Plants & People

The Jones & Bartlett Learning Topics in Biology Series

We are pleased to offer a series of full-length textbooks designed specifically for your special topics courses in biology. Our goal is to supply comprehensive texts that will introduce non–science majors to the wonders of biology. With coverage of topics in the news, emerging diseases, and important advances in biotechnology, students will enjoy learning and relating science to current events.

AIDS: The Biological Basis, Fifth Edition
Benjamin S. Weeks & I. Edward Alcamo

Alcamo's Microbes and Society, Third Edition
Benjamin S. Weeks

Evolution: Principles and Processes
Brian K. Hall

Human Embryonic Stem Cells, Second Edition
Ann A. Kiessling & Scott C. Anderson

The Microbial Challenge: Science, Disease, and Public Health, Second Edition
Robert I. Krasner

Plants & People
James D. Mauseth

Plants & People

James D. Mauseth
University of Texas at Austin

World Headquarters
Jones & Bartlett Learning
25 Mall Road
Burlington, MA 01803
978-443-5000
info@jblearning.com
www.jblearning.com

Jones & Bartlett Learning books and products are available through most bookstores and online booksellers. To contact Jones & Bartlett Learning directly, call 800-832-0034, fax 978-443-8000, or visit our website, www.jblearning.com.

Substantial discounts on bulk quantities of Jones & Bartlett Learning publications are available to corporations, professional associations, and other qualified organizations. For details and specific discount information, contact the special sales department at Jones & Bartlett Learning via the above contact information or send an email to specialsales@jblearning.com.

Production Credits
Chief Executive Officer: Ty Field
President: James Homer
SVP, Editor-in-Chief: Michael Johnson
SVP, Chief Technology Officer: Dean Fossella
SVP, Chief Marketing Officer: Alison M. Pendergast
Publisher: Cathleen Sether
Senior Acquisitions Editor: Erin O'Connor
Senior Associate Editor: Megan R. Turner
Editorial Assistant: Rachel Isaacs
Production Manager: Louis C. Bruno, Jr.
Senior Marketing Manager: Andrea DeFronzo
V.P., Manufacturing and Inventory Control: Therese Connell
Composition: CAE Solutions
Cover Design: Kristin E. Parker
Text Design: Scott Moden
Associate Photo Researcher: Lauren Miller
Cover Image: © iStockphoto/Thinkstock
Title Page Painting: *Three Birches* by Felipe Garcia
Printing and Binding: Strategic Content Imaging (SCI)
Cover Printing: Strategic Content Imaging (SCI)

To order this book, use ISBN 978-1-4496-5717-8.

Library of Congress Cataloging-in-Publication Data
Mauseth, James D.
Plants & people / James D. Mauseth. — 1st ed.
p. ; cm.
Includes index.
ISBN 978-0-7637-8550-5 (alk. paper)
1. Botany. 2. Plant ecology. 3. Human-plant relationships. 4. Plants, Useful.
I. Title. II. Title: Plants and people.
QK47.M384 2012
580—dc23
2011016284

6048

Printed in the United States of America
26 25 24 23 22 10 9 8 7 6 5 4 3

Brief Contents

Contents

Chapter 14 Spices and Herbs: Plants that Make Eating Fun 339

Chapter 15 Plants as Sources of Medicines, Drugs, and Psychoactive Compounds. 356

Preface

This book has four main objectives:

1. to introduce you to the basic principles and concepts of plant biology,
2. to discuss the roles of plants and people in issues of global importance such as climate change, endangered species, genetically modified organisms, pollution, and so on,
3. to describe the plants we use in our daily lives as food, drink, clothing, drugs, and ornamental plants, and
4. to encourage an interest in plants, an interest that you probably have already if you are reading this preface.

Most of us know a great deal about biology simply because we observe our own lives. We people are animals, and like any other animal—or any other organism for that matter—we must take in energy and nutrients (food), we must be aware of our environment and respond to it (is it hot, cold, storming, dangerous?), we must protect ourselves from diseases and predators, and so on. We can use our knowledge of our own animal biology to think about and understand the biology of plants. People and plants are similar in so many ways, ways that are fundamental to all life: Our bodies are composed of proteins, fats, enzymes, cells; we grow, develop, die; we interact with other organisms in our environment. And yet people and plants differ in many ways: We animals move, eat, and produce waste; we have a limited number of each organ (two eyes, two lungs, one heart, one brain); and we have limited life spans. In contrast, plants do not move, eat or go to the bathroom; they have hundreds or thousands of leaves, flowers, and other organs; and many plants live much longer than animals, never experiencing old age. These fundamental concepts of biology are discussed in the first section of this book.

The second section focuses on global issues. The world—Earth itself as well as the organisms that live on it—is a dynamic place that is constantly changing. Volcanoes build mountains, which are then eroded away by rain and snow; continents drift from place to place on the planet's surface; and changes in Earth's orbit cause ice ages to come and go. Such changes would occur even if no life existed on Earth, but living organisms also cause changes by their own metabolism, by eating things, moving things, generating waste. All the oxygen we breathe is produced by green plants as they photosynthesize in the sunshine; no free oxygen at all was in the air before photosynthesis evolved. Plants have altered Earth's entire atmosphere, and we are now doing the same thing by burning coal, oil, and natural gas.

We must be aware of the changes that occur and think about which are inevitable, which are natural, and which are caused by us. We need to think carefully about the consequences of our actions: Are we helping or harming our future existence? Are we helping or harming other organisms that share Earth with us? Organisms other than we humans have no control over their future or their fate, but we do, and we can use our intelligence to analyze problems and ensure a better future.

How did you start your day today? Got out of bed (the sheets were probably cotton), turned on the lights (most electricity is generated from coal), brushed your teeth (many toothpastes contain algae), washed with soap (soaps are made with plant oils), dried with a towel (cotton), put on clothes (more cotton), ate breakfast (wheat toast? oatmeal? granola? fruit? tacos? coffee? tea? cocoa?) while reading the newspaper (plant fibers). We have considered no more than an hour of your day, and almost everything you have done relied on plants and plant products. Many of us—you, too—are experienced botanists without realizing it. We recognize many plants as tasting wonderful (strawberries, apples, blueberries) or not really (brussels sprouts), as being nutritious (whole grain bread) or no so much (sugary foods, cigarettes). We love to run our hands through fragrant rosemary or mint but not through poison ivy, cacti, or stinging nettles. We give flowers on special occasions, holidays, and for friendship, and we know which flowers are suitable for each. After thinking about all the plants we use in our daily lives, try to guess how many were discovered in your lifetime: Probably none. Almost all our familiar foods and fibers have been gathered and used by people for millennia; most were being cultivated on farms or orchards three or four thousand years ago. Some were domesticated in Asia, others in Europe, Africa, or the Americas and have since been distributed to most parts of the world where people live. Even processed plant foods such as beer and wine have ancient origins and long histories, as is also true of fibers that are spun into thread and woven into cloth. These plants and their stories constitute the third section of this book.

The fourth objective of this text—to encourage an interest in plants—does not have its own separate section. This objective has been achieved, I hope, by having all parts of the book focus on plants and processes that are familiar to you. I have also tried to tell the story of plants and people in ordinary language as much as possible, avoiding technical terms that you will probably not need in most aspects of your life. On the other hand, I have assumed that anyone reading this book will want to learn new things, including basic technical words that are needed to express biological concepts accurately.

I just mentioned "the story of plants and people" and that itself is an important concept. We are not dealing with isolated facts and figures, with tables and charts of data, or with lists of bullet points. The biology of people is a story that is intimately related to the biology of plants and all other organisms, a story that began long ago with our pre-human ancestors who ate plants and lived in forests or grasslands. The story became more complex as our ancestors learned to cultivate plants, to transport seeds with them as they migrated, to pull weeds, and to cut down forests to make fields. Such practices had both beneficial and harmful consequences. For example, increased crops permitted increased human populations as well as increased pollution and more rapid spread of diseases; the discovery of medicinal plants improved human health but crop failures resulted in mass starvations. The story of plants and people continues even now as we realize that our increasing numbers and consumption are causing great harm through habitat destruction, the extinction of species, and climate change. Fortunately, we have become aware of the ways in which our biology adversely affects the environment, and we realize that we can take corrective action. Plants cannot think or plan or see the consequences of their biology, but people can. It may be that right now, the story of plants and people is entering its best, most compassionate, and most hopeful phase, and we are part of it.

Ancillaries

For Students

To further enhance the learning experience, Jones & Bartlett Learning offers the following ancillary materials:

The *Companion Website* (http://biology.jbpub.com/plantsandpeople) provides content exclusively designed to accompany *Plants & People*. The site hosts an array of study tools, including chapter outlines, study quizzes, an interactive glossary, animated flashcards, crossword puzzles, and web links for further exploration of the topics discussed in this book.

For Instructors

An *Instructor's Media CD*, compatible with Windows® and Macintosh® platforms, provides instructors with the following resources:

- The PowerPoint® ImageBank contains all of the illustrations, photographs, and tables (to which Jones & Bartlett Learning holds the copyright or has permission to reproduce electronically). These images are inserted into PowerPoint slides. Instructors can quickly and easily copy individual images into existing lecture slides.
- The PowerPoint Lecture Outline presentation package, prepared by Tharindu Weeraratne of the Unversity of Texas at Austin, provides lecture notes and images for each chapter of *Plants & People*. Instructors with the Microsoft® PowerPoint software can customize the outlines, art, and order of the presentation.

To receive a copy of the *Instructor's Media CD*, please contact your sales representative.

The following materials are also available for qualified instructors to download from the Jones & Bartlett Learning website, www.jblearning.com:

- Text files of the *Testbank*

Acknowledgements

This book has benefited from the generous, conscientious thoughts of many reviewers. They provided numerous suggestions for ensuring thoroughness of topics covered, improving clarity of presentation, and identifying illustrative examples that will enhance the reader's understanding and interest. I am deeply indebted for their generosity and patience. It was humbling to see how freely they shared their experience and careful thoughts with me, writing out advice in detail. I thank them all.

Maren E. Veatch-Blohm, Loyola University Maryland
Margaret Carroll, Framingham State College
Joseph Darbah, Ohio University
Susan Dunford, University of Cincinatti
Robert C. Evans, Rutgers University-Camden
Scott Heckathorn, University of Toledo
Professor Steven J. Karafit, Hendrix College
Sarah MacDonald, Missouri Valley College
Chintamani S. Manish, Midland Lutheran College
James D. Metzger, The Ohio State University
Edgar Moctezuma, University of Maryland
Richard Mueller, Utah State University
Donald Pfister, Harvard University
Kumkum Prabhakar, Nassau Community College
Manfred Ruddat, University of Chicago

The initial production of a textbook is not by any means the sole effort of the author. I am fortunate to benefit from the many contributions of numerous talented individuals at Jones & Bartlett Learning; the current editorial staff is one of the best and most skillful. I especially thank Molly Steinbach for her insightful input at every stage. After consulting with many teachers at numerous colleges, she mapped out the basic concepts of this book and then approached me about writing it. Over the next several months, she and I worked as a team, having many discussions about the needs of biologists in this area, the level of the text, the organization of the text, and the use of boxes, tables, and illustrative material. Molly was an active participant at all stages and was always willing to consult with reviewers to determine what teachers might want and what students might need. I thank her for her endless patience, unlimited energy, and insightful judgment. It was a pleasure to work with Rachel Isaacs as she prepared the manuscript for production; she tactfully asked for clarification of various points about the manuscript without actually implying that I had made any errors or omissions. I was so happy to work with Lou Bruno during the production of this book; we have worked together on others, especially *Botany: An Introduction to Plant Biology, 4th Edition*. Production is the stage during which every letter, period, space, and figure must be correct, and everything must be on schedule. Lou always has everything under control and seems to know each part of the book by heart, recognizing when certain parts are not in agreement. His calmness, knowledge, and care were reassuring.

I also thank my good friend Chris McCoy, certainly not only for his wonderful art but also for many conversations about plants as they are used in art, food, and furniture. He provided insights and details on diverse aspects of spices, beverages, and cuisines of several countries. Last but not least, I especially thank my partner Tommy Navarre for generously sharing with me his knowledge of food safety, food processing, the humane treatment of farm animals, and concepts of sustainable agriculture. But most of all, I appreciate his never-ending support, encouragement, and confidence.

Jim Mauseth

About the Author and the Artist

Jim Mauseth has been a professor of botany at the University of Texas at Austin since 1975, specializing in plant anatomy and evolution. His research focuses on plants whose bodies have become highly modified as they adapted to extreme environmental conditions: for example, the succulent, spiny bodies of cacti allow them to survive in desert habitats, and many parasitic plants (relatives of mistletoes) are able to spend most of their lives completely inside the bodies of host plants because the parasites now lack roots, stems, and leaves. Other books by Jim include *Botany, An Introduction to Plant Biology; Plant Anatomy; A Cactus Odyssey: Journeys into the Wilds of Bolivia, Peru and Argentina* (with Roberto Kiesling and Carlos Ostolaza); and *Plant Structure, A Color Guide* (with Bryan Bowes).

Chris McCoy has taught art at St. Andrews Episcopal School in Austin since 2005, with 29 years teaching in public and museum art schools. His students study drawing from life observation, painting, sculpture, architecture, and design. He assists in guiding students on field trips to Chicago, Los Angeles, Marfa, and France including Paris, Giverny, and Pont Aven where he teaches his students about art history and living life fully as an artist. Chris specializes in drawing and photography, creating collages of vintage photographs combined with modern materials. Throughout his life in Iowa and Texas, he has encouraged young artists at the dawning of their creativity. From an early age, he was a careful observer of birds, insects, plants, and people. Current interests employ those observations in gardening, collecting and creating art, photography, and making good food with friends.

Metric and Imperial Units for Measuring

There are two sets of units used to measure things. The metric system has units such as meters and grams; it is also used throughout the world. The imperial system is a hodge podge of oddball units such as inches, feet, miles, ounces, quartz, gallons, and so on, and it is used only by Americans. The metric system is logical and orderly: each unit is one thousand times larger than the next smaller unit. For example, length is measured in meters, and we have 1000 nanometers (nm) = 1 micrometer (μm); 1000 micrometers = 1 millimeter (mm); 1000 millimeters = 1 meter (m); and 1000 meters = 1 kilometer (km) (the abbreviation for micrometer uses the Greek symbol μ, pronounced "mu"). The imperial system by contrast has 12 inches = 1 foot; 3 feet = 1 yard; 5280 feet = 1 mile. People teaching the metric system often claim that its advantage over the imperial system is that it is easy to convert measurements from one unit to another; for example, if a plant cell is 20 μm (micrometers) wide, we know immediately that it is 20,000 nanometers wide or 0.02 millimeters wide. In contrast, if something is 20 feet long, we need a calculator to figure out its length in inches or yards or miles. But if you think about it briefly, that doesn't really matter, we almost never do conversions like that, either between feet and miles or micrometers and nanometers, and, if we do, we use a calculator or computer.

The real advantage of the metric system, and this is a very important advantage, is that it has a convenient unit for almost anything we want to measure whereas the imperial system has no units for really small things. As mentioned, many plant cells are about 20 μm wide, so it is convenient to measure cells in micrometers: small parts of cells are 1 or 2 μm and large cells are 100 to 200 μm; all these numbers are easy to write, read, and verify. We know immediately if we have made a mistake by saying that a part of a cell is 200 μm long or that an entire cell is only 2 μm long. But because the smallest imperial unit of length is the inch, that same cell would be 0.000787 in. wide. Small parts would be 0.0000393 in. (1 μm), and large ones would be 0.00787 in. (200 μm). It would be extremely easy to make a mistake with the number of zeros after the decimal. For whole plants, both the meter and the foot are convenient: many bushes are about 3 feet tall (1 meter); big ones are 10 ft (3 m) tall, and short ones only 1 ft (0.3 m) tall. We could use either system easily and not have to worry about making a mistake with zeros by the decimal. Similar examples could be given for measuring volume and weight.

Because the metric system has convenient units for anything no matter how small or large (**FIGURES 1–3**), it is the system used in science throughout the entire world: no scientist uses the imperial system for anything. Engineers also use the metric system worldwide, unless they are designing a product to be used only in the United States, and then they might use inches and ounces, but more often engineers use metric for everything.

Because this is a science book, it should use the metric system. But because most of you readers are probably Americans with little or no experience with the metric system, you will understand me best if I use the imperial system. The compromise I decided on is to give all measurements in metric units followed by the imperial equivalent in parentheses. For example, I might write that "a tree is 5.3 m (17.4 ft]) tall." But microscopic distances and very tiny weights do not mean much to most of us; for most readers, knowing that a plant cell is 20 μm (0.000787 in.) wide means nothing. But after you read several chapters about cells and tissues, the concept of 20 or 2 or 200 µm will start to mean something to you, and this will then let you read other books or articles about cell biology and understand them. If you read those same chapters with measurements of 0.000787 in. or 0.0000787 in. or 0.00787 in., you would also become familiar with cell sizes in inches, but there would be no point...there would be almost no other book or article that would use inches for cells. So for very small quantities, only metric units are used.

TABLE 1 contains metric and imperial conversions. And you can use a computer or phone to access websites that convert all types of units.

FIGURE 1. Most of us have a set of standard metric weights readily available. A dollar bill (of any denomination) weighs exactly 1 gram (g), 2 pennies weigh 5 grams, 15 quarters and 3 pennies weigh almost exactly 100 grams, and a small Post-it Note® weighs one-tenth (0.1) of a gram.

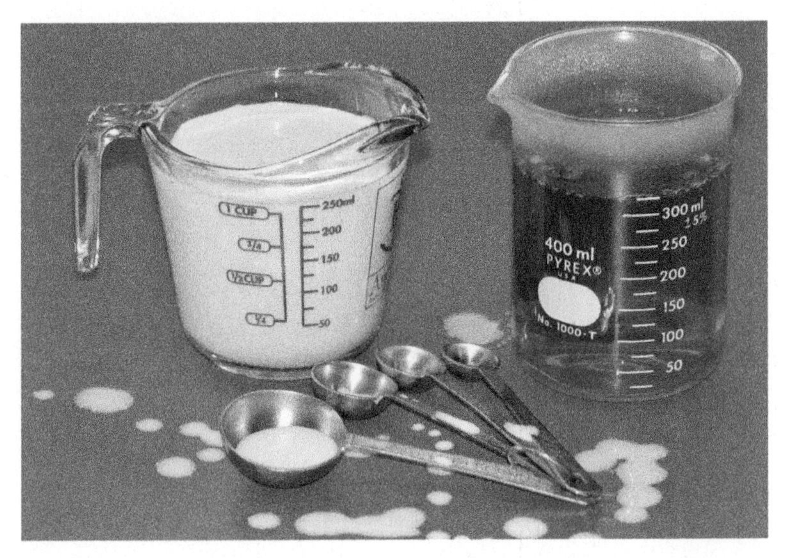

FIGURE 2. The metric system measures volumes in liters (L) and milliliters (mL). A tablespoon (foreground) is 14.7 mL, and a teaspoon (the next one back) is exactly 5 mL. One cup is slightly less than 250 mL. Drops of milk are irregular, but 20 drops of water equal 1.0 mL. The measuring cup and the beaker use the abbreviation "ml" but the correct abbreviation is "mL."

TABLE 1 Imperial Unit to Metric Units Conversions

Length	
1 in = 2.5 cm	1 mm = 0.04 in
1 ft = 30 cm	1 cm = 0.4 in
1 yd = 0.9 m	1 m = 40 in
1 mi = 1.6 km	1 m = 1.1 yd
	1 km = 0.6 mi
Weight	
1 oz = 28 g	1 g = 0.035 oz
1 lb = 0.45 kg	1 kg = 2.2 lb
Volume	
1 tsp = 5 mL	1 mL = 0.03 fl oz
1 tbsp = 15 mL	1 L = 2.1 pt
1 fl oz = 30 mL	1 L = 1.06 qt
1 cup = 0.24 L	1 L = 0.26 gal
1 pt = 0.47 L	
1 qt = 0.95 L	
1 gal = 3.8 L	
Temperature	**Interval Equivalents** (°C °F)
$°C = (°F - 32) \times \frac{5}{9}$	1° = 1.8°
	5° = 9°
$°F = °C \times \frac{5}{9} + 32$	10° = 18°

FIGURE 3. Many rulers have centimeters (cm) along one side and inches along the other. Here, we can see that this book, chosen entirely at random, is 193 cm (about 7.5 in.) wide.

Introduction

1

Plants Are Important to People

Plants Provide Our Oxygen, Energy, and the Atoms of Our Bodies

Could you live without plants? There is an easy way to find out: hold your breath. As we take in a lungful of air, our bodies absorb the oxygen, a chemical that is essential to our life. Without oxygen we simply cannot live. And all the oxygen we breathe, all the oxygen that sustains our lives was produced by plants. The green parts of plants contain a pigment called chlorophyll, which captures sunlight and makes its energy available for photosynthesis (FIGURE 1.1). During photosynthesis, plants force carbon dioxide and water to react with each other, converting them into sugar and oxygen. The oxygen is released into the air, where it sustains our life. Without the oxygen produced by plants and photosynthesis, we could not live more than a few minutes. No animal or fungus can perform photosynthesis.

Plants supply all the fruits shown here, as well as the wood for the table, the paper in the sketchbook, and the handle for the knife. Bacteria and fungi convert milk to cheese. Of the fruits here, some come from trees, some from herbs, some come from plants that live for years, others that live for only a few months. This lunch will provide necessary carbohydrates, proteins, vitamins, and so on, but just as importantly, it will be delicious and enjoyable. (Courtesy of Chris McCoy.)

FIGURE 1.1. The green parts of plants carry out photosynthesis, which captures energy, creates organic compounds, and releases oxygen. The cyclists are breathing the oxygen, and their food is based on plants, whether they eat the plants as salad, pasta, vegetables, or as milk or meat.

The sugar produced by photosynthesis also is critically important to our lives. First, plants use it to build all parts of their bodies; plant metabolism combines simple sugar with various minerals from the soil and constructs all the chemicals found in the plant body. All the proteins, starches, oils, flavors, aromas, and other chemicals begin as sugar produced by photosynthesis. If we tried a similar feat—eating nothing but sugar and mineral supplements—we would die quickly. We must eat more complex foods, foods that are parts of a plant body or that come from animals or fungi that have themselves consumed plants. Our digestive systems break these complex chemicals down into simpler forms such as amino acids, fatty acids, and so on, and then we absorb them through our small intestine and use them to build our own bodies. The chemicals that make up our bodies all started out as parts of the bodies of many plants. Even if you mostly eat meat and never eat fruit and vegetables except as a last resort, the meat you consume came from an animal that ate plants. Look at your hands, arms, any part of your body you can see: all of it started off as sugar produced by photosynthesis in a plant. Without plants, our very bodies would not exist.

Photosynthesis captures the energy of sunlight and is the source of our own energy. When we eat food, part of it is used to construct our bodies, to make more skin (we shed skin cells constantly, which is especially noticeable in dry weather or after a sunburn), more hair, more muscle, and more fat. But most of our food is "burned," or more accurately, it is combined with oxygen in a process called respiration that gives us the energy we need to move, to be warm, to pump our blood: all the energy we need just to stay alive comes from the energy that plants captured during photosynthesis. We may sit in the sun to warm ourselves, but that warmth cannot be used to make our muscles move or to drive our metabolism.

It is a profound thought: all the material of our bodies, all the energy we use, all the oxygen that keeps us alive, all of this comes from plants. And it is not just we people; all other animals are just as dependent on plants as we are.

Plants Protect Us

We use plants to protect us from the environment. Think of yourself at this very moment, while you are reading this. If you are indoors in a building, imagine what would happen if all the wood—wood comes from plants—suddenly disappeared: many buildings would either collapse or vanish except for the nails. Sitting in a concrete building? Wood forms were used to hold the concrete as it hardened. There is a good chance your chair and table would vanish if wood suddenly ceased to exist. Now think of your clothing: without plants, your cotton shirts, pants, dresses, coats, socks, sneakers, and underwear would all vanish—most of us would be naked without plant products. To get a feeling for how much we rely on plants to protect us from the environment, imagine yourself somewhere with no plants—the Sahara Desert, for example, or the North Pole—and imagine you are naked.

Plants protect us in other ways. We get warmth from firewood, coal, and charcoal. In many parts of the world all heating and cooking are done with fires fueled by hand-gathered twigs, sticks, and dry grass that people collect themselves. Since ancient times, people have obtained medicines from plants, and even today some of our drugs are extracted directly from plants, while others are produced from starting materials obtained from plants.

Plants protect us on a much larger scale and in ways we rarely think about. They alter many factors in our environment such that our lives are possible. For example, plants stabilize soil. It is easy to assume that this has little to do with the lives of most people, but remember that all parts of Earth's surface are exposed to winds and rains that can move soil (FIGURE 1.2). Without a stabilizing layer of vegetation, soil would erode away until all of it had been blown or washed into one of the oceans and the land's surface was just barren, uninhabitable rock. Any bare patch of soil will produce a dust cloud as wind blows over it, and this has many consequences: blinding dust storms cause traffic accidents that kill people; the dust chokes people and animals, making breathing difficult; it covers the leaves of downwind plants, inhibiting their growth; and so on. The soil carried off by a dust storm is the uppermost, richest topsoil, and without it, the remaining soil is less fertile and more rocky, less capable of sustaining revegetation. Rain and melting snow also erode soil, washing it lower on hillsides and carrying it into streams, lakes, and rivers, making them muddy, damaging the gills of fish and invertebrates, and polluting the water we drink. Almost all of us drink water pumped out of rivers. Along coastlines, plants stabilize sand dunes, preventing them from blowing away. Such dunes lessen the impact of storm surges and high water that accompany hurricanes and tropical storms. Without coastal dunes, inland flooding would be much worse. Coastal wetlands and marshes offer similar protection, as well as being a habitat for millions of fish, birds, crustaceans, and other animals (FIGURE 1.3). Trees and shrubs that grow along rivers hold their banks in place, reducing erosion and flooding and helping to keep mud and silt out of the water. Without vegetation covering its surface, soil would be in constant motion, mostly moving downslope and downstream, and land would be mostly bare rock.

FIGURE 1.2. The vegetation at the top of the hill was disrupted by farming, and now without roots to stabilize the soil, erosion occurs with every rain, washing soil particles down into a river below.

FIGURE 1.3. These dune and marsh plants along the Texas Gulf coast stabilize sandy soil and help hold it in place when hurricanes and tropical storms come through this area. In the years between storms, habitats like this support a large diversity of plants (palms, grasses, and sedges are visible), as well as many sorts of animals. Without these plants, the dunes would erode and salt water would be pushed inland during storms, damaging large areas of habitat.

Plants Affect Us Psychologically

Plants affect our mood and emotions. Just think of these foods, all of which come from plants: chocolate, sugar, cinnamon, mint, peppermint, and other spices; cakes, pies, pastries, doughnuts made from wheat flour; fresh apples, peaches, cherries, oranges, walnuts, pecans, almonds, cashews; tortillas (both corn and flour), pasta. If your tastes are focused more on fast-food hamburgers, then probably at least the catsup, mustard, tomatoes, lettuce, pickles, french fries, and onion rings are things you like. It is probably safe to say that everybody has some plant-based food they really enjoy.

Other foods are more calming. It is hard to imagine relaxing without a cup of tea or coffee, or perhaps a glass of wine or beer. If we extend the list to hard liquors (ones that have been distilled to increase their alcohol content), the flavors and aromas are derived from the plants that were used in the fermentation mix. Marijuana and tobacco are other plant products that calm us, but tobacco also appears below under the heading Plants Can Kill Us.

Even plants we don't eat affect us emotionally. Think of flowers given as gifts: roses on Valentine's Day, flowers children give to their parents, the flowers we give to friends who are ill, or just a simple bouquet or potted plant from one friend to another. We use flowers to express our emotions and feelings, in celebrations and in mourning: weddings always involve flowers, as do funerals. Gardening is a hobby that gives pleasure to many people, whether they have an extensive garden or just a plant on a windowsill. Perfumes and fragrances, which are so important on many social occasions, are extracted from flowers, fruits, or other plant parts. Many religious services are still accompanied by the burning of incense, especially frankincense and myrrh, just as they have been for thousands of years.

Plants Can Kill Us

Plants are rooted to the soil, unable to move but surrounded by animals that, given the chance, will eat them, lay eggs in them, break them apart to use as nesting material, or maybe just trample them without notice. Numerous defense mechanisms have evolved in plants; most plants are far from helpless. Many plants are so toxic that eating just a few leaves or seeds will kill an adult within a day or two. And the poisons are diverse, some affecting our nervous system, others our heart, and others our general metabolism. Probably the plant poison most familiar to people is that in poison ivy, and stinging nettles are painfully familiar to many people. The sap of many spurges (the genus *Euphorbia*) is not only poisonous if eaten, just touching the sap can cause blisters on the skin. An especially insidious plant poison is nicotine: it is slow acting, but smoking tobacco usually results in cancer of the lungs, throat, or mouth. Worldwide, more than 1.3 million people die from lung cancer each year.

Some plants might not kill us, but they have spines that make our lives miserable. Sharp stickers have evolved numerous times in many different groups of plants and can be found on leaves, stems, and even fruits. Most likely everyone has been cut by a rose thorn at least once in their life, and most people in arid areas have first-hand experience with the spines of cacti, agaves, yuccas, and other desert plants. Spiny vegetation occurs in all kinds of habitats, not just deserts; many palms and bamboos of wet areas are spiny, as are prairie plants such as prickly poppies and thistles. Although spines are usually merely painful for us, they can be deadly to animals, either injuring them badly and leading to infection or blinding them. On a small scale, many tiny plants that are preyed upon by tiny insects also have spines, too small for us to notice, but effective against the insects.

And just as we occasionally trample and harm plants without intending to hurt them, plants may do the same to us. Wind-pollinated plants such as ragweed, cedars, oaks, and others release copious amounts of pollen into the air, causing misery for millions of people who suffer from hay fever, allergies, and asthma. The irritation these plants cause is entirely incidental; it is not a defense mechanism, it does not protect the plants, and they derive no benefit from our suffering. It is simply that our immune systems are unnecessarily sensitive to chemicals on the pollen grains.

People Are Important to Plants

We Cultivate and Protect Plants

At present, people are important to plants, both helping and harming them. We help certain plants by cultivating them as crops for food, lumber, cloth, oils, and ornamentals. Wheat, rice, corn, potatoes, cotton, and soybeans dominate vast areas of Earth's land area, but they could not do this without human help (FIGURE 1.4). If we became extinct, these crop plants would soon disappear, unable to compete against wild plants without our help. We plow ground for them, sow their seeds, water and weed them. Afterward, we harvest them but save some of their seed to propagate them the next year. For some crops, we provide them with pollinators by keeping hives of honeybees close to the fields. For vanilla, people actually pollinate each flower by hand. This leads us to wonder whether we have domesticated plants or if they have domesticated us.

FIGURE 1.4. These oats are thriving under artificial conditions created by a farmer: the soil was plowed to make it soft, seeds were sown, weeds have been eliminated, fertilizers supplied, and insect pests have been eliminated. Without all this care, weeds would quickly take over this field, and animals, insects, and fungi would attack the oats.

We realize now that natural areas on Earth are not unlimited and they cannot restore themselves if damaged too badly. Conserving natural habitats has become a goal that many people have accepted, and large areas of land have been protected in the form of national and state parks, nature preserves, and wildlife refuges owned and controlled by government agencies. In addition, a great deal of habitat is protected by nongovernmental organizations such as The Nature Conservancy, World Wildlife Fund, Natural Resources Defense Council, Sierra Club, and many others. In some cases, conservation is brought about by arranging conservation easements with landowners: the land remains in private hands but the owners agree to restrict hunting, grazing, farming, construction, and other practices that would damage the plants and animals. Conservation easements are a relatively new method in conservation biology, and they have shown that many landowners are eager to be involved in protecting the natural aspects of the land.

As the concept of conservation biology has developed, we realize it is important to preserve tracts of land that are large enough to include self-sustaining populations of plants and animals. If our goal is to protect an endangered plant species, then it is best to have a preserve that supports the animals that pollinate those plants, as well as any other animals such plants depend on. Many plants are adapted to natural cycles of flooding or fires; those too should be included in conservation efforts.

Many habitats were already severely damaged before people developed the concept of conservation ethics. For example, much of the midwestern region of the United States was tall grass prairie before it was cleared for farming; virtually no natural patches of tall grass prairie have survived. However, numerous efforts are underway to restore some of these areas, and an entire field of science—restoration ecology—is developing.

Fortunately, it is possible for each of us to protect plants and animals in our own daily lives. We can consume less and make thoughtful choices about what we do.

FIGURE 1.5. Urban sprawl causes large amounts of habitat destruction. When people live widely separated from each other, more freeways are needed, but if more people would live in high-density housing, cities could be more compact and fewer highways and parking lots would be needed.

We can use less food, paper, chemicals; walk or bicycle more and drive less; keep our homes and classrooms less heated in winter and less air-conditioned in summer. Rather than use either paper or plastic bags at stores, carry a reusable cloth bag. If each of us consumes less, there will be less pollution, less natural land converted to shopping malls, parking lots, and farmland; less overfishing of oceans. Each of us can make a difference every day.

We Damage the Habitats of Plants

People harm plants in many ways, both intentionally and inadvertently. Probably the greatest damage is **habitat destruction** caused by clearing land for construction projects, cutting forests for lumber, draining wetlands for farming, and flooding valleys behind dams (FIGURE 1.5). Smog and air pollution damage plants downwind, and dumping wastes into rivers harms aquatic life. Oil spills from drilling platforms and oil tankers has damaged thousands of miles of coastal habitats.

What Are Plants?

Being a book about plants and people, we should consider just what plants and people are. Technical details are a bit complicated, but almost certainly you already have an accurate idea of which organisms are plants and which are not.

How Plants Are Related to All Other Organisms

There are three basic types of organisms, called **Bacteria**, **Archaea**, and **Eukarya**. Bacteria are probably already familiar to you: they are microscopic, usually have a body consisting of just a single simple cell shaped like a rod, a spiral, or a sphere. Bacteria occur in almost every conceivable place, such as soil, water, air; in forests, grasslands, deserts, and snow-covered areas; on and in our bodies, on plant surfaces, and in our food unless it has been sterilized. Some bacteria cause the diseases that are treated with antibiotics. One group of bacteria is the Cyanobacteria, which are green and able to photosynthesize.

Archaeans are less familiar to most people, but they are also microscopic and unicellular. They are not as widespread as bacteria, mostly being located in habitats that are extremely acidic or hot, such as the geysers and thermal pools of Yellowstone National Park. Cells of both bacteria and archaeans have no nucleus, and thus the two groups are known as **prokaryotes** (this means "before nuclei").

The third group, Eukarya, consists of all plants, algae, animals, fungi, and many microscopic organisms. This diverse group has an important unifying feature: their cells each contain a nucleus ("eukarya" means "true nucleus"). In addition, the cells of **eukaryotes** are much more complex than those of prokaryotes, as described in Chapter 9. When eukaryotes first originated billions of years ago, perhaps by the modification of some archaeans, they were all organisms whose bodies consisted of just a single microscopic cell. At this early stage, some evolved to be capable of photosynthesis, and these then became more numerous. Many variations evolved and some gradually evolved into algae and plants. Meanwhile, some of the other early eukaryotes evolved into animals and fungi (such as mushrooms, puffballs, and yeast). Until recently, most biologists assumed that fungi and plants were closely related because both groups reproduce by spores, neither moves, and neither has any of the internal organs that typify most animals. But many key aspects of the metabolism of fungi resemble those of animals and differ from those of plants.

In the photosynthetic line of evolution, **algae** diversified into several types. At present, there are green algae, red algae, brown algae, and several other types, each characterized by their pigments and other features (**FIGURE 1.6**). About 420 million years ago, one group of green algae evolved to have features that allowed them to live on land rather than in water. Members of this group are the **true plants** (see Chapter 9 for more about the differences between algae and plants). The first true plants were small and simple, but as time passed, several new features evolved in some of them, and gradually they diversified into the plant groups that exist today. Following are brief descriptions of the major groups of plants, most of which are probably familiar to you (**TABLE 1.1**).

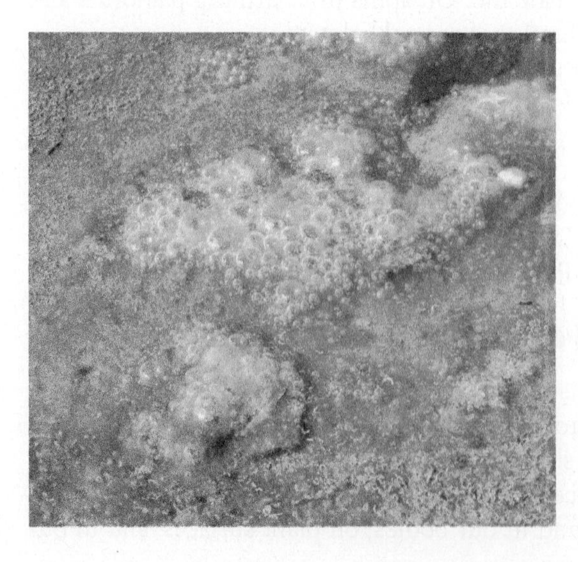

FIGURE 1.6. Slow-moving streams are good habitats for many green algae, such as these. They are green because they contain chlorophyll and other pigments like true plants, and they carry out photosynthesis. The bubbles are oxygen liberated by photosynthesis.

TABLE 1.1. The Major Types of Plants and Their Relatives

I. **Green algae,** in particular a group called Charophyta, are the group of organisms most closely related to plants.

II. **Bryophytes** are plants that have no vascular tissue and no seeds. Members are mosses, liverworts, and hornworts.

III. **Vascular plants** are plants that have vascular tissues, tissues that conduct water and nutrients from one part of a plant to another.

A. **Ferns and fern-allies** are plants that have vascular tissue but do not make seeds during sexual reproduction.

B. **Seed plants** are vascular plants that produce seeds during sexual reproduction.

1. **Cycads** are the "sago palms" and their relatives.
2. **Conifers** are the pines and their relatives.
3. **Angiosperms** are the flowering plants. Angiosperms are the most abundant of all types of plants. There are more types (species) of angiosperms than all other plants combined, and they occur in more habitats and cover more of the Earth's surface than all other plants. Almost all our food, spice, medicinal, crop, and ornamental plants are angiosperms.

The Major Groups of Plants

1. Flowering plants are technically known as **angiosperms**. Flowers are reproductive structures; some parts produce pollen grains, which carry sperm cells, other parts have egg cells. During reproduction, new embryos form as part of a seed, and all seeds occur inside fruits. Plants with obvious flowers, such as roses, petunias, snapdragons, and lilies are flowering plants (FIGURE 1.7). But in many other angiosperms, the flowers are tiny and inconspicuous; for example, the flowers of grasses are so small and pale you might never notice them, and the same is true for those of oaks, elms, cattails, and palms (FIGURE 1.8). Other angiosperms have large, obvious flowers, but they bloom so rarely—at least in cultivation—that many people mistakenly assume that they never flower. Cacti are a good example of this: when kept on a windowsill, potted cacti may survive and grow without being healthy enough to flower, but in nature they produce spectacular flowers that no one could miss. All angiosperms have **vascular tissues**, that is, tissues that conduct water and nutrients from one part of the plant to another. Plants with vascular tissues are **vascular plants**.

FIGURE 1.7. The name "flowering plant" sometimes confuses people. It is meant to indicate plants that reproduce with flowers rather than cones or some other means. In the winter, when this rose bush has no flowers on it, it will still be a flowering plant. The more proper term "angiosperm" is less confusing but is not used as often except in technical writing. (Courtesy of Chris McCoy.)

2. Conifers are the pines, spruces, firs, and their relatives (FIGURE 1.9). They produce cones rather than flowers, and are always trees that live for years; they are never small herbs or plants that live for less than a year. Their leaves are needles or scales, they are never broad and flat like those of most flowering plants (never like a geranium or maple leaf). Like angiosperms, all conifers are vascular plants and all produce seeds.

3. Cycads might not be familiar to many of you because they only survive where winters are mild; even a brief, light frost kills most species of them (FIGURE 1.10). Cycads are vascular plants that produce seeds in structures that superficially resemble the cones

FIGURE 1.8. The flowers of some flowering plants are so small or inconspicuous it is easy to be familiar with the plant but not realize that it ever makes flowers. Sandburs (several species of the genus *Cenchrus*) are familiar to all of us: the burs that hurt our feet are flowers with stickers on them. Later the flowers develop into fruits that are larger and even more painful.

of conifers. But they differ from conifers in significant details (particularly their very soft wood) and are considered to be only distantly related to conifers.

4. Ferns are probably very familiar plants. They tend to have large leaves subdivided into many parts with complex shapes (FIGURE 1.11). They have vascular tissue but never produce seeds during their sexual reproduction. All plants produce spores, even

FIGURE 1.9. This cedar (*Cedrus*) is a conifer and is related to pines, firs, and other plants that reproduce with cones instead of flowers. Cedars grow to be large trees with short needle-like leaves that occur in clusters. Their cones produce resins (often called "pitch"), some of which have oozed out and are visible as white patches.

FIGURE 1.10. Cycads (also called "sago palms") often look like small palm trees with large, leathery leaves. They produce seeds in very large cones, each with many large seeds.

FIGURE 1.11. Many ferns grow best in low, filtered light in the shade of large trees although some are adapted to the intense sunlight of deserts. Here, we see only leaves of bracken fern (*Pteridium*), their shoots grow entirely underground, spreading widely; it is possible that there is only a single fern plant in this image. Ferns never produce flowers or cones, instead some of their leaves produce spores that blow away and later grow into new plants (see Chapter 7 for more details).

the seed plants, but spores are especially easy to see on the underside of many fern leaves. Several groups of plants are so similar to ferns—and so unfamiliar to most people—that they are just called **fern allies** in most wildflower guides and other non-technical books. Fern allies include horsetails (also called scouring rushes; FIGURE 1.12), spike-mosses, and club-mosses (neither is a true moss; FIGURE 1.13). Like ferns, these are vascular plants that never make seeds.

5. Mosses, liverworts, and **hornworts** are often grouped together and informally called **bryophytes** (FIGURE 1.14). Bryophytes are tiny plants, and their bodies have fewer tissues and organs than do the bodies of all other plants: they have none of the vascular tissues that all other plants have, and they never make seeds, cones, or flowers. Many do not have anything equivalent to leaves, roots, or stems; the bodies of many liverworts are just flat sheets of green cells.

It is easy to think of some plants as "advanced" and others as "primitive," but we must be careful when we think about this. Mosses and liverworts do have simple bodies and they lack many features present in angiosperms, but that is not the same as being primitive or poorly adapted. All plants that are alive today must be reasonably well-adapted to their habitats, otherwise they would have been crowded out by other plants, they would have become

FIGURE 1.12. Horsetails (also called scouring rushes; *Equisetum*) are usually small plants but these, *Equisetum giganteum*, are the largest ones known. Horsetails are rather common in moist areas, especially along railroad tracks. Equisetums have hollow jointed stems (they can be pulled apart easily), and their surface contains silica and is very rough. (Courtesy of Chad Husby.)

FIGURE 1.13. Club mosses (*Lycopodium*) prefer shady, moist habitats such as this forest in North Carolina. Although they have vascular tissues, club-mosses never become very large, rarely more than a meter tall.

FIGURE 1.14. Mosses and liverworts often grow together, as shown here. The larger, ribbon-like plants with a netted pattern on their surface are liverworts in the genus *Conocephalum*. The smaller, leafy plants are mosses. Unlike the other plants illustrated in this chapter, these do not have vascular tissue and never grow to be very large. Being so small, mosses and liverworts often live in small cracks in rock where seed plants could not fit.

extinct. The tiny simple body of a moss or liverwort requires only a few resources to build, and just a short period of photosynthesis will make enough sugar to synthesize all the chemicals needed to build their small bodies, whereas weeks of photosynthesis are necessary before a flowering plant has enough energy and chemicals to bloom. Each plant is well-adapted to its own particular habitat. Many people like to think of ourselves, human beings, as being the pinnacle of evolution, the most advanced animal to ever exist, but there are millions more worms than there are people, living in a greater number of places, even inside some of us. Rather than thinking about which is more or less advanced, it is more productive to think about how each is adapted to its environment, and how the lives of various organisms impact those of others.

An Individual Plant

It is important to distinguish the individual from the group. Think about "people"—we are the species **Homo sapiens**—and think about yourself. Much of your biology is like mine or that of anyone else, but there are aspects of your life that differ from those of all other people. Each of us is unique in some way. Each of us may have a particular strength—some are smarter, stronger, more dexterous, more talented, more artistic, more resistant to disease—and each of us has a particular weakness, but our entire species, all of us together, have all the strengths and all the weaknesses. Whereas each individual has a set of characteristics, the whole species is more varied because it has all the characteristics of all the individuals. At some point, an individual person must die, but our species does not have to. If at least some of us are reproducing and some are adapted to conditions somewhere on Earth, *Homo sapiens* should be able to persist as a species for millions of years.

The same is true of plants. Individual plants of a particular species have certain characteristics, but the entire species has all the characteristics of all its members. The species is variable, and with enough variability, it may be able to evolve rapidly enough that it can survive even if its habitat changes. A critical point here is that larger populations tend to have more genetic diversity and thus are more likely to survive changes in the environment. When we impact plants, usually by reducing their populations by damaging their habitats, we reduce the number of individuals and the variability of the species. If we cause plant populations to be too small, we put the entire species at risk of extinction.

Things that Are Not Plants

Many creatures resemble plants in being fixed to the soil and unable to move from place to place. Don't be embarrassed if they have fooled you into thinking they are plants; all of the following have been mistakenly classified as plants at some point in the past. **Lichens** are not plants, they are not even individual organisms (FIGURE 1.15). Each consists of a network of fungal cells associated with hundreds of cells of algae, the two growing together. Some lichens grow as a simple crust on the surface of rocks or wood, but others have what appear to be stems and leaf-like structures. But despite the similarity, lichens are not plants. The corals that grow in tropical oceans are stationary, unable to move from place to place, and some have a fan-like shape that resembles a leaf. But corals are entire colonies made up of thousands or millions of tiny animals; they are not plants. Fungi have many plant-like characters, but they are their own group, the Kingdom Fungi. Examples are mushrooms (often called toadstools), puffballs, bracket fungi, morels, and chanterelles (FIGURE 1.16).

FIGURE 1.15. This rock in Zion National Park is covered with several lichens, each with different pigments: one is orange, others are dark gray, light gray and chartreuse. None is an actual plant; instead, each is a set of fungus cells and algal cells growing together, and each helping each other (this is a mutualistic relationship). The algae could not live on bare rock in full sunlight like this without the fungi, and similarly the fungi must have the algae to grow here.

FIGURE 1.16. This mushroom is the deadly fly agaric (*Amanita muscaria*). Many people mistakenly assume that fungi are plants but actually they are more closely related to animals. Mushrooms are often called toadstools.

BOX 1.1. The Scientific Method

Background

People have always tried to understand the world and how it works. At first, people believed that the world was filled with many spirits—some good, some malicious—that control weather, crops, floods, diseases (and recovery from disease), plagues of grasshoppers, and so on. Gradually, these were replaced by a multitude of gods and goddesses, and then later, the concept developed that there was a single god controlling all aspects of the world. Starting just before the 1400s, a new method, called the **scientific method**, slowly began to develop, a method that could help us understand details of how the world functions. Several fundamental principles were established as follows.

Principles of the Scientific Method

Source of information. *All accepted information can be derived only from carefully documented and controlled observations or experiments*. Explanations made by priests, prophets, scientists, or anyone else cannot be accepted automatically; they must be subjected to verification and proof. For example, for hundreds of years, medicine was taught using texts based on the work of Galen, a Roman physician who lived in the second century CE (common era = AD). In the early 1500s, Andreas Vesalius began dissecting human corpses and noticed that in many cases, Galen had been mistaken. Vesalius promoted the idea that observation of the world itself (in this case, human bodies) was more accurate than accepting undocumented claims, even if the claims had been made by an extremely famous, respected person and were preserved in ancient, highly revered books.

Examples from our own modern times are important. Various drugs and herbal supplements are advertised as curing illnesses or improving health. Under the scientific method, such claims themselves, even if made by famous and respected drug companies, are not sufficient. The actual data must be made available and the research must be described so carefully that other, independent scientists can repeat the studies so as to determine whether the drugs or supplements are indeed effective or whether these are cases of false advertising.

Phenomena that can be studied. *Only tangible phenomena and objects are studied*, such as heat, plants, minerals, and weather. We cannot see or feel magnetism or neutrons but we can construct instruments that detect them reliably. In contrast, we do not see or feel ghosts, leprechauns, elves, trolls, or unicorns, and no instrument has ever detected any of these reliably: if such things do exist, they must be intangible and cannot be studied by the scientific method. Anything that cannot be observed cannot be studied.

Constancy and universality. *Physical forces that control the world are constant through time and are the same everywhere*. Water has always been and always will be composed of hydrogen and oxygen; gravity is the same now as it has been in the past. The world itself changes—mountains erode, rivers change course, plants evolve—but the forces remain the same. Experiments done at one time and place should give the same results if they are carefully repeated at a different time and place. Constancy and universality allow us to plan future experiments and predict what the outcome should be: if we do the experiment and do not get the predicted outcome, it must be that our theory was incorrect, not that the fundamental forces of the world have suddenly changed. This prevents people from explaining things as miracles or the intervention of evil spirits. For example, if someone claims that a new drug cures a particular disease, we can check that by testing the same drug against that disease. If it does not work, the first person (1) may have made an innocent mistake, (2) may have tested the drug on people who would have gotten better anyway, (3) may have been committing fraud. But we do not have to worry that the difference in the two experiments is due to the fundamental laws of chemistry and physics having changed, or that the first experiment outcome was altered by benevolent spirits, the second by evil spirits.

Basis. *The fundamental basis of the scientific method is skepticism*, the principle of never being certain of a conclusion, of always being willing to consider new evidence. No matter how much evidence there is for or against a theory, it does no harm to keep a bit of doubt in our minds and to be willing to consider more evidence. For example, there is a tremendous amount of evidence supporting the theory that all plants are composed of cells, and there is no known evidence against it. All our research, all our teaching assumes that plants indeed are composed of cells, but the concept of skepticism requires that if new, contrary evidence is presented, we must be willing to change our minds. As a further example, consider people who have been convicted of crimes, then later—often years later—DNA-based evidence indicates they are innocent: skepticism is the willingness to consider new evidence.

Scientific studies take many forms, but basically they begin with a series of observations, followed by a period of experimentation mixed with further observation and analysis. At some point, a hypothesis, or model, is constructed to account for the observations: a **hypothesis** (unlike a speculation) must make predictions that can be tested. For example, scientists in the Middle Ages observed that plants never occur in dark caves and grow poorly indoors where light is dim. They hypothesized that plants need light to grow. This can be formally stated as a pair of simple alternative hypotheses: (1) plants need light to grow, and (2) plants do not need light to grow. The experimental testing may involve the comparison of several plants outdoors, some in light and others heavily shaded, or it may involve several plants indoors, some in the normal gloom and others illuminated by a window or a skylight. Such experiments give results consistent with hypothesis 1; hypothesis 2 would be rejected.

A hypothesis must be tested in various ways. It must be consistent with further observations and experiments, and it must be able to predict the results of future experiments: one of the greatest values of a hypothesis or theory is its power as a predictive tool. If its predictions are accurate, they support the hypothesis; if its predictions are inaccurate, they prove the hypothesis is incorrect. In this case, the hypothesis predicts that environments with little or no light will have few or no plants. Observations are consistent with these predictions. In a heavy forest, shade is dense at ground level and few plants grow there. Similarly, as light penetrates the ocean, it is absorbed by water until at great depth all light has been absorbed; no plants or algae grow below that depth.

If a hypothesis continues to match observations, we have greater confidence that it is correct, and it may come to be called a **theory**. Occasionally, a hypothesis does not match an observation; that may mean that the observation is wrong (we scientists do make mistakes), or that the hypothesis must be altered somewhat, or that the whole hypothesis has been wrong. For instance, plants such as Indian pipe or *Boschniakia* grow the same with or without light; they do not need light for growth (**FIGURE B1.1**). These are parasitic plants that obtain their energy by drawing nutrients from host plants. Thus our hypothesis needs only minor modification: all plants except parasitic ones need sunlight for growth. It remains a reasonably accurate predictive model.

Note the four principles of the scientific method here. First, the hypothesis is based on observations and can be tested with experiments; we do not accept it simply because some famous scientist declared it to be true. Second, sunlight and plant growth are tangible phenomena we can either see

FIGURE B1.1. This *Boschniakia* is a parasitic plant; it has a modified root system that attaches to surrounding plants and obtains nutrients from them. It does not have chlorophyll, and it cannot perform photosynthesis itself, but it relies on the photosynthesis occurring in its host plants. Mistletoes are partially parasitic, but unlike *Boschniakia*, they carry out their own photosynthesis and just obtain water and mineral from the host.

directly or measure with instruments. Third, if we repeat the experiment anytime or anywhere we expect to get the same results. And fourth, we interpret the evidence as supporting the hypothesis but we keep an open mind and are willing to consider new data or a new hypothesis.

Before the 1900s, theories that had overwhelming support were called "laws." These were statements considered to be universally true and proven for all time. But biologists do not use the concept of "law" any longer; instead, we realize that some of what we know is incomplete and some is wrong. So we keep an open mind and continue making observations and experiments. Occasionally people will try to discredit evolution because it is only a theory and not a law; what they do not realize is that there are no laws in biology.

The concept of intelligent design has recently been proposed to explain many complex phenomena. Its fundamental concept is that many structures and metabolisms are too complicated to have resulted from evolution and natural selection. Instead they must have been created

(continued)

by some sort of intelligent force or being. This may or may not be true, but this does not help us to analyze and understand the world; instead, the concept of intelligent design is an answer in itself, an answer that discourages further study. Photosynthesis is certainly complex, and it may have been designed by some intelligent being, but believing that does not help us to understand it at all, it does not help us plan future experiments. In contrast, the scientific method is a means through which we are discovering even the most subtle details of photosynthesis.

Areas Where the Scientific Method Is Inappropriate

Certain concepts exist for which the scientific method is inappropriate. We all believe that it is not right to wantonly kill each other, that racism and sexism are bad, and that things such as morality and ethics exist. But both morality and ethics have no chemical composition, no physical structure, no temperature; they are not tangible and thus cannot be studied by the scientific method. Science can study, measure, analyze, and describe the factors that cause people to kill each other or to be racist or sexist, and it can predict the outcome of these actions. But science cannot say if such actions are right or wrong, moral or immoral. Consider euthanasia: many types of incurable cancer cause terrible pain and suffering in their final stages, which may last for months. We have drugs that can arrest breathing so that a person dies painlessly and peacefully. Science developed the drugs and can tell us the metabolic effects of using them, but it cannot tell us if it is right to use them to help a person die and avoid pain. Biological advances have made us capable of surrogate motherhood, of detecting fetal birth defects early enough to allow a medically safe abortion, and of producing insecticides that protect crops but pollute the environment. These advances have made it more important than ever for us to have a well-developed ethical philosophy for assessing the appropriateness of various actions.

Types of Interactions Between Organisms

All organisms live close to other organisms, interacting with multiple other creatures throughout their lives. Any place on Earth that has life will have several types, not just one. If you see plants, then there almost certainly are fungi present as well, along with bacteria, soil algae, and numerous animals, perhaps most of them microscopic. And of course, if you are seeing the plants, that means that at least one person is there, interacting with the other organisms.

Most interactions between organisms can be classified as one of four basic types:

1. Both members benefit. A relationship is **mutualistic** if both organisms benefit from the relationship, if both organisms grow and develop better together than if one of the partners were missing (Figure 1.15). Many plant and people relationships are mutualistic, for example when we cultivate crops or establish and maintain a garden. Placing hives of honeybees near crops is beneficial to the crop, the bees, and to us. Establishing national and state parks as places of both wildlife preservation as well as recreation is beneficial to ourselves and the organisms in the parks.

2. One member is unaffected, the other is helped. In a **commensal** relationship one partner is unaffected and the other benefits. As we build towns, cities, and farms, we also build vacant lots and small pieces of land that are no longer natural but neither are they cultivated. These are ideal for plants such as dandelions, chickweed, plantago, tumbleweeds, and other weeds that we often just ignore. These plants benefit from the disturbances that we create, but their presence has almost no impact on us at all. When we build artificial lakes, reservoirs, and canals, many water plants benefit but we are not affected. Have you ever gone hiking then pulled sticky seeds out of your socks and shoes? You have helped distribute the seeds of some plants (so the plants benefit), but you probably didn't even notice the seeds. Unfortunately, this happens on a larger and more serious scale: seeds and animals are unintentionally moved long distances, even from one continent to another as materials are shipped around the world. This usually has no direct impact on us, but it introduces the organisms into new habitats, habitats where they may have never before existed. If they survive (or even worse, if they thrive), they become introduced invasive weeds that benefit from an expansion of their range, but which harm the native plants and animals.

3. Both members are harmed. A relationship is a **competition** when two or more species compete for the same resource. For example, roots of separate plants compete for water in the soil, leaves compete for light, and plants compete just to have enough space for their bodies. A competition tends to harm both members because neither one grows as well as it could if it had the entire resource all to itself. People compete with plants and other organisms for space and water, and people usually win. We eliminate all vegetation and animals from huge areas of land so that we can build towns, factories, farms, ranches, ski areas, marinas, airports, highways, sport facilities with huge parking lots, and so on. We dam rivers and divert water to our needs, leaving downstream ecosystems deprived of water. We even compete with plants for their pollinators and seed dispersers when we spray insecticides to kill insects and when hunters kill birds and animals.

There are not many examples of competitions in which plants win against people, but there are a few. Certain weeds, for example witchweed (*Striga*), are so invasive and difficult to control that they have forced farmers to abandon cropland. If ranchers allow pastures to be overgrazed in drier areas, prickly pear cactus (*Opuntia*) can invade and spread so aggressively that the pasture becomes useless.

4. One member is harmed, the other benefits. In a **predatory** relationship, one member is harmed and the other benefits, and usually this term is reserved for specific pairs of organisms in which the actions occur repeatedly. For examples, lions prey on zebras, and parasitic plants such as mistletoe prey on their host plants. Animals such as buffalo that graze on grasses and others, like deer that browse on leaves and twigs, are described as predators. In this respect, we prey on plants when we pick the flowers or fruits of wild plants: we take something from them but we are not cultivating them, we are not helping them. Similarly, we prey on rare orchids, bromeliads, parrots, and tropical fish when we collect them from their natural habitats rather than cultivating them. I know of no examples in which plants prey on people.

Important Terms

algae
angiosperm
Archaea
Bacteria
bryophyte
commensal
competition
conifer
cycad
Eukarya
eukaryote
fern
fern ally
flowering plant
habitat destruction
Homo sapiens
hornwort
hypothesis
lichen
liverwort
moss
mutualism
predation
prokaryote
scientific method
theory
true plant
vascular plant
vascular tissue

Concepts

- All the material of our bodies, all our energy, and all the oxygen we breathe are produced by plants.
- Plants provide us with necessities such as clothing, housing, heat, medicine, and other essentials.
- Plants provide us with nonessential foods, flavors, drinks, aromas, and beauty that add richness to our lives.
- The natural areas of Earth are not unlimited and we must preserve wilderness areas, species and their habitats; in some cases the habitats have been damaged and must be restored.
- Many activities by all sorts of people destroy or damage habitats, endangering plants, animals, and other species.
- All plants are eukaryotes (their cells have true nuclei); most plants are vascular plants (they have vascular tissues); most plants we use are seed plants, either conifers or angiosperms (flowering plants).
- Relationships between organisms can be mutualistic, commensal, competitive, or predatory.
- The scientific method is a set of standards for the type of information and logic that can be used when trying to understand material aspects of the world. One principle is skepticism, that we must keep our minds open to new possibilities.

Plants Themselves

I

The rancher who put up this sign is expressing an important concept: almost everything we eat has been cultivated on a farm or ranch. Much of our seafood is tended in "farms" rather than being caught in the open ocean. The sign could go farther, because most of the things we drink (coffee, tea, beer, wine) and many other things we use, such as cloth, paper, medicines, and ornamental flowers, are produced by agriculture.

2 Whole Plants: Introduction to Plant Bodies, Growth Forms, and Life Spans

Plant bodies are organized differently than animal bodies, and in general are much simpler, which allows many to survive to very great age and size. This live oak tree (*Quercus virginiana*) was considered worthy of such respect as an individual living organism that it was transplanted, at great effort and expense, to save it from urban development.

The bodies of all plants are extremely simple; almost all are composed of just three fundamental **organs:** stems, leaves, and roots (**TABLE 2.1**). A few familiar plants seem exceptional: cacti appear to have no leaves, and carrots look as if they are just leaves attached directly to a root, but all the parts are present (cactus leaves are tiny and carrot stems are the thin bit of tissue where leaves come together at the top of the root). The simplicity and uniformity of plant bodies allow us to easily understand almost any plant we encounter.

Whole Plant Body

TABLE 2.1. Plant Organs

Stem
- nodes
- internodes
- axillary buds
- terminal bud (in perennial plants)

Leaves

Root

Flower (a flower is considered to be a modified shoot)

Shoots: Leaves and Stems

Leaves and stems are always associated with each other: Stems produce leaves, and all leaves occur on stems. We speak of the two together as a **shoot** (FIGURE 2.1). Stems are made up of **nodes** (the points where leaves are attached) and **internodes** (areas between nodes). Internodes are almost always narrow, only a few tenths of an inch (a few millimeters) in diameter while growing (for example the stems of seedlings or the new twigs at the ends of branches on trees), but they can be broad (as in cacti and other plants with succulent stems).

If you examine a shoot while it is still elongating, you will notice that the leaves and internodes at its base (its **proximal** end) are larger and more mature than those at its tip (its **distal** end) (FIGURE 2.2). Each shoot apex has a set of cells (its **apical meristem**) that grow and multiply by cell division, producing new cells for the stem as well as cells that become part of new leaves. While young, the shoot is short and has only a few leaves, nodes, and internodes, but as it grows it becomes longer and bears more of each. Some shoots, such as those of vines and lawn grasses, grow for many months, becoming long and with numerous leaves, whereas those of trees or irises grow only briefly and produce only a few

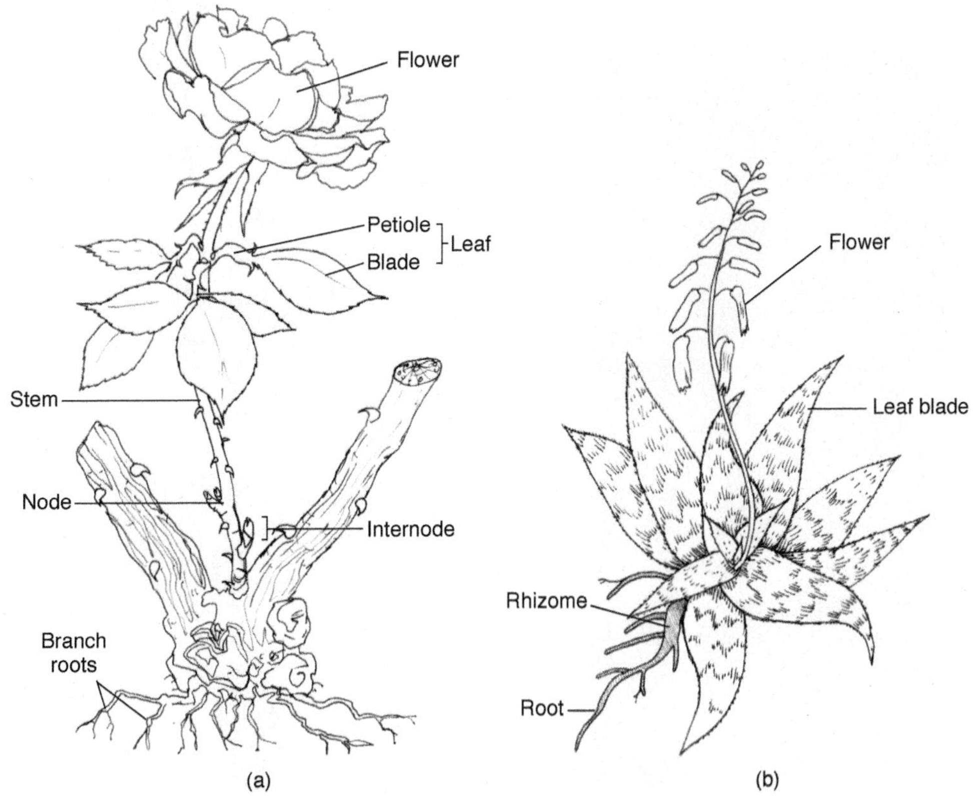

FIGURE 2.1. The body of almost all plants consists of roots, stems, and leaves. **(a)** Most flowering plants called dicots (broadleaf plants) have bodies similar to this rose. **(b)** Many of the flowering plants called monocots can be recognized by their long, strap-shaped leaves. This aloe is adapted to dry habitats: it has thick, succulent leaves and a fleshy root.

FIGURE 2.2. The two uppermost leaves are smaller and younger than the next two down, and those are smaller and younger than the next two. The apical meristem of this shoot is producing leaves in pairs, and if the youngest two that are visible here could be pulled apart, an even younger pair of leaves would be seen.

new leaves each year. Once finished growing, shoots of perennial plants (those that live longer than 1 year) will produce a dormant **terminal bud**: the apical meristem cells stop dividing and become protected by a set of tightly overlapping, tough, waterproof **bud scales** that keep the meristem safe from winter cold, insects, and fungi (FIGURE 2.3). In the following spring, the terminal bud becomes active again, allowing the shoot to resume its growth.

Each node on a stem usually has one or several **axillary buds** located in each **leaf axil**, the upper side of the point where a leaf attaches to the stem. Like terminal buds, axillary buds consist of dormant meristematic cells protected by bud scales. Under ordinary circumstances some axillary buds grow out as flowers (these are floral buds), others (vegetative buds) grow out as new shoots, allowing the shoot to branch and grow more rapidly than a single stem could (FIGURE 2.4). Most axillary buds remain dormant and act as a safety mechanism, giving plants a remarkable capacity to survive injury: even if large parts of a shoot

(a)

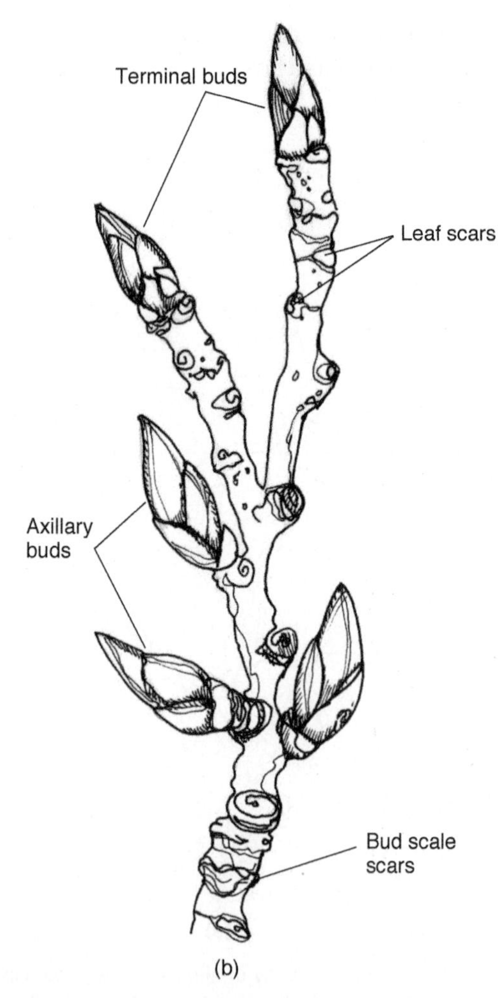

(b)

FIGURE 2.3. This twig of cottonwood became dormant in autumn; its shoot apical meristem is covered in bud scales, making this a terminal bud. Three large lateral buds are present, and the leaf scars indicate where the terminal bud had been last winter.

are damaged, as long as one or several axillary buds survive, they might be able to grow out as a new replacement shoot and keep the plant alive.

Leaves have evolved to carry out many functions in a shoot (**BOX 2.1**) but one of their most important functions is to obtain energy and building material. We animals do this by eating food, but plants have **foliage leaves** that absorb carbon dioxide from air and convert it to sugar using the energy of sunlight. The sugar is then transported from foliage leaves to stems, roots, flowers, and other parts of the plant where it is needed. Some of the sugar is converted into various chemical compounds that are needed to build the plant's cells and tissues, and some of the sugar is respired for energy, just as we obtain energy from the sugars and other carbohydrates we eat.

FIGURE 2.4. This node of jasmine (*Jasminum*) has two leaves, but the one on the right was removed to show its axillary bud more clearly. The bud is beginning to grow as a branch; the bud in the other axil, on the left, is also growing.

With foliage leaves, we can explore the concept that structures should have features that aid their functions. Being broad and flat, foliage leaves have a large surface area that is good for collecting light energy and carbon dioxide. Foliage leaves are also thin because neither light nor carbon dioxide penetrate deeply: if a foliage leaf were thick, cells in its lower side would have too little light for photosynthesis, and cells in its center would not have enough carbon dioxide. Furthermore, a leaf needs a transport tissue (**phloem**) that collects the newly synthesized sugars and transports them to the stem. It also needs another transport tissue (**xylem**) to carry water from the roots and stem and distribute it throughout the leaf such that leaf cells do not die from dehydration in dry air (leaves lose water just as our skin does, and both must be kept moist). The two transport tissues, called **vascular tissues**, are easily visible in most leaves, forming **netted** or **reticulate veins** in leaves of **dicots**, and **parallel veins** in **monocots** (**FIGURES 2.5** and **2.6**). Both patterns have such a high number of veins per square inch (about 6 square centimeters) that every photosynthetic cell is close to a vein, and sugar can move quickly into the phloem, and water is carried to every part of the leaf. Most leaves also have one or several especially large veins, called **midribs**, that provide strength to support the leaf.

Another structure/function relationship is seen in adaptations that minimize self-shading (**TABLES 2.2** and **2.3**). Most leaves have a **petiole** (stalk) that holds the **leaf blade** (also called a **lamina**) away from the stem such that it is more exposed to sunlight and is not shaded by leaves above it. In very sunny habitats such as deserts and mountain tops, this is not so important and leaves often lack a petiole. Also, many monocots have such long, strap-shaped leaves that petioles are not necessary (**FIGURE 2.7**).

Leaf blades may be either simple or compound. In a **simple leaf**, the blade consists of just one piece, but in a **compound leaf**, the blade is composed of several pieces, called **leaflets**, each attached to a rachis. If all leaves attach to the rachis at the same point, it is a **palmately compound** leaf, but if they are attached in two rows, the leaf is **pinnately compound** (**FIGURES 2.8** and **2.9**).

Simple and compound leaves allow us to consider structure/function relationships further. Whenever a plant produces a new structure, it must pay a cost, the new structure is not free; the same is true for animals as well. As a shoot's apical meristem makes a new foliage leaf, the plant must move some of its existing sugars,

BOX 2.1. Several Common Types of Modified Shoots

Bulbs. Onions, shallots, leeks (**FIGURE B2.1a**). Bulbs have a short vertical, subterranean stem with thick, fleshy leaves that completely encircle the stem, each leaf being more or less tubular. Each leaf stores water and nutrients (onions store both starch and oils). The stem of an onion is the short central basal portion that we usually throw away; onion rings are cross sections of the leaves. The large yellow or white onions in stores are harvested after the tops of the leaves have died and shriveled; while growing, each tubular "onion ring" leaf extends upward above ground as a green, tubular photosynthetic leaf.

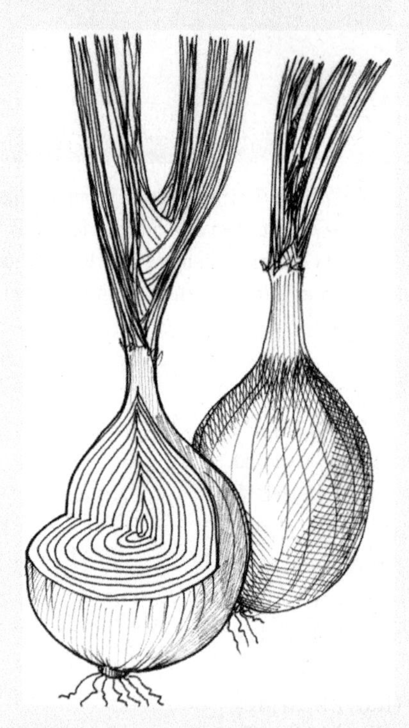

FIGURE B2.1a. An onion (*Allium*) is a bulb; it has an extremely short, vertical stem surrounded by thick, fleshy leaves. While onion leaves are growing, they have a long green portion that extends above ground and photosynthesizes whereas the lower portion of each leaf remains below ground and stores nutrients as part of the bulb. The organization of an onion is very similar to that of the orchid in Figure 2.10.

Tubers. Potatoes (**FIGURE B2.1b**). Tubers are short horizontal, subterranean shoots with extremely broad stems (the entire potato). The "eyes" of a potato are its axillary buds; its leaves are such tiny scales below each eye that they are difficult to see. A potato tuber grows from a slender subterranean shoot that then expands into a tuber. You may be able to see the slender shoot on one end of a potato, and the apical meristem on the other end. The word "tuber" is often used for any irregular, oddly shaped stem.

FIGURE B2.1b. This "eye" of a potato tuber (*Solanum tuberosum*) is really a vegetative axillary bud; it has become active and will grow out as a branch that looks and acts like a complete potato plant. Although some potato plants are grown from seeds, most are grown from pieces of tuber.

Rhizomes. *Iris, ginger, asparagus, bamboo* (**FIGURES B2.1c** and **B2.1d**). Rhizomes are thick subterranean shoots that grow horizontally, but unlike tubers, rhizomes grow indeterminately: they can become extremely long, branch profusely, and spread very widely (especially bamboo). The leaves of the rhizome are small and brown; because they are underground, they cannot photosynthesize. Some of the axillary buds become active and grow upward as vertical shoots with large green, photosynthetic leaves; these shoots often appear to be separate plants, but each is just a branch of the same plant. The asparagus shoots we eat are young vertical branches that are harvested before they can become tall, leafy, and too tough to eat.

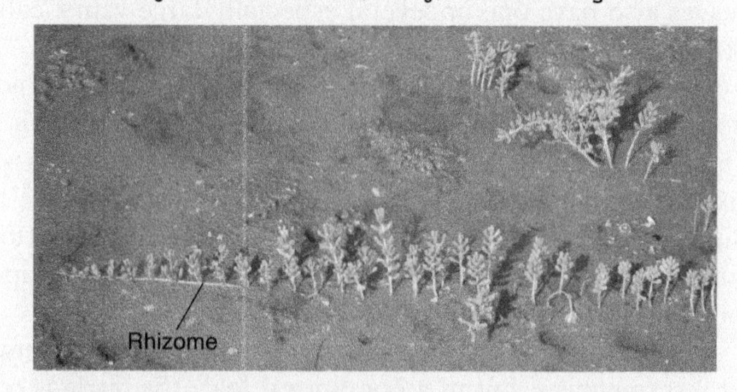

FIGURE B2.1c. Most rhizomes grow so deeply underground it is difficult to illustrate them, but the rhizome of this pickleweed (*Batis*) grows near the surface of mud in a shallow bay. The rhizome is the horizontal shoot, and each of the upright shoots is a branch growing from an axillary bud. Some axillary buds can grow out as horizontal rhizomes, allowing the plant to spread extensively.

FIGURE B2.1d. Rhizomes of ginger in a grocery store. These rhizomes are thick and fleshy with very short internodes. All of their roots and vertical branches were trimmed off when they were harvested and prepared for sale.

Tendrils. Grape vines, cucurbits (squash, cucumber), passionflower, garden peas (**FIGURE B2.1e**). Tendrils are produced on certain climbing plants. Tendrils are long and slender, and when they touch something, they wrap around it, harden, and thus provide support to the rest of the shoot. Some tendrils are modified leaves; others are modified branch shoots that grow from axillary buds.

FIGURE B2.1e. This tendril has wrapped around a wire support that allows the plant to grow very tall without having a strong woody trunk. Many garden plants such as peas and squash support themselves with tendrils.

Stem succulents. Cacti, cactus-like spurges (**FIGURE B2.1f**). Stem succulents have exceptionally broad stems composed of cells able to store large amounts of water. They are adapted to deserts, and during a rain, their roots gather water and conduct it up to the succulent stem where it is stored. If the plant is able to store enough water this way, it can survive many months without rain. All plants must store water in cells; none has any large cavity equivalent to an animal's stomach. Ordinary, non-succulent plants have such narrow stems with so few cells that they have little water storage capacity and scant ability to survive drought. The capacity to store water has evolved independently many times in varied plant families, and the term "succulent" can be applied to any of them; only the succulents in the cactus family (Cactaceae) are cacti.

FIGURE B2.1f. This *Euphorbia polygona* is a stem succulent; it has a thick fleshy stem that stores water and carries out photosynthesis. The plant does have green chlorophyll present, but that is hidden by a very thick layer of bluish-white wax. The plant is a little more than 2 inches (6 cm) tall.

Leaf succulents. *Agave, Aloe, Yucca* (**FIGURE B2.1g**). Leaf succulents have exceptionally thick leaves rather than succulent stems. They too are desert plants, but they store water in their leaf cells rather than those of their stems.

FIGURE B2.1g. This *Aloe brevifolia* is a desert plant that stores water in thick, succulent leaves. Many species in the genus *Aloe* are beautiful and easy to cultivate in a garden or flower pot. The sap of *Aloe vera* is excellent for treating minor burns.

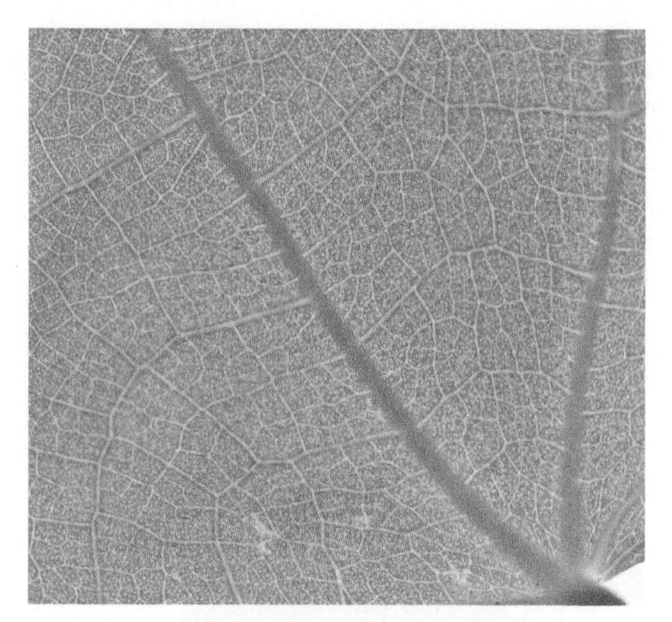

FIGURE 2.5. This is a close-up view of a dicot leaf (the leaf of a broad-leaf plant, *Malvaviscus*) showing its netted venation. The larger veins are dark because they are thick and have many dense cells, the smaller veins are bright because they are so fine that sunlight passes through them.

FIGURE 2.6. Close-up view of a monocot leaf (banana, *Musa*), showing its parallel venation (veins are light, areas with chlorophyll are dark). Compare with Figure 2.5: all the veins here run parallel to each other whereas those in a dicot form a network.

amino acids, fats, vitamins, and minerals into the developing leaf to support its growth. The leaf must then live long enough such that its photosynthesis produces more sugars than the plant spent to make the leaf. Consider a plant that needs to make one square yard (about one square meter) of new leaf surface: it could make one large leaf or many small ones. A large simple leaf blade acts like a sail or kite: it catches wind and pulls on the shoot, or in a strong wind, the blade might be damaged, shortening its life and reducing the amount of photosynthesis it can do. Simple leaves that are small may survive and function longer than large ones.

TABLE 2.2. Phyllotaxy: The Arrangement of Leaves on a Stem

The position of leaves and nodes on a stem is not random. Each leaf will have a particular orientation with regard to the leaves above and below it, and in some cases more than one leaf is attached to a node. This arrangement is called phyllotaxy.

Alternate leaves. A single leaf is attached at each node; this occurs in the greatest number of plant species. The most common type of phyllotaxy with alternate leaves is spiral phyllotaxy; each next higher, younger leaf is not directly above the leaf below it but instead is positioned about one-third of the way around the stem, and the next higher is about one-third further. The leaves are arranged in a spiral that winds upward around the stem. Alternate leaves can also have distichous phyllotaxy; each leaf is located half way around the stem relative to the one below, so the leaves are arranged in two rows.

Opposite leaves. Two leaves are attached at a node, one on either side of the stem. Typically, the next set of leaves higher on the stem is located one-quarter of the way around the stem, so the stem has four rows of leaves; this is decussate phyllotaxy.

Whorled leaves. Three (rarely more) leaves are attached at each node.

TABLE 2.3. Leaf Parts

Petiole (with an abscission zone in deciduous leaves)
Lamina (blade)
- simple lamina
- compound lamina with leaflets
  - palmately compound
  - pinnately compound

When identifying plants, it is often useful to examine the edge of the lamina (its margin) to see if it is smooth, toothed, lobed, and so on. The lamina tip may be long and pointed or short, even blunt. The two sides of the lamina base usually attach to the petiole at the same point, but in the elm tree one side attaches to the lamina slightly farther out than does the other side.

In contrast, compound leaves can be large because each leaflet is small and flutters independently of the others; the leaf acts less like a sail and there is less chance of being damaged. Almost all large leaves are compound. Other factors are also important: a compound leaf can abscise an infected leaflet, thus keeping the other leaflets safe, but if a fungus infects a simple leaf, it may spread to the entire leaf, killing all of it. On the other hand, if two leaves have the same surface area and

FIGURE 2.7. Many monocots, like this yucca (*Yucca thompsoniana*), have such long, narrow leaves that they do not shade each other significantly and do not need petioles.

FIGURE 2.8. This stem of jasmine has two leaves at each node (opposite leaves) and each leaf is palmately compound, having three leaflets that all attach to the petiole at the same point.

FIGURE 2.9. This leaf of *Mimosa* is pinnately compound: its leaflets are attached to the rachis in two lines. Each leaflet itself is pinnately compound.

other features, it is less expensive to make a simple leaf compared to a compound one. A plant can increase the life span of its leaves (and thus the length of time they produce sugars) by making the leaves poisonous or spiny or so fibrous that animals are less likely to eat them, but each of those adaptations increases the cost of the leaf and requires that each leaf must function longer. As a result, leaves vary considerably in texture, size, shape, edibility, and longevity. Many pine needles are tough, resinous, and inedible and they remain healthy and photosynthetic for years. On the other hand, leaves of beans, lettuce, and many other food plants are so soft and nontoxic that insects and deer eat them immediately unless protected by a farmer or gardener.

Most leaves are attached to the stem at one small spot, but in monocots, the base of the leaf is wrapped completely around the stem and extends down the stem for some length before it actually is attached. This portion is called a **sheathing leaf base**, and it may be as much as several internodes long (FIGURE 2.10). It gives the leaf a very firm attachment to the stem and often provides extra strength and protection to the stem itself.

Unless eaten by animals or killed by disease or frost, most foliage leaves die a controlled death. Certain minerals and vitamins are transported out of dying leaves and stored in the stem where they can be used again by the plant. **Deciduous leaves** (those that fall off the plant as they die) produce an **abscission zone** with two layers of cells: the outer cells die and break down, releasing the petiole and lamina from the stem. Cells of the inner layer convert themselves to cork cells, which create a protective **leaf scar** that prevents fungi and bacteria from entering the stem (FIGURE 2.11). Because the sheathing leaf base of monocots completely encircles the stem, when they abscise a leaf, its scar forms a complete ring around the node. Nondeciduous leaves lack an abscission zone and remain on the plant after they die, often hanging down as a layer of dry, inedible,

FIGURE 2.10. The base of each leaf of this orchid (*Dendrobium*) completely ensheaths the stem; the leaves do not attach to the stem in one small spot as a leaf with a petiole would. Parallel venation is visible in the lowest leaf base.

FIGURE 2.11. This shoot of chinaberry (*Melia azedarach*) was photographed in winter after the leaf had abscised. The leaf scar and dormant axillary bud are visible.

FIGURE 2.12. Leaves of Spanish dagger (*Yucca torreyi*) have no abscission zone, so they remain on the plant after they die. Not only are they are hard, dry, and non-nutritious, but each has a sharp point that deters animals from climbing the stem and eating the fresh leaves and shoot apex.

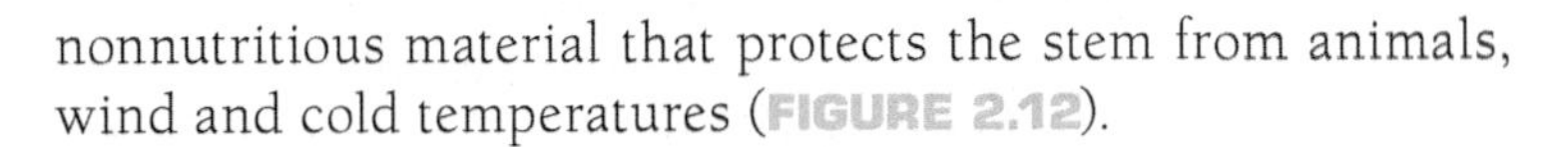

nonnutritious material that protects the stem from animals, wind and cold temperatures (FIGURE 2.12).

Root Systems

Roots perform many tasks simultaneously. They support the shoot, absorb water and minerals, and they produce hormones necessary for shoot growth. If you have pulled out a weed or transplanted a plant, you know that if soil around the roots is loosened, the plant is unstable and must be propped up until its roots reestablish a firm grip on the soil. One consequence of being well rooted and stable is that leaves and flowers are held high and in the proper position. But another consequence is that plants are not mobile like animals; they cannot flee from herbivores or diseased neighbors; they cannot move to a place with more sunlight or water or better soil. Plants must cope with life in one spot, defending themselves, if they can, from animals, fungi, summer's heat, and winter's cold. Imagine standing naked in one spot for an entire year, or like some trees, for hundreds of years. Even if someone brought you food and water, there is little chance you would survive. But, of course, plants must because they are rooted there. Roots have a profound effect in determining which features will be beneficial to a plant's survival.

Unlike us, plants never take a drink of water. Instead they absorb water through their roots and do so slowly and constantly as long as the soil has enough moisture. And they cannot choose their water; they always absorb the water present in soil. This is actually beneficial because soil water contains minerals such as potassium, calcium, iron and so on, and plants obtain these necessary nutrients by absorbing them through their roots.

It is important that a plant have a root system that is just the right size. It should be large enough to collect all the water and nutrients needed for the shoot but not so large that the plant wastes resources building an unnecessarily extensive root system. The balance between shoot and roots is maintained by hormone production

TABLE 2.4. Root Parts

Root tip (with root cap and root apical meristem)
Zone of elongation, maturation
Root hair zone
Mature region where lateral roots can emerge

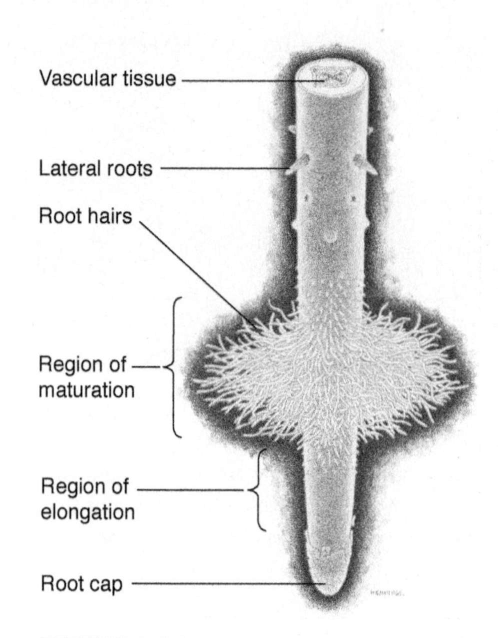

FIGURE 2.13. The tip of a root consists of a root cap, the root apical meristem, a zone of elongation, and a region where root hairs are formed.

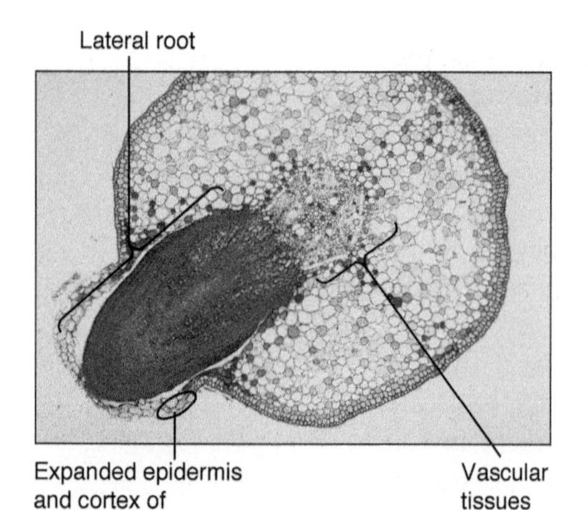

FIGURE 2.14. This young lateral root of willow (*Salix*) was initiated deep within the root tissues. Its growth caused considerable damage to the parent root cortex, and it broke the surface open (×50).

in each: healthy roots produce hormones that promote shoot growth whereas unhealthy or dormant roots do not. Similarly, shoots control root growth by providing hormones and sugars necessary for root metabolism.

An individual root is simpler than a stem. It has no leaves, nodes, internodes, or axillary buds (**TABLE 2.4**). Instead, a root is a narrow cylinder with a **root apical meristem** at its far (distal) end that produces new root cells (FIGURE 2.13). Cells at the root tip are younger, smaller, and less mature than those at the root's base (closer to the shoot): basal cells, being mature, do not elongate and so older parts of a root are not pushed through the soil. But cells that have just been produced by the root apical meristem are young and small and must enlarge. As they do, they push forward into the soil; this region is the **zone of elongation**. These delicate cells are protected because the root tip excretes a slippery, lubricating substance called mucigel. Also the apical meristem is covered by a **root cap**, a set of cells that can be replaced easily if they are damaged by being pushed against soil particles. Roots are able to grow only in soil that has adequate space between the individual soil particles, spaces wide enough for the root tip to push through. Roots cannot grow into solid rock or tightly packed clay.

Just behind (proximal to) the zone of elongation is the **root hair zone**. In this area, some of the root's surface cells grow outward as root hairs, which, being only one cell wide, are extremely slender and able to penetrate into very fine spaces between soil particles, spaces filled with water and mineral nutrients. Root hairs absorb these and transfer them into the root, which in turn transports them to the shoot.

No plant has a single, long unbranched root. Instead, roots branch and bear **lateral roots** that also branch and so on. A root system is usually a highly branched set of roots that somewhat resembles the shoot. Because roots have no axillary buds, they need an alternative way to branch. Some of the root's internal cells form themselves into a cluster that functions as a new root apical meristem. It grows outward through the parent root's outer tissues until it emerges and then pushes into the soil (FIGURE 2.14).

Two fundamentally different types of root systems occur within plants: taproots and fibrous roots. **Taproot systems** have one main root, and familiar examples are carrots and beets. The swollen size of these taproots is striking, but the critical feature is that when a seed germinates, its seedling root produces the entire root system, so only one single root attaches to the base of the shoot. In plants with **fibrous root systems**, the seedling root dies quickly, and new roots must be produced in the stem. Consequently these roots attach to the shoot in various places. Roots that arise in a stem or leaf rather than in another root are called **adventitious roots** (FIGURE 2.15). Fibrous root systems composed of adventitious roots are typical of monocots.

Certain plants are adapted to live on the branches of trees or floating on lakes, and as might be expected, their roots must be adapted to those different "soils." Vines like ivy and *Philodendron* climb trees or walls because they produce adventitious roots that adhere to rough surfaces like brick or bark. **Epiphytic** plants like *Tillandsia* live on the branches of trees, holding themselves there with wire-like roots that grip the branch's bark. Epiphytic orchids have both gripping roots as well as other roots that dangle into the air, absorbing water when they are moistened by rain, fog, or dew. Water hyacinth and several other plants float on the surface of lakes; their roots dangle into the water but do not reach the lake bottom. Because their roots attach to nothing, they provide no anchorage and the plants drift from place to place.

FIGURE 2.15. This shoot of poison ivy (*Toxicodendron radicans*) has climbed up a tree trunk and is attached to it by numerous adventitious roots. Adventitious roots form in a part of the plant that is not a root.

Reproductive Organs: Flowers, Fruits, and Seeds

Some trees live for thousands of years but none is immortal. The same is true of animals and all other organisms: because individuals have a limited life span, if a species is to persist through time, some of its individuals must reproduce before they die. Like animals, plants reproduce sexually, but plant sex organs differ from those of animals. In angiosperms (flowering plants), the sex organs are located within flowers, those of conifers are in cones, and ferns reproduce sexually with spores. Despite the differences in structures, several basic principles are universal in all organisms: genetic information from a paternal (male) parent is combined with information from a maternal (female) parent such that the offspring (seeds or babies) have some features of each parent. The significance and details of this are covered in Chapter 7 but the basic structures of flowers are presented here.

In general, each flower has four types of organs (**TABLE 2.5**) attached to a **receptacle** (a specialized section of stem) supported by an internode called a **pedicel** (**FIGURE 2.16**). Just like vegetative shoots, each flower is produced by an apical meristem, the floral apical meristem. **Sepals** are outermost and are usually green and somewhat leaflike; they enclose and protect the rest of the flower as it develops. When a flower is ready to open, the sepals spread apart allowing the inner organs to expand. **Petals** are typically the most colorful, fragrant, and distinctive portions of flowers; we distinguish between roses, larkspurs, morning glories, and so on by the color, size, and shape of the petals. More importantly, bees, birds, and other animals that carry pollen from flower to flower also identify the correct flower by petals characters. **Stamens** produce pollen, and within each pollen grain is a microscopic cell that produces two sperm cells. **Carpels**, located in the center of a flower, contain structures that produce egg cells. Stamens and carpels are often referred to as the male and female parts of a flower, but technically this is not quite

TABLE 2.5. Flower Parts
Sepals: protect flower parts as they develop
Petals: attract pollinators
Stamens: produce pollen, which contains sperm cells
Carpels: produce structures that contain egg cells; carpels develop into fruits

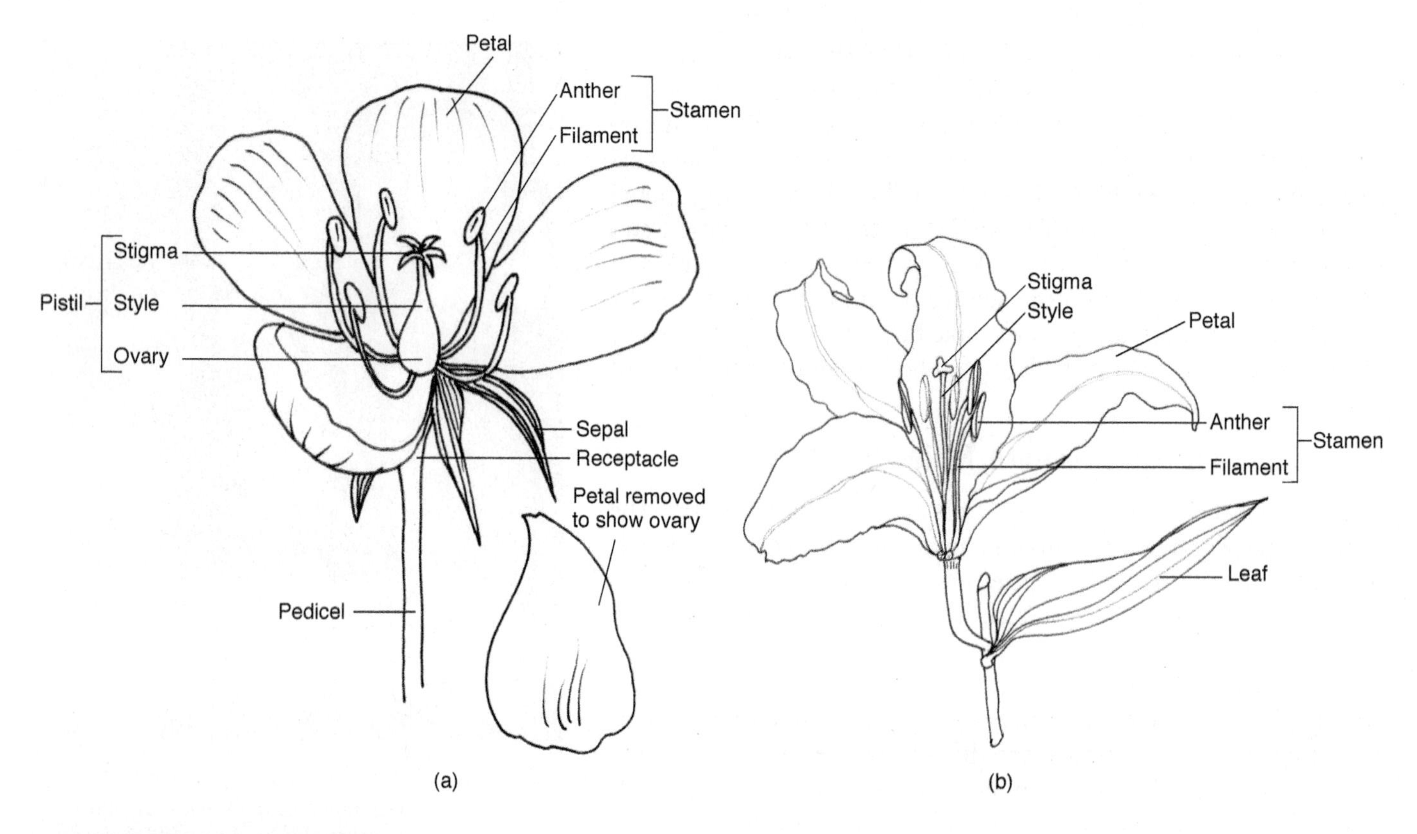

(c)

FIGURE 2.16. Most flowers have four types of structures: sepals, petals, stamens, and a pistil. **(a)** Dicot flowers often have five of each organ, that is, five sepals, five petals, and so on. The carpels are often joined together as a single structure. One petal has been drawn as being detached so that the carpel could be seen. **(b)** Monocot flowers usually have three of each organ instead of five, and very often the sepals and petals greatly resemble each other. **(c)** In most flowers, the ovaries at the base of the carpels are too small to be seen easily, but they are readily visible in lilies. On the top is the stigma with three lobes, in the center is the style, and on the bottom is the green ovary (this is actually three ovaries fused together into one structure). Inside the ovary are the ovules that will develop into seeds if the flower is pollinated. The ovary will develop into a fruit.

correct (only the cells that produce the sperms and eggs should be called male and female in plants).

For sperm cells to meet egg cells, pollen must be moved from stamens to carpels; this is usually done by pollinators such as birds, bees, and bats, or by wind. After a sperm cell fuses with an egg cell, the resulting cell grows into an **embryo** (an embryonic plant) within a seed. Simultaneously, parts of the carpel develop into a **fruit** (FIGURE 2.17). With humans, an embryo grows continuously, then is born and continues to grow. But in plants, the embryo and seed become dormant while still small, and then must be dispersed away from the parent plant by wind or by animals eating the fruits and then either spitting the seeds out or by defecating them. Later, sometimes years later, the seeds absorb water, the embryo resumes activity and grows into a new plant with roots, stems, leaves, and flowers of its own.

Reproduction in Other Plants

Conifers (pines and their relatives) reproduce sexually by making sperm cells and egg cells inside cones, then wind carries pollen from one plant to another. After fertilization, seeds develop inside cones rather than inside fruits. This is similar to sexual reproduction in flowering plants, but because of certain technical details, cones are not considered to be either flowers or fruits.

Sexual organs in ferns are less like those in flowers. Ferns produce spores that blow away, then germinate and grow into new, very small plants that look nothing like a typical fern. However, these small plants make the necessary sperm cells and egg cells, and the fertilized egg grows into a plant that is an ordinary fern. The fern embryo never becomes dormant and never is part of a seed. See Chapter 7 for details.

FIGURE 2.17. Each of these tomato fruits (*Lycopersicon esculentum*) is the base of a carpel that began developing into a fruit after fertilization caused seeds to begin developing inside (each tomato has many seeds). The flower stalks (pedicels) are developing into fruit stalks, and the five sepals of each flower are still present. The petals and stamens withered and fell off shortly after fertilization.

Comparisons Between the Bodies of Plants and Those of Animals

Plant bodies differ greatly from animal bodies, and it is worthwhile to think about these differences to understand several underlying principles.

Complex Versus Simple Bodies

We have already discussed that plants are stationary rather than mobile, that they absorb water and mineral nutrients through their roots rather than by drinking, and that they gather energy and carbon with their foliage leaves rather than by eating. Plants do not need to detect food, nor do they need to hunt, catch, or digest it. They do not need to defecate indigestible parts of food as we must. The complex, sophisticated sense organs, central nervous system, muscles, and skeleton that we must have are unnecessary to plants. We must be careful to eat healthful, nutritious foods to maintain not just our basic metabolism, but especially all our complex organs: our organs are expensive but plants do not need to either construct or maintain such organs. Furthermore, even though our organs keep us alive, they also make us susceptible: a wild animal that goes blind or deaf, or that loses its teeth or breaks a leg will soon die even if all the rest of its body is healthy. In contrast, large parts of most plants can die but the plant will survive as long as some axillary buds and roots are functional. Our complex organs are adaptive for our methods of obtaining food and avoiding danger, and a plant's simple organs are adaptive for its biology.

Localized Growth and Diffuse Growth

Plants produce new cells and grow only in their shoot and root apical meristems and leaf primordia; this is **localized growth**. In contrast, all parts of an animal's body grow (**diffuse growth**). From the time we are embryos until we are about 18 years old, all parts of our body grow simultaneously; our small feet and hands grow larger, as do our heart, lungs, liver, and so on, and all are always the same age. Due to a plant's localized growth, an individual plant can have older, completely mature and highly functional leaves at the same time that it is still producing new leaves, leaves that are too young to function well. This is especially important for trees: the trunk and branches can be mature and so hard that they cannot grow taller while at the same time the tree has numerous twigs that are young and soft enough to grow easily. If a tree had to grow to its full size while its wood was immature and soft, the tree would collapse long before it reached full size.

There are some exceptions. Individual leaves and fruits grow diffusely, whereas in us humans our finger nails and hair have localized growth, each grows only at its base. Our skin too is constantly being renewed by cell divisions localized in its deepest layers.

Indeterminate and Determinate Organogenesis

An especially important aspect of localized growth in plants is that it permits **indeterminate organogenesis**. A shoot apical meristem can continue to produce new leaves, nodes, internodes, and axillary buds as long as it is healthy. There is no predetermined number of leaves, branches, roots, or flowers a plant can have. In contrast, most animals have **determinate organogenesis:** the number of organs we will ever have is predetermined by our genes. During a brief period while we are embryos, all our organs are initiated then no new ones are ever established again in our lifetimes. No matter how long we live, we will never have more than two arms, two legs, two eyes, one liver, and so on. Even when our permanent teeth replace our baby teeth, that is not new organogenesis; the primordia for our permanent teeth were established before we were born.

Having indeterminate organogenesis causes plants to differ from animals in many ways. Plants have numerous temporary organs whereas animals have a set number of permanent ones. When we are 80 years old, so are our eyes, teeth, stomach, heart, and so on, and many of those will probably not be as healthy as they were when they were only a few years old. But an 80-year-old tree will be using new foliage leaves that are only a few months old, its root hairs will also be new, and so will its flowers. As we get older, so do all of our organs; as plants get older, they always make new organs and throw the old ones away.

This also affects the concept of "growth." As we animals grow, all of our organs grow but we do not make additional organs. If an animal could somehow continue to grow throughout its entire life, it would get taller, heavier, maybe faster, maybe able to catch more prey or be better at escaping predators, but it would not be able to reproduce more. It would still have the same number of sex organs and would produce the same number of offspring. But plants are different. Each year a tree becomes larger by adding more leaves, branches, and roots. It can gather energy, carbon dioxide, water, and minerals faster than the year before. And the important thing, it will have more flowers, more pollen, more eggs, more seeds, and more offspring. A 20-year-old oak might produce a few dozen acorns each summer, but when it is 200 years old, it will be able to produce thousands every year. For plants, localized growth and indeterminate organogenesis permit an ever-increasing reproductive capacity.

Life Spans

An important concept of life is death, the cessation of life. Death can be caused not only by external factors such as disease, predators, and accidents but also by internal factors that act as part of normal development. Plant life spans vary from just a few months to thousands of years; many insects also live for just a few months and some animals live more than 100 years but none lives for thousands. Plants that live less than a year are **annual plants**, those that live longer are **perennials**.

Annuals

Despite having the potential for indeterminate organogenesis, annual plants automatically die while only a few months old, while the environment is still benign and while the plant is otherwise healthy (FIGURE 2.18). Examples are cereal grasses like wheat and oats, vegetables like peas and beans, and many garden flowers and weeds. The seeds of annual plants usually germinate in spring, grow rapidly into a small plant, then flower, produce seeds and fruits, and then die. All the thousands of plants in a wheat field will die once their seeds are mature, even though conditions are mild enough that nearby plants continue to thrive. One of the shortest life spans known in plants is that of *Arabidopsis thaliana*: it is only 6 weeks. Its seeds germinate quickly and the seedlings grow for about 3 weeks, then they produce a set of flowers (FIGURE 2.19). Within another 3 weeks, the flowers have been pollinated, formed seeds and fruits, then die. *Arabidopsis thaliana* is used in numerous experiments to study inheritance of genetic traits and their influence on development and the plant's health. In about a month and a half after beginning an experiment, the results can be obtained and planning for a new experiment can begin. Such studies in perennial trees would take years.

Perennials

Perennial plants have no specific life span, instead their longevity depends on environmental conditions (FIGURE 2.20). Their death is just an accident of nature, caused by wind, fire, floods, pathogens, or rare freezes. If well protected, they may live for hundreds of years. A long-lived plant must protect itself for a long time: if it lives

FIGURE 2.18. This was photographed in July, and most plants are still alive and healthy, but the annual plants—the fields of wheat (*Triticum*)—have died as a natural part of their own development and life cycle.

FIGURE 2.19. This plant of *Arabidopsis thaliana* is fully mature: it has some flowers still open and some older flowers are already developing into seeds. This annual plant will complete its life cycle and die naturally when only about 2 months old. (Courtesy of Tharindu Weeraratne.)

FIGURE 2.20. This redwood (*Sequoia*), planted just after the home (chateau de Cheverny) was built in 1630, is almost 400 years old. Unlike annuals and biennials, perennials such as this do not have a life span determined precisely by their genes.

for several years, it is certain to be noticed by an herbivore or by some insect that can lay eggs in it, or a pathogenic spore will land on it. Perennials must use some of their energy to construct defensive chemicals or protective surfaces. Annuals, however, live such a short time that with some luck, they will not be attacked by predators or pathogens. It is less important for them to use any of their resources to build defenses.

Biennial plants are perennials that live for 2 years. These plants germinate and grow vegetatively for 1 year, then become dormant during winter. The following spring, they grow briefly then flower, produce seeds, then die (FIGURE 2.21). Under natural conditions, biennials do not flower in their first year nor do they survive their second. Examples are beets, carrots, and cabbage.

Reproductive Cycles

Plants vary in the number of times they reproduce in their lifetime. Annuals and biennials have only one reproductive cycle: they reproduce once and then die after their seeds and fruits are mature. Because perennials live for many years, they can have many reproductive cycles. Once old enough, trees and shrubs usually bloom every year for many years. Plants that reproduce only once are **monocarpic**; those that reproduce several times are **polycarpic**.

A few perennials are unusual because they are monocarpic. They live for many years without ever flowering, instead merely becoming larger. Then at some point the plant undergoes a massive flowering, using all the energy and resources it has been storing for years to produce thousands of flowers and seeds, after which the entire plant dies (FIGURE 2.22).

The bamboo is an important example of monocarpic perennials. After a bamboo seed germinates, it grows and spreads, sending out underground rhizomes, that, after many years may fill an entire valley. A single plant can live, in some species, for 120 years without ever flowering. When it reaches the correct age, it flowers: the thousands of bamboo shoots of the single plant produce flowers, then millions of seeds and fruits. Afterward, the entire plant dies; every single bamboo stalk in an entire valley may suddenly die in 1 year, after having been healthy for more than a century. Giant pandas eat only bamboo leaves, nothing else. In the past, when a bamboo plant would die, the pandas would simply migrate to a new valley with young, healthy bamboo. But now people live in much of the panda habitat, and between one valley and the next, there may be roads, rail lines, canals, towns, and fields. It is no longer easy for pandas to migrate, and often when their bamboo dies, the pandas starve.

Several species of trees undergo a reproductive cycle called masting: all the trees in an area bloom and produce seeds at irregular periods several years apart. In the years when they are not flowering, animals that depend on their seeds for food have little to eat and many starve. Consequently, when the trees do bloom, they produce more seeds than the few animals can possible eat, so many seeds survive unharmed.

(a)

(b)

FIGURE 2.21. **(a)** In its first year, a biennial plant has only a very short stem with all its leaves attached close together and is unable to flower. **(b)** In its second year, a biennial plant produces a large group of flowers (an inflorescence), almost always on a tall stalk. After the fruits and seeds are mature, the plant dies. This is mullein, *Verbascum thapsus*.

FIGURE 2.22. This giant agave (*Agave americana*) has been cultivated in an Austin park for probably 20 years and never flowered until the year this photograph was taken. The entire central stalk grew to its full height in just a few weeks and produced thousands of flowers. Each is now developing into a fruit, which will open in autumn and release the seeds. All the leaves at the plant's base are dead or dying, and by autumn the entire plant will be dead, except for its seeds and the four or five "plantlets" (each more than a meter tall) that it produced from axillary buds.

Juvenile and Adult Phases

Newly germinated seedlings, just like newborn babies, cannot immediately undergo sexual reproduction. Instead, all their energy and resources are directed to growth, development, and survival. This non-reproductive period is the **juvenile phase**. Later, the plant enters its **adult phase** during which it is able to reproduce. The juvenile phase lasts just a few days or weeks in annual plants, but it may be several years long in perennials; some trees do not produce their first flowers until they are at least

FIGURE 2.23. The top part of this cactus (*Melocactus*) is a cephalium, which is the adult stage of its growth. The bottom part of the plant is the juvenile stage and it was (and still is) unable to flower. When the plant became old enough to flower (when it became adult), its growth changed completely; since that time it has grown as a cephalium. Four small pink flowers are open at the top of the cephalium. In almost all other cacti, the adult plant looks like the juvenile, and the shoot apex continues to make a body that looks like an ordinary cactus.

15 years old. In many animals, at least some features of an individual change when they pass from juvenile to adult, and this is also true of a few species of plants. In ivy, citrus, junipers, and some cacti (*Melocactus*, *Backebergia*), the leaves and stems produced when they are juvenile are obviously different from those produced when they are adult (**FIGURE 2.23**). But in most plants, there is no change in outward appearance; adult plants can be distinguished from juveniles only when they are flowering or bearing fruit.

Herbs and Woody Plants

The plant kingdom contains two fundamental types of body: herbaceous and woody (**TABLE 2.6**). An **herbaceous plant** (an **herb**) grows only by means of its apical meristems and leaf primordia, as described earlier in this chapter. In contrast, **woody plants** contain two additional meristems that produce additional cells: a **vascular cambium** that produces xylem and phloem, and a **cork cambium** that produces cork cells that cause bark to be waterproof and resistant to microbes (xylem produced by a vascular cambium is called wood, whereas xylem produced by apical meristems or leaf primordia is just called xylem). The word "herb" is often used to mean a plant with a savory flavor used for cooking; many culinary herbs such as peppermint and spearmint are from herbaceous plants, while others like oregano and rosemary are leaves of woody plants.

TABLE 2.6. Types of Plant Body

Herbaceous. An herb grows only at its root, shoot, and flower apical meristems and its leaf primordia. All tissues in the body of an herb are derived from these meristematic regions, and technically they are known as primary tissues. They make up the plant's primary body.

Woody. A woody plant also always has root, shoot, and flower apical meristem as well as leaf primordia. But in addition, they also produce a vascular cambium and cork cambium which produce wood, phloem, and bark. Tissues derived from the vascular and cork cambia are technically known as secondary tissues and compose the plant's secondary body.

Herbs

By definition, herbs are plants whose stems and roots never make wood, no matter how long they live. Familiar examples of herbs are snapdragons, African violets, wheat, corn, and squash. All their cells are produced by apical meristems. Because they lack wood, the shoots of most herbs are so slender and weak they cannot grow very tall without falling over; herbs are usually small. They are often annuals but some are perennial; ferns, gladiolus, and ginger are examples of herbs that live for many years.

Several exceptional herbs such as bamboos and palms are so large, hard, and tough they appear to be woody plants. But they are strong due to having large amounts of fibers, not wood. What is the difference? These fiber cells are produced by shoot apical meristems, not by vascular cambia. Bamboos and palms are big, perennial herbs.

Woody Plants

All woody plants begin life as herbs. During its first several weeks or months of life, a seedling is completely herbaceous, but at a certain stage, some cells in the seedling convert themselves into vascular cambium cells, and others convert to being cork cambium cells. These two new meristems then begin producing wood, phloem, and cork (FIGURE 2.24). The vascular cambium allows stems and roots to add new conducting cells each year: as a woody plant grows larger, with a greater number of leaves and photosynthetic capacity, and a more extensive root system, the plant needs a greater amount of xylem and phloem if it is to conduct the extra sugar down to the roots and the extra water up to the shoots. Because of the vascular cambium, the trunk, branches, and roots have more conducting capacity every year and are also stronger. Vascular cambia are another example of localized growth that allows for indeterminate growth. The giant redwoods in California exemplify how massive woody plants can become.

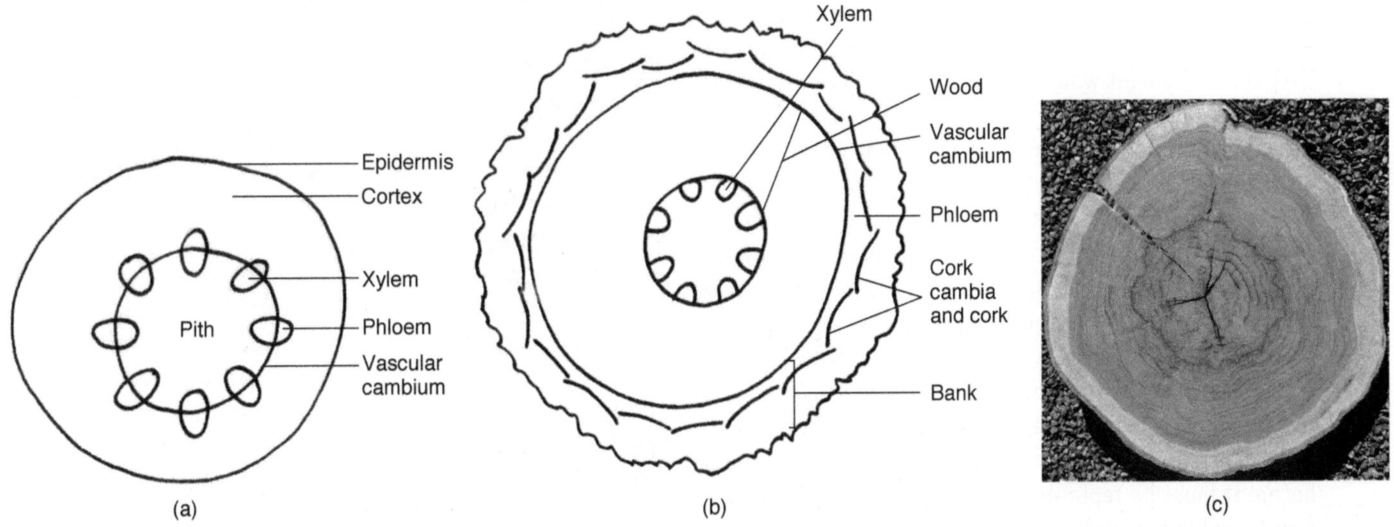

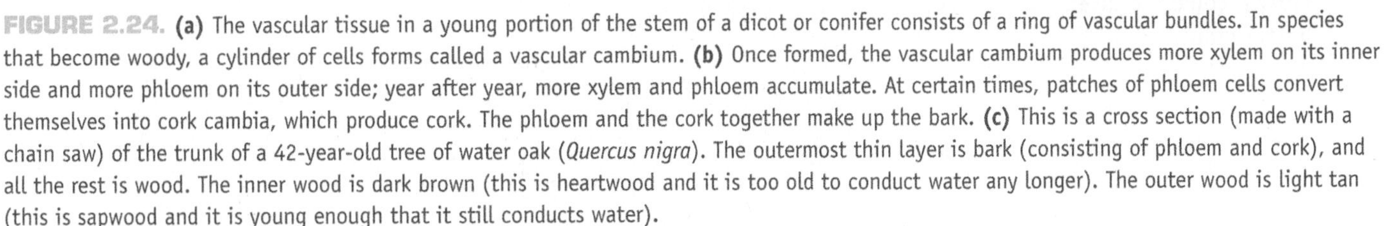
FIGURE 2.24. **(a)** The vascular tissue in a young portion of the stem of a dicot or conifer consists of a ring of vascular bundles. In species that become woody, a cylinder of cells forms called a vascular cambium. **(b)** Once formed, the vascular cambium produces more xylem on its inner side and more phloem on its outer side; year after year, more xylem and phloem accumulate. At certain times, patches of phloem cells convert themselves into cork cambia, which produce cork. The phloem and the cork together make up the bark. **(c)** This is a cross section (made with a chain saw) of the trunk of a 42-year-old tree of water oak (*Quercus nigra*). The outermost thin layer is bark (consisting of phloem and cork), and all the rest is wood. The inner wood is dark brown (this is heartwood and it is too old to conduct water any longer). The outer wood is light tan (this is sapwood and it is young enough that it still conducts water).

BOX 2.2. Food Plants as Examples of Plant Types

Several of the concepts of this chapter will be more understandable if we relate them to plants that are familiar to us, especially food plants. Our diet must supply us with energy, vitamins, minerals, and interesting flavors. Some animals obtain all these from a single species of plant but we humans eat a variety of foods.

We obtain most of our energy by eating foods rich in carbohydrates, usually starch: wheat (as bread and pasta), rice, corn (either as kernels or as corn syrup that has been added to other foods to sweeten them), potatoes, and sugar (from sugar cane or sugar beets). Wheat, rice, and corn are grasses that are annual, monocarpic, and herbs: each year a new crop must be planted, allowed to grow for several months, then be harvested after the fruits have matured and the plants have died. In each of these, we harvest the seeds and fruits (kernels of wheat, rice, and corn are often called seeds, but each is actually a single dry fruit that adheres tightly to a single seed). We discard all the rest of the plant, or merely use it for straw for livestock such as horses and cattle. Potatoes are also annual herbs but we harvest the tubers not the seeds, fruit, or any other part.

Sugar beets are biennial, but we harvest them during autumn of their first year; at that time they have stored large amounts of sugar in their taproots. If we did not harvest them, they would use that sugar to produce a tall shoot bearing hundreds of flowers in their second year. At that time, the beet would be empty of sugar and useless as a food.

Sugar cane is a rhizomatous perennial grass. Pieces of rhizome are planted and allowed to grow. Some of the axillary buds grow out as more rhizomes, causing the plant to spread horizontally; other axillary buds grow upward as photosynthetic shoots. Once these are mature, the aerial shoots are harvested by cutting them off at their bases, then their sugary sap is squeezed out of their nodes and internodes. The subterranean rhizome is unharmed and it will send up more aerial shoots the following spring. A field of sugar cane can be harvested for many years before it must be replanted with fresh pieces of rhizome (seeds could be used, but pieces of rhizome give a better crop).

Most of us obtain a significant part of our protein by eating meat, fish, eggs, and milk, but members of the legume family (Fabaceae) provide us with many protein-rich seeds: beans, peas, lentils, peanuts, and soybeans. These plants too are annual, monocarpic herbs and must be replanted every year.

In contrast, perennial, polycarpic woody trees and shrubs provide us with many of the foods we eat for vitamins, sweetness, and flavors. Trees of the rose family (Rosaceae) provide apples, peaches, pears, apricots, almonds, and cherries. The citrus family (Rutaceae) provides oranges, lemons, limes, grapefruits, and kumquats. Blueberries and cranberries are the fruits of small shrubs. Olives, harvested mostly for their oil, are obtained from trees, as are avocados. Tree crops, being perennial and polycarpic, do not need to be replanted each year, and a single tree will produce a harvest every year for many years.

Notice that almost everything mentioned above is a seed or fruit. Seeds must be rich in energy and nutrients because they support the growth of the embryo when the seed germinates. The flavors and nutrients in fruits fulfill a very different role. It is best if a plant's seeds are dispersed away from the parent plant; if all the seeds fall to the ground below the parent tree, they will compete with each other—and with the parent tree—for water and minerals when they germinate and grow. Some seeds and fruits, like those of maples and dandelions, blow away, but juicy, flavorful, nutritious fruits encourage animals to eat them; if the seeds fall out during sloppy eating, or are spit out or later defecated, then they will have been dispersed away from the parent tree.

We consume the leaves of only a small number of plants, as salads of lettuce, cabbage, spinach, and onions, or in drinks like tea. Leaves are typically low in carbohydrates, proteins, and sweetness but they do provide minerals, vitamins, and other beneficial chemicals. These plants are small herbs and are most easily harvested by cutting the entire shoot (a head of lettuce or cabbage) away from its root. The tea we drink is made from leaves of small perennial woody shrubs, and each leaf must be picked by hand. Tea provides no significant carbohydrate, protein, or vitamins, but it does offer unique flavors and factors that calm us.

Certain exceptional woody plants produce so little wood they are easily mistaken for herbs. Many large annuals like sunflowers produce a small amount of wood before they die. Also, barrel cacti and many other succulents have a small amount of wood in their center. But the amount of wood is not important: if a plant makes any wood at all, it is a woody plant, not an herb.

Important Terms

adult phase
adventitious roots
angiosperm
abscission zone
annual plant
apical meristem
axillary bud
biennial plant
bud scale
bulb
carpel
compound leaf
cork cambium
deciduous leaf
determinate organogenesis
dicot
diffuse growth
embryo
epiphyte
fibrous root system
foliage leaves
fruit
herb
herbaceous plant
indeterminate organogenesis
internode
juvenile phase
lamina
lateral roots
leaf axil
leaf blade
leaflet
leaf scar
localized growth
midrib
monocarpic
monocot
netted venation
node
organ
palmately compound leaf
parallel venation
pedicel
perennial plant
petal
petiole
phloem
phyllotaxy
pinnately compound leaf
polycarpic
primary growth
primary plant body
proximal
receptacle
reticulate vein
rhizome
root apical meristem
root cap
root hair zone
sepal
sheathing leaf base
shoot
simple leaf
stamen
taproot system
tendril
terminal bud
tuber
vascular cambium
venation
woody plant
xylem
zone of elongation

Concepts

- Bodies of almost all plants are composed of roots, stems, and leaves.
- Most axillary buds remain dormant and act as a safety mechanism in case the shoot apex is damaged.
- Structures should have features that aid their function.
- Every structure or metabolism has a cost; nothing is free.
- The adaptive value of structures and metabolisms must be evaluated in terms of the type of biology an organism has.
- Plants have indeterminate organogenesis; animals have determinate organogenesis.

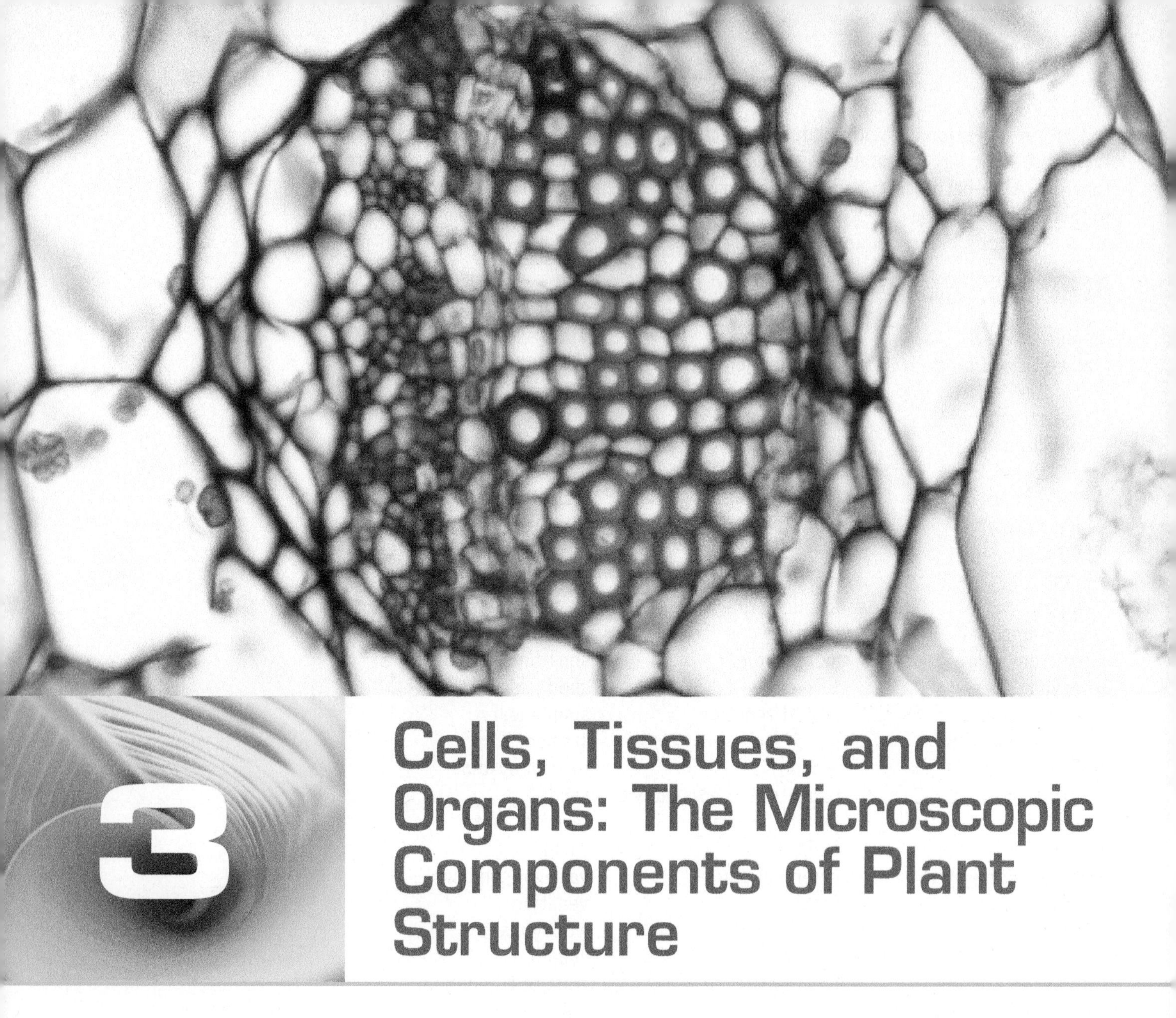

3 Cells, Tissues, and Organs: The Microscopic Components of Plant Structure

Plants are composed of a variety of cells and tissues that must work together in a coordinated fashion if the plants are to survive. These are cells that conduct water (red) and sugars (blue) in the veins of a leaf.

All organisms contain a material called **protoplasm**, which is composed of water, numerous organic compounds, and mineral nutrients. The nature of protoplasm was not well understood until the early 1900s. For hundreds of years it was thought to be a unique substance with an extraordinary property not found in other substances such as rock, water, air, iron, and so on: protoplasm could be alive. One of the first experiments to actually analyze protoplasm was reported in 1644 by Jean-Baptiste van Helmont. He carefully weighed a small willow plant and a pot of soil. After planting the willow in the soil, he cultivated it for 5 years, then unpotted it, washed the roots to obtain all remaining soil, and discovered that the plant had increased in weight by 164 pounds (73 kg [kilograms]) and the soil had lost only 2 ounces (57 g [grams]). He concluded that the plant and its protoplasm must be composed mostly of altered water and a small amount of minerals. He concluded that protoplasm might be special, but it was composed of ordinary chemical compounds.

Other chemists confirmed that protoplasm is indeed composed of water, organic compounds, and mineral nutrients, but the belief still persisted that protoplasm must differ from other substances by having a component called "vital force," which if present allowed it to be alive and if absent caused it to be dead. Only in the late 1800s did Louis Pasteur and others prove that vital force does not exist: protoplasm is alive when its many complex chemical reactions are running properly, it dies if some of these reactions fail.

The protoplasm of all organisms is divided into small bodies called cells, but our understanding of cells was also slow to develop. Early Greek philosophers knew that living creatures have organs: plants have roots, stems, and leaves; animals have hearts, stomachs, and so on. But even with the best eyesight, organs looked merely homogeneous or at most mealy or fibrous. After the first microscopes were invented, in 1665 Robert Hooke discovered that plant organs are composed of small boxes, which he named cells. Plant cells were easy to study even with crude early microscopes because plant cells are large and each is surrounded by a cell wall that appears as a fine line when viewed by a microscope. It was quickly realized that all parts of a plant's body are composed of cells. Also, early plant microscopists realized that plant cells are composed of small units (**organelles**): all green cells in leaves, stems, and unripe fruits hold their green photosynthetic pigment chlorophyll in small, bright green dots named chloroplasts (FIGURE 3.1). Starchy foods like potatoes have starch grains in their cells, and colorful petals and fruits display their pigments as particles or droplets in the cells.

Progress was much slower in the study of animal cells. They lack cell walls, and neighboring cells fit together so tightly that even today it is almost impossible to identify individual animal cells with even the best modern light microscopes. Furthermore, most organelles within both plant and animal cells have no natural color, so they remained unseen until microscopists began experimenting with artificial dyes and stains.

We now know that all organisms are composed of cells, and that in plants, each cell consists of a cell wall surrounding a small bit of protoplasm. The protoplasm itself contains many diverse organelles, and cells specialize for particular tasks by adjusting the number and metabolic activity of their organelles. Cells that specialize for photosynthesis develop large numbers of chloroplasts whereas flower petal cells develop various pigments. The same is true of us: our skin cells produce large amounts of the pigment melanin whereas our red blood cells fill themselves with hemoglobin instead; our liver cells have an abundance of organelles that break down toxins whereas muscle cells can contract because of the proteins they contain.

The body of every large organism, whether tree or human, is composed of organs that are in turn composed of cells. It is possible to imagine a bit of protoplasm or a single cell growing to the size of a tree or a human but it is difficult to imagine it being able to form

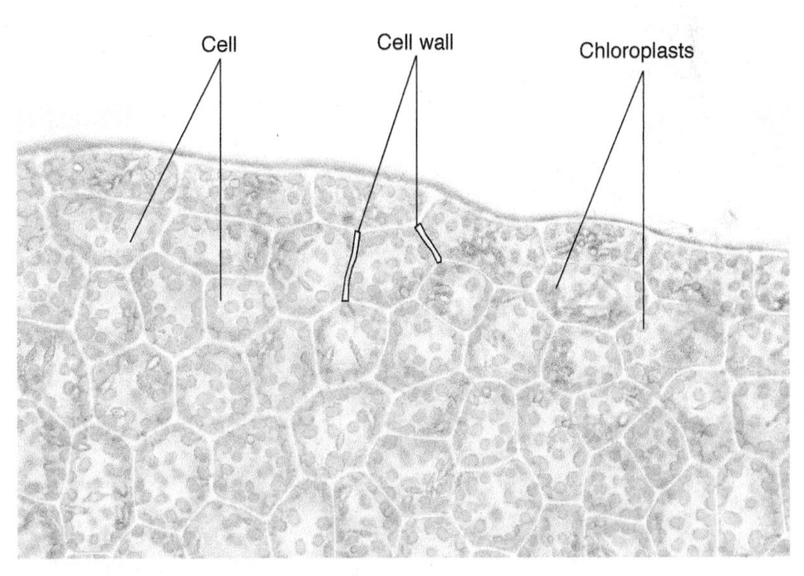

FIGURE 3.1. This is a view of a liverwort, *Pallavicinia*, a plant that has a body only one or two cells thick, thus allowing us to see details that are difficult to view in thicker plants. The white lines are cell walls, and each space enclosed by walls is a distinct cell. Each cell has numerous green spots, organelles called chloroplasts that contain chlorophyll and carry out photosynthesis (×200) (each cell is about 40μm wide; a strip of tissue 1 inch long would have about 625 cells side by side).

specialized parts, each with a particular structure and function. Because organisms are composed of cells and because cells can specialize as they grow and develop, organisms can become both large and complex.

Cells

It is important to ask ourselves "Where do cells come from?" "How do cells grow and develop?" "Do cells die?" All cells originate in just one of two ways: either by (1) cell fusion or by (2) cell division. Cell fusion is the rarer type, occurring when one sperm cell fuses with one egg cell, resulting in one zygote (a fertilized egg). The zygote then grows and develops and at some point it will divide into two new daughter cells. These in turn grow then divide, and the process is repeated many times. Every cell in a plant's body and in our own bodies can be traced to its parental cell and the one that preceded that and so on back to the zygote. None of our cells is ever produced in any other way.

Because we animals have diffuse growth, most of our cells retain this capacity to grow and divide, thus allowing us to grow to adult size (our brain cells are exceptional and stop dividing while we are very young). Plants are different though, and have localized growth. While a plant embryo is still tiny inside an immature seed, cells in the center stop dividing and instead grow and differentiate into mature cells of shoot and root; these cells almost never divide again and do not contribute any further to the growth of the plant. Cells at the shoot and root tip, however, organize themselves into the root apical meristem and shoot apical meristem, and it is these cells that will continue to divide and produce new cells for the plant for the rest of its life (**FIGURES** 2.2 and **3.2**). All plants therefore have some small dividing cells, some developing cells, and some full-sized, fully mature cells. This is important to remember because plant cells usually change dramatically as they mature. Think of a flower's apical meristem, that of a cherry for example: while a flower bud is still tiny all its cells are small, soft, and some are dividing. The cells produced by the meristematic cells develop in various ways. Those that will be part of the sepals produce large numbers of chloroplasts and turn green; others produce a delicate pink pigment and become petals; still others will become enormous, filling themselves with sugar and red pigment and becoming the fruit we eat; and finally those that will make up the pit grow to their adult size, then alter their walls to become rock hard, then they die. These last cells, just like those of our own hair, nails, and skin, perform their protective function while dead.

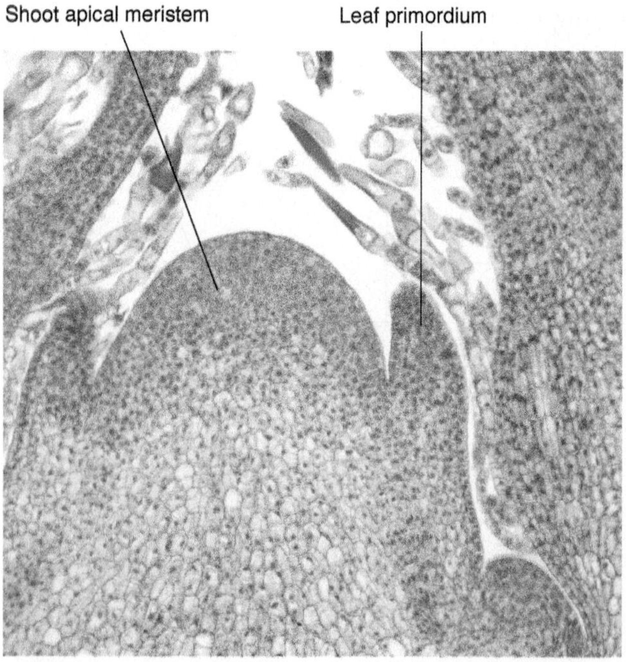

Shoot apical meristem
Leaf primordium

FIGURE 3.2. This shoot apex has been cut vertically down its center (a longitudinal section) and magnified enough to see its very small cells. Each cell divides into two new cells, then both grow back to their original size then divide again, thus producing new cells for the shoot. Along the sides of the meristem, some cells grow more rapidly than others, producing bulges that develop into leaves (×100).

Organelles

The **cell wall** is the outermost organelle; it surrounds and protects the rest of the cell (**TABLE 3.1**). All plant cells have a **primary cell wall** that is thin and flexible but strong enough to give the cell its shape and to control its size

TABLE 3.1. Organelles of a Plant Cell

Cell Part	Function
Cell wall	Provides strength.
Plasma membrane	Controls which chemicals enter or leave the cell.
Central vacuole	Storage, recycling, part of turgor pressure system of support.
Plastids	
Chloroplasts	Perform photosynthesis.
Chromoplasts	Store pigments other than chlorophyll.
Amyloplasts	Store starch.
Proplastids	These are plastids in meristematic cells; they differentiate into the other types.
Mitochondria	Carry out respiration.
Endoplasmic reticulum	A set of tubular membranes involved in moving material throughout the cell; ER also synthesizes certain lipids.
Dictyosomes	Package material into vesicles.
Ribosomes	Carry out protein synthesis.
Microtubules	Provide a framework inside the cell; involved in moving organelles through the cell.
Cytosol	The liquid (nonmembranous, nonparticulate) part of cytoplasm.

(**FIGURE 3.3**). Most of a wall's strength comes from one of its components, **cellulose**, which is a long, chain-like carbohydrate molecule. Cellulose molecules aggregate side by side into fibrils that wrap around the cell, reminiscent of the bandages wrapped around a broken arm to make a cast. Also, just like a cast, the cellulose must be held in place, and a plant uses other carbohydrates called **hemicelluloses** for this. Cell walls also contain proteins, pectin, and water (**TABLE 3.2**). Primary cell walls are extremely permeable, that is, water and small molecules like sugars, vitamins, and amino acids pass through them easily, allowing cells to obtain needed materials from neighboring cells.

When a cell divides, two new cell walls form in its center, separating one new cell from the other. These two new walls must adhere to each other and to the preexisting wall of the parent cell; otherwise the plant would fall apart into many single cells. The adhesive plants use to glue one cell to another is surprising—jelly, or more exactly, pectin (calcium pectate). Where two cells adjoin each other, a thin layer of pectin, called the **middle lamella**, holds the cells to each other. All the cells of the wood in a baseball bat or a piano or a giant redwood tree are held in place by the principle ingredient of jelly.

Where neighboring cells meet, primary cell walls have thin areas, called **primary pit fields**, and these in turn have extremely fine holes (**plasmodesmata**; singular: plasmodesma) that allow the protoplasm of one cell to actually be in contact with that of the neighboring cells (**FIGURE 3.4**). All cells of a plant are interconnected by plasmodesmata. **Secondary walls** are produced in sclerenchyma cells after the cell has grown to its proper size and shape; these are discussed below.

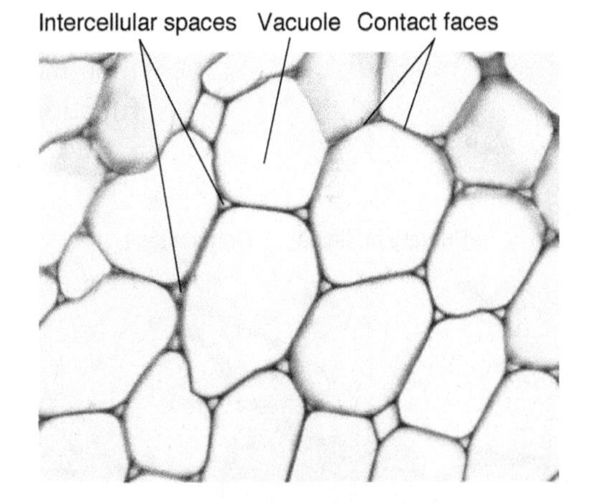

FIGURE 3.3. Each line here is a cell wall, and where cells touch each other, there are two walls pressed against each other. The region where cells touch is called a contact face, and it is where the cells are glued together by a middle lamella. Although it appears as if each cell is empty, each actually has a thin layer of organelles too small and sparse to see with a light microscope (these cells do not have chloroplasts that are easy to see as in Figure 3.1). The large white center of each cell is its central vacuole. The small spaces between adjacent cells are intercellular spaces. These are mature cells in the center of a sunflower stem (×300).

TABLE 3.2. Components of Cell Walls

Basic components found in all cell walls	Components found in specialized walls
Cellulose	Casparian strips: lignin and suberin
Hemicellulose	Cork: suberin
Cell wall proteins	Epidermis: cutin and wax
Pectins	Sclerenchyma: lignin
Water	

The small mass of protoplasm surrounded by a cell wall is called the cell's **protoplast**, and its outermost layer is a membrane, the **plasma membrane** (sometimes called the plasmalemma). Like all cell membranes, this is composed of lipids and proteins, and is **selectively permeable**; it controls which molecules pass through it and which do not. The plasma membrane then determines which molecules are allowed into the cell (usually nutrients) and which are not (usually toxic substances).

Each cell, at least while young, has a single **nucleus** that contains genetic material (DNA) that stores most of the information needed for plants to grow, develop, and reproduce (FIGURE 3.5). The nuclear DNA guides the synthesis of enzymes that carry out the cell's metabolism. There is a limit to the rate at which one nucleus and its DNA can guide a cell's metabolism; consequently, a cell must not grow so large that the nucleus can no longer control it. Rarely, some cells allow their nuclei to divide such that the cell is binucleate or tetranucleate (or more), and this allows the cell to become larger or have a more rapid metabolism. Certain phloem cells (sieve tube members) destroy their nucleus as part of their development, and then must function while being enucleate.

Each nucleus is surrounded by two selectively permeable membranes, the **nuclear envelope**, which keeps the nuclear material separate from the rest of the cell. The nuclear envelope allows the cell to have two distinct compartments, each with a specialized metabolic environment: one inside the nucleus and adjusted for the type of chemical reactions that must occur there, and a second environment suitable for the rest of the protoplasm (the portion of the protoplasm that is not nucleus is called **cytoplasm**). This is called **compartmentation**. Nuclear pores control passage of material between the nucleus and cytoplasm.

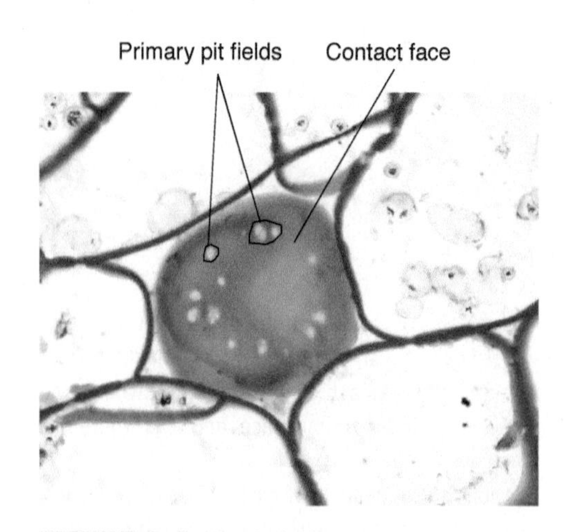

FIGURE 3.4. The dark blue area is a contact face (two cell walls glued together) seen in face view (contact faces are in side view in Figure 3.3). The white areas in the contact face are primary pit fields, areas where the two walls are exceptionally thin. Each contains many plasmodesmata, but they are too narrow to be visible with a light microscope (×200).

The **central vacuole** is an organelle that occurs in plants, algae, and fungi but never in animals, and it affects many aspects of plant biology. It is surrounded by a selectively permeable membrane (the **vacuole membrane** or tonoplast), so it is a compartment distinct from that inside the nucleus and in the rest of the cytoplasm. The vacuole membrane is able to forcefully pump certain molecules—for example waste products, sugars, and potassium—into the central vacuole. Unlike animals, plants do not dump their wastes outside their bodies (they have no kidneys or colon) but instead store them in their central vacuole (FIGURE 3.6). Storing waste inside your own body seems counterintuitive but waste products have little nutritional value for animals or fungi, so as they accumulate inside the central vacuole, the cell's value as food decreases. Animals prefer to eat young, nutritious leaves and stems rather than old, mature waste-filled ones.

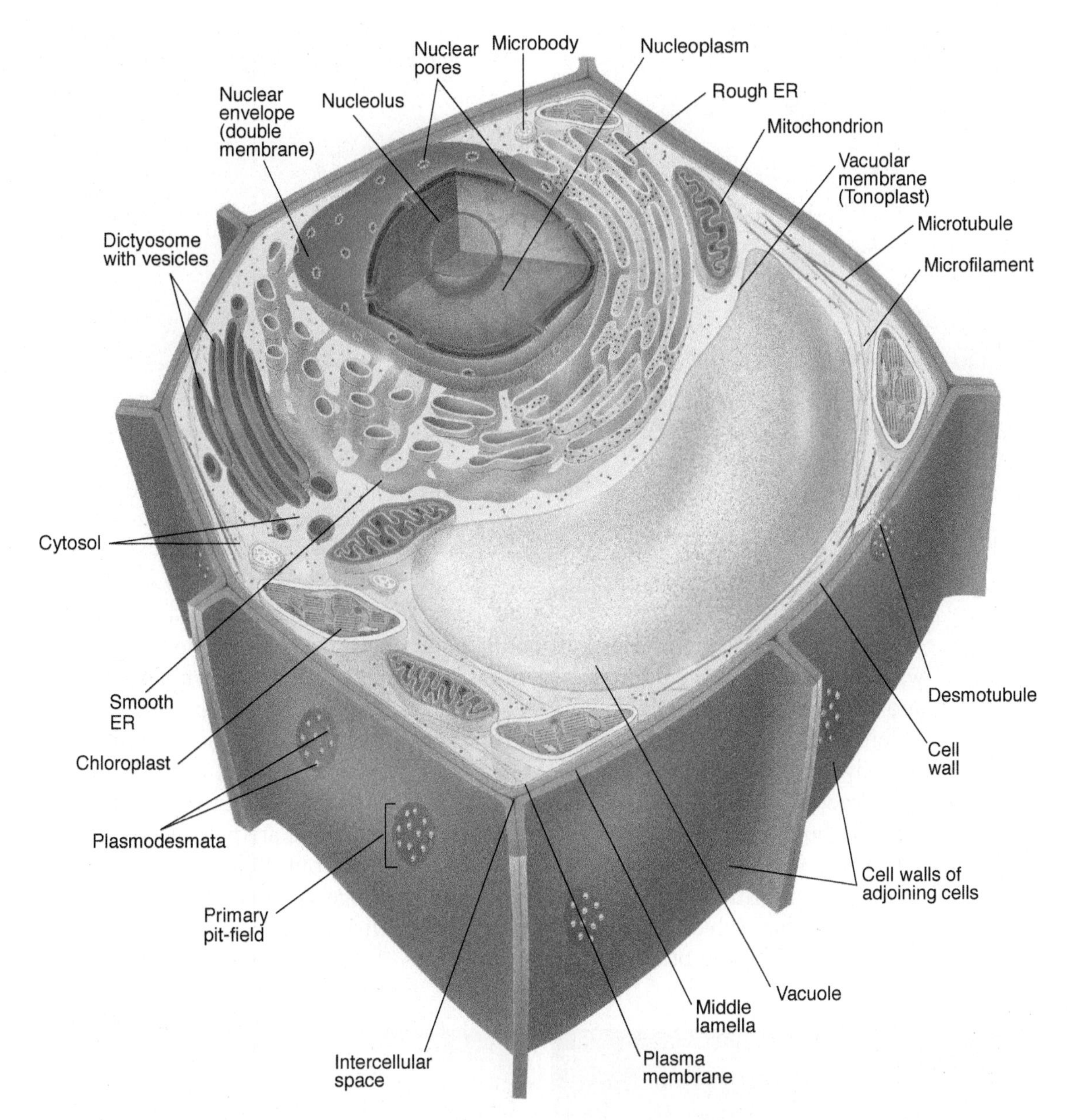

FIGURE 3.5. Diagram of a generalized plant cell. The vacuole has been drawn exceptionally small here to save space, otherwise the diagram would occupy several pages and be mostly vacuole.

Plants obtain defense against predators merely by storing their waste rather than throwing it away. This defense mechanism would not work for us: we generate such a large quantity of waste that if we tried to store even one month's worth, we would be too heavy to move. In addition to wastes, central vacuoles are used to display pigments in flower petals and fruits in many plants (FIGURE 3.7).

Plant cells contain small **plastids**, a type of organelle that can change its structure and metabolism, thus allowing cells to also alter their physiology. In shoot and root apical meristems, plastids are small, nondescript **proplastids**, bodies with an outer and inner membrane; some plastid DNA; and a small amount of water, chemicals, and other components. As meristem cells grow and divide, their proplastids also grow and divide such that each new daughter cell receives an adequate supply of proplastids. If some of the meristem cells become part of a leaf primordium, their

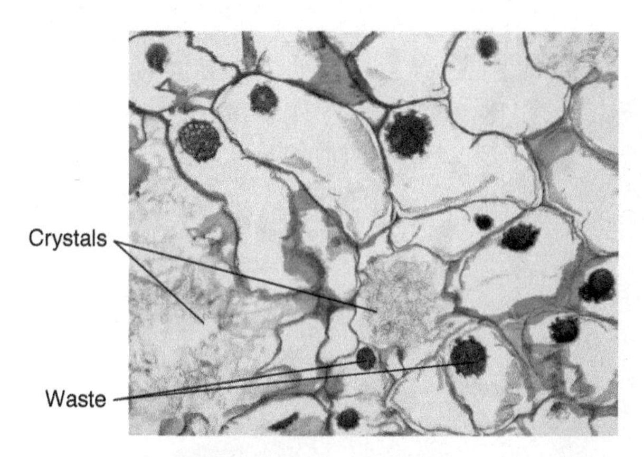

FIGURE 3.6. The dark red materials in these very old cells are accumulated waste products. Many crystals are also present, and in general these cells appear old (×300).

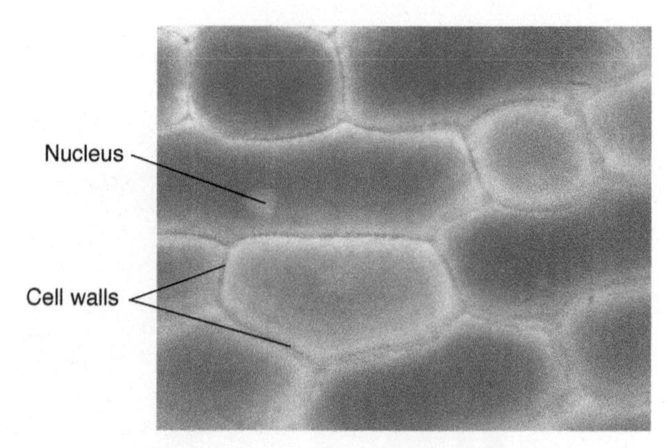

FIGURE 3.7. These are fresh cells from the skin of a purple onion. The color is pigment stored in each cell's central vacuole, and we can see that the vacuole occupies almost the entire volume of each cell. The dark lines are cell walls and clear areas are nuclei. These cells fit together tightly with no intercellular spaces (×200).

proplastids convert themselves into **chloroplasts** by synthesizing chlorophyll along with other factors needed for photosynthesis (Figure 3.1). Each chloroplast enlarges and its inner membrane folds into a complex three-dimensional shape. Once fully differentiated, carbon dioxide can pass through the outer chloroplast membrane and be converted into sugar by the chloroplast's enzymes and the light energy captured by chlorophyll. Chloroplasts are abundant in cells of dark green leaves, like those of spinach, but are more sparse in leaves, stems, and fruits that are pale green.

Many flowers and fruits are green while immature but change to some other color when mature. In many cases, this color change is caused by chloroplasts converting themselves to **chromoplasts** as they dismantle their chlorophyll and replace it with some other pigment (FIGURE 3.8). As starchy organs like potato

(a)

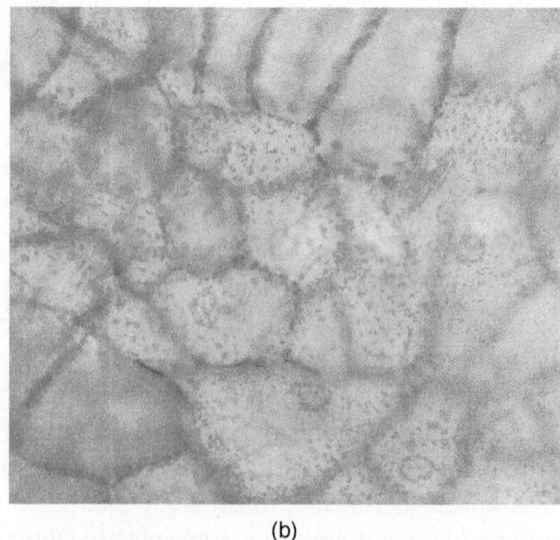

(b)

FIGURE 3.8. **(a)**The red color of these peppers (*Capsicum*) is due to pigments in chromoplasts whereas the stalks are green because they contain chloroplasts. **(b)** Each orange particle is a chromoplast in cells of red pepper; although they look orange here, altogether they produce the red color of the peppers in part a (×80).

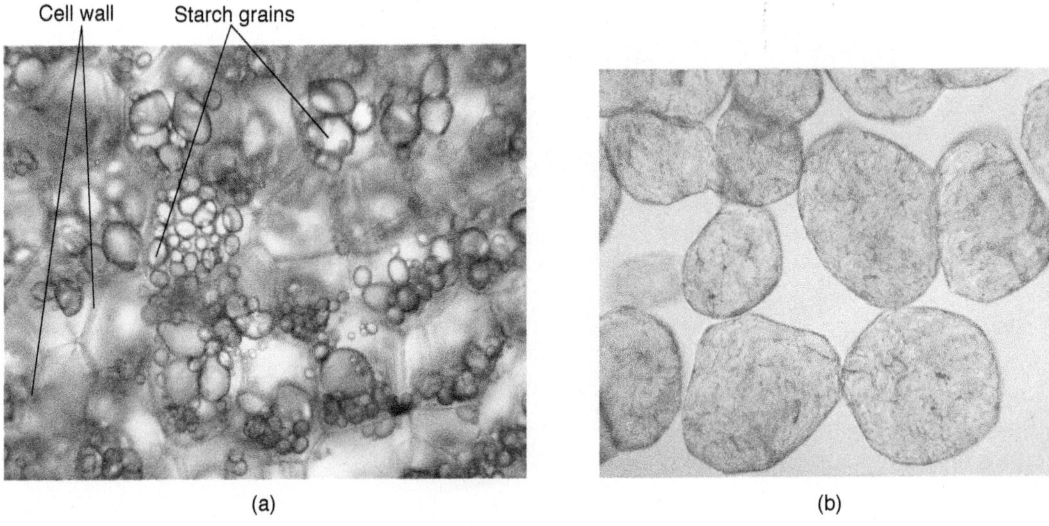

FIGURE 3.9. **(a)** This is a view of living potato cells, each of which contains many oval amyloplasts (starch grains). Cell walls are so thin they are just barely visible. Most cells are about half filled with amyloplasts, most of which are so heavy they have settled to the bottom of the cell (×80). **(b)** These are boiled potato cells: the only thing visible is greatly swollen starch grains (same magnification as part a). Boiling has caused the starch grains to absorb water and swell, making them more digestible for us (×80).

tubers, yams, and rice grains develop, their proplastids turn into **amyloplasts** (starch grains) (FIGURE 3.9). Amyloplasts absorb sugars and polymerize them into starch, which is so stable it can be stored in the cells for months or years. Most plant cells have at least a few amyloplasts at some point in their life, but starchy cells have numerous amyloplasts and develop in roots, stems, wood, bark, fruits, and especially seeds. When the plant needs the energy stored in the starch, the amyloplast breaks it back down into sugars, then loads them into phloem where they can be transported to tissues and organs that need them. As a seed germinates, starch is broken down and sugars are moved to the root and shoot apical meristems. Similarly, when a century plant finally blooms, amyloplasts in cells throughout all its massive leaves release sugar that is used to build its giant flower stalks (see Figure 2.22). Some of the cells and their amyloplasts may be more than 20 years old. We animals never store starch; we use fats in our adipose cells for long-term energy storage.

Two other critical roles of plastids are that they synthesize certain lipids and amino acids that no other organelle can make. Only algae and plants have plastids; no other organisms do.

Starch, sugar, and fats are energy-rich molecules, but that energy must be transferred to a smaller molecule, called ATP, by the process of respiration (see Chapter 5). Some steps of respiration occur in cytoplasm but the last steps, which involve oxygen, occur in **mitochondria** (singular: mitochondrion) (FIGURE 3.10). All cells of all organisms (except bacteria and archaeans) have mitochondria, and cells that require large amounts of energy (rapidly growing meristems, germinating seeds, muscles, liver) have large numbers of mitochondria. Like plastids, mitochondria have an outer and inner membrane, their own mitochondrial DNA, and they can divide to produce more mitochondria.

Within the cytoplasm are many other small organelles, composed mostly of membranes, that are involved in moving materials from place to place within the

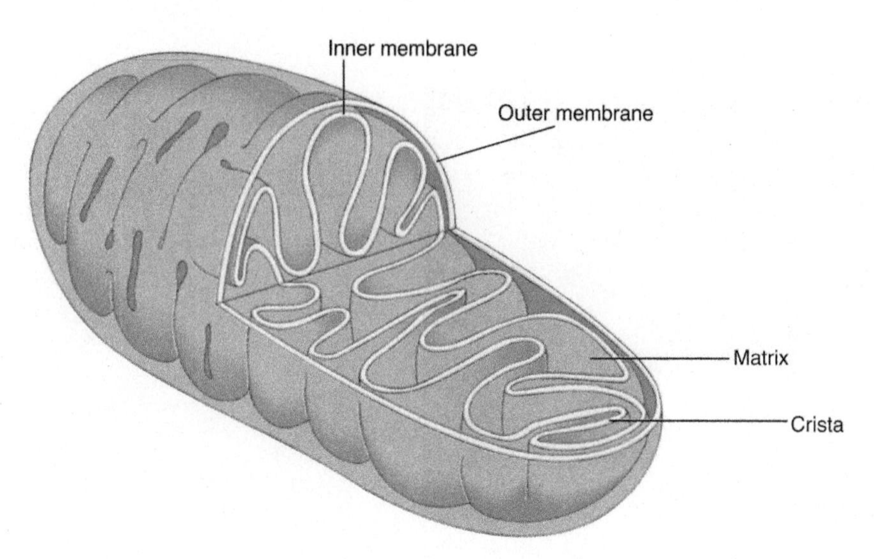

FIGURE 3.10. Diagram of a mitochondrion showing the smooth outer membrane and highly folded inner membrane.

cell. The **endoplasmic reticulum** is a network of tubes and sacs that extends throughout the cytoplasm. Like the plastids, it also synthesizes certain lipids that are made nowhere else. **Dictyosomes** are sets of flattened sacs that process materials that will be exported from the cell. Raw materials are absorbed by the dictyosomes, converted to finished products, then are packaged into a vesicle (a small bag composed of membrane), which moves through the cell. When a vesicle touches the cell membrane, the two fuse and the vesicle's contents are released to the exterior of the cell. This is especially common in us, and is the method by which we secrete our digestive enzymes into our stomach. Carnivorous plants like Venus's flytraps and sundews also secrete digestive enzymes by means of dictyosome vesicles. In plants that are slimy (okra, prickly pears), the slimy mucilage is synthesized in dictyosomes and then stored in the central vacuole.

Many of the membranes mentioned above are interconnected, at least from time to time. The outer envelope of the nucleus connects to the endoplasmic reticulum, which in turn can form vesicles that accumulate to produce a dictyosome. Dictyosome vesicles can fuse with plasma membrane or vacuole membrane. When plastids and mitochondria wear out and stop functioning, they enter the central vacuole by fusing with the vacuole membrane; once inside they are digested into individual molecules that can be re-used to build new organelles. Membranes, vesicles, and organelles are constantly moving in plant cells; a plant may stand still but its protoplasm is in constant motion. This set of interconnected membranes is called the endomembrane system.

Finally, plant cells contain several organelles that are not composed of membranes. **Ribosomes** are small particles that construct proteins from amino acids, guided by molecules produced by DNA. **Microtubules** and microfilaments are linear aggregates of proteins that act like a scaffold (the **cytoskeleton**) and give the cytoplasm structure, while at the same time being responsible for pulling organelles from site to site within the cell. Finally, many plant cells contain crystals of calcium salts; these may be spiky balls (druses), a single large spear-like crystal (styloids), tiny needles that occur in bundles (raphides), or numerous tiny cubic crystals (crystal sand). The liquid portion of cytoplasm, the part not composed of membranes or particles, is **cytosol**.

Intercellular Spaces

In addition to cells, plants are also composed of **intercellular spaces**. Imagine filling a room with beach balls that are so fully inflated they are almost inflexible. Once the room contains as many balls as possible, there will still be a great deal of empty space between the balls. Also, each ball will touch its neighbors in just a small area (called a **contact face**) (FIGURE 3.11). Now imagine filling the room with balls that are just a little underinflated. You can pack more balls into the room, there is less space between them, and each ball has a greater contact area with its neighbors. Now imagine using balls that contain so little air that each is extremely flexible; the room can be completely filled with balls because each can bend and flex to match the contours of its neighbors. There is no longer any air space left in the room, and the entire surface of each ball contacts one neighbor or another. Animal cells and plant meristematic cells are like the last group: their surfaces are so delicate the cells mostly fit snuggly together. But as plant cells grow, the pressure of the central vacuoles often causes them to be more rounded and they push away from their neighbors, partially tearing the middle lamella, and creating intercellular spaces at their corners. If they expand more, the cells become almost spherical and the intercellular spaces become larger.

Intercellular spaces allow oxygen and carbon dioxide to diffuse quickly throughout a plant. Because plants have no lungs, gills, or circulatory system, they cannot move these gases forcefully through their bodies the way we animals do, and plants must rely on diffusion. Gases diffuse thousands of times more rapidly through intercellular spaces than through cells, and without intercellular spaces, the innermost cells of thick plant parts would suffocate.

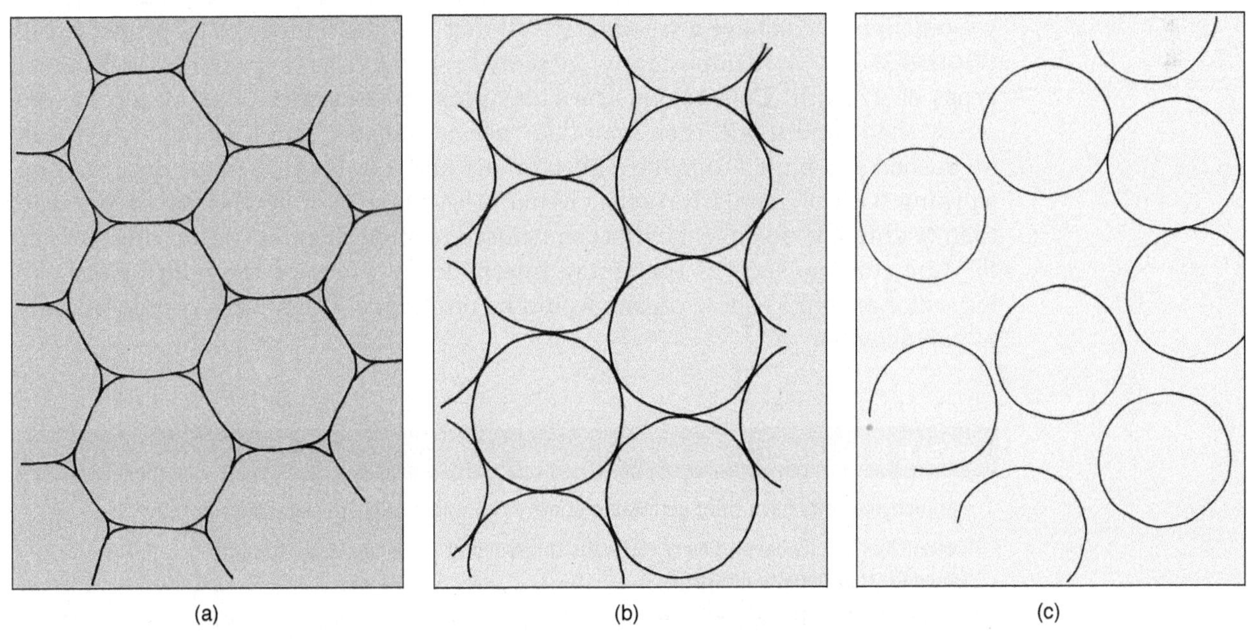

FIGURE 3.11. If plant cells are not very turgid, or if they are glued together strongly by their middle lamellas, then each has extensive contact with adjacent cells and intercellular spaces between them are small **(a)**. But if cells are very turgid, they swell and pull away from each other **(b)**, causing the intercellular spaces to expand, which permits air to diffuse more rapidly through the tissue. With even more expansion **(c)**, some cells pull completely away from some of their neighbors.

Tissues

There are just three types of plant cells: parenchyma, collenchyma, and sclerenchyma, and they never occur in a disorderly jumble (**TABLE 3.3**). Instead numerous cells develop in groups that allow them to perform certain functions. For example, certain cells work together as the epidermis (or skin) of the plant, others function in water conduction, and so on. If all the cells of a tissue are similar, the tissue is a simple tissue, but if several different types of cells work together, they are called a complex tissue.

Simple Tissues

Parenchyma is a simple tissue composed of just parenchyma cells, and these are defined as cells that have a uniformly thin primary wall and also lack a secondary wall (**FIGURE 3.12**, and see Figures 3.1 and 3.3). Because primary walls are permeable to small molecules, parenchyma cells are well adapted for many metabolic functions: material can be moved easily into parenchyma cells, chemically altered, and then be moved back out to other parts of the plant. Photosynthetic cells are parenchyma (often called chlorenchyma): carbon dioxide, water, and light enter the cells easily and newly synthesized sugars leave easily. Similarly, starch storage cells in tubers and seeds are parenchyma cells, as are cells that store oils (avocado, sunflower seeds), pigments (petals, fruits), or water (cacti and other succulents). Meristematic cells are also parenchyma; their thin walls allow the cells to grow.

Collenchyma and **sclerenchyma** are both cells that are specialized such that they provide much more strength and support than can be offered by the thin walls of parenchyma cells. Collenchyma cells are defined by having primary walls that are especially thick at each corner of the cell (Figure 3.12), whereas sclerenchyma cells have a secondary wall that lies just interior to its primary wall (**FIGURE 3.13**). More importantly, these two cell and tissue types provide different types of strength. Collenchyma provides **plastic strength**: it can be forced into a new shape and it will then maintain that new shape. Think of clay, which can be pushed and pulled to make a bowl or cup and it holds that shape once we stop working with it. Similarly, collenchyma can grow into a new shape or size and then maintain it, so collenchyma can be used to strengthen long, thin organs *while they are growing*, such as the young internodes of vines or the midribs of large leaves for example. These organs would be too fragile if they were composed only of parenchyma.

TABLE 3.3. Basic Types of Plant Cells
Parenchyma cells have thin, permeable primary cell walls; cells are metabolically active.
Collenchyma cells have primary cell walls thickened at corners; plastic strength (strength that permits growth).
Sclerenchyma cells have both primary and secondary cell walls; elastic strength (strength that restores mature size and shape).
Fibers provide flexible elasticity.
Sclereids provide inflexible elasticity.

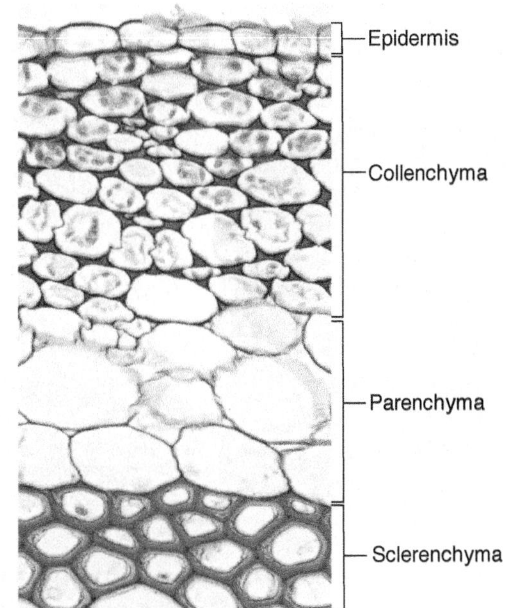

FIGURE 3.12. A stem of cucumber has been cut in transverse section and the outer part photographed. The uppermost layer of cells is the epidermis, and just below it is a band of collenchyma about eight to ten cells thick. The corners of collenchyma cells are thickened. The large cells with thin walls in the center are parenchyma, and the cells at the bottom are sclerenchyma: they have secondary walls (×150).

BOX 3.1. Parenchyma, Sclerenchyma, and Food

Most of the fruits, vegetables, and other plant parts we eat as food consist of almost pure parenchyma. With their thin walls, parenchyma cells are soft and easy for us to chew if eating our food fresh or cooked, and they can be ground, mashed, and sliced for processed food. Whereas collenchyma and sclerenchyma are used for strength, parenchyma cells are the sites for synthesis and storage of an amazing variety of organic compounds: carbohydrates, fats, proteins, vitamins, pigments, flavors, and other nutritious materials essential to our health.

Parenchymatous foods are easy to recognize because we can bite through them easily: apples, pears, strawberries, blueberries, potatoes, lettuce, spinach, and so on. All seeds such as popcorn, beans, rice, and wheat are parenchyma as well, but in their dry, ungerminated, uncooked condition, they are too hard to be edible, despite having thin cell walls. Boiling allows water to loosen hard starch grains and protein bodies, and to soften the walls.

Collenchyma is not present in all plants, and is never really abundant. It is almost always just a minor fraction of all the cells in leaves and stems, and virtually never occurs in roots. The most familiar collenchymatous food is celery: we eat the petioles of leaves, and each ridge along the surface is a mass of collenchyma cells. Also, each vascular bundle inside the petiole has a cap of collenchyma along one side; these bundles cause celery's stringiness. Other leafy vegetables with thick petioles, like rhubarb and bok choy, also have abundant collenchyma. Most of us eat these for their flavor and nutrients, not for the pleasure of chewing endlessly on collenchyma strands.

Plants can store starch and proteins in fibers, but they do so only rarely. With the low nutrition and difficulty in chewing, we never use truly fibrous material like mature bamboo shoots and wood for food. However, many vascular bundles contain fiber cells, so even when eating parenchymatous foods we still consume some sclerenchyma. The fibers may be very noticeable, as in asparagus that is a bit too old, green beans, snow peas, artichokes, pineapples, and mangoes. In others, the fibers are a bit softer, especially after being cooked, but you may still notice them in things like squash, pumpkin, and zucchini. You have probably noticed when carving jack-o-lanterns that cleaning a pumpkin is really just a matter of removing seeds and fibers.

Sclereids also occur in some of our foods. Clusters of them cause the grittiness in pears. The seed coats of beans, peas, and most other seeds are also made up of sclereids, as is the covering on seeds of corn. The pieces that get stuck in our teeth while eating corn on the cob are composed of sclereids (and the inedible cob is mostly fibers). In small seeds that are ground into flour (wheat, rye, barley), the sclereid-rich seed coats are broken down into small pieces (the bran) that are easy to eat. Sclereids also make up the brown covering of unpopped popcorn, and popping breaks up the covering. In all cases, we do not digest any part of the secondary walls of sclerenchyma cells; all the walls simply pass right through us.

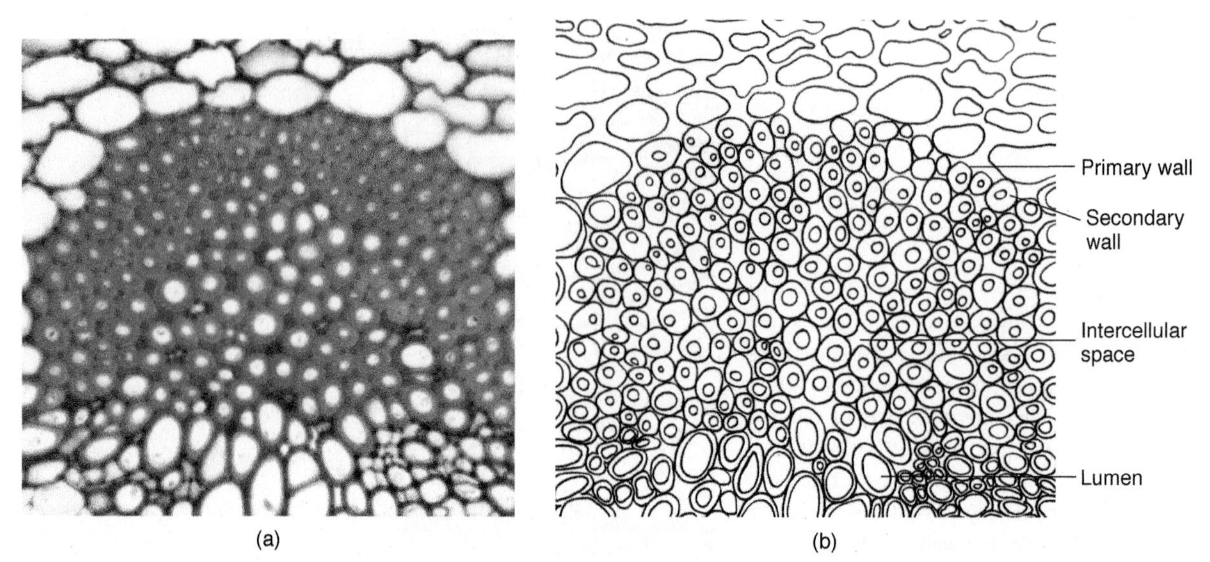

FIGURE 3.13. Sclerenchyma cells are recognizable because they have a thick secondary wall just inside their primary wall. The secondary wall is so thick in each cell that it occupies most of the cell's volume: by the time each cell had finished constructing its secondary wall, the cell's protoplasm had been squeezed down to fit inside the white area (the lumen) in the center of each cell. These cells appear round, but they are very long fiber cells that have been cut in transverse section. It is unusual to have such large intercellular spaces in a mass of fibers (×200).

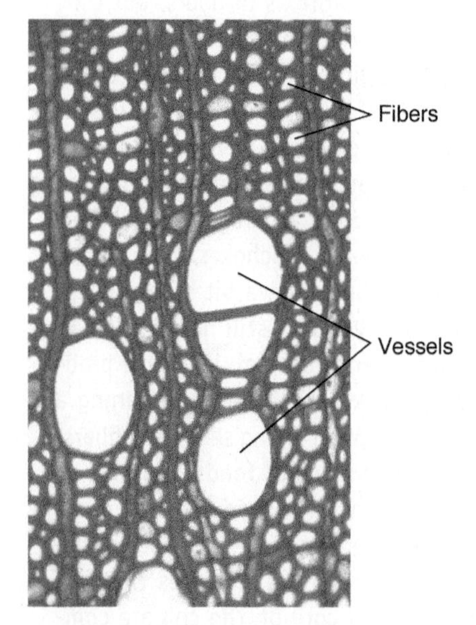

FIGURE 3.14. This is a transverse section of wood of American hornbeam (*Carpinus caroliniana*), showing that almost all the cells are fibers (all the cells with small white centers). Fibers give the wood both strength and flexible elasticity. The several wider cells are vessel elements, which conduct water through the wood (×200).

Collenchyma cells occur together as bands with no other types of cells mixed in with them (Figure 3.12). This is collenchyma tissue, a simple tissue. Collenchyma does not occur in all plants, but in those that have it, it usually occurs as a band several cells thick located just interior to the epidermis of stems, and just below or above the largest veins of leaves. Our most familiar example is celery: its stringy masses are whole vascular bundles, each of which has a mass of collenchyma alongside it. Each ridge on the surface of a celery stalk is a bundle of just collenchyma cells.

In contrast, sclerenchyma provides **elastic strength**. Like an elastic waistband in clothing, it can be pulled to a new shape but once we stop pulling, it returns to its original size and shape. Sclerenchyma differentiates in large organs *after they have already grown* to their proper size and shape. Woody branches are good examples of sclerenchyma's elasticity: the branch grows to a particular form and then is occasionally bent by wind or snow or even heavy fruit or birds (FIGURE 3.14). The branch should not stay in the new shape; it should spring back to its original shape once the wind has stopped. The secondary wall is so strong it cannot be stretched by the central vacuole, so a fully differentiated sclerenchyma cell cannot grow; instead, it begins its life as a parenchyma cell, grows to its proper size and shape, and only then can it build its secondary wall. If you have cleaned and cut asparagus for cooking, you know that the top green young part is soft and edible; the basal white part is old, tough and fibrous. The green part is still growing longer and all its cells have just thin primary walls. The lower white part has finished elongating, all its cells have reached their proper length, and some have deposited a secondary wall and transformed themselves from parenchyma cells into sclerenchyma cells.

The secondary wall is made of the same components as the primary wall, but it is much thicker and stronger. Also, sclerenchyma cells add another component, **lignin**, which hardens into a solid matrix around all the other wall components, making the secondary wall extremely tough and resistant to decay. Lignin also makes the wall impermeable to water and almost everything else, so if a cell encased itself completely in a secondary wall, it would soon die of starvation or asphyxiation. Instead the secondary wall has small tunnels (called **pits**) that provide a passageway for nutrients (FIGURE 3.15). In sclerenchyma tissue, in which all cells have secondary walls (for example the shells of nuts, the masses of fibers in bamboo, or the spines of cacti), the pits of neighboring cells must interconnect.

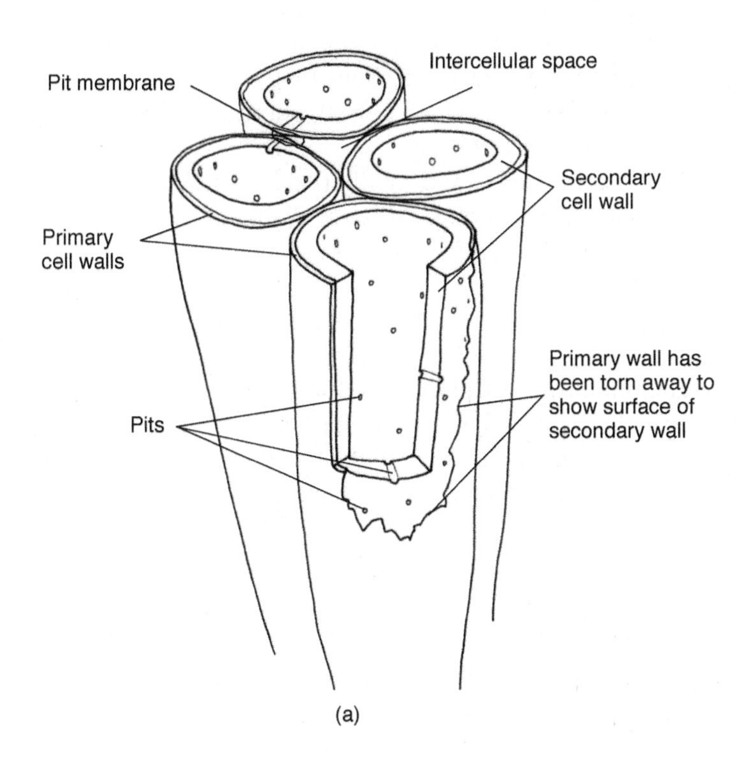

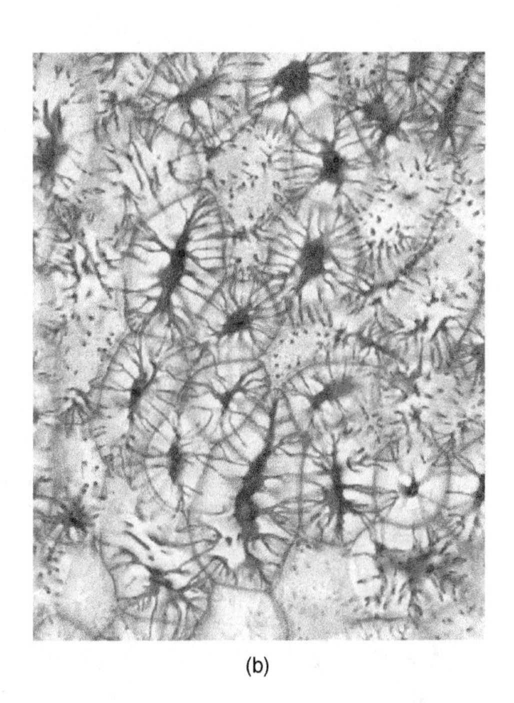

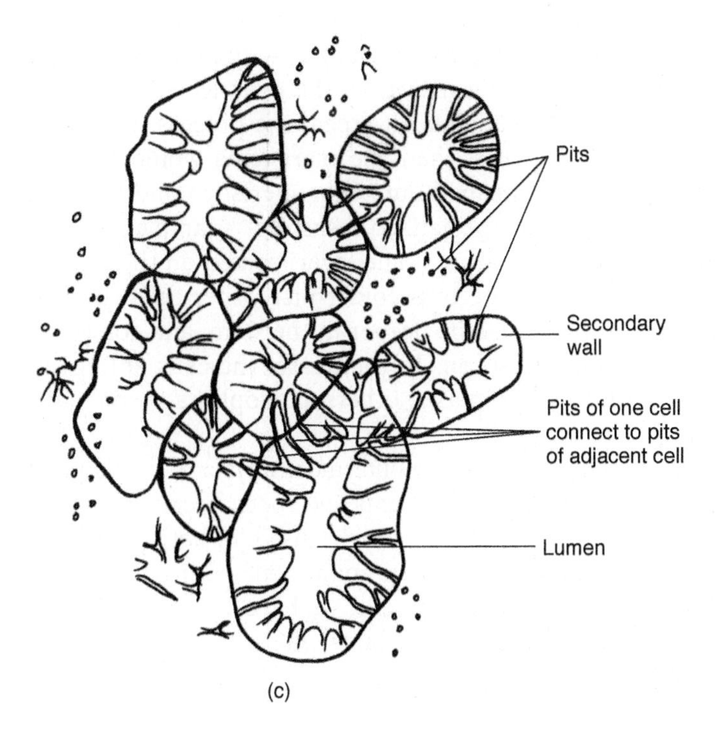

FIGURE 3.15. Sclerenchyma cells, both fibers (**a**) and sclereids (**b, c**) are interconnected by pits. A pit is an area where there is no secondary wall, but the primary walls of each cell are still present (**a**), so particles cannot pass from one cell to another through pits. Fibers are usually interconnected by very few pits whereas sclereids have many. (**b**) is a section through the pit of a peach (×250).

Sclerenchyma provides plants with two types of elastic support: flexible and inflexible. Flexible support is provided by long sclerenchyma cells called **fibers**. These allow woody stems and fibrous leaves to bend and flex without breaking. If the fiber cell walls are extremely thick and heavily lignified, as in wood of oak or hickory, it may take a great deal of force to flex the wood, but with thinner walls and less lignin, as in willow wood, the fibers are more supple. Inflexible support results from spherical or cube-shaped sclerenchyma cells called **sclereids**. Sclereids make up the shells of seeds like walnuts, pecans, coconut, and so on; inside each shell is a delicate seed and its embryo, both made of fragile parenchyma cells. A shell made of fibers might flex without breaking, but the embryo inside would be crushed; inflexible sclereids are needed. We have bones with these same properties: our long bones provide flexible elasticity (our ribs are more flexible than our leg bones) whereas our skull must provide inflexible elasticity if it is to protect our brain.

Many sclerenchyma cells die after they make and lignify their secondary wall. The strength is provided just by the wall, a living protoplast is not needed. In fact, keeping the sclerenchyma cells alive after they have completed their secondary wall would be an unnecessary metabolic expense.

Sclerenchyma cells often occur as large groups of just fibers or just sclereids, with no other cells mixed in (FIGURE 3.16). This is sclerenchyma tissue, a simple tissue. The majority of sclerenchyma in most plants occurs as part of complex tissues, mixed together with parenchyma cells. Wood and bark are good examples of complex tissue that contain sclerenchyma and parenchyma together. Fibers are used to make cloth, fabrics, and paper, and are discussed in Chapter 16.

Complex Tissues

Meristems are small discrete groups of parenchyma cells that do at least two things: they divide to make new cells, and they make those cells in specific layers and patterns (Figure 3.2). Their thin walls allow these cells to easily absorb nutrients that have been transported to them by xylem and phloem from the rest of the plant. The nutrients are converted into more complex compounds such as carbohydrates, lipids, proteins, DNA, and these in turn are assembled into microtubules, membranes, proplastids, and all the other organelles that make up protoplasm. The accumulation of newly synthesized protoplasm causes the cells to enlarge, then the cells divide, producing new, smaller cells which repeat the process. If all new cells continued to act as meristematic cells, then the meristems would become larger, but instead, some of the cells are pushed out of the meristem and begin to differentiate such that they can carry out specialized tasks. For example, cells on the surface of the shoot apical meristem form a single layer located on the surface of stems and leaves; these cells differentiate into epidermis. Other cells, located deeper in the shoot apical meristem will also be deeper in the stem, and some will mature as xylem, others as phloem, and so on. In each case, the cells enlarge as they differentiate, and some produce chloroplasts, others amyloplasts, some make secondary walls, and so on.

The **epidermis** is a complex tissue that forms the outermost surface of young plant organs (FIGURE 3.17). If an organ, such as woody stems and roots, becomes old, the epidermis will be replaced by bark. The word "epidermis" is used for both plants and animals but a plant epidermis differs greatly from an animal one. Our epidermis is thick, consisting of many layers of dead and dying cells, we have sweat glands, hair follicles, nerves, and numerous other features. In plants, the epidermis is almost always just one single layer of parenchyma cells.

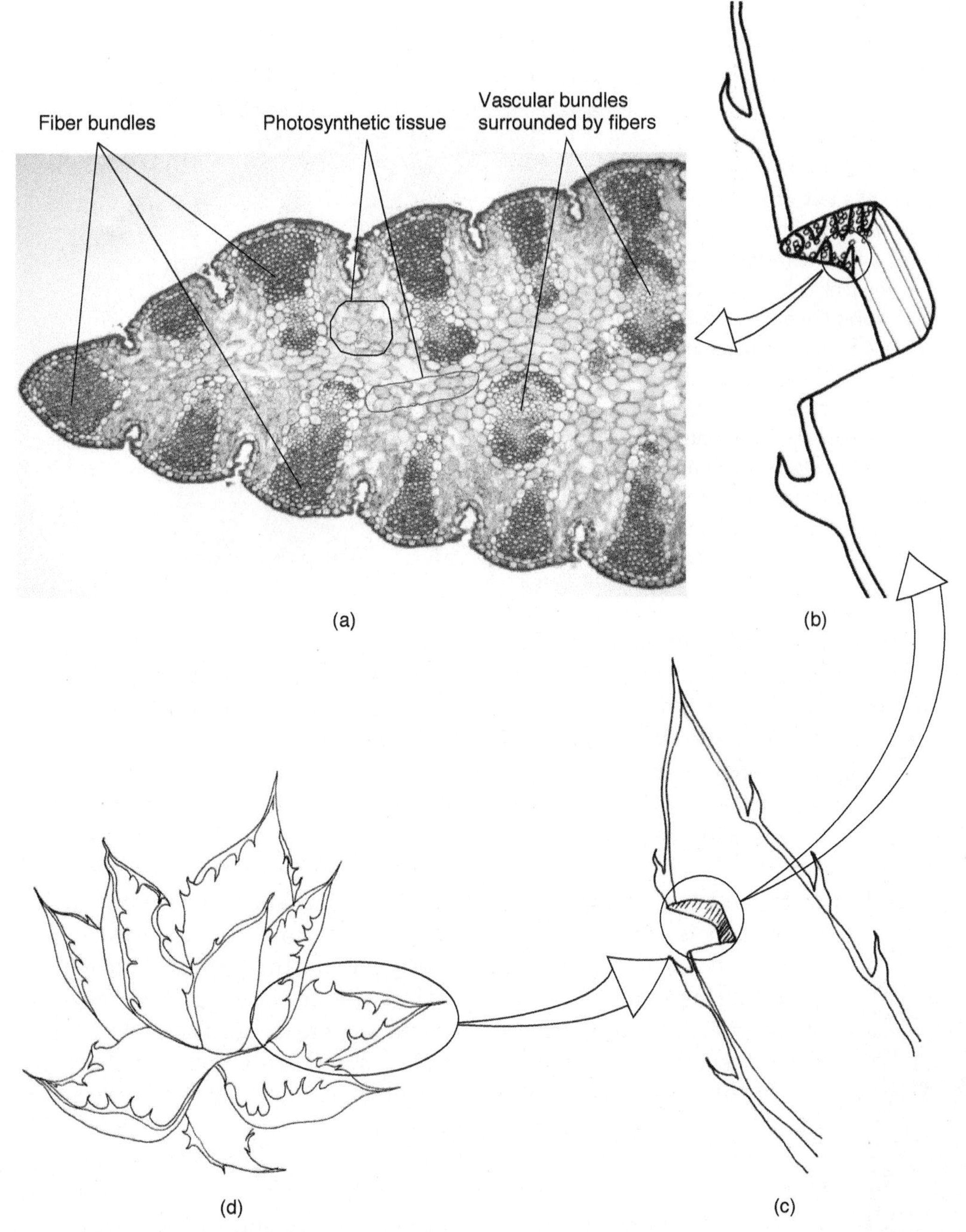

FIGURE 3.16. The long strap-shaped leaves of many monocots such as yuccas and agaves often have bundles of fibers along each surface, making the leaves tough, flexible, and resistant to being torn by wind and eaten by animals. The fiber bundles are often extracted and used for rope or weaving coarse cloth. Some of the bundles shown in transverse section in the micrograph would be as long as the entire leaf shown in the drawing (×70).

As epidermis cells differentiate, the endoplasmic reticulum produces several kinds of fatty acids that are transferred across the plasma membrane and into the cell wall that is exposed to the plant's environment. The fatty acids accumulate in and on the wall, and polymerize into an impermeable substance called **cutin**; a layer of cutin is called the **cuticle**. Wax is also present in the cuticle of most plants. Being impermeable, the cuticle prevents water from evaporating rapidly out of the plant on dry days.

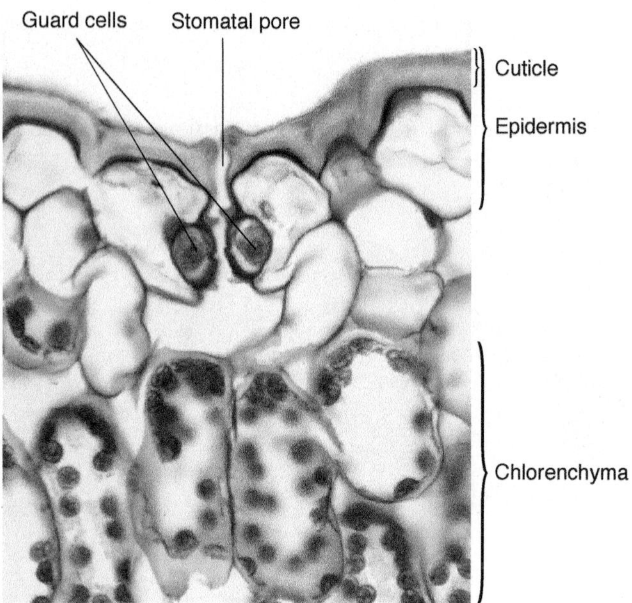

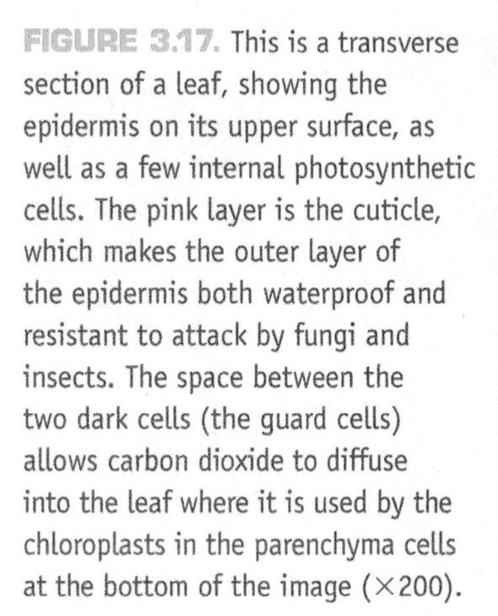
FIGURE 3.17. This is a transverse section of a leaf, showing the epidermis on its upper surface, as well as a few internal photosynthetic cells. The pink layer is the cuticle, which makes the outer layer of the epidermis both waterproof and resistant to attack by fungi and insects. The space between the two dark cells (the guard cells) allows carbon dioxide to diffuse into the leaf where it is used by the chloroplasts in the parenchyma cells at the bottom of the image (×200).

It also greatly hinders the movement of oxygen and carbon dioxide. Certain organs, such as petals and roots, do not carry out photosynthesis, do not need carbon dioxide, and their metabolism is so low they need little oxygen. Organs like this can have an epidermis in which the cuticle is uninterrupted and has no holes in it.

Foliage leaves have a more complex epidermis. Leaves must allow large amounts of carbon dioxide to pass from the air across the epidermis and into their photosynthetic cells while simultaneously allowing oxygen to escape. Consequently, foliage leaf epidermis has numerous small holes, called **stomatal pores**, that are not blocked by cuticle and that allow the plant to exchange gases with the air (**FIGURE 3.18**). There is a problem however: water vapor can also escape through the pores, and this could dehydrate the plant. While the plant is photosynthesizing during sunlight hours, there is little that can be done, and if the soil is moist enough the roots absorb adequate water to compensate for the water lost by leaves. But at night, when photosynthesis is not possible, water would still be lost even though the plant would not need to absorb carbon dioxide. This problem was solved when guard cells evolved. **Guard cells** are curved cells that occur in pairs, one on either side of a stomatal pore (**FIGURE 3.19**). At sunrise they increase their turgor pressure, swell and bend slightly more, opening the stomatal pore. At sunset, they lose turgor pressure, shrink and close the pore, preventing water loss at night. The term **"stoma"** (plural: stomata) is often used to refer to either the pore or both the guard cells and the pore (for example, an epidermis might have 100 stomata per square millimeter [an area about the size of the period at the end of this sentence]).

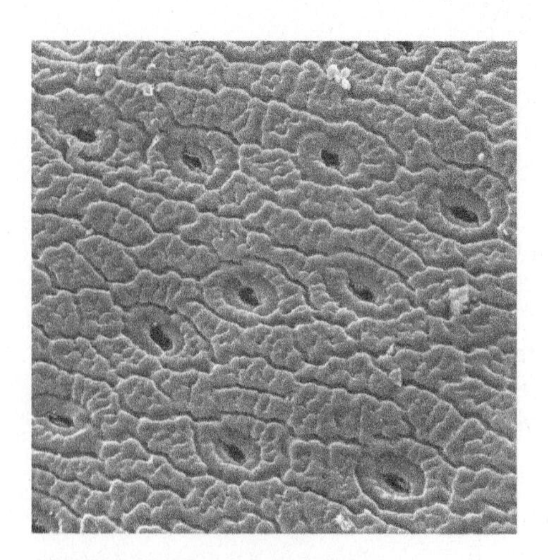
FIGURE 3.18. A scanning electron micrograph of the outer surface of an epidermis, showing the numerous stomata. Every photosynthetic cell (out of view on the other side of the epidermis) is close to a source of carbon dioxide (×50). (Courtesy of Urs Eggli.)

The epidermis of at least some parts of plants may produce **hairs (trichomes)** (**FIGURE 3.20**). Plant hairs may be as simple as just a single epidermis cell growing outward as a long cylinder (a unicellular trichome), or the cell may subdivide such that it is several cells long or wide (multicellular trichomes). They may be unbranched or branched, and most are parenchyma, but some hairs build a secondary wall and become sclerenchyma. In many cases, plant hairs die immediately after reaching their full size.

FIGURE 3.19. Each stomatal pore is surrounded by two guard cells and then by other epidermis cells. This specimen was prepared while the guard cells were swollen and pushed apart, opening the pore (×800). (Courtesy of Urs Eggli.)

FIGURE 3.20. The hairs on this leaf of lambs ear (*Stachys byzantina*) affect its biology in many ways: they make it difficult for insects to walk on the leaf, they keep dirt and dust away from the stomata, they reflect excess sunlight away from chloroplasts, and, when dew forms, the droplets form on the tip of the hairs where they will not block stomatal pores.

Hairs affect plant biology in many ways. Bright habitats like deserts, seashores, and high alpine areas may have too much sunlight, and a layer of dead hairs prevents the leaves from being sunburned and having their chlorophyll damaged. A thick layer of hairs, especially if they are curly or sharp, make it difficult for animals to walk around on a plant's surface, so they are less likely to eat the plant or lay eggs in it. Also, a layer of hairs creates a zone of calm air near the leaf's surface, which reduces the rate at which wind draws water out of the leaf through open stomatal pores.

Glandular hairs secrete substances, almost always defensive compounds. Those of stinging nettle secrete the irritating compounds formic acid and histamine, those of mala mujer have toxins that will blister the skin. Glandular hairs on *Pavonia odorata* secrete such sticky compounds that they entangle insects on the plants where they starve to death without getting a chance to harm the plant. These plants have been planted in gardens and plantations as a natural way to control insect pests.

Phloem is a complex tissue that conducts sugars throughout the plant (**TABLE 3.4**). In many plants, it also has cells that store sugars and other compounds, and has fibers that strengthen the vascular bundles in stems and leaves. The conducting cells of phloem in flowering plants are called **sieve tube members**, and they are basically parenchyma cells that have grown to be long, slender tubes (**FIGURES 3.21** and **3.22**). When young, they have plasmodesmata in primary pit fields, but as the cells mature, their plasmodesmata become wider (about 2.0 µm in diameter) and we call them **sieve pores**. The primary pit fields are also given a new name, **sieve areas**. Sieve tube members are aligned end to end with their sieve pores interconnected.

TABLE 3.4. Vascular Tissues

Phloem: conducts water, minerals, sugars, and other organic material; conducting cells are living parenchyma cells but they destroy their nuclei before they begin to conduct.

Sieve cells have small sieve areas and narrow sieve pores only; they do not have sieve plates with a large sieve area and especially wide sieve pores.

Sieve tube members have small sieve areas and narrow sieve pores on their side walls, but their end walls are sieve plates with a large sieve area and especially wide sieve pores.

Xylem: conducts water, minerals (and rarely sugars); conducting cells are sclerenchyma cells and must digest their protoplasm and die before water can be pulled through them.

Tracheids have no perforations; all water must enter or leave through pits and must cross pit membranes. Tracheids provide a very safe method of conduction, but cause high friction.

Vessel members have perforations; water moves from member to member within a vessel by passing through perforations. Water passes from one vessel to another through pits and pit membranes. Vessels provide low-friction conduction, but when an air bubble forms, it incapacitates many vessel members.

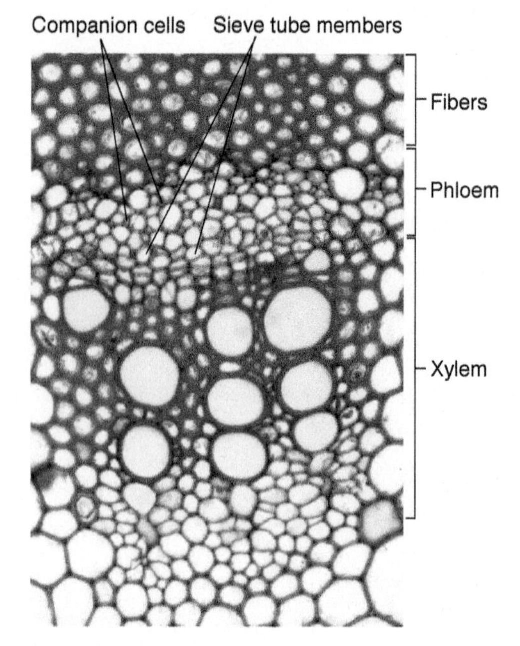

FIGURE 3.21. Vascular bundles have both xylem and phloem running side by side, parallel to each other. The xylem in this vascular bundle contains both narrow fibers and wide vessel elements; the phloem consists of sieve tube members and companion cells. A large mass of fibers protects the outer side of the bundle. This is from a stem of sunflower (×220).

A long series of sieve tube members is a **sieve tube**. The nuclei break down during development, so sieve tube members lack a nucleus, and instead their metabolism is controlled by the nucleus of an adjacent cell, a **companion cell**. The vacuole membrane also breaks down, causing most of the cytoplasm to become a watery solution called **phloem sap**.

Technically, the terms "sieve tube member" and "companion cell" should be used only for angiosperms. For all other vascular plants, the terms "sieve cell" and "albuminous cell" should be used.

Xylem is a complex tissue that conducts water and dissolved mineral nutrients, usually upward from roots, through stems and into leaves and flowers (Figure 3.21). Like phloem, it has conducting cells as well as storage cells and fiber cells. There are two types of xylem conducting cell: tracheids and vessel members.

Tracheids begin life as small parenchyma cells and they grow to be long, narrow cells with tapered ends; they resemble very long fibers (**FIGURE 3.23**). At maturity, they deposit a lignified secondary wall that has many broad pits; water flows from one tracheid to the next through the interconnected pits. Once the wall is complete, the tracheids die and digest all their protoplasm: when functional, a tracheid consists only of a cell wall. It is important that tracheids occur in groups, with many tracheids side by side and with their tapered ends interdigitating, so any mature tracheid is surrounded by many others; they do not function by themselves.

The pits that interconnect tracheids are like the pits of fibers and sclereids, but much wider and more numerous. An important point about pits is that each is a tunnel just through a secondary wall of a cell, not through the primary wall. Thus, a set of two aligned pits is not a completely open passage between two tracheids: there are two primary walls (called the **pit membrane**). Even though pit membranes are permeable, water moves more slowly when it has to cross the pit membrane as opposed to when it is merely being pulled through the pit or through the tracheid lumen. Thus, as water is pulled upward it passes from one

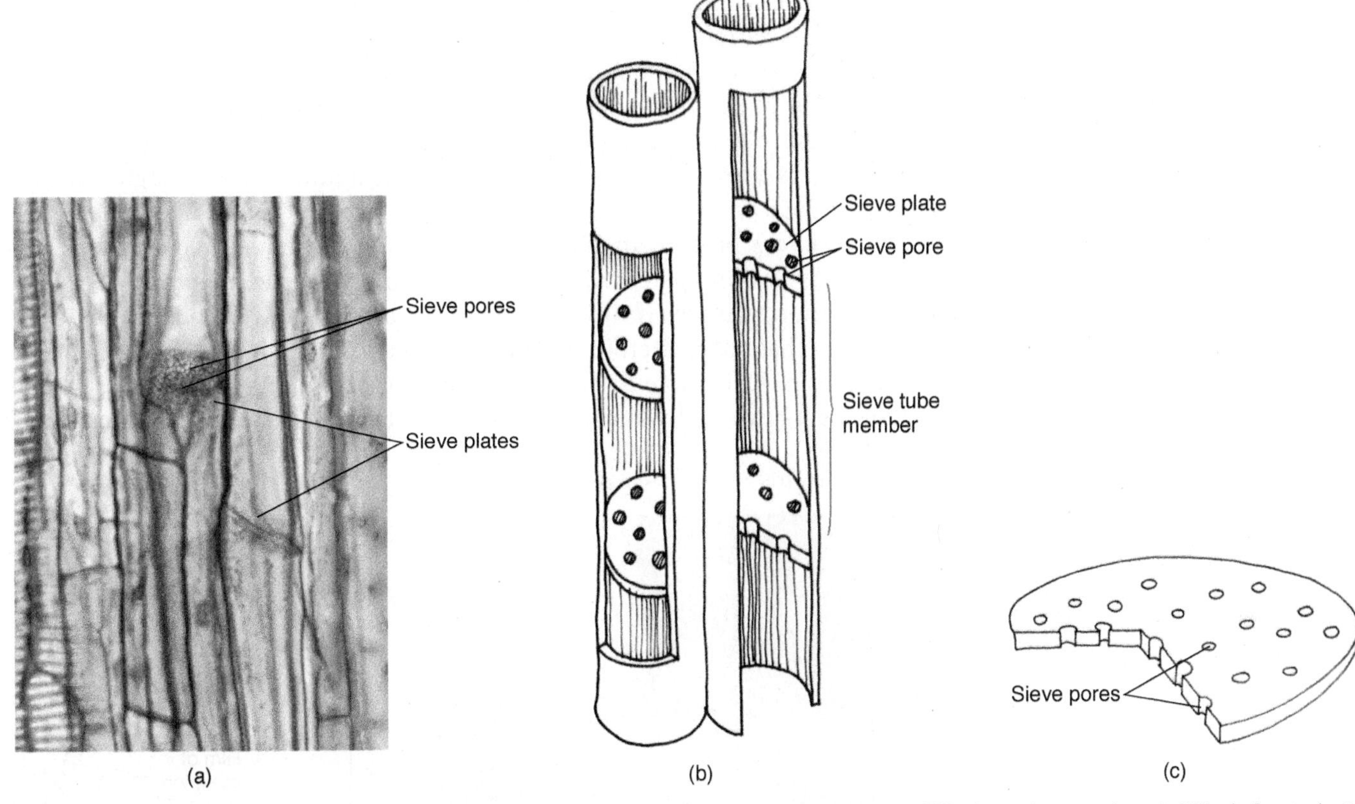

FIGURE 3.22. **(a)** Sieve tube members are long, slender, and delicate, and consequently are difficult to photograph and difficult for students to see in lab. Two sieve tubes are visible here, both with tilted sieve plates. One sieve plate is in side view, the other is in face view such that its sieve pores (white dots) are visible (×300). **(b)** and **(c)** are diagrammatic representations.

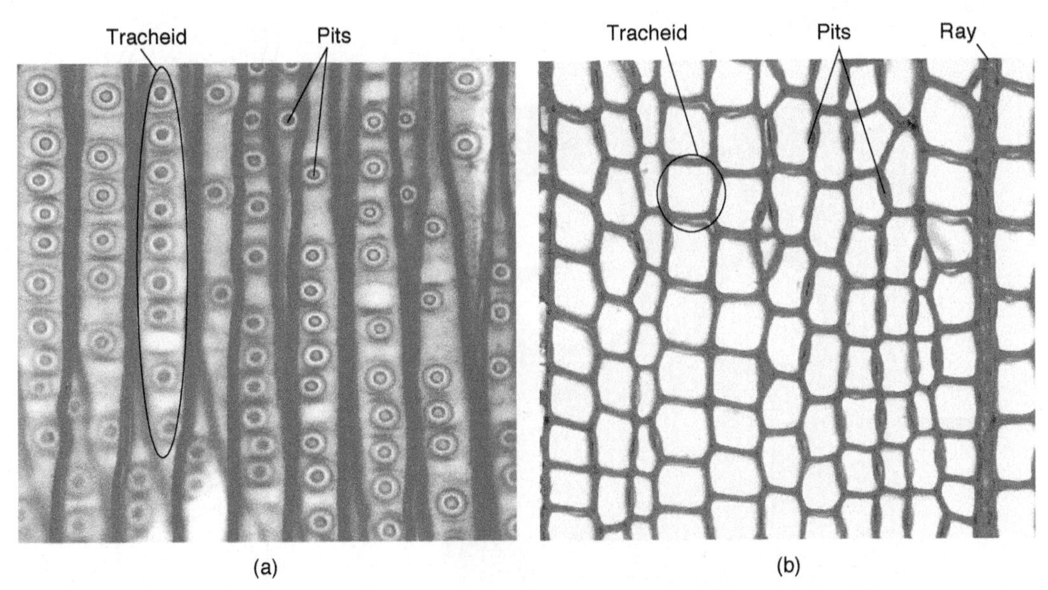

FIGURE 3.23. Tracheids are long, slender conducting cells that occur in masses, each interconnected to its neighbors by many broad pits. As water passes from one tracheid to another, it must pass through the narrow pits as well as the pit membranes. In **(a)**, the tracheids were cut lengthwise so that we could see their side walls; the pits are visible in face view; in **(b)**, the tracheids were cut in transverse section and we see the pits in side view (×300).

tracheid to another going through the pits of both. Tracheids are exceptionally long cells, but none is as long as an entire plant. If a plant is 1.0 m tall (about 3 feet) and its tracheids are each about 1 cm long (about 2.5 inches), each water molecule would need to be pulled through at least 100 tracheids and pit membranes. Because tracheids are narrow and each pit is even narrower, transporting water through tracheids involves a lot of friction (imagine drinking a milkshake through a narrow straw rather than a wide one).

Vessel members provide an alternative, low-friction method of conducting water in xylem. As a young parenchyma cell grows to be a vessel member, it becomes a very wide, short cylinder (FIGURE 3.24). Like a tracheid, it deposits a lignified secondary wall that has numerous wide pits aligned with the pits of vessel members lying alongside it. Just before dying and degrading away all its protoplasm, it completely digests its two end walls: it produces two large holes, called **perforations**,

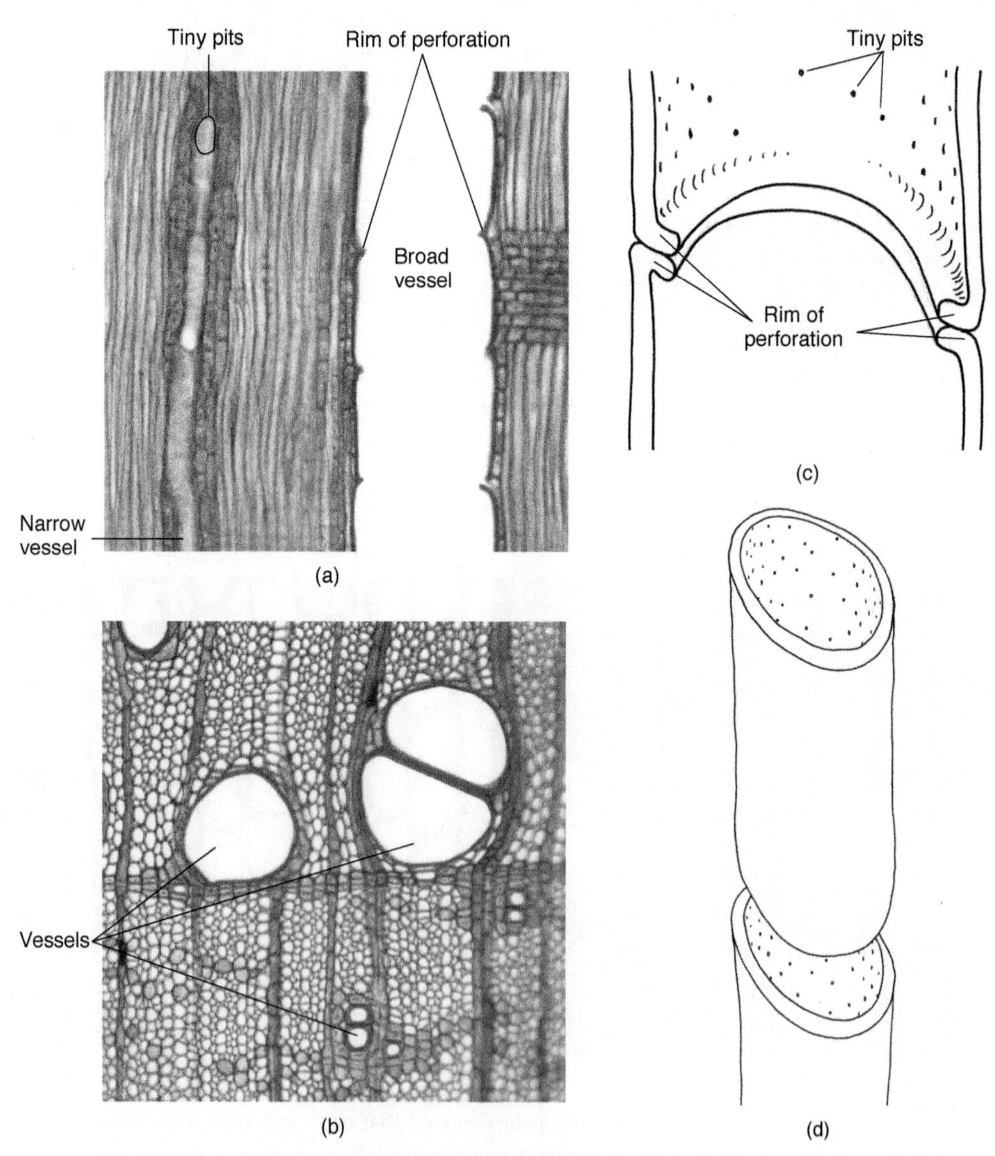

FIGURE 3.24. Vessel members are usually broader and shorter than tracheids, and each has a large hole (a perforation) on each end where it connects to the adjacent members of the same vessel. Pits along the sides of vessel members allow water to transfer from one vessel to another. **(c)** and **(d)** show how two vessel elements fit together.

one on each end of the vessel member. Once mature, it looks basically like the cement pipes used to build sewer lines, and indeed vessel members fit together the same way: the perforation on each of its ends is aligned with those on the vessel members above and below it. Hundreds or thousands of vessel members become interconnected this way, forming a long tube called a **vessel**. As water is pulled through vessels, it moves from one member to another through perforations—wide, open holes—rather than through narrow pits with pit membranes. Water moves much more easily and with less friction through a set of vessels as compared to a set of tracheids. Although vessels are often very long (often more than a meter [3 feet]), each does have a top and bottom end where the terminal vessel member has only one perforation, not two. Water must enter and leave a vessel through the pits on the sides of the vessel members.

The Internal Organization of the Primary Plant Body

Individual organs such as roots, stems, leaves, and flower parts are each composed of distinctive arrangements of tissues. Stems vary from bulbs to rhizomes to tendrils and others (see Box 2.1), but as mentioned before, they all have the same stereotyped parts: internodes, nodes, axillary buds, and so on. The same is usually true of the internal organization of plant organs: despite an outward appearance of diversity, internally, most stems are remarkably similar in the organization and patterns of their tissues, and the same is true of roots, leaves and flower parts.

Stems

If we cut across the internode of a young stem and examine it in transverse section, the pattern of tissues will be the same for almost all species. The outermost layer will be epidermis, followed by cortex, a set of vascular bundles, and then pith in the very center (FIGURE 3.25). The epidermis of a stem varies from one species to another, but it always is similar to the epidermis described above. The cuticle and waxes may be especially thick in plants of desert regions and rainforests; in deserts, the extra cuticle helps to keep water in the plant, but in rainforests it helps to keep all that rain from leaching chemicals and nutrients out of the plant. Stomata may be abundant in the stem epidermis of some species but completely absent from that of others.

Interior to the epidermis is the **cortex**, by definition the entire region between epidermis and the vascular bundles. In many species, the cortex is only about a millimeter or two thick (less than one-quarter inch) but in cacti and other succulents, the cortex may be several centimeters thick (from one to several inches) and consist of many layers of giant parenchyma cells that store water in their exceptionally large central vacuoles. The outermost layers of cortex are often a band of collenchyma cells that provide plastic strength to the internode; in other plants, this region remains parenchymatic while the internode grows, but once it reaches its full length, the cells differentiate into a band of fibers that provide elastic strength. Any band of specialized cells just below the epidermis is called a **hypodermis**. Cortex is usually green because at least some cells have chloroplasts. It is not unusual to find various types of defensive cells in the cortex: cells filled with poisons, irritating compounds or crystals, all of which make the stem less palatable to animals. In milkweeds and spurges, the milky sap that exudes when the plant is cut is produced in cortex cells (FIGURE 3.26).

In the center of the stem of most plants is the **pith**, which is similar to cortex. It is very narrow and almost always consists of just parenchyma cells, some with crystals

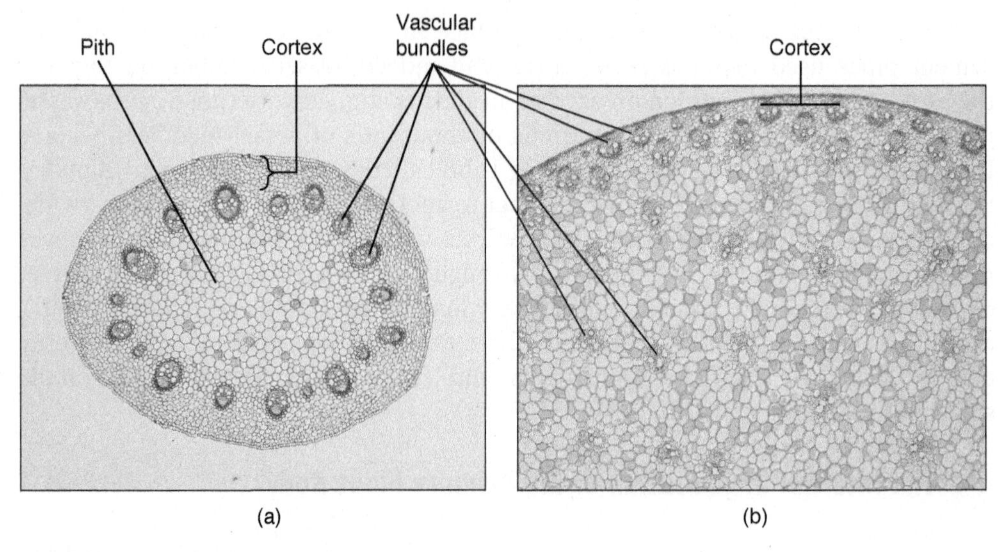

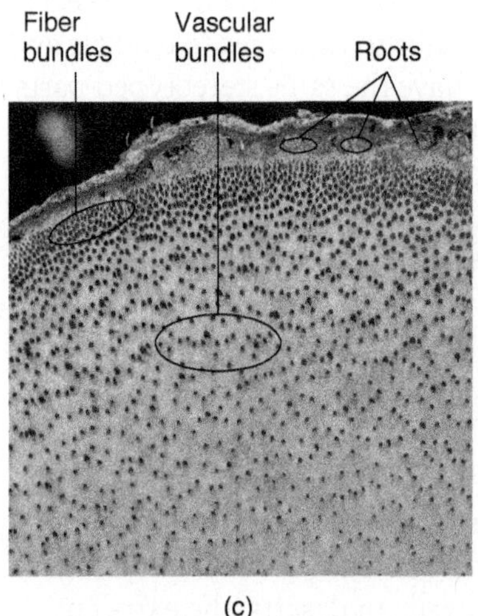

FIGURE 3.25. **(a)** Stems of this buttercup, like those of all vascular plants other than monocots, have a single ring of vascular bundles located between the cortex and pith (×20). **(b)** Stems of corn, like those of all other monocots, have many vascular bundles, not just a single ring (×35). **(c)** Palm trees are monocots, so they too have many vascular bundles; this large palm trunk has hundreds of vascular bundles. The outermost, very dark bundles consist of just fibers, without xylem or phloem (about one-quarter life size).

or defensive compounds. Starch grains may be present, but the pith is so slender it contains too few cells to provide much storage volume.

Between the cortex and pith is a ring of closely spaced **vascular bundles**, each having phloem on the outer side and xylem on the inner side (Figure 3.25a). When viewed in an internode cut in transverse section, as few as 20 or 30 vascular bundles might be visible in narrow stems, but there can be hundreds in wider stems. The phloem and xylem in these bundles is like that described above. Phloem contains sieve tubes and companion cells as well as other parenchyma cells that store material. Fibers may be mixed in as well, but it is most common for the outer margin of each bundle to have a cap of fibers or collenchyma cells running upward parallel to the bundle. This cap not only strengthens the bundles and the stem, it protects the phloem from animals like aphids that suck sap from sieve tubes (FIGURE 3.27).

The xylem of each bundle usually has 10 or more (maybe many more) vessels accompanied by parenchyma cells and fibers (FIGURE 3.28). Even a very small plant with narrow stems might have 20 vascular bundles each with about 20 vessels; thus there would be at least 400 vessels visible anywhere the stem is cut in transverse section. Each vascular bundle usually runs straight up the stem from one node to the next, but at the nodes, most bundles will merge with one or more of their neighbors, then continue up into the next internode. The separate bundles visible in internodes are really

FIGURE 3.26. A leaf was pulled off this milkweed (*Asclepias tuberosa*, also known as butterfly bush) causing milky sap to flow out of secretory cells in its cortex. Milky sap is often poisonous and can cause serious harm if it contacts your eyes.

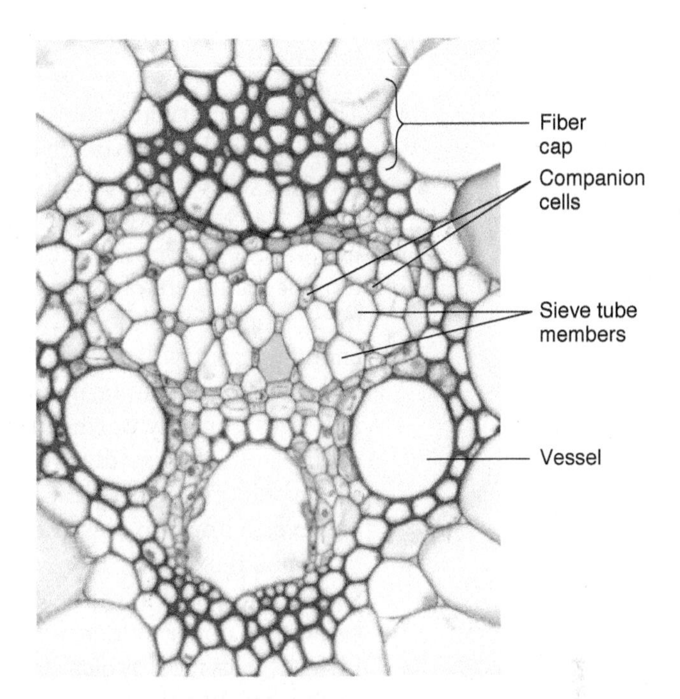

FIGURE 3.27. This vascular bundle of corn is protected by a cap of fibers between the phloem and the cortex. Companion cells and sieve tube members are especially easy to identify in this specimen (×200).

parts of an interconnected network of bundles whose vessels can exchange water with each other, just as their sieve tubes can exchange sugars.

Having a single ring of vascular bundles is by far the most common pattern; it occurs in ferns, conifers, and broadleaf plants (dicots). In monocots, there are numerous vascular bundles distributed throughout the center of the stem, not just in one ring (Figure 3.25b, c). Also, in most monocots, rather than just having a cap of fibers protecting the phloem, each vascular bundle may be completely encased in a **bundle sheath** composed of many fibers.

At nodes, where leaves and axillary buds are attached to stems, are sets of vascular bundles called **leaf traces** and **bud traces**. They are attached to the stem bundles at one end, cross the cortex and enter the leaf and axillary bud at the other: typically, several leaf trace vessels run side by side with several stem vascular bundle vessels, picking up water through the pits that interconnect the two. Then the leaf trace vessels carry the water out to the leaves. Similarly sieve tubes carry phloem sap from leaves into the stem then they transfer the sugars to the stem vascular bundle sieve tubes.

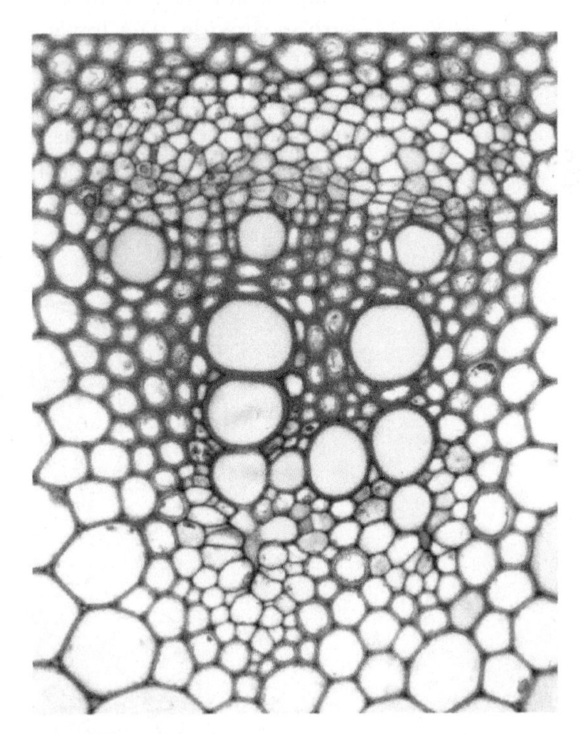

FIGURE 3.28. This is a very young vascular bundle, near the shoot tip of a sunflower. It has not yet formed all its xylem, but already almost a dozen vessels are present. The stem had many more vascular bundles: even at the shoot tip where leaves are young and small, about 100 vessels are mature and conducting water (×200).

Leaves

The internal structure of the blade of a foliage leaf facilitates photosynthesis. It has adaptations that maximize the absorption of light and carbon dioxide and a dense array of vascular bundles (often called leaf veins) that distribute water to all parts of the leaf while exporting sugars. The epidermis on the lower side of a foliage leaf (the lower epidermis) has numerous stomata, permitting rapid absorption of carbon dioxide. Hairs are often present, keeping the air calm around the stomata and thus increasing the likelihood that water that escapes out of the stomata may remain long enough that it will, by chance, reenter the leaf. The upper epidermis usually has many fewer stomata or none at all. Sunlight warms leaves, so air rises off them just as it rises from any warm surface. If the upper epidermis has stomata, any water that diffuses out is immediately swept away and has no chance to form a humid layer that would protect the leaf. Also, when dew forms, it covers the upper leaf surface, and even a fine film of water would block a stomatal pore, preventing the absorption of carbon dioxide.

The interior of a leaf is called **mesophyll**. Leaves sometimes have a hypodermis below the upper epidermis, but this would block light and is not very common. Instead, it is most typical to have one or two layers of columnar cells that project downward, like the bristles of a brush from the upper epidermis (FIGURE 3.29). This is **palisade mesophyll** and it is always photosynthetic parenchyma, with each cell having numerous chloroplasts. Palisade mesophyll cells lie close together but only rarely touch each other; instead of being glued tightly together by middle lamellas, almost every part of the cell wall faces intercellular space, from which it can absorb carbon dioxide. The lower portion of foliage leaves is occupied by **spongy mesophyll**, a tissue in which cells are widely separated, touching only one or two neighbors; intercellular spaces make up more than half the volume. The arrangement of foliage leaf mesophyll thus allows carbon dioxide to enter through numerous stomata in the lower epidermis, diffuse rapidly throughout the spongy mesophyll, quickly reaching all parts of the leaf, then diffusing up and around all sides of the palisade

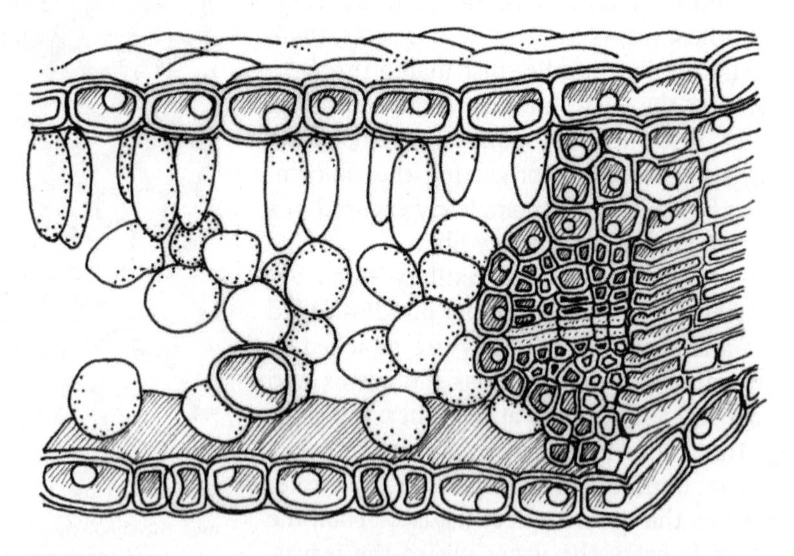

FIGURE 3.29. The interior of a leaf has many air spaces that allow carbon dioxide to diffuse from the stomata in the lower epidermis up to the photosynthetic cells of the palisade parenchyma. Xylem in the vascular bundle carries water to leaf cells; phloem brings sugar out of the leaf to the stem.

mesophyll cells. At the same time, light is passing through the upper epidermis and being captured by palisade cell chloroplasts where it powers the conversion of carbon dioxide into sugar.

Vascular bundles run in a single layer between the palisade and spongy mesophyll. Each has phloem on its lower side, xylem on the upper side (FIGURE 3.30 and see Figures 2.5 and 2.6). In transverse section, the midrib and larger vascular bundles have relatively large amounts of xylem and phloem, and typically also have caps of collenchyma or fibers. The smallest, finest bundles may have only one or two vessels and sieve tubes, and no caps.

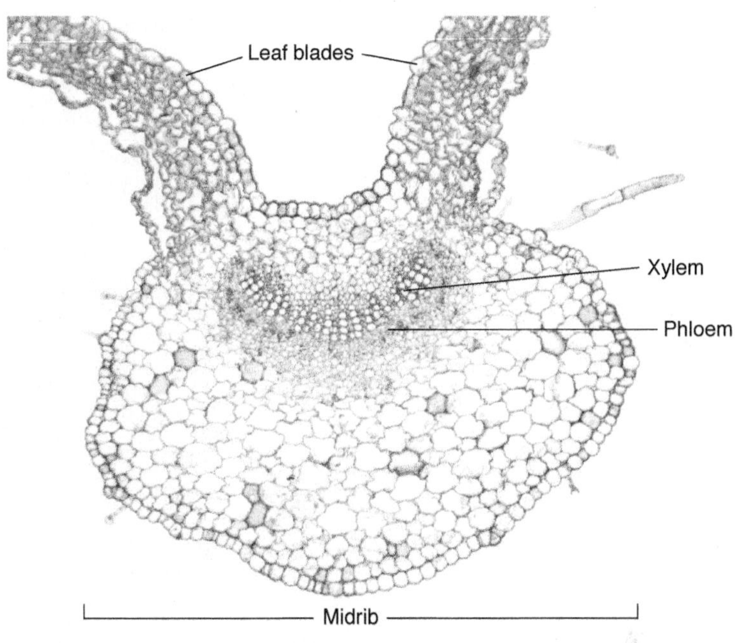

FIGURE 3.30. The midrib of this foxglove leaf contains a large amount of phloem and xylem. Whereas the leaf blade is thin, the midrib is thick and provides strength that holds the leaf into the sunlight (×50).

Roots

Roots evolved from stems and still resemble them in many ways, but several internal features are adaptations that increase a root's ability to absorb water and minerals (Figures 2.13 and 2.14). Because roots are subterranean and do not photosynthesize, stomata are not necessary in their epidermis. Instead, root epidermis produces thousands of special unicellular hairs, **root hairs**, that greatly increase the root's absorptive surface area and that are narrow enough to enter very tiny soil spaces where water and minerals are most likely to be found. Root cortex is typically narrow, just a few layers of rounded parenchyma cells. Intercellular spaces are prominent at their corners; roots need to allow oxygen to diffuse through them so that they don't suffocate. Some root cortex cells may contain crystals or defensive compounds, but typically a root's cortex lives only briefly. In many plants, it dies not long after the root hairs have stopped absorbing material.

The **endodermis** is a cylinder of parenchyma cells that lie between the cortex and vascular tissues in roots (FIGURE 3.31). It is a critically important barrier between the outside world and the plant. Because cell walls are permeable, any minerals dissolved in the soil water can diffuse through the root epidermis and cortex, just by moving through walls and intercellular spaces, without ever actually entering a cell's protoplasm. This is true both for beneficial minerals as well as harmful ones like lead and other toxic metals. But the radial walls (the sides, top, and bottom walls) of endodermis cells have a **Casparian strip**, a band where the walls are incrusted with waterproof chemicals. Nothing can diffuse past the Casparian strip; at this point, water and anything else must be accepted by a plasma membrane, enter an endodermis cell and then be released on the inner side of the endodermis. The plasma membrane allows beneficial elements in but blocks harmful ones.

Root vascular tissues do not occur in bundles. Instead xylem occurs in the center as a star-shaped column, the star having three or four arms in most roots (FIGURE 3.32). Alternating with the xylem arms are columns of phloem. Between the xylem, phloem, and endodermis are parenchyma cells called **pericycle**. Some pericycle cells divide and form the primordia of lateral roots, most simply store a little starch. Monocot roots are typically much wider than those of dicots and all other plants. The central column of xylem in monocot roots is exceptionally wide and may have dozens of arms alternating with an equal number of phloem columns. Roots never have pith.

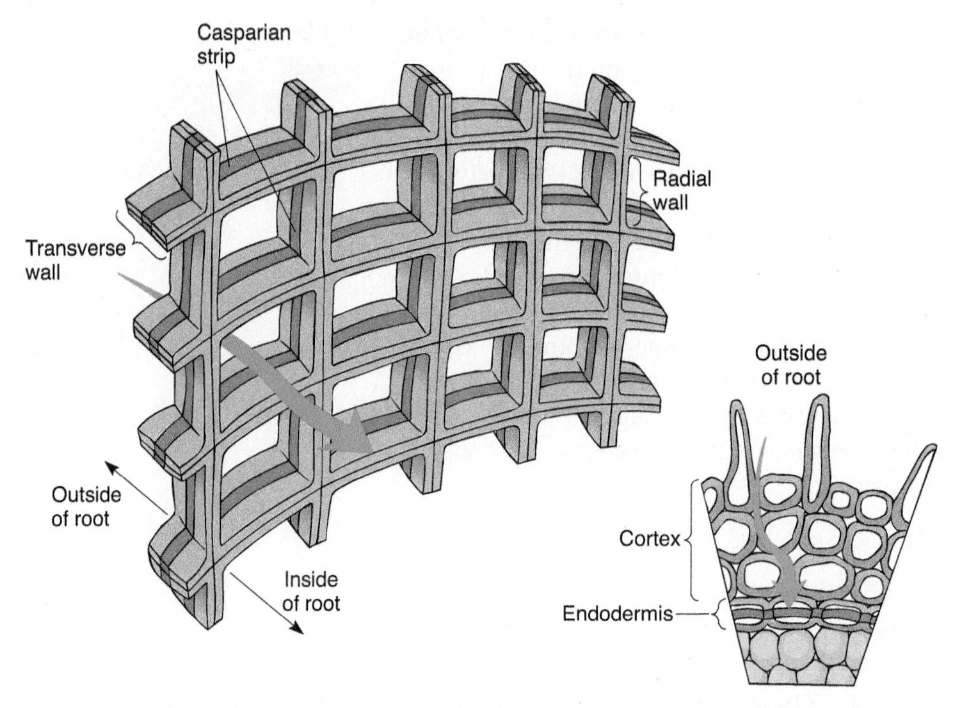

FIGURE 3.31. The endodermis is a cylinder, one cell thick, with Casparian strips on the radial walls of all cells. Chemicals dissolved in soil water cannot get past the endodermis by simply moving through the walls. Any chemical that is not accepted by a cell plasma membrane cannot get into the xylem.

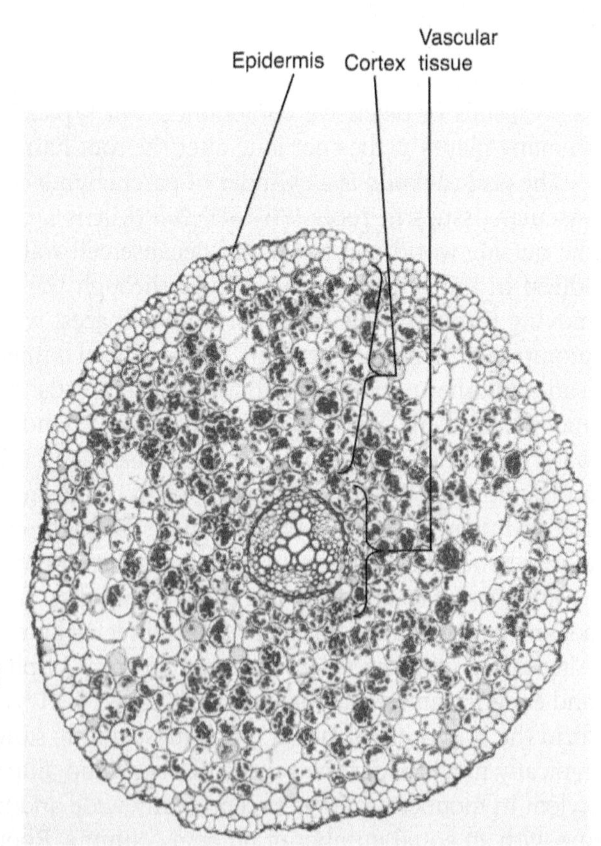

FIGURE 3.32. Transverse section of a root of buttercup. The cortex is broad, and the xylem in the center is shaped like a star rather than being part of vascular bundles as in stems. Columns of phloem are located between the arms of the xylem star (×50).

The Internal Organization of the Secondary Plant Body

In stems and roots of an herb, all cells of epidermis, cortex, xylem, phloem, and pith stop dividing, then they differentiate and mature within a few hours or days after having been produced by an apical meristem. But in woody plants, some cells located between the xylem and phloem do not stop dividing, instead they continue to act as meristematic cells and they constitute the vascular cambium. Vascular cambia produce cells to both the interior and exterior of the cambial cylinder. Cells on the inner side develop into **wood (secondary xylem)**; those on the outer side become **secondary phloem** (see Figure 2.24). These are complex tissues, each containing several types of cells, but wood, being a type of xylem, will have tracheids or vessels or both, and secondary phloem will have sieve tubes in angiosperms, sieve cells in other types of plants. Both secondary xylem and phloem in angiosperms usually have fibers, which give wood and bark their strength. Because fibers are present, angiosperm wood is called **hardwood** (FIGURE 3.33). In contrast, wood of conifers, even giant Douglas firs and coastal redwoods, have no fibers and so are called **softwoods** (FIGURE 3.34). In some cases, the tracheids of conifers have such thick, heavily lignified walls they constitute a stronger wood than the fibrous wood of some angiosperms. Wood of the conifer bald cypress (a softwood) is much harder and more rot-resistant than is wood of the angiosperm balsa (a hardwood).

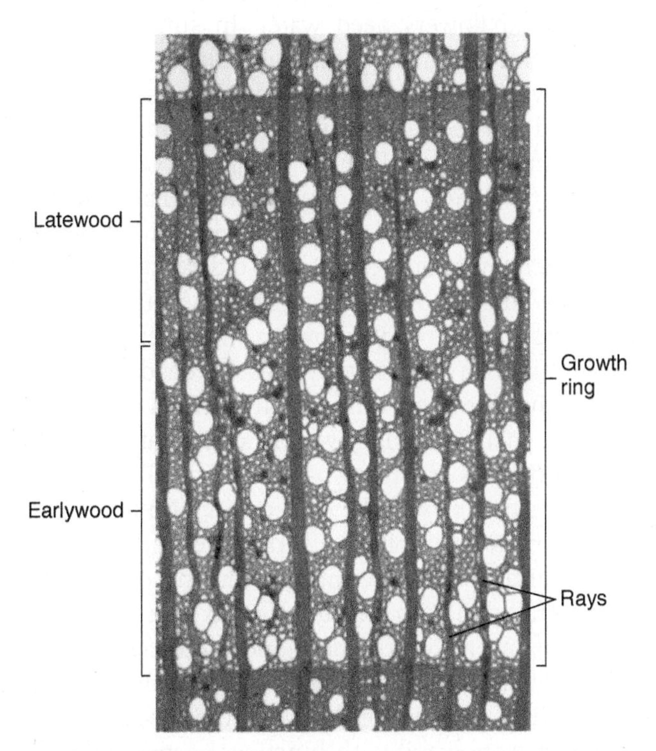

FIGURE 3.33. This is a transverse section of wood from an apple tree (*Pyrus malus*). Vessels are abundant but not very wide, and most of the cells between the vessels are fibers: because this has fibers, it is a hardwood. Rays are abundant, and one entire growth ring is visible, along with the latewood of the previous year and the earlywood of the following year. The vascular cambium and bark would be located above the top of the image (×50).

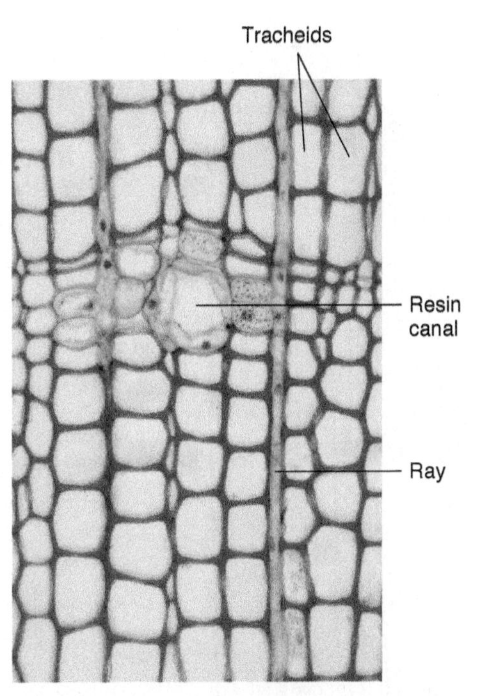

FIGURE 3.34. This is a transverse section of wood from a pine tree (*Pinus*). No fibers are present, so this is a softwood. All the cells with red-stained walls are tracheids. A resin canal is present, and several rays are visible (×200).

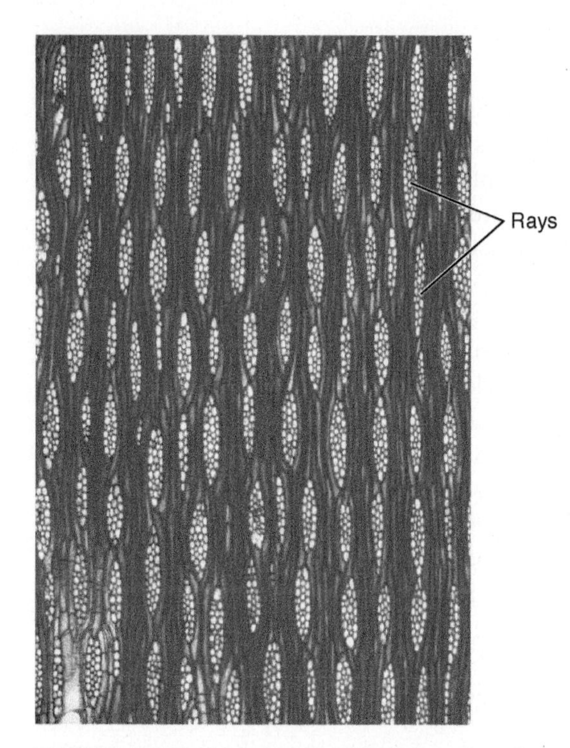

FIGURE 3.35. This is a tangential section of wood, cut near the surface of the tree trunk, parallel to the surface. The rays are masses of living parenchyma cells that store water, minerals, and starches in the wood. While they are alive, ray cells help resist fungi and insects that invade the wood (×40).

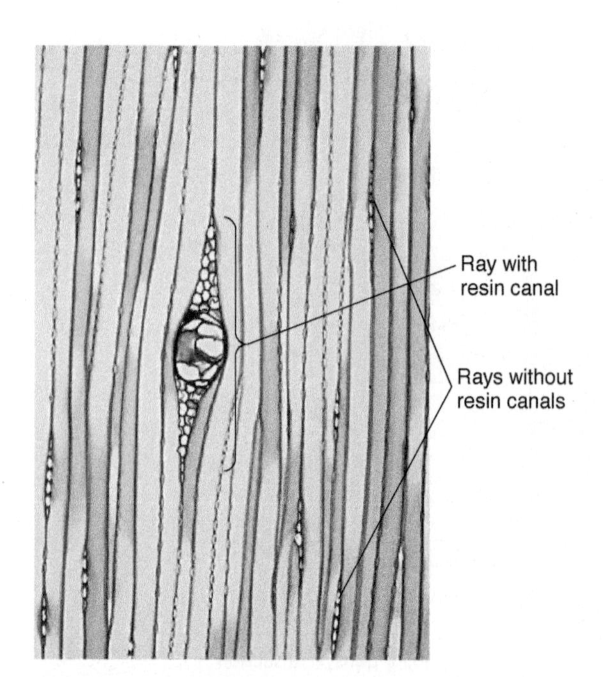

FIGURE 3.36. Tangential section of pinewood, showing that most rays are tall and narrow, but one has a resin canal in it. All the elongate, vertical cells are tracheids (×40).

Wood and secondary phloem have many parenchyma cells that store water and starch during winter when trees are leafless. Wood parenchyma cells also defend the rest of the wood from boring insects and wood-rotting fungi. All these are elongate cells produced by long, slender vascular cambium cells called fusiform initials. Both wood and secondary phloem have **rays** (FIGURES 3.35 and 3.36), groups of short parenchyma cells that store starch and water and that also help protect wood from invasive organisms. Ray cells are produced by short ray initials in the vascular cambium.

Vascular cambia are another example of localized, indeterminate growth. A vascular cambium can remain active for hundreds, even thousands of years, causing the trunk, branches, and roots to become wider each year. The most spectacular examples of this are the giant redwood trees of California, some of which have trunks 17 meters (57 feet) thick. Some trees become wide very quickly because their vascular cambia produce large numbers of wood cells each year, whereas in others, the vascular cambia are less active and even very old tree trunks are slender.

In springtime, vascular cambia usually produce wood with a high percentage of wide vessels. This provides high-volume conduction while soil is moist and leaves are young; also many trees flower in spring, and flowers need water. In summer, cambia usually produce wood that has a greater abundance of fibers and only a few, narrow vessels; this wood provides extra strength. The two types of wood are called **earlywood** (or springwood) and **latewood** (or summerwood). All the wood produced in a single year is called a **growth ring** (or an annual ring). The outermost growth ring (nearest to the vascular cambium and bark) is the newest one, and each one deeper in is 1 year older (FIGURE 3.37).

While wood is young, it contains living cells. All the tracheids and vessel elements are alive while growing and differentiating. It is only when they are mature that they must be dead and devoid of protoplasm so that they can conduct. In addition, most wood fiber cells die once their secondary wall is formed and lignified. But cells of the wood rays are living parenchyma cells, and often there are living cells next to vessels or mixed in with fibers. Also, when first formed by the cambium, all tracheids and vessels are filled with water and are conducting, but during dry seasons, some cells accidentally fill with air bubbles and stop conducting. Over the course of several years, more and more cells fill with air until finally all the tracheids or vessels of a growth ring are useless for conduction. But during this time, the vascular cambium will have made more wood with functional conducting cells. In many species, after vessels fill with air, the parenchyma cells next to them fill them with plugs called **tyloses** (singular: tylosis), which prevent fungi from growing in them. Once a growth ring stops conducting, its parenchyma cells synthesize numerous compounds that are antimicrobial and rot-resistant, then the cells die. At this point, all cells in the wood are dead, but it is still strong and helps hold up the trunk, branches, and roots. The outermost wood, that which has living cells and some conduction, is **sapwood**, whereas the inner wood that is not conductive and not alive is **heartwood** (see Figure 2.24c).

Although secondary phloem is also produced annually and cyclically, it typically does not have recognizable growth rings. Also, because it is produced to the outside of the wood and vascular cambium, it is pushed outward every time a new ring of wood is added. This causes the secondary phloem to be stretched and to crack, and bits and pieces of it fall off the tree. Whereas wood remains permanently on the tree (unless a branch falls off), secondary phloem remains attached only temporarily.

When a young woody stem or root sheds its first bark, the epidermis is shed with it, and the plant needs a new protective layer. Within the secondary phloem, band-like regions of parenchyma cells begin to divide and produce new parenchyma cells to their outer side. The dividing cells are the cork cambium. The cells they produce differentiate into **cork cells** by enlarging, putting a chemical called suberin into their walls, then dying. Suberin not only makes cork cells waterproof, they become resistant to fungi, bacteria, and other microbes. Cork is also indigestible, and because it is dead, it has no nutritional value, so animals do not seek it as food. The mixture of secondary phloem and cork is **bark**. Bark may be hard if the secondary phloem is very fibrous and the layers of cork are thin, or bark can be soft and spongy if cork is abundant

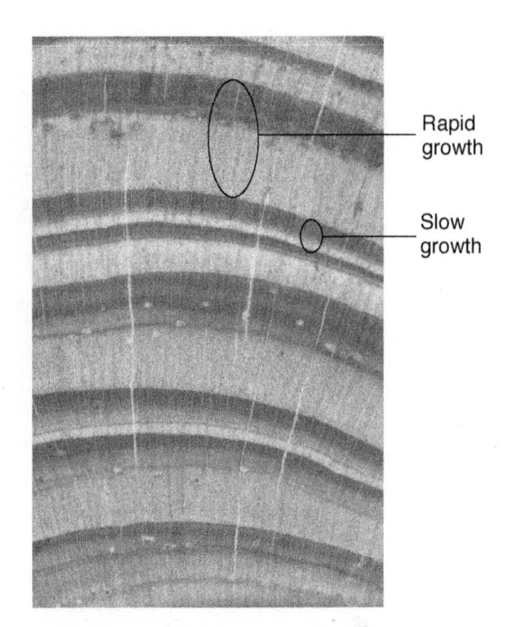

FIGURE 3.37. Growth rings are easy to see in this transverse section of a limb from a pine tree. Each growth ring consists of a light-colored band of earlywood and a darker band of latewood. Thick rings were formed when the tree (or at least the limb) was growing vigorously, thin rings indicate years when the limb's vascular cambium made only a few cells. White streaks are rays; dots are resin canals (about twice life size).

BOX 3.2. Dendrochronology: Analyzing Past Events with Tree Rings

The growth rings in wood carry information about a plant's past environment. Each year, a vascular cambium produces a new growth ring, and if the tree is healthy and vigorous, if it has enough rain, sunlight, and warmth, the cambium will produce many xylem cells and make a thick ring. On the other hand, if the summer is cool, dry, or cloudy, the plant will not photosynthesize well and there will be few resources for the cambium, and it might make only a thin ring. By examining a tree trunk cut in transverse section, we can usually see, even without a microscope, that some rings are thick, others are thin. By cutting or sanding the surface carefully, and then using a microscope, we can actually count the number of cells produced each year. If the wood was taken from a living tree, we know that the outermost ring is this year's growth, the next one deeper in is last year's and so on. Many trees live to be 500 or more years old, and we can identify which ring corresponds to each year for several centuries into the past.

Identifying good years (with wide rings) and bad years (narrow ones) is only the beginning. A single narrow ring surrounded by ones of ordinary size would indicate that just a single poor year occurred. However, if several narrow rings occur together, that would indicate many years of poor conditions for the tree. Perhaps there was an exceptional drought that lasted for many years, or a prolonged cool period. Long periods of bad weather in the tree's past might indicate that conditions were so poor that people in the region would have had poor crops and may have faced starvation. In contrast, a large number of wide rings together might indicate that weather was good for the region and any people there would have been prosperous. In areas where people have kept no written history, such as the American Southwest during the time of cliff dwellings, the dendrochronology record is invaluable.

Tree rings record more than just weather. If the tree survived a forest fire, one of its rings should show a scar and charcoal. Similarly, signs of unseasonal freezes, volcanic eruptions, and floods can be detected in tree rings, and from the age of the ring, we know when those events occurred.

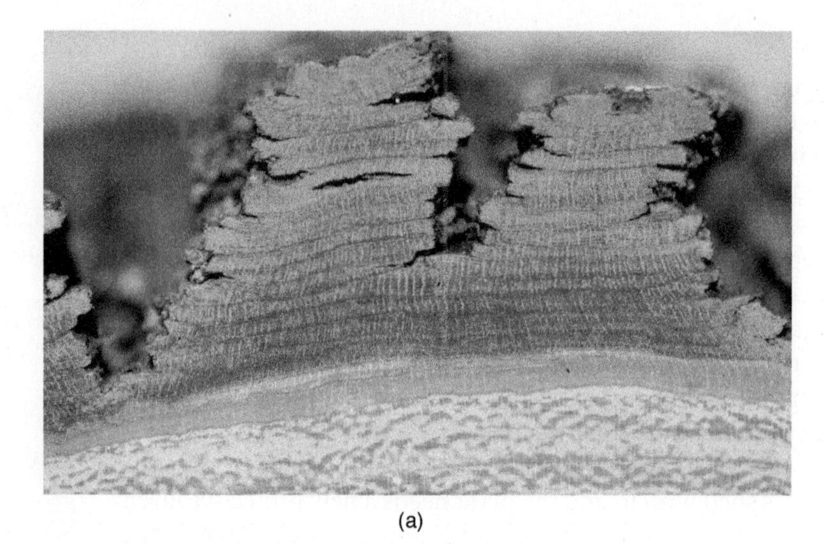

(a)

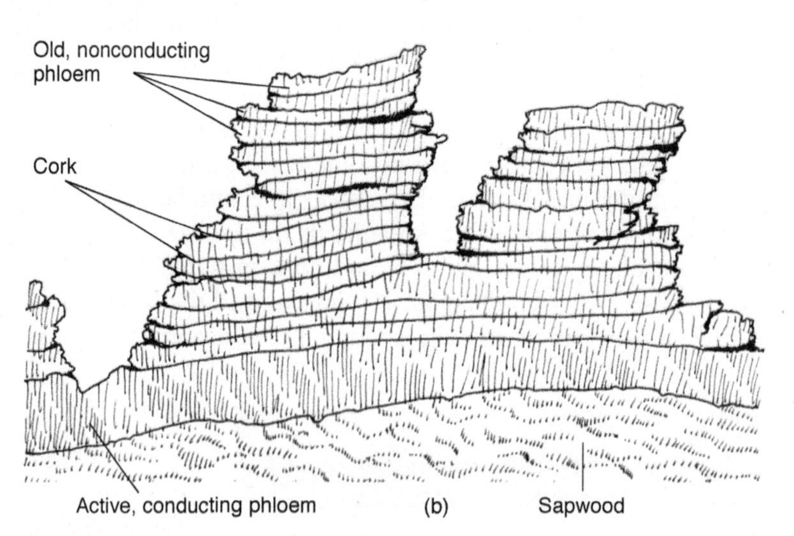

FIGURE 3.38. It is rare for bark to stay on a sample of wood; usually the weak cells in the region of the vascular cambium break and the bark separates from the wood. **(a)** The bark of this sample of honey mesquite (*Prosopis gladulosa*) contains many small patches of cork, each produced by a cork cambium that produced only a few cells then stopped functioning. **(b)** Because the bark does not stretch, it forms fissures as the tree forms more wood and pushes the bark outward.

and the secondary phloem has few or no fibers. Because bark is constantly being pushed outward and shed, new cork cambia must be produced periodically, each one deeper in the secondary phloem than the previous one (FIGURE 3.38).

Bark has **lenticels**, small regions of loose cork cells with intercellular spaces that allow oxygen to diffuse into the plant (FIGURE 3.39). Ordinary regions of cork are impermeable to gases because their cells fit together tightly, but in lenticels, cork cells are rounded just enough to create intercellular spaces at their corners, permitting oxygen to diffuse into the inner regions of phloem, cambium, and sapwood, all of which have living cells that need oxygen for respiration. This does increase the risk of invasion by bacteria, but passageways for oxygen are essential.

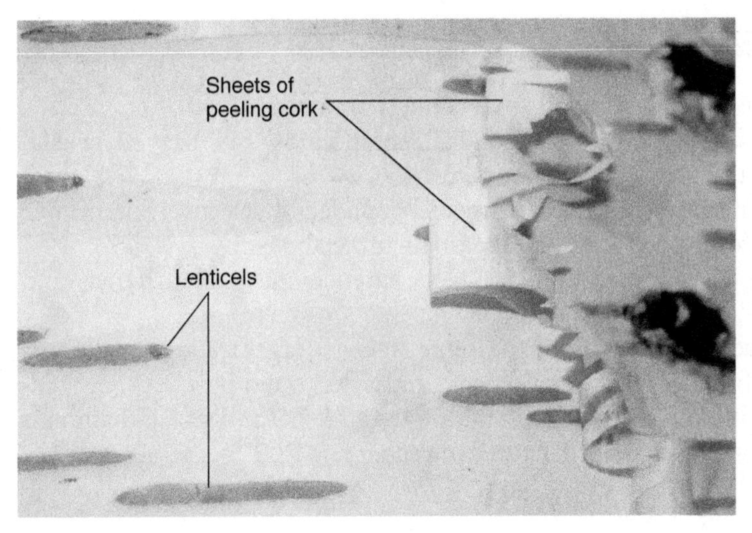

FIGURE 3.39. Unlike the bark of mesquite in Figure 3.38, birch bark (*Betula*) is very smooth and peels off in thin sheets. Its smoothness makes its wide, short lenticels visible. Oxygen diffuses through intercellular spaces in the lenticels and permits the living cells inside the phloem, cambium, and sapwood to respire.

Important Terms

amyloplast
bark
bud trace
bundle sheath
Casparian strip
cell wall
cellulose
central vacuole
chloroplast
chromoplast
collenchyma
companion cell
compartmentation
contact face
cork cell
cortex
cuticle
cutin
cytoplasm
cytoskeleton
cytosol
dictyosome
earlywood
elastic strength
endodermis
endoplasmic reticulum
epidermis
fiber
growth ring
guard cell
hair
hardwood
heartwood
hemicellulose
hypodermis
intercellular space
latewood
leaf trace
lenticel
lignin
mesophyll
microtubule
middle lamella
mitochondrion
nuclear envelope
nucleus
organelle
palisade mesophyll
parenchyma
perforation
pericycle
phloem sap
pit
pit membrane
pith
plasma membrane
plasmodesma
plastic strength
plastid
primary cell wall
primary pit field
proplastid
protoplasm
protoplast
ray
ribosome
root hair
sapwood
sclereid
sclerenchyma
secondary phloem
secondary wall
secondary xylem
selectively permeable
sieve area
sieve pore
sieve tube member
sieve tube
softwood
spongy mesophyll
stoma
stomatal pore
tracheid
trichome
tylosis
vacuole membrane
vascular bundle
vessel
vessel member
wood

Concepts

- All organisms are composed of cells that contain protoplasm.
- Protoplasm consists of ordinary chemical compounds and has no special properties, no vital force.
- Each cell consists of numerous compartments, bounded by selectively permeable membranes, and each specialized for a particular metabolism.
- There are three basic types of plant cell: parenchyma, collenchyma, and sclerenchyma.
- Almost all herbaceous stems have the same internal organization: epidermis, cortex, vascular bundles (each with phloem and xylem), and pith.
- Some terms, such as epidermis and hair, are used for both plants and animals but may refer to very different structures in each.

Basic Metabolism, Translocation of Water, and Mineral Nutrition

4

Death is a natural part of life for both plants and people. By shedding these leaves in autumn, the parent tree was able to minimize the amount of tissue it would need to protect in winter. Furthermore, as these leaves are decomposed by fungi, bacteria, and tiny animals, their minerals will be released and may be taken up by the roots of some plant, perhaps even the plant that shed the leaves. (Courtesy of Byran Hoyt.)

The bodies and metabolism of all plants—indeed of all organisms—are based on the fundamental principles of chemistry and physics. Although organisms can be alive, can reproduce, and can evolve, they have no special substances or powers not present in rocks or air or any other ordinary substance. Biology is easier to study because the same rules and equations you learned in chemistry class can be used to understand organisms.

Because biology is based on chemistry and physics there are two consequences we should take time to think about in a Plants and People book. First, living creatures are natural aspects of the world of rocks, minerals, mountains, air, and so on; we are not set apart from it, not detached from it. And second, we humans have the capacity and intellect to be aware of our world and to study it, to understand it, to appreciate and love it. We have the capacity to realize that our actions can either harm the world or sustain it. During the last several hundred years, especially since the Industrial Revolution, we have emphasized our ability to domesticate plants and animals and to use the world and its living beings as sources of raw materials that can be manufactured into things without worrying greatly about the pollution we have caused. But now we do see the harm we are doing and realize we must alter some of our actions.

We share Earth with all living creatures, and we must consider the consequences of our actions. We must also value our humanity, our human spirit, and our human responsibilities, and just as we take pride in our technological accomplishments, we can also take pride in our recent and ongoing efforts to minimize the pollution we cause and the steps we are taking to restore damaged parts of the environment. The principles introduced in this chapter help us understand the chemical and physical methods we can use to live more harmoniously in our world.

To whom much is given, much is expected (Luke 12:48).

Part 1: Atoms and Molecules

All substances are made up of **atoms**. Each atom is composed of a central nucleus containing positively charged protons and neutral neutrons, surrounded by clouds of negatively charged electrons (**FIGURE 4.1**). The various types of atoms differ in the number of protons present in each nucleus: hydrogen has one, helium has two, and so on (**TABLE 4.1**). The number of neutrons in each nucleus varies, but is similar to that of the protons. If the number of electrons in each atom is exactly equal to the number of protons, then the negative charges of the electrons balance those of the protons and the atom is neutral. If there are extra electrons, the atom is said to be a **negative ion** (an anion); if there are too few electrons, the atom is a **positive ion** (a cation; pronounced cat eye on). If all atoms of a substance are the same, the substance is an element; for example, hydrogen is an element, and all atoms of hydrogen have just one proton in each nucleus. Other elements important to biology are carbon (6 protons), nitrogen (7 protons), and oxygen (8 protons).

Atoms combine with each other to make **molecules**. For example, each molecule of water is made up of two atoms of hydrogen and one of oxygen (abbreviated H_2O), each molecule of carbon dioxide has one atom of carbon and two of oxygen (CO_2), and the sugar glucose has six carbon atoms, twelve hydrogens, and six oxygens ($C_6H_{12}O_6$) (**FIGURE 4.2**). Because each molecule of water, carbon dioxide, and glucose contains more than one kind of atom, they are compounds, not elements. Some molecules are composed of several identical atoms: the air we breathe is composed mostly of molecules of nitrogen (N_2) and oxygen (O_2). The atoms in a molecule are held together by interactions of the electrons of each; if the electrons hold the atoms together tightly, we say that they are strongly bonded, but if some part of the molecule can separate from another part easily, we say it has a weak bond. Molecules may be neutral or either positively or negatively charged.

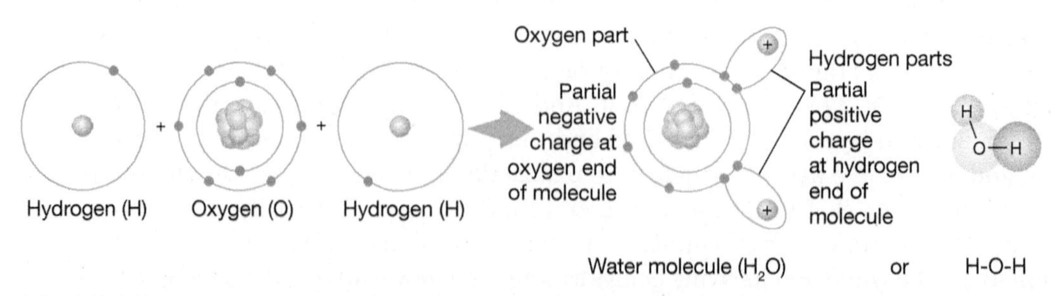

FIGURE 4.1. An atom of hydrogen has one proton and one electron, whereas an atom of oxygen has eight of each (plus eight neutrons in its nucleus). If two atoms of hydrogen combine with one of oxygen, they give off energy and become a molecule of water.

TABLE 4.1. The Number of Protons in Several Elements Important for Life

Element	Symbol	Number of protons (atomic number)*
Hydrogen	H	1
Carbon	C	6
Nitrogen	N	7
Oxygen	O	8
Magnesium	Mg	12
Phosphorus	P	15
Sulfur	S	16
Potassium	K	19
Calcium	Ca	20
Iron	Fe	26

* The number of protons in the nucleus of each element is known as the element's atomic number.

Water

Water has many characteristics important for life, and many of them are due to water molecules being small and "sticky." The two hydrogen atoms lie more or less on the same side of the oxygen atom, so each molecule is slightly positive on one side but slightly negative on the other; water molecules are **polar**, with a positive pole and a negative one. Opposite charges attract, so each water molecule attracts nearby water molecules, as well as any other charged molecule. Water's attraction to itself is called cohesion. Cohesion makes it difficult to boil water or to melt ice, and difficult for water to evaporate, so lakes and rivers tend to be very stable and safe for living organisms, remaining liquid unless temperatures are extremely cold, and not evaporating away unless droughts are prolonged (FIGURE 4.3).

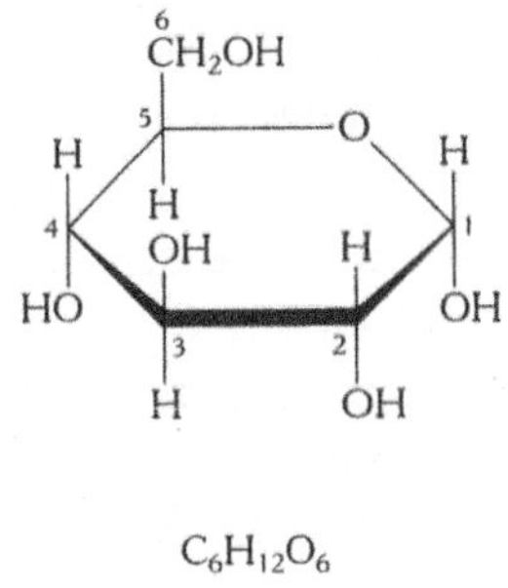

$C_6H_{12}O_6$

FIGURE 4.2. A molecule of the simple sugar glucose has six atoms of carbon, twelve of hydrogen, and six of oxygen. It is possible for this number of carbons, hydrogens, and oxygens to combine in many different ways and to also have the formula of $C_6H_{12}O_6$; when they are arranged as in this diagram, they are glucose.

FIGURE 4.3. Snowshoeing in Zion National Park. Because water molecules are cohesive (they stick to each other), water is extremely stable and does not melt, freeze, evaporate, or condense easily. Water is present here in its solid form (snow and ice), its liquid form (inside the bodies of the living organisms), and as a gas (water vapor in the air, which is invisible to us). A great deal of solar energy is needed to melt the snow to liquid and then cause it to evaporate; and similarly, temperatures must be very cold for water vapor to condense to rain or snow.

Water's attraction to substances other than itself is called adhesion. Because each water molecule is very small, water easily interacts with many other molecules that have positive or negative regions. The positive sides of many water molecules surround negative portions of substances and cause them to dissolve, and the same is true for the negative sides causing positive regions to dissolve. For example, sugar molecules also have both positive and negative regions, and consequently sugar quickly dissolves in water. Substances that interact with water this way are said to be **hydrophilic** (water loving). In contrast, substances in which all parts are neutral, such as oils and waxes, are **hydrophobic** (water fearing): they do not interact with water and do not dissolve in it.

Acids and Bases

As a first step to understanding acids and bases, we can say that **acids** are compounds that break down and release a **proton** (an H^+) into a solution, whereas a **base** is any compound that releases a **hydroxyl ion** (an OH^-) when it breaks down. This is important because other compounds in the same solution may then pick up a proton and become positively charged (or neutral if they were already negative); others may pick up a hydroxyl and become negatively charged. Adding an acid or base to a solution can potentially change many other molecules in the same solution. Protoplasm is an extremely complex solution of thousands of types of molecules, most of which will be affected by the presence of acids or bases.

In a weak acid, the proton is held tightly so only a few of the acid molecules in a solution release a proton. The solution does not become very acidic and has only a low concentration of protons able to affect other molecules; examples are vinegar (acetic acid) and lemon juice (citric acid). Strong acids break down extensively and many protons are released, strongly affecting the solution; examples are sulfuric acid and hydrochloric acid. There are also weak bases (bleach, sodium hypochlorite), and strong ones (lye, sodium hydroxide), depending on how many hydroxyls are released. **pH** is a measure of acidity: solutions with a pH near 7 are neutral; those between 0 to 6.9 are acidic; and those with pH of 7.1 to 14 are basic. Limestone is basic and causes ground water, streams, and lakes in an area to be basic (**alkaline**), whereas soils rich in leaf litter and humus are acidic as are waters that flow from them. Pure water, unaffected by soil, plants, or animals, is neutral.

Acids and bases are important in biology for several reasons. First, many common organic compounds are acids and bases. Examples of common acids are amino acids in proteins, fatty acids in oily and greasy foods, and nucleic acids in cell nuclei. In most of these, the neutral molecule is relatively inactive, but when the proton or hydroxyl comes off in solution, the remainder of the molecule has a charge and can react with another molecule that has an opposite charge. Also, many mineral elements that roots must obtain from soil are soluble in acidic solutions but not alkaline ones. If the soil is too alkaline, the minerals do not dissolve and cannot be absorbed by roots.

Water is unusual in being both an acid and a base: in a body of water (in a drinking glass, in a lake, snow, ocean) almost all water molecules are complete, that is, they are H_2O, but a tiny fraction spontaneously break down into H^+ and OH^-. Because one acidic H^+ is produced every time one basic OH^- is, the solution remains neutral (pH 7), but still there are protons and hydroxyls available to affect other molecules in the solution.

Organic Molecules

Millions of organic compounds—those that contain at least one carbon atom—are possible, but fortunately they can be classified into just a few families of compounds,

TABLE 4.2. Main Groups of Macromolecules in Cells

Carbohydrates composed of sugars	Lipids composed of fatty acids
Proteins composed of amino acids	Nucleic acids composed of nucleotides

such as carbohydrates, proteins, lipids, and so on (**TABLE 4.2**). This is because most carbon atoms in an organic compound carry a chemical group, known as a **functional group**, and these give the compound its particular properties. A carbon compound that has an acid functional group is an acid; those with an –OH functional group are alcohols, and so on. Because most organic compounds have several or many carbons, each compound can have several functional groups and simultaneously be an acid and an alcohol or other things.

Another thing that simplifies biological chemistry is that many important compounds are **polymers**. Each is made up of smaller units (**monomers**) that tend to be similar to each other. For example, simple sugars can be joined together to make complex sugars; the cellulose molecules in cell walls are giant molecules with thousands of carbon atoms and thousands of functional groups, but each is really a very simple polymer of the simple sugar glucose. Similarly, proteins are polymers of amino acids.

Carbohydrates

Carbohydrates are familiar organic compounds such as sugars, starches, and cellulose. The basic unit of carbohydrates are **simple sugars (monosaccharides)**, sugars that cannot be broken down into smaller sugars. Monosaccharides can be polymerized into **disaccharides** (two sugars) and **polysaccharides**. There are hundreds of monosaccharides, most which are made only by plants and algae, not by animals. The great diversity of monosaccharides makes it possible to have thousands of types of disaccharides and polysaccharides, but only a few need to be mentioned here.

Monosaccharides are classified by the number of carbon atoms they contain. Simple sugars called tetroses have four carbon atoms, pentoses five, hexoses six, and so on. One important pentose is deoxyribose, the D in DNA (deoxyribonucleic acid). Familiar hexoses are glucose (refer back to Figure 4.2), fructose (often called corn syrup on food labels), and galactose. All hexoses have more or less the same formula ($C_6H_{12}O_6$), but they differ in the arrangement of their atoms such that each hexose differs from all others in its shape and the position of side groups. This allows enzymes to easily distinguish one hexose from another. The enzyme that polymerizes glucose into starch (amylose synthase) only interacts with glucose and never accidentally puts fructose or galactose into the growing starch molecule.

It is difficult to overstate the importance of glucose. Too much sugar is bad for our health, but glucose is what photosynthesis produces, and using only glucose and some inorganic ions like ammonium, sulfate, and a few others, plants build their entire bodies, everything. Glucose can be converted not only into all the other hexoses, but into all the other sugars. From these, plants make many types of acids, including amino acids, fatty acids, and nucleic acids. Every organic molecule in a plant began as glucose in a chloroplast. And because all of our own food is also ultimately derived from plants (meat comes from animals that eat grass, oats, and corn), all our organic molecules began as glucose as well. Plants also polymerize glucose into various long-chain polysaccharides. Bonded end to end in a particular way (with a beta-1,4 bond), glucoses become cellulose, a structural material used to build cell walls. Bonded in a different way (with an alpha-1,4 bond), glucoses become starch (amylose), a means of safely storing glucose for many days, months, or years.

Beyond converting glucose to other materials, almost all organisms have enzymes that break down starch, releasing the glucose molecules when they are needed. The glucose is then converted into other molecules or respired for energy. Much of our own energy comes from digesting and respiring starch in our food. Surprisingly, very few organisms can break cellulose down to get its glucose molecules; instead cellulose passes right through us as "dietary fiber." Glucose itself cannot be stored in large quantities because it interacts with water and would cause a cell to swell. This interaction does not occur with starch; starch grains can be stored in cells with no adverse effects.

When a plant needs to transfer energy from one area to another, such as from leaves to growing tissues or developing fruits, it uses energy-rich molecules, typically the disaccharide sucrose. A cell converts part of its glucose to fructose, then bonds one glucose to one fructose in such a way that the resulting sucrose molecule is very stable and has little tendency to react with other molecules. Sucrose can be safely transported from one end of a plant to the other in phloem. Phloem sap of white lupine has as much as 154 grams (about 5.4 ounces) of sucrose per liter. Other small polysaccharides are also transported: raffinose (a trisaccharide of galactose, glucose, and fructose) and stachyose (a tetrasaccharide similar to raffinose but with two galactose molecules instead of just one).

Amino Acids and Proteins

Amino acids are small organic molecules that are both an acid and a base. One end of each is a carboxyl group (acidic), the other end is an amino group (basic), in between is the rest of the molecule with a side group that is a functional group with some special property. Some side groups are hydrophilic, others hydrophobic, some are long, others short, and so on. Only 20 different amino acids are used to construct all proteins. The carboxyl group of one amino acid forms a chemical bond (called a peptide bond) with the amino group of a second amino acid (FIGURE 4.4). Even after an amino acid has reacted this way, it still has a reactive end, so more amino acids can be added, creating a long, unbranched molecule called a **protein** (very short proteins are sometimes called peptides). Any particular organism has thousands of types of proteins; we know that humans have at least 30,000 types, and when all organisms on earth are considered, there must be millions of types of protein. Each differs from all others in the number of amino acids it contains and in their sequence.

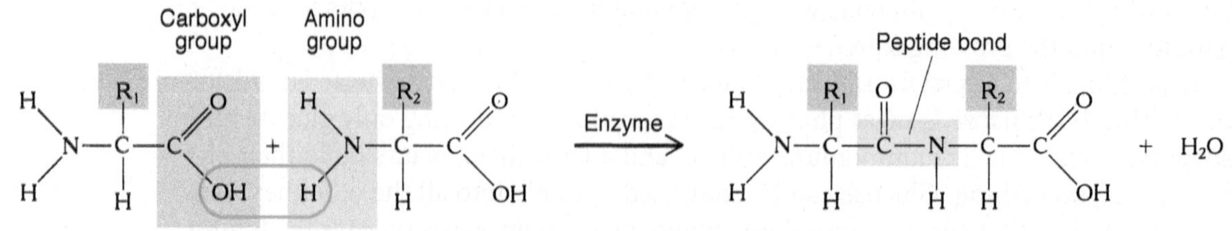

FIGURE 4.4. Proteins are composed of amino acids that bond to each other by having a carboxyl group on one react with an amino group on another. After the reaction, another amino acid can be added, then another, and so on until the protein is complete. The bonding of amino acids like this occurs only in ribosomes and only when guided by messenger RNA. The boxes marked R_1 and R_2 are side groups, and each of the twenty amino acids has a unique side group and thus unique chemical characteristics.

The amino acids in any particular type of protein occur in extremely precise sequences, and this gives the protein distinctive properties. Each protein molecule spontaneously folds into a distinctive shape because of its sequence of amino acids. Positively charged amino acids repel each other but are attracted to negatively charged ones, and this may cause one part of a long protein to fold back, bringing oppositely charged regions together. Similarly, various hydrophilic regions attract each other or interact with the water of the protoplasm whereas hydrophobic regions cause the protein to fold such that as many hydrophobic regions as possible are brought together to form a pocket. If part of a protein has amino acids that make it hydrophobic, such regions will tend to dissolve into the lipids of membranes. By organizing the 20 amino acids in various sequences, and by making proteins of various lengths, a huge number of proteins, each with distinctive properties, is possible.

The sequence of amino acids is said to be the **primary structure** of the protein. The particular shape caused by the attraction and repulsion between amino acids is called its **tertiary structure** (in between, short regions form an alpha helix, others form a beta pleated sheet; these are its **secondary structure**) (FIGURE 4.5). In many cases, once a protein has folded into its tertiary structure, its surface has the correct shape and characteristics to adhere to other proteins; the entire grouping is **quaternary structure**. This ability to automatically organize a shape simply due to characteristics of the molecule itself is called **self-assembly**.

Proteins have many roles in biology. Some are simply storage forms of amino acids. As a flower produces seeds, it fills them with nutrients that will allow the embryo to grow rapidly when the seed germinates. Amino acids are polymerized into seed storage proteins that fold into such a compact mass (their tertiary structure) that almost all water is excluded from it; the protein is dry, stable, and lightweight (FIGURE 4.6). We eat storage proteins in beans, peas, whole wheat, and other protein-rich plant-based foods. Other proteins play a structural role; they interact with each other (forming a quaternary structure) to form long strong microtubules or microfilaments that act as

FIGURE 4.5. This is a model of a protein that has folded into its tertiary structure due to the interactions of the side groups on each of its amino acids. When folded like this, the protein has the correct structure to carry out a particular function. (Structure from Protein Data Bank 2AAS. J. Santoro, et al., *J. Mol. Biol.* 229 [1993]: 722–734. Prepared by B. E. Tropp.)

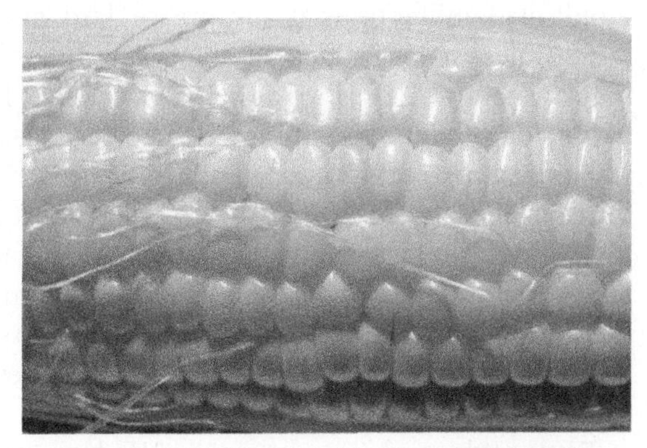

FIGURE 4.6. As these corn kernels mature, their cells synthesize and store large amounts of a protein called zein. Protein makes corn nutritious for us: after we eat corn, our stomach and small intestine digest the zein down to its amino acids, which we then absorb and use to synthesize the proteins our bodies need.

a skeleton inside each cell. Microtubules are composed of two proteins that associate into pairs, then the pairs aggregate into microtubules.

Some proteins act as **hydrophilic channels** in membranes. Their primary structure has alternating regions of hydrophilic and hydrophobic amino acids; the protein folds in such a way that the hydrophilic amino acids form a channel in the center whereas hydrophobic ones face outward. Such a protein does not remain in water for long but instead sinks into a hydrophobic membrane. Once there, the central channel acts as a passageway allowing hydrophilic molecules to pass through the membrane.

Many proteins act as biological catalysts called **enzymes**. Catalysts enter a chemical reaction and cause it to proceed faster or more easily or at a lower temperature than the reaction would without the catalyst. The catalyst itself emerges unaffected by the reaction, so it can then catalyze another and another, often participating in several hundred per second. To be an effective catalyst, an enzyme must have an **active site**, a set of amino acid side groups that are brought together by the protein's folding. The combined properties of each of the active site's side groups create a chemical environment that speeds up a reaction. The active site must have exactly the right size, shape, and set of side groups, which means that the protein must be folded into the correct tertiary shape. If the cell's pH is too acidic or alkaline it might cause the protein to fold improperly and no active site will form. Too much heat or cold, or too many ions like calcium (Ca^{++}) and magnesium (Mg^{++}) can also affect the active site for better or worse. A protein that is inactivated is denatured.

Each enzyme catalyzes only one type of reaction because each active site must be precise. The enzyme that polymerizes glucose into cellulose is called cellulose synthase, and its active site accepts only glucose, no other monosaccharide. Proteases depolymerize proteins back into amino acids, and lipases digest lipids. The fact that each enzyme catalyzes only one type of reaction is **substrate specificity**, and it allows a cell to control its metabolism by controlling the types of enzymes it produces: if it makes all the enzymes necessary for photosynthesis, it will develop into a chlorenchyma cell; if it makes the ones for secondary walls and lignin, it becomes a sclerenchyma cell and so on.

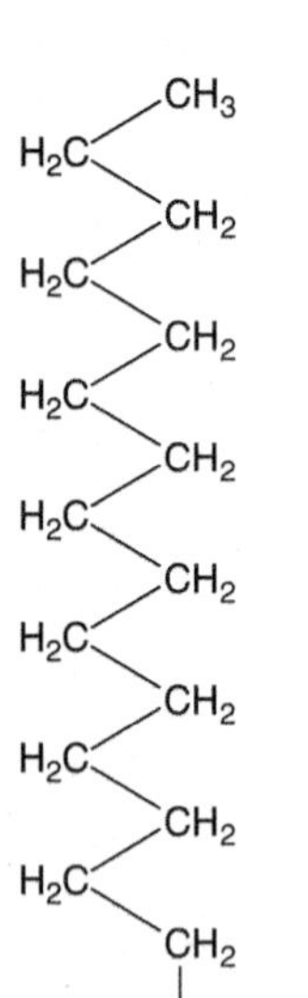

Lipids

Lipids are familiar to us as oils, grease, lard, and butter: things that do not dissolve in water unless soap is added. Unlike water, lipids have no positive and negative regions; they are neutral, described as being **nonpolar**. The basic units of many lipids are **fatty acids**, long chains containing up to 26 carbons and having an acid group (a carboxyl, just as in amino acids) at one end (FIGURE 4.7). If every carbon is bonded to two hydrogens, the fatty acid is **saturated** and tends to be a straight molecule that lies closely to neighboring fatty acids. In a group, saturated fatty acids fit together so well they tend to crystallize and be firm or hard at room temperature (butter, grease). If some hydrogen is missing, the fatty acid is **unsaturated** and has a kink in it, and it cannot lie tightly against its neighbors. It has less tendency to crystallize, so unsaturated fatty acids are soft or even liquid at room temperature (olive oil, corn oil) (**BOX 4.1**).

Fatty acids tend to polymerize. We have seen two examples in earlier chapters. Epidermis cells secrete fatty acids on to their outer wall where they polymerize into cutin if they are moderately long, or into wax if they are very long fatty acids. Almost all other fatty acids react with glycerol, which can hold three fatty acids, forming a **triglyceride** (FIGURE 4.8). This is the main storage form found in lipid droplets in cells of oily seeds and fruits, such as avocado, peanuts, and sunflower seeds. Being nonpolar and hydrophobic, triglycerides have even less tendency to attract water than does starch, and because most enzymes are water-soluble not lipid-soluble, a lipid droplet in a cell is almost inert.

Palmitic acid
(saturated)

FIGURE 4.7. All fatty acids have a carboxyl group at one end; the rest of the molecule is just carbon and hydrogen. If no double bond is present, the molecule is a saturated fatty acid and it can be straight. This allows them to align side by side; these are often solid at room temperature (a fat).

BOX 4.1. Lipids: Oils, Fats, and *Trans*-fats

The term "lipid" covers both fats and oils and several other compounds (see previous page and page 85). **Fats**, by definition, are solid at room temperature whereas **oils** have a lower melting point and thus are liquid at room temperature. If oils are cooled sufficiently, they solidify, and if fats are heated, as when we use them for frying, they liquefy. Fatty acids are long, unbranched backbones of carbon atoms with hydrogen attached and with an acid group at one end. In a saturated fatty acid, each carbon is attached to two others by what is called a carbon-carbon single bond, and each carbon also has two hydrogens attached to it (**FIGURE B4.1**). In an **unsaturated fatty acid**, at some point in the backbone, two adjacent carbon atoms are attached to each other by a carbon-carbon double bond, and each of these two carbons has only a single hydrogen attached to it. The fatty acid is not saturated with hydrogens. If an unsaturated fatty acid has one double bond, it is **monounsaturated**; if two or more double bonds, it is **polyunsaturated**.

Saturated fatty acids tend to align easily with each other and have an orderly, stable packing. They have to be heated to disrupt their orderliness and make them move around as a liquid, so saturated fatty acids make up the solid lipids, the fats. The double bond of an unsaturated fatty acid causes a kink in the backbone, so neighboring fatty acids cannot align well; instead they make a jumble and move around even when cool; unsaturated fatty acids are the oils.

Within the bodies of plants and animals, the relative proportions of saturated and unsaturated fatty acids determine whether a membrane or a lipid droplet is solid, soft, or liquid. The proportions can be changed by the cells as the surrounding temperatures change. In plants and cold-blooded animals, body temperature is similar to environmental temperature and changes with the seasons. Membranes must remain fluid at all times, so these organisms often add oily unsaturated fatty acid in winter, and then replace them with saturated fatty acids in summer because the saturated fatty acids will not become too fluid (too "runny") in the heat.

Two types of double bond are possible in unsaturated fatty acids. If the two parts of the backbone lie on the same side of the double bond, this is a ***cis*-unsaturated fatty acid**, but if the two parts lie on opposite sides, it is a ***trans*-unsaturated fatty acid**, usually just called a ***trans*-fat**. A polyunsaturated fatty acid can have both types of double bond. Neither plants nor animals ever make *trans*-fats naturally in their bodies. Whereas natural dietary fats are generally beneficial, *trans*-fats are not essential and are never healthful in our diets. They increase the risk of heart disease; they raise the level of "bad" LDL cholesterol and reduce our "good" HDL cholesterol.

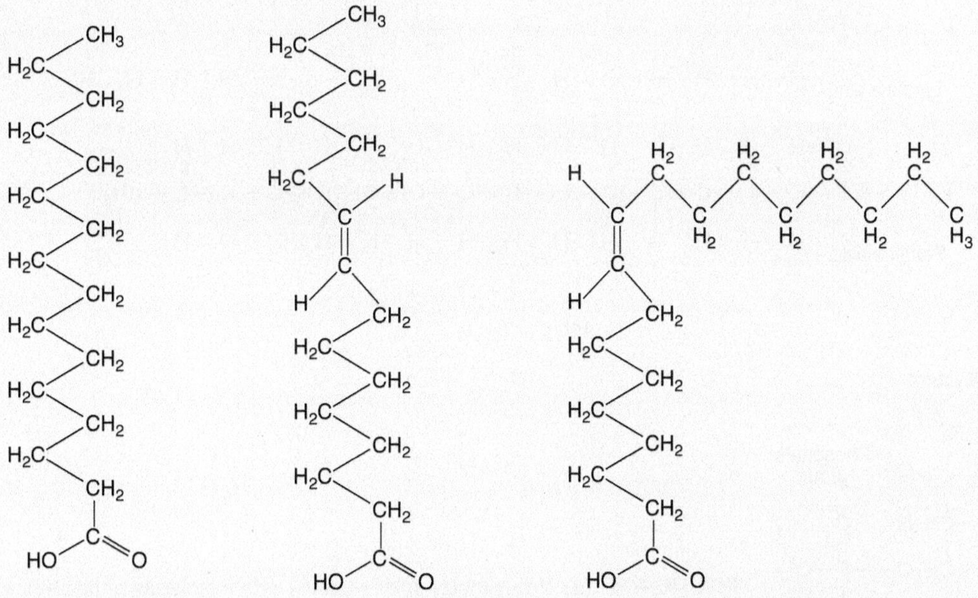

(a) Saturated fatty acid (b) *trans*-Unsaturated fatty acid (c) *cis*-Unsaturated fatty acid

FIGURE B4.1. The various types of fatty acids. **(a)** Saturated, **(b)** *cis*-unsaturated, and **(c)** *trans*-unsaturated.

(continued)

The only source of *trans*-fats in our diets is through food processing: we synthesize them artificially in factories. Unsaturated fatty acids, being oils, often cannot be heated enough to use for frying because they tend to smoke, scorch, and develop a "burnt" flavor. Also, in baked goods like cakes, cookies, health food bars, and so on, the food tastes better if it is not "oily" at room temperature. In contrast, many fats can be heated to high temperatures; they melt into a liquid at about the temperature needed for cooking. And in baked goods, they give the food a good texture. Furthermore, unsaturated fatty acids (oils) and foods made with them do not have a shelf-life as long as that of saturated fatty acids (fats) because oxygen reacts with double-bonded carbons, and we perceive this as the fat becoming rancid and inedible. If unsaturated oils are used in foods that need to be stored (such as most commercially prepared, packaged foods), then artificial antioxidants such as BHA and BHT must be added (you will see them in many labels, near the end of the ingredients).

Although it would be possible to find adequate supplies of natural fats needed for baking (such as lard), it is easier and cheaper to start with plant oils and **hydrogenate** them (chemically force hydrogen onto their double bonds). If the oils were hydrogenated to saturation, they would become extremely hard, too hard. Instead, the oils are **partially hydrogenated**, which merely saturates some double bonds and leaves others. This results in fats that have a longer shelf-life than the original oil, and they have a better texture for baking and frying. The problem is that partial hydrogenation converts some of the natural *cis*-fatty acids into unnatural *trans*-fatty acids. The danger of *trans*-fats has only recently become known, and most food processors have redesigned their products so that they no longer use *trans*-fats, but this is not true of all processors or of all processed foods. It is always best for you to check the label and avoid anything with ingredients listed as "*trans*-fats" or "partially hydrogenated."

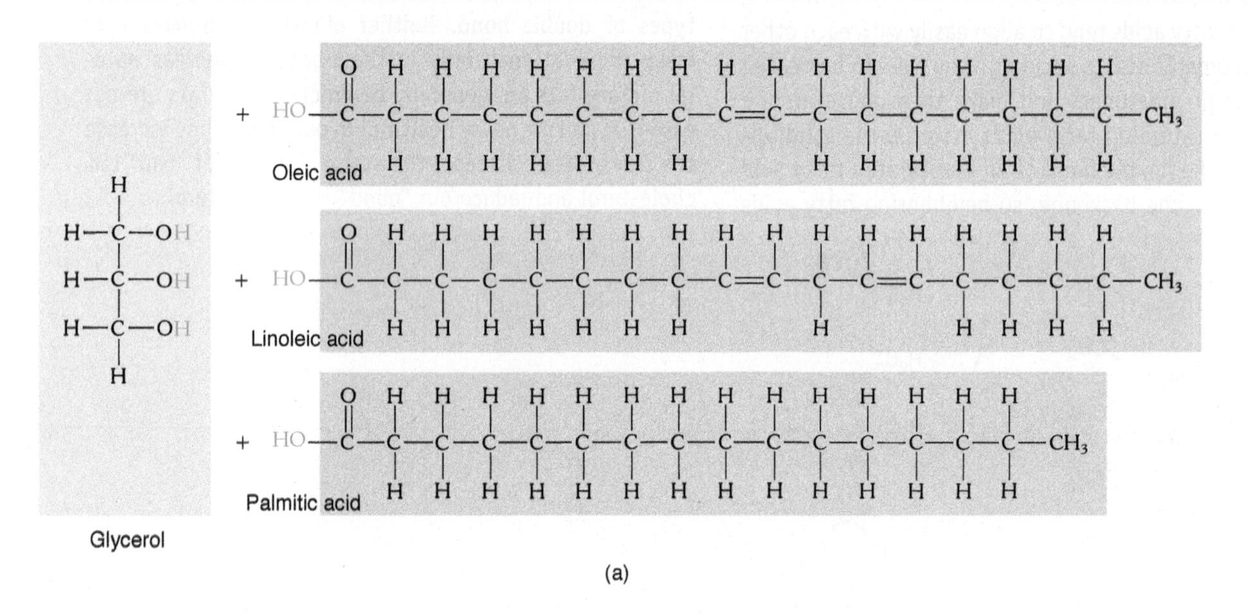

FIGURE 4.8. **(a)** If three fatty acids combine with a molecule of glycerol, the result is a triglyceride; no part of such a molecule can dissolve easily in water, so triglycerides form droplets in cells. **(b)** In a phospholipid, the phosphate group is hydrophilic whereas the fatty acids are hydrophobic: the phosphate associates with water and the fatty acids associate with other lipids. Cell membranes are composed of phospholipids.

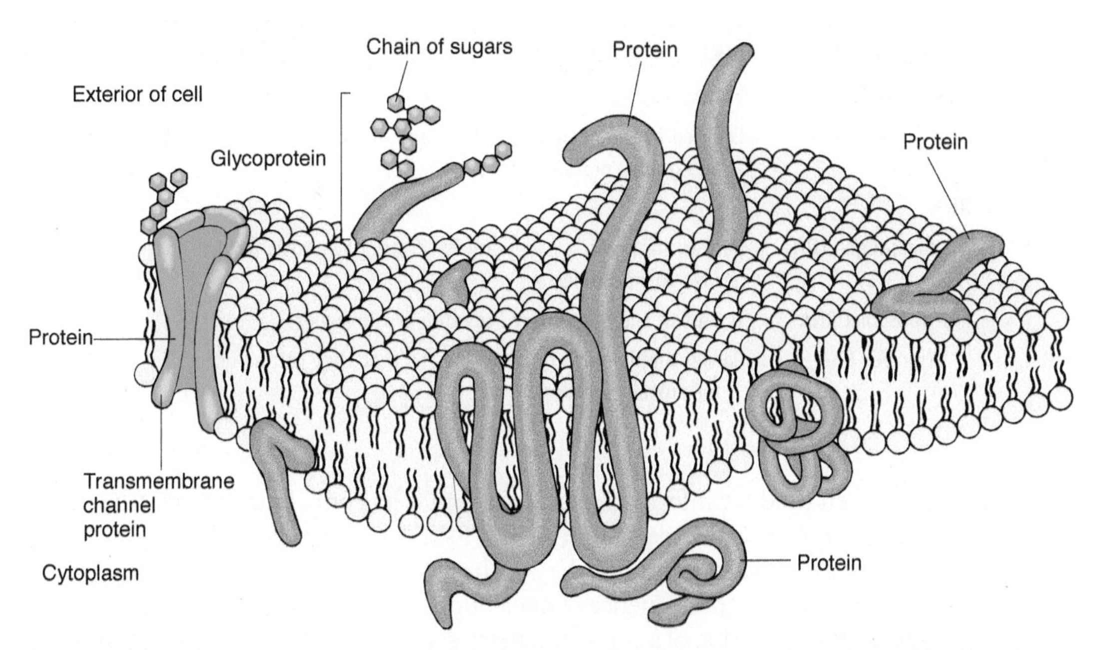

FIGURE 4.9. Cell membranes are composed of two layers of phospholipids with the fatty acids of one layer facing those of the other. Cell membranes also contain many proteins. The parts of the proteins that are inside the membrane must contain amino acids that have lipid soluble (hydrophobic) side groups.

Phospholipids are formed when glycerol combines with two fatty acids and a phosphate group. Phosphate routinely gives off a proton (an H^+) in water, so the rest of the phosphate group has a negative charge. Like other lipids, phospholipids are neutral and hydrophobic over most of their surface but are charged and hydrophilic at the end with the phosphate group; the phosphate causes the molecule to be polar. If a small amount of phospholipid is poured onto water, it will spread into a film one molecule thick, with the phosphate end of each lipid facing the water and the rest of the lipid protruding upward, as far away from the water as possible. In cells, which have water everywhere, phospholipids form themselves into membranes two molecules thick (a bilayer): the phosphate groups in each layer face outward and interact with water, the hydrophobic fatty acids face inward and interact with each other (FIGURE 4.9). This is the basic arrangement of all membranes in cells. Each membrane is completed by having proteins associate with them: proteins that have hydrophobic surface regions sink into the membranes, those with hydrophilic regions lie on its surface. Membranes must be flexible because they produce and receive vesicles, and different organelles fuse with each other. In cold climates, where lipids tend to solidify, plants add more unsaturated fatty acid to their membrane phospholipids; in hot climates they use more saturated fatty acids.

Nucleic Acids

Nucleic acids are organic polymers involved in storing and transmitting information. How does a molecule store information? The monomers of nucleic acids are **nucleotides**, and we can think of each of them as being similar to a letter in an alphabet: with the 26 letters in our alphabet, we can write sentences, articles, and books (information storage) and then make copies of those to send that information to various places where it is needed. Someone reads that information and uses it to guide his or her actions or thoughts. **Deoxyribonucleic acid** (**DNA**) is the nucleic

TABLE 4.3. Nucleic Acids

Nucleic Acid	Function
DNA	Located in the nucleus, plastids, and mitochondria; stores information and guides protein synthesis.
RNA	
messenger RNA	Carries information necessary for protein synthesis from DNA to ribosomes.
ribosomal RNA	Is part of the quaternary structure of ribosomes.
transfer RNA	Carries amino acids to ribosomes so that they can be used for protein synthesis.

acid most often used for information storage, and it has only four "letters" (the deoxyribonucleotides A, T, G, C; see **TABLE 4.3**) (**FIGURE 4.10**).

The information in DNA guides the synthesis of proteins, but DNA never participates directly. Instead, the sequence of nucleotides in a certain section of DNA, called a **gene**, guides the formation of a complementary sequence in **messenger RNA (mRNA)**, which leaves the nucleus, moves to the cytoplasm, and, working with ribosomes, provides the information needed to assemble amino acids into the proper sequence (primary structure). A ribosome has two subunits, each consisting of several molecules of nucleic acid (**ribosomal RNA**, **rRNA**) and enzymes. When the proper proteins and rRNAs encounter each other, they automatically self-assemble into a large complex with elaborate quaternary structure. The presence of mRNA then causes the two ribosome subunits to come together around it (**FIGURE 4.11**). Amino acids are brought to the ribosome by **transfer RNA (tRNA)** molecules; one part of each tRNA can "read" three "letters" of the mRNA. Whichever tRNA has the amino acid that corresponds to those three letters is allowed to enter the ribosome, bind just long enough for the amino acid

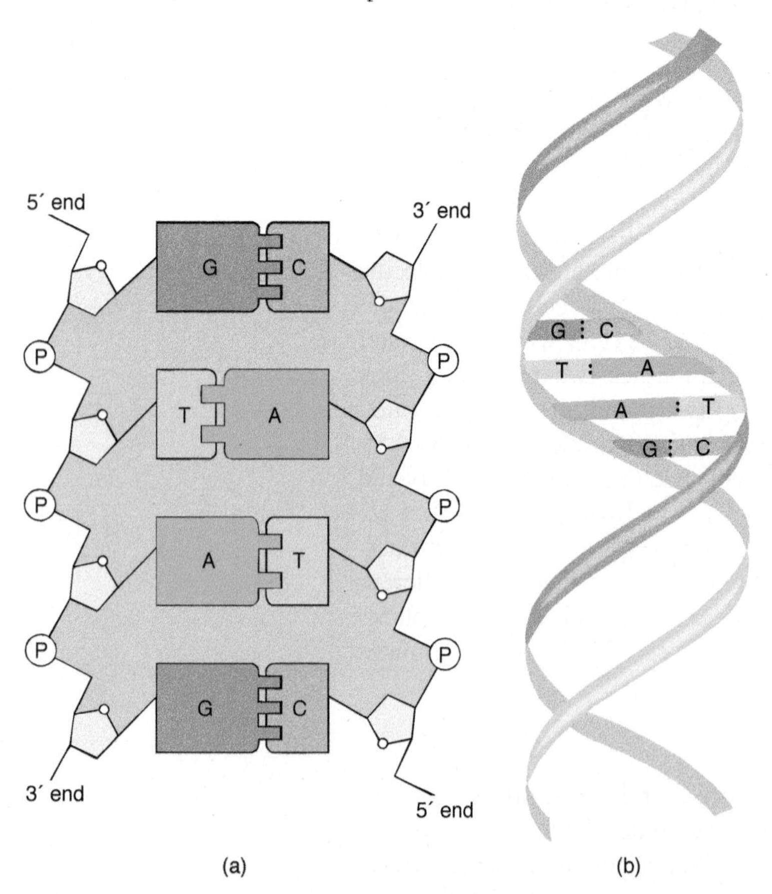

FIGURE 4.10. A molecule of DNA consists of a linear sequence of nucleic acids abbreviated as A, T, G, and C. Each molecule of DNA is paired with another such that each A is paired with a T and each G is paired with a C. The pairing is done by the enzyme that synthesizes DNA.

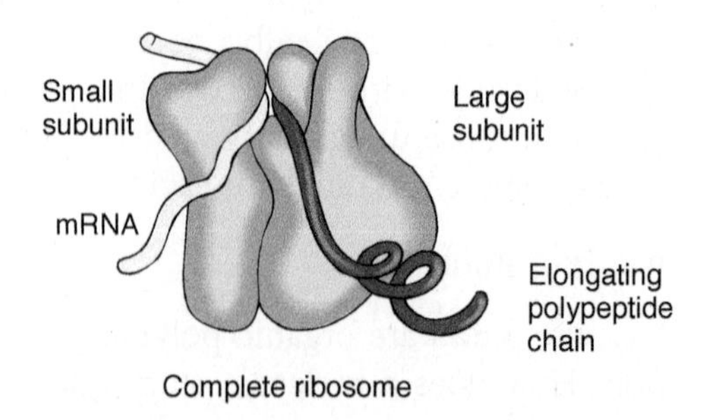

FIGURE 4.11. Ribosomes consist of two parts that have several grooves in them: messenger RNA fits into one, transfer RNA brings amino acids into another, and the protein that is being synthesized emerges from a third groove.

to be attached to the growing protein, then the empty tRNA is ejected. The ribosome pulls the next three mRNA "letters" into the active site and waits for the proper tRNA to bring the proper amino acid. As the growing protein emerges from the ribosome, it folds into its tertiary structure, forming active sites, or clustering with other proteins.

DNA also guides the formation of small pieces of RNA that do not do any of the above. These have regulatory roles: some complex with mRNA and inactivate it, others might bind to parts of DNA and either active or repress it. These small regulatory RNAs have many names, but in general are referred to as **microRNAs**.

Plastids and mitochondria also have DNA and ribosome and all the other components necessary for protein synthesis.

Secondary Metabolites

Many other types of compounds are synthesized by plants and play distinctive roles in their metabolism. These are often given the general label of **secondary metabolites**, and they include diverse compounds such as pigments, fragrances, poisons, and chemicals that give many of our foods their flavors. The taste of vanilla, chocolate, mint, and cinnamon are due to secondary compounds, as are the burning chemicals in poison ivy and hot peppers. An extremely common type of secondary metabolite is a class of chemicals called tannins; they are bitter and astringent, they cause our mouths to pucker when we eat unripe fruit. More importantly, tannins denature proteins, so when animals eat tannin-rich plants, their mouths and stomachs are damaged, and the animal either dies or learns to avoid that plant. We use tannins in several ways; in high concentrations, we use them to tan hides, that is, to denature the proteins in animal skins so that our leather shoes and belts do not rot. In lower concentrations, tannins give tea and red wine their mild astringency.

Several groups of secondary metabolites are familiar and easy to see: pigments. There are several distinct types of pigments, each type being made by its own particular metabolic pathway. Carotenoids are a group of lipid-soluble pigments similar to each other in their structure and in the enzymes that make them. Carotenoids are mostly yellow and orange and occur in chromoplasts and chloroplasts. Animals need carotenoids but do not synthesize them, and must get them in their diet, from the plants they eat. The yellow of egg yolks is entirely plant-derived carotenoids, and chickens are often fed yellow plant material so that their egg yolk will be more appealing to us. People need carotenoids as a necessary material for our vision and the synthesis of vitamin A. Anthocyanins are water-soluble pigments, predominantly blues and reds. For some reason, an entirely new type of water-soluble flower pigments, betalains, evolved in cacti, four-o'clocks, amaranths, and their relatives. Plants that have betalains do not have anthocyanins. Chlorophylls are a very small family of pigments: true plants have only two, chlorophyll *a* and chlorophyll *b* (both are green and both capture the light used in photosynthesis), but algae have a few other types of chlorophyll as well.

Part 2: The Movement of Water Throughout a Plant

Water molecules, just like all other molecules, are in constant motion. In liquids and gasses, molecules move a short distance until they strike another molecule; then, unless they combine chemically, the two bounce off each other and move in new directions. The hotter the material, the more rapidly molecules move, but even when frozen into ice, water molecules vibrate in place and occasionally break away from their neighbors and move away.

Water Movement Across Membranes

If two substances are mixed together, their constant motion causes each to gradually mix with the other. This is **diffusion**, technically defined as the movement of molecules from areas where they are more concentrated to areas where they are less concentrated; this is an important principle to remember. If water is added to a material that is also hydrophilic and interacts with it, water diffuses into that substance and tends to stay there. For example, a central vacuole might contain a mix of water and glucose. If water is more concentrated in the cytoplasm, then water will move from the cytoplasm, across the vacuole membrane and into the central vacuole (FIGURE 4.12). Movement across a membrane like this is **osmosis**. Water will continue to move into the central vacuole until either of two things happens: so much water enters the vacuole that the concentration of water is equal in both the cytoplasm and the vacuole, or until the vacuole swells so much that its pressure against the wall is so great that no more water molecules can force their way in. It is important that the membrane be selectively permeable; that is, some materials can diffuse through it whereas others cannot. In this case, the vacuole membrane is permeable to water but not to glucose. If it were permeable to glucose, glucose would diffuse out and there would be no build up of pressure.

The tendency of glucose or any other hydrophobic molecule to attract and hold water is its **osmotic effect**. A dilute solution of glucose will have only a weak osmotic effect, a weak ability to attract more water to itself. A concentrated, syrupy solution has a stronger osmotic effect and a stronger ability to attract more water. Dry saltine crackers and table salt have such strong osmotic effects that they can pull water out of humid air.

Plants control osmotic effects very precisely. If a cell needs to be turgid or to swell and grow, it can pump solutes (especially sugars and potassium) into the vacuole and cause water to diffuse in and build up pressure. If the cell needs to shrink, it can pump the solutes out and water will follow, causing turgor pressure to drop.

Using osmotic effects, central vacuoles and walls function together to provide strength to cells, leaves, flowers, and other soft organs. As the vacuole membrane pumps material into the vacuole, the vacuole becomes "osmotically drier"; that is, its tendency to absorb water increases. As water enters, the vacuole swells and presses the rest of the protoplasm outward against the wall, but the wall resists, becomes stretched and taut. We say the cell becomes **turgid**; examples are fresh lettuce and crisp apples

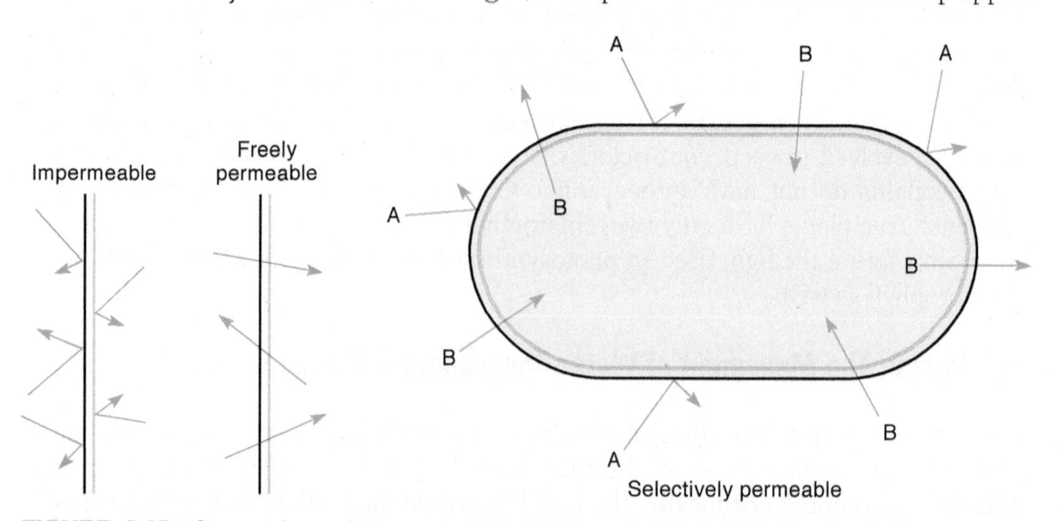

FIGURE 4.12. If a vacuole membrane pumps material into a central vacuole, the vacuole becomes "drier," causing water to move into the vacuole, which creates turgor pressure. Diffusion of material through a membrane is osmosis.

FIGURE 4.13. These leaves are being supported by turgor pressure in all their cells. They are able to support themselves only if they have enough solutes and water in their central vacuoles.

FIGURE 4.14. These leaves are wilted because the leaves are losing water to dry air faster than the roots are able to bring water in from the dry soil: there is not enough water in the plant to fill all the central vacuoles full enough to generate turgor pressure.

(FIGURES 4.13 and 4.14). If there is not enough water, the vacuole does not put pressure on the wall and the plant is **wilted**. This is similar to putting air in a tire: when air is pumped into the tire, the tire becomes strong enough to support a bicycle or car. Turgor pressure in plant cells is not strong enough to hold up entire trees (wood is necessary for that), but it does maintain the shape of leaves, flowers, and many stems: if something can wilt, it is being supported by turgor pressure. We animals use this type of hydrostatic pressure to maintain the size and shape of our eyeballs, and males use it when sexually aroused.

When a cell needs to grow, it softens the cell wall just enough that vacuole pressure causes the wall to stretch. Once the cell has reached its proper size, it reinforces the wall enough to stop further growth. Familiar examples of this are flowers: just before a flower opens, its petals are fully formed but all their cells are extremely tiny with miniscule central vacuoles. The vacuole membranes pump enough material into the vacuoles so quickly that they absorb water, swell and stretch the cell walls causing the petal to grow to its full size in just a few hours. Once the petal is fully expanded, the central vacuoles make up almost the entire cell volume, and the walls are stretched so thin that the petals are delicate, easily damaged, and usually die within a few days. In contrast, leaves, stems, and roots must be permanent and must not be so delicate; as they grow, the cell adds new cellulose, hemicellulose, and other wall components at a rate just fast enough to strengthen the walls but not stop growth.

Osmotic effects are also the basis for opening and closing stomatal pores. At sunrise, guard cells pump potassium into their central vacuole, causing water to follow it. The guard cells swell into a kidney bean shape, push away from each other, and the stomatal pore opens, allowing carbon dioxide to enter the leaf (FIGURE 4.15). At sunset, the cells reverse

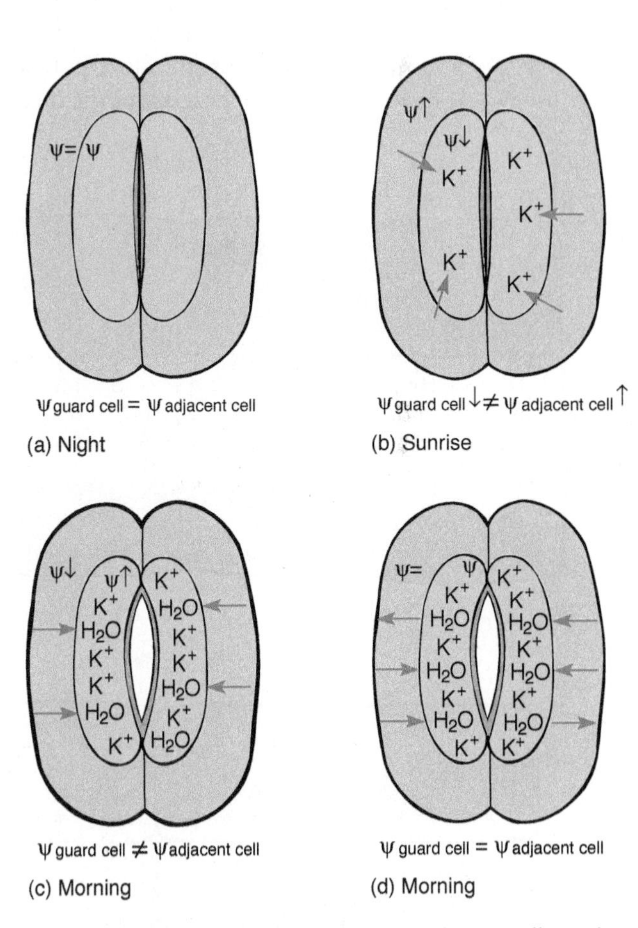

FIGURE 4.15. At night, guard cells and adjacent cells are in hydraulic equilibrium **(a)**, but at sunrise, potassium is pumped into guard cells **(b)** and water follows **(c)**, generating turgor pressure. This causes the guard cells to swell, bend, and open the stomatal pore **(d)**. At sunset, when the pores must be closed, the process is reversed. The symbol Ψ (pronounced "sigh") is a measure of a solution's tendency to absorb more water or to lose water.

this: they pump potassium out of the vacuole, and again water follows it, turgor pressure drops, the cells relax, and the stomatal pore is closed.

Some leaves have motor cells in their petioles. When the leaf blade needs to be held out into the sun, the motor cells pump potassium into their central vacuoles, swell, and become turgid. But at night or if the sunlight becomes too intense, the motor cells wilt and the leaf blade folds downward.

Remember that all of this depends on selectively permeable membranes. If glucose or potassium or other solutes could leak out of the central vacuole, no pressure would build up and these processes would not be possible.

Sugar and Water Are Pushed Through Phloem

Plants, just like animals, must move water and nutrients from certain parts of their body to other parts. Conduction through sieve tubes is easy to understand. In organs that produce sugars (called **sources**, such as leaves), sugars are forced into the sieve tube members by molecular pumps in the plasma membrane. As sugar increases in the cells, they become osmotically drier and water follows automatically. If these were ordinary parenchyma cells, they would become turgid or would grow, but these cells have holes (the sieve pores) interconnecting them. The sugar water (the phloem sap) is just squeezed into the next cell in the sieve tube and a flow begins (FIGURE 4.16). In areas that need sugars (called **sinks**), membrane pumps extract sugar from the sieve tubes, and water follows again. Compare this with the water supply of a city: many large tanks put water into the city's network of pipes, and many faucets let water out. The volume of flow depends on which tanks and faucets are open at any moment.

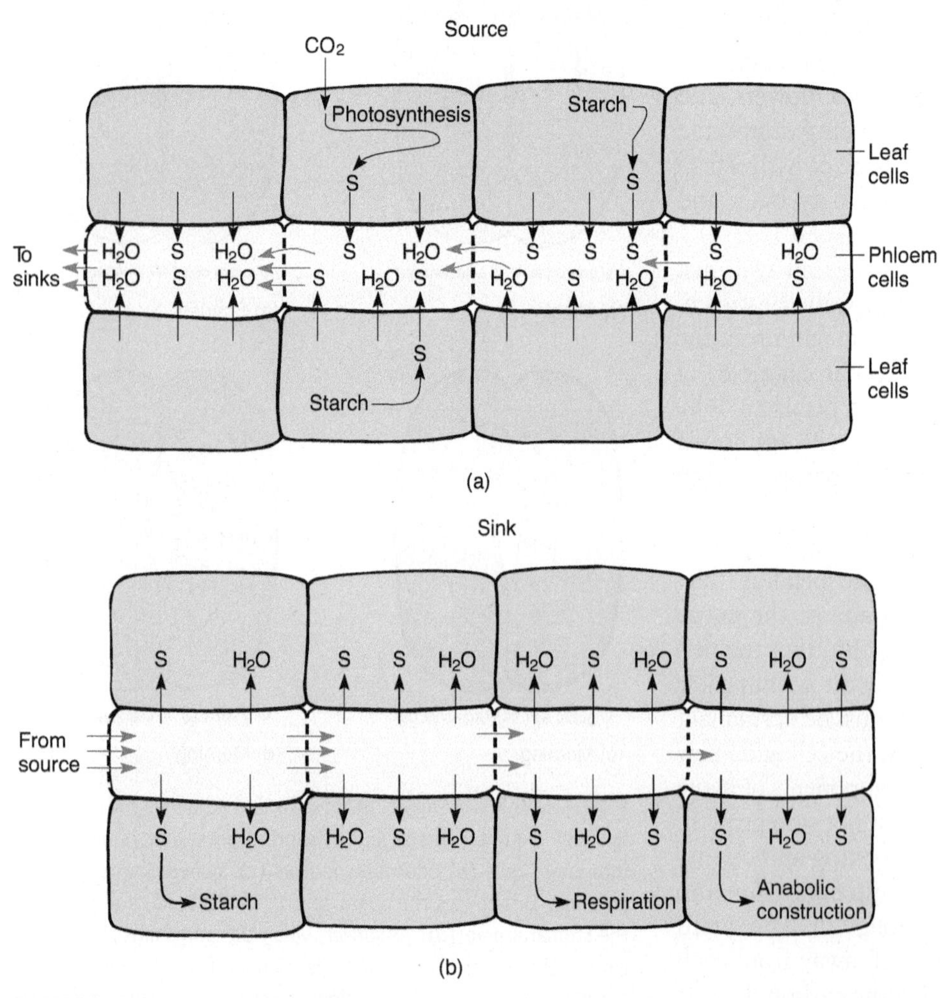

FIGURE 4.16. This diagram illustrates sugar transport through phloem; the details are given in the text.

Any plumbing system will have leaks and must have a repair mechanism. Phloem is always in danger of leaking if an animal bites into it, if a stem breaks and tears the phloem open, or even whenever leaves, flowers, or fruits fall off a plant. Sieve tubes contain two chemicals, **callose** and **p-protein**, that lie quietly in the sieve tubes if they are functioning normally. But if the phloem is broken open, the high turgor pressure inside it causes a sudden rush of phloem sap toward the leak. This sweeps the callose and p-protein toward the damaged area, where both form sticky masses (a **callose plug** and a **p-protein plug**) that instantly seal the tube and prevent any further leakage (FIGURE 4.17). These act instantly, much more rapidly than the way our blood vessels seal themselves.

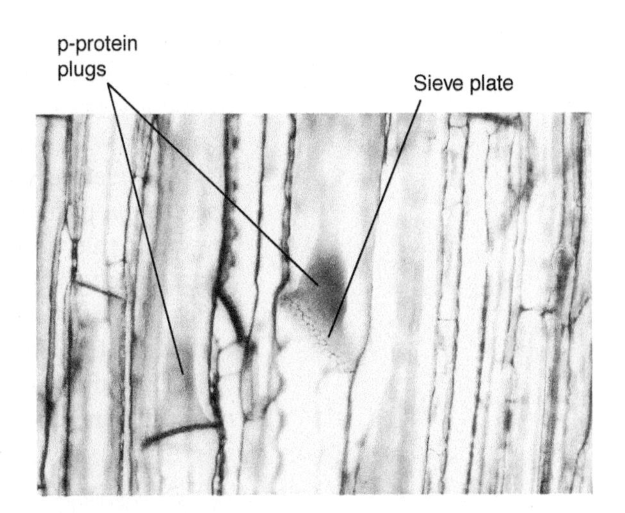

FIGURE 4.17. When this material of squash was being prepared for microscopy, it was cut open, causing the phloem sap to surge toward the cut, sweeping p-protein along and forming p-protein plugs, visible as dark brown masses. Sieve pores in the sieve plates are also visible (×500).

Water and Minerals Are Pulled Through Xylem

The movement of water and minerals upward through xylem is completely different from the transport of phloem sap or blood. Whereas water is under pressure and being pushed through phloem (and our blood vessels), water is *pulled* through xylem. Water molecules cohere (stick to each other); this is easy to see when water freezes into ice. Imagine grasping just the top of a long icicle and lifting: you will lift all the water molecules, not just the ones you are touching. A similar thing happens in plants: as water molecules escape from leaves, the leaf cells become osmotically drier, so they pull water into themselves from the xylem. Each water molecule pulls on the ones below themselves, and these pull on the next and so on down to the roots. This pull causes water to flow through the mass of tracheids or vessels, passing from one to the next as the water molecules move upward. Notice that this is completely different from blood flow in our body: our blood flows in the space between the living cells of our blood vessel walls; xylem sap is pulled through a mass of dead cells.

Safety and Flow Volume Are Important in Water Transport

Recall that xylem has two kinds of conducting cells, tracheids and vessel elements. Vessel elements tend to be very wide and are interconnected by perforations, which are complete holes: vessels allow relatively large volumes of water to be conducted with very little friction. Tracheids tend to be narrow and they do not have perforations, so each water molecule must move across pit membranes every time it goes from one tracheid to another. Tracheids provide a conducting system that does not allow high volume flow, and it has considerable friction, but they provide greater safety. Imagine the icicle again: if you pull on it and its lower end is frozen to something or if it is so long that it is very heavy, it will break and the top and bottom portions separate. The same can happen to water in xylem: if both air and soil are dry, both pull on the water and the water column (the water column is the mass of water itself, not the cells) may break (**cavitate**): the pull on the water has overcome its cohesion (FIGURE 4.18). Water in the upper part will continue to be pulled upward, but now there is nothing pulling on the lower water, so it sinks back down due to its weight, and an air bubble (an embolism) forms between the two. If this happens in a tracheid, the air bubble will expand only until it fills the tracheid: it cannot cross the pit membrane so it cannot damage any other tracheid. But if the cavitation occurs in a vessel member, the embolism expands right through the perforations because they have nothing similar to pit membranes to stop the expansion. The embolism can fill all the hundreds or thousands of vessel members in a vessel. Thus, a cavitation in a tracheid causes the plant to lose

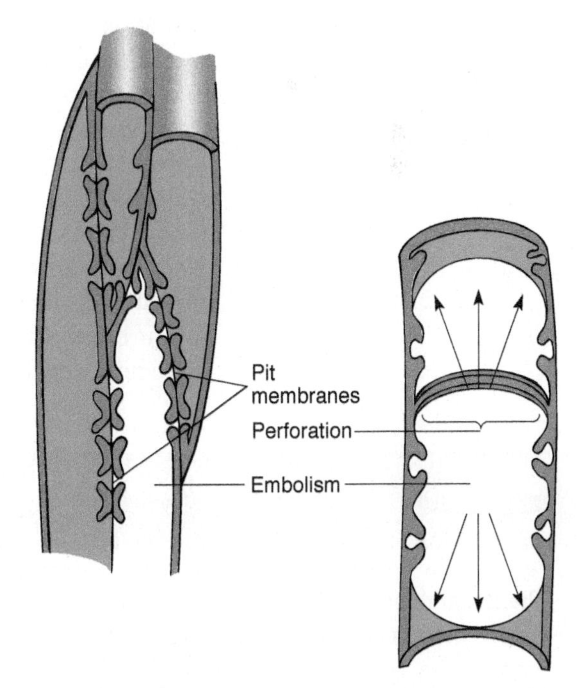

FIGURE 4.18. Severe tension can overcome the cohesion of water molecules and cause an air bubble to form and expand. The air bubble can pass through holes such as perforations but is stopped by pit membranes. If an air bubble forms in a tracheid, it cannot spread beyond the tracheid, but if a vessel element cavitates, the air bubble spreads throughout the entire vessel.

the use of just that one tracheid, whereas a cavitation in a vessel causes the plant to lose the use of thousands of vessel members. In dry habitats where cavitations occur frequently, tracheids offer a safer means of conduction, but in moist or even moderate habitats, cavitations are so rare that the benefits of low friction in vessels outweigh the risk of spreading embolisms.

It is important to compare our water conduction with that of plants. We have a hollow heart that uses a great deal of energy pumping our blood out through our wide aorta, which then branches out to smaller arteries, arterioles, then capillaries. Capillaries return blood to venules, to veins, and then back to the heart in a closed loop: each water molecule in the blood, and each red blood cell follows this circuit over and over thousands of times a day. We have an extremely low safety system with almost no redundancy: failure of our single heart or our single aorta, or one of our few major arteries is usually fatal. In contrast, plants pull water through many, many tracheids or vessels; even small plants have thousands, and large trees have billions. Plants do not have anything like an aorta or artery; there never is just one or a few very wide tracheids or vessels that all or most water must pass through. Xylem, and phloem as well, are extremely redundant and have a high margin of safety.

Consequences of Water Movement

Plants and animals handle water very differently, and the consequences are important. We animals take in water as a liquid, either by drinking it or eating foods that contain moisture. We also lose most of our water as a liquid, such as urine or sweat, and some of it as a gas from our lungs when we exhale. Liquid water can contain dissolved minerals, so a danger we face is that of losing minerals by too much sweating, urination, or diarrhea. But plants don't urinate or sweat; they almost never lose water in the liquid form, it is mostly just as water vapor escaping through the epidermis, and water vapor cannot carry mineral or other solutes. What is a consequence? Plants trap minerals: they bring them in from the soil as they absorb liquid water, move them upward through the xylem, but then cannot lose them as the water goes out through the epidermis. Day after day, minerals accumulate in the plant body. What happens to them? Plants need many of them for their own metabolism, as described below. And when animals eat plants, the minerals end up inside the animal and are essential for the animal's metabolism. Think of the iron in our blood, the calcium and phosphate in our bones, and so on...it all comes in as part of our food and it all came in to the biological world by means of plants. We never have to eat dirt to get minerals.

BOX 4.2. Plants, People, and Irrigation of Farms

Plants and people interact in many ways because of water. The earliest civilizations quickly learned to master two types of large construction projects: walls around their cities and canals to irrigate their fields. Even today, irrigation projects are massive and redirect the flows of large rivers. At Grand Coulee Dam, much of the Columbia River is diverted into canals that distribute irrigation water across hundreds of square miles of eastern Washington. Huge canals also irrigate the massive Central Valley in California as well extensive areas in Arizona. Lesser projects have been built almost everywhere in the western United States. In many areas, for example in west Texas, farms are irrigated by water pumped from aquifers beneath the farms.

The redirecting of water has many consequences for both plants and people. Rivers have less water after irrigation water has been removed, so there is slower flow, waterfalls are less vigorous, river bottoms are not as agitated. All of these areas are natural habitats for many plants and animals, and removing water for irrigation disturbs the creatures in these habitats. So much water is removed from the Rio Grande River in Texas and from the Colorado River in California and Arizona that neither is actually a river at its mouth, where it meets the ocean. Instead, both are just slow moving brackish marshes.

Irrigation water must go somewhere. Only a small fraction is actually absorbed by roots and then transpired out through leaves. The rest just moves past the plant, through the soil. Much of it sinks downward, especially in sandy soils, which do not hold water tightly. The water enters aquifers and then flows horizontally underground until it emerges as a spring or empties into a river. For many of us, a new spring or increased flow from an existing one would be a pleasant sight; there would be a small creek or pond, some aquatic plants and animals, some waterfowl. But however pleasant, these all are habitats that have been altered from their natural state due to extra water, we have interrupted a natural ecology.

If the irrigation system is in a deep valley that has no river draining it, the excess irrigation water has no means of draining away. It flows to the lowest part of the valley, accumulates there as a lake or marsh, and evaporates. Examples are California's Imperial Valley, the Great Salt Lake area in Utah, and the "basin and range" area of Nevada. Because water enters these valleys as a liquid and leaves as water vapor, minerals are trapped and accumulate. Also the irrigation water has picked up fertilizers from fields, so it is very rich in minerals. As the water evaporates, the minerals become more concentrated, the lake becomes saltier, osmotically drier, and at some point plants can no longer grow in it: its minerals attract water molecules more strongly than roots can. The lake becomes a salt marsh or salt lake, or if all water evaporates, it becomes a salt desert in which few things can live. Making otherwise fertile soil salty is called **salinization** (**FIGURE B4.2**), and it is an important problem in many irrigation projects.

In areas where fields are irrigated with ground water that has been pumped up, aquifers become depleted. Because the aquifer has lost some of its water, the natural springs and seeps that it feeds may have reduced flow or may dry up completely, and again natural habitats are damaged.

FIGURE B4.2. Rain water accumulates in this low-lying area and forms a temporary pond. But it evaporates before the pond becomes deep enough to drain away by a stream. Consequently, all the salts that were in the water remain here after the water evaporates, and the soil becomes too salty for any plants to grow here. This pasture is being damaged by salinization.

Part 3: Mineral Nutrition

All organisms need a small number of chemical elements for their metabolism and structure. Our own bodies need iron for our blood; calcium and phosphorus are necessary for our bones and teeth. All in all, our bodies need about 17 different chemical elements, and if our diets lack even one of them, we become ill. In the same way, plants too have a small number of chemical elements that are essential for their life, and these are therefore called the **essential elements** (**TABLE 4.4**). For the most part, the elements that are essential for plants are also essential for us animals; a noteworthy exception is that plants do not need sodium (Na^+) at all whereas we can't live without it.

Carbon, hydrogen, and oxygen come from air and water, but other chemicals come from the minerals present in soil. The ways that plants obtain and process these minerals is called **mineral nutrition**. As rocks weather and break down, they release minerals, which roots absorb. Nitrogen is an important exception; it is never part of a rock matrix. It has to be captured from air and will be discussed separately below.

TABLE 4.4. Essential Elements in Plants

Element	Function
Macro Essential Elements	
Carbon	Almost all organic compounds
Hydrogen	Almost all organic compounds
Oxygen	Many organic compounds
Nitrogen	All amino acids; all nucleic acids; chlorophyll
Potassium	Osmotic balance; enzyme activator; movement of guard cells and motor cells
Calcium	Controls the activity of many enzymes; component of the middle lamella; affects membrane properties
Phosphorus	Phospholipids; nucleic acids; many sugars have phosphate attached to them during certain reactions
Magnesium	Chlorophyll; activates many enzymes
Sulfur	Some amino acids
Micro Essential Elements	
Iron	Chlorophyll synthesis; enzymes involved in respiration
Chlorine	Unknown; possibly involved in photosynthetic reactions that produce oxygen
Copper	Plastocyanin, a compound involved in transporting electrons
Manganese	Chlorophyll synthesis; necessary for the activity of many enzymes
Zinc	Activates many enzymes
Molybdenum	Involved in nitrogen reduction
Boron	Unknown

All of these except for boron are also essential elements for us humans; we cannot live without these. Unlike plants, however, we also need fluorine, iodine, cobalt, selenium, chromium, and sodium. We obtain fluorine by adding it to our drinking water (fluoridation) and iodine is obtained by adding it to salt or by eating seafood. Notice that sodium is not an essential element for plants; even though our lives depend on it, and we can become sick just by losing too much sodium by sweating, plants do not need it at all.

Essential Elements

An element is an essential element if a plant cannot live normally without it. There are three criteria that determine if an element is essential:

1. The element must be necessary for complete, normal plant development through a full life cycle. This includes surviving stresses such as droughts, freezes, insect attacks, and so on, not just an easy life in a greenhouse.
2. There must be no substitute for the element.
3. The element must be acting within the plant, not outside it. This third criterion sometimes causes confusion. For example, iron is an essential element, but in alkaline soils, iron is mostly present as an insoluble compound that roots cannot absorb. If we add other elements to the soil, they might acidify the soil and thus release the iron, making the plant flourish. It would be easy to conclude that the elements we added were essential, even though all they are doing is making iron more readily available.

Certain elements do not need to be tested because they are so obvious. No organism at all can live without carbohydrates, lipids, or proteins, so carbon, hydrogen, oxygen, and nitrogen are automatically known to be essential. Also, magnesium is part of the chlorophyll molecule so it will be essential to any green, photosynthetic plant, and because phospholipids make up cell membranes, the phosphorus of phosphate groups must be essential.

We use **hydroponic experiments** to actually test if an element is essential. Seedlings or cuttings of a plant are grown in solutions with numerous chemicals that are thought to be necessary (and without any that are believed to be poisonous). If the plants die, we obviously made a bad guess, and either some element must be added or something must be removed. Once we have a solution that supports growth, we can prepare a second solution that is identical except that one single element is left out. If the new plants grow in that new solution, then the eliminated chemical was not essential. But if the plants die or show disease symptoms, then the element needs to be examined more closely.

At present, we believe there are 16 elements that are essential to all plants (Table 4.4). However, when we prepare the hydroponic solutions, we know that the chemicals we are using, as well as the water and the glass containers, are contaminated with miniscule amounts of various elements, and it may be that some of those are essential in extremely low amounts.

Some essential elements are needed in rather large amounts, for example the nitrogen needed to build the amino acids of proteins, the nucleic acids, and many other compounds, or the phosphorus used in the phosphate groups of many organic compounds. These are called the **macro essential elements**. But certain essential elements, called **minor essential elements** or **trace essential elements**, are needed in tiny amounts.

Soils Provide Essential Elements for Most Plants

Soils are derived from the breakdown of rocks. Volcanoes deposit basalt; upwelling magma cools into granite; old sea floors may become exposed as limestone, marble, or sandstone. These rocks have a hard, crystalline structure that roots cannot penetrate, but as these rocks are broken down (**weathered**) by rain, wind, freezing, or acids from lichens and mosses, the crystal matrix releases various elements (FIGURE 4.19). Some were part of the matrix itself, some were merely trapped there as the rock formed.

FIGURE 4.19. These rocks will last for thousands of years but not forever; they are gradually being broken down by rain, snow, and especially by the action of acids formed by the lichens growing on their surface. As bits and pieces break off, they form a thin layer of soil that allows mosses and small flowering plants to grow, which accelerates the weathering of the rocks.

Initially, rock may be broken down into boulders and other large pieces, but gradually it is weathered into finer and finer pieces. An important classification of soil is based on the size of soil particles: coarse sand has particles that range from 2.0 mm down to 0.2 mm; fine sand is 0.2 to 0.02 mm; silt is 0.02 to 0.002 mm in diameter. Soils in which most particles are finer than 0.002 mm are clay soils.

Soil particle size is important because it affects water-holding capacity and mineral availability. In coarse and fine sand, much of a soil's volume is occupied by spaces between soil particles. Such soils have plenty of room for air (necessary for keeping roots alive) and water. However, sandy soils do not hold water for long, because it moves through the large soil pores easily and either evaporates into the air or drains away into an aquifer. Even though water is adhesive, there is just not much soil particle surface area for water molecules to adhere to in sandy soils. Once a rain stops and water has drained away, the water that does remain, held by adhesion to soil particles, is called the soil's field capacity (FIGURE 4.20). In silty soils, the tiny soil particles occupy more of the soil volume yet still there is plenty of room for air and water. Silt particles have more surface area, so they hold water better than sandy soils, and have a higher field capacity. Silty soils tend to be very good for most plants. In clay soils, particles are so small that there is an overabundance of surface: clay particles hold water so tightly it cannot drain away, but roots are not very successful at pulling the water away from the clay particles' surface. With so much water, there is little room for air, and roots may suffer from lack of oxygen.

FIGURE 4.20. The very slight depression in the center of this slope is an area where fine soil particles accumulate, giving the soil more water-holding capacity (a greater field capacity). This area stays moist longer after a rain, and the plants that grow in the depression do not have to be as drought-tolerant as the sagebrush on the hilltops.

As a rock's matrix breaks down and mineral nutrients are released, most come off as positively charged cations (K^+, Cu^{++}, Mg^{++}, Ca^{++}) and they remain close to the rock's surface, which has a negative charge. This is beneficial because otherwise rains would wash the minerals away as quickly as they are released by weathering. However, roots face the problem of pulling the nutrients away from the rock. This is done by cation exchange. As roots respire, they release carbon dioxide, which is converted to carbonic acid. Being an acid, carbonic acid gives off one or two protons, and if this occurs next to a soil particle, the proton's positive charge might loosen a nutrient from the soil's surface. With luck, the nutrient will diffuse in the direction of the root and be absorbed. One cation (a proton) has been exchanged for another (a mineral nutrient).

Once water and minerals reach a root hair, they may cross the cell wall and enter the protoplasm, or they may just penetrate deep into the root by diffusing through the water and pectin in the cell walls of the epidermis and cortex. As mentioned in Chapter 3, they can only do this until they reach the Casparian strip; at that point, the wall is incrusted with hydrophobic material and further diffusion through the wall is impossible. The minerals must be accepted by an endodermis cell plasma membrane if they are to reach the xylem and phloem. Casparian strips prevent unwanted, harmful minerals from reaching the rest of the plant.

Nitrogen Is Essential and Unique

Nitrogen is an essential element that differs from the mineral elements: it is found in air but not as part of a rock matrix. Also, it is in a form, N_2, that plants cannot use. Surprisingly, even though all plants, animals, fungi, and all other organisms must have nitrogen for many critically important molecules, only a tiny number of bacteria have an enzyme, nitrogenase, that can bind N_2 and use it. These are called nitrogen-fixing bacteria. The processing of nitrogen is called nitrogen metabolism, and it involves three steps: fixation, reduction, and assimilation. In nitrogen **fixation** and **reduction**, nitrogenase binds N_2 and forces electrons onto it (adding electrons to a chemical is a reduction). This converts the nitrogen into an ammonium ion, a state similar to that of nitrogen in amino acids. During nitrogen **assimilation**, the bacterium attaches ammonium ions to various acids, converting them to amino acids; at this point the nitrogen is part of the bacterium's body, it has been assimilated. Some of the amino acids are then used to build proteins; others are converted into nucleic acids and other compounds. The bacterium might be eaten by tiny soil animals or digested by soil fungi, or if it dies and decays, the nitrogenous compounds that are released might be absorbed by roots. Whatever happens, the nitrogen that was assimilated by this bacterium is now part of another organism. Some soil microbes convert the nitrogenous compounds back to ammonium, and others respire it, much like we respire carbohydrates. This converts the nitrogen to nitrate (NO_3^-) or nitrite (NO_2^-). Roots can absorb all three forms, ammonium, nitrate, and nitrite. If roots have ammonium available, they can immediately use it to make amino acids just as the bacterium did. If the roots absorb nitrate or nitrite, they must force electrons onto it, reducing it, but roots have no problem with this, it is just N_2 that they cannot process. Once the nitrogen is assimilated, some is used by the roots themselves for their own metabolism, but a large part is loaded into phloem and conducted up to the shoots.

Most microbes that fix nitrogen are **free-living**; that is, they live on their own and are not intimately associated with other organisms. One of the most common free-living microbes is *Nostoc*, a cyanobacterium that forms colonies large enough to see easily: they look like dark pieces of cellophane when dry, or like rubbery sheets

FIGURE 4.21. Nitrogen-fixing bacteria live in the nodules on the roots of this cowpea (a legume). The bacteria obtain sugar and other nutrients from the cowpea, and use part of their energy to convert atmospheric nitrogen to organic nitrogen. Root cells absorb part of the nitrogen from the bacteria. (© Nigel Cattlin/Alamy.)

when wet. They are common on desert soils. Other nitrogen-fixing microbes are **symbiotic**, they live in a close association with plants. Roots of legumes like alfalfa, peanuts, and soybeans form nodules that contain millions of cells of the nitrogen-fixing bacterium *Rhizobium* (**FIGURE 4.21**). The roots actually provide the bacteria with sugars and other nutrients, and they also protect the bacteria from oxygen (nitrogenase is poisoned by oxygen). The bacteria secrete nitrogenous compounds to the roots. Symbiotic nitrogen-fixing cyanobacteria also associate with alders, the water fern *Azolla*, and with many liverworts and hornworts. Because the plants live symbiotically with nitrogen-fixing bacteria, they can grow on very poor soils deficient in nitrogen. These plants are often the first to colonize bare, rocky areas.

Mycorrhizae and Phosphorus Absorption

Although roots can absorb phosphate from soil on their own, they cannot do it as well as many fungi can. Roots of most plants form a symbiotic relationship with certain soil fungi, a relationship in which fungi absorb phosphate and pass much of it on to roots while roots provide the fungi with sugars. This symbiotic relationship is called a **mycorrhiza** ("fungus root"), and there are several types. In ectomycorrhizae, the fungi form a dense sheath of fungal cells around the surface of the root. In endomycorrhizae, the most common type, the fungi actually penetrate into the root, all the way to the endodermis, and even penetrate root cells. Within the root

BOX 4.3. Fertilizers, Pollution, and Limiting Factors

Plants in nature usually do not grow as vigorously as they potentially could. For example, desert plants typically grow more rapidly if given extra water, plants in shady areas grow better if given a bit more light, whereas some extra nitrogen fertilizer usually helps prairie grasses, which already have enough water and light. Any plant grows at a particular speed and vigor because it is limited by some factor, such as too little water or light or nitrogen fertilizer. An important concept is that there is only one single **limiting factor** at a time for any plant. If we give desert plants more light or fertilizer, they will not grow faster, it is water that is limiting them. But if we do give them extra water, their growth rate will increase until some other factor becomes limiting, perhaps lack of nitrogen. While growing slowly, the plants could get nitrogen quickly enough, but now that they are growing faster, their ability to obtain nitrogen from the soil may be limiting. If we give the plants both water and nitrogen fertilizer, they may grow even more rapidly, but finally, some other factor will become limiting. On many farms, plants are irrigated and fertilized, and they are planted far enough apart that they do not shade each other and their roots do not interfere with each other. Such plants grow much more rapidly than they would in nature, but they are still limited, in this case, by their own genetics, their own innate metabolic capacity. There is always a limiting factor.

The concept of a limiting factor is important in understanding techniques for reducing the damage caused by pollution. Under natural conditions, the water in rivers and lakes has so few nutrients that algae grow slowly, and they are so sparse that the water is blue. In the middle of the twentieth century, pollution from farms and cities fertilized rivers and lakes and allowed algae to grow more vigorously. As rain drained from fields, lawns, gardens, and golf courses, it carried much of the fertilizers that had been applied to stimulate the growth of crops, flowers, and grass. Also, most household waste that is flushed down toilets is an excellent organic fertilizer. With all these extra inputs of nutrients, populations of algae became so dense that "pond scum" floated near the surface of rivers and lakes, and the water was green because of the abundance of microscopic algae (**FIGURE B4.3**).

FIGURE B4.3. When livestock such as cattle, pigs, or poultry are kept in outdoor pens, rain washes manure into streams, fertilizing them and causing algae to grow profusely.

It would have been difficult to stop all pollution, but people realized that it was only necessary to control one single pollutant so as to create a limiting factor. Phosphate was chosen as the target. Phosphorus is an essential element, and it is naturally low in pure water. A large amount of the phosphate pollution in the rivers was coming from laundry detergent and dishwashing soap. With a little effort, phosphate-free detergents were invented, and now they are used almost universally, so there is much less phosphate pollution. The concentration of phosphate in rivers dropped so low that algae could no longer thrive, and their populations fell to more normal levels. Even though the water is still heavily polluted with nitrates, sulfates, and other nutrients, the algae cannot use them as long as phosphate is kept low enough to limit their growth. If we could reduce the phosphate runoff from farms and lawns, the levels of algae would drop even more and rivers and lakes would be even cleaner.

It would be better to control and reduce all types of pollutants, but by keeping one at limiting levels, we can at least minimize some of the damage caused by pollution.

cell, fungus cells branch into tiny tree-like structures called arbuscules. The fungi fill these with phosphorus, which is then transferred to the plant. The fungi benefit from this by receiving sugars and other nutrients from the roots.

Diseases Caused by Lack of Essential Elements

If a plant grows in a soil that has too little of one of the essential elements, the plant will suffer from a **deficiency disease**. Such diseases are most likely to be encountered in cultivated plants if they are grown in unsuitable soils. One of the most frequent situations that cause deficiency disease is cultivating plants that need acid soils in areas with alkaline soils; such plants have difficulty absorbing enough iron.

Deficiency of certain minerals causes specific symptoms. Lack of iron causes chlorosis, a yellowing of leaves due to inability to synthesize chlorophyll (**FIGURE 4.22**). Necrosis is the death of patches of tissue; if the leaf tips and margins die, it is probably caused by a lack of potassium. Deficiency of manganese causes necrosis of tissues between leaf veins even though the veins themselves remain alive and green.

FIGURE 4.22. This azalea leaf is from a plant growing in alkaline soil. Azaleas require acidic soil and suffer iron deficiency in alkaline soils.

Important Terms

acid
active site of enzyme
alkaline
amino acid
atom
base
callose
callose plug
cavitate
deficiency disease
deoxyribonucleic acid
diffusion
disaccharide
DNA
enzyme
essential element
fatty acid
free-living organism
functional group
gene
hydrophilic
hydrophilic channel
hydrophobic
hydroponic experiment
hydroxyl ion
limiting factor
macro essential element
messenger RNA (mRNA)
microRNA
mineral nutrition
minor essential element
molecule
monomer
monosaccharide
mycorrhiza
negative ion
nitrogen assimilation
nitrogen fixation
nitrogen reduction
nonpolar molecule
nucleic acid
nucleotide
osmosis
osmotic effect
p-protein
p-protein plug
pH
phospholipid
pit membrane
polar molecule
polymer
polysaccharide
positive ion
primary structure of protein
protein
proton
quaternary structure of protein
ribosomal RNA (rRNA)
salinization
saturated fatty acid
secondary metabolite
secondary structure of protein
self-assembly
simple sugar
sink
source
substrate specificity
symbiotic organisms
tertiary structure of protein
trace essential element
transfer RNA (tRNA)
triglyceride
turgid
unsaturated fatty acid
weathered
wilted

Concepts

- The bodies and metabolisms of all organisms are based on the fundamental principles of chemistry and physics.
- Millions of organic compounds are possible, but most can be classified into just a few families of compounds.
- Many important biological compounds are polymers composed of monomers.
- Using only glucose, water, and a few inorganic minerals, plants build all the molecules of their bodies.
- Self-assembly is the tendency to automatically organize shape due to characteristics of a molecule itself.
- Wherever water is mixed with a hydrophilic substance, water tends to move from where it is more concentrated to where it is less concentrated.
- Because all plants absorb water as a liquid and lose it as a gas, they accumulate minerals.
- The growth rate of all organisms is determined by a limiting factor.

Energy Metabolism: Photosynthesis and Respiration

5

All organisms must take in energy. Plants and algae obtain their energy from sunlight, while other organisms obtain energy from the food they consume. This prickly pear cactus will use the light energy to power chemical reactions that build its body and make its flowers, fruits, seeds, and so on. Animals will eat the fruit and pads, and fungi will decompose the body after it dies. The two processes of photosynthesis and respiration distribute the sun's energy throughout all living organisms. (Courtesy of Byran Hoyt.)

All living organisms must obtain energy; there is no alternative. Principles of physics tell us that there is no such thing as perpetual motion, so as soon as an organism runs out of energy, its metabolism must stop. Plants obtain energy through photosynthesis: special pigments capture light energy and then, through a series of metabolic steps, they place that energy onto molecules that can carry it to places that need the energy.

In addition to energy, photosynthesis gathers two other things: carbon and electrons. The organic compounds in a plant's body, as well as our own bodies and those of all organisms, are built up of many carbon atoms, and these are obtained by means of photosynthesis, from carbon dioxide in the environment. Because organic compounds are more complex than carbon dioxide, each carbon atom needs more electrons to form the various bonds present in carbohydrates, proteins, and so on that are not present in a molecule of carbon dioxide. Consequently, photosynthesis supplies plants—and all organisms that use plants as food—with these three essentials: energy, carbon, and electrons (FIGURE 5.1). Plants are **autotrophs** (they feed themselves).

All nonphotosynthetic organisms, such as animals, fungi, and most microbes, must obtain these three essentials by eating foods that contain them. They are **heterotrophs** (they feed on others). But they are faced with the problem of converting the food molecules into the various molecules they need for their own metabolism, and to

FIGURE 5.1. When wood burns, the heat given off is the energy that the plants had captured by photosynthesis and used to make the wood cells. Burning converts the organic compounds of the wood (mostly cellulose) back to carbon dioxide and water, the two chemicals the plant had used in photosynthesis.

converting much of their food into energy for movement, keeping their bodies warm, and controlling their metabolism. This is accomplished through **respiration**.

Photosynthesis and respiration are complimentary processes. Photosynthesis brings in energy, carbon, and electrons, but it makes only one product, the simple sugar **glucose**. Respiration on the other hand, takes glucose and converts it to dozens of other compounds, many of which can be used immediately, others of which are further modified by other metabolic reactions. Both processes are essential: without photosynthesis, all organisms would starve, and without respiration, there would be an abundance of glucose but little else. The topics of this chapter involve these three processes: manipulating energy, carbon, and electrons.

Reactions

Oxidation/Reduction Reactions

When one compound donates electrons to another, we say it **reduces** that compound, and at the same time, it becomes **oxidized** (it loses electrons). Unlike electricity that comes through wires in our homes, electrons on biological compounds are always associated with atoms, so oxidations and reductions occur in pairs: one partner is oxidized, the other reduced as electrons are transferred. Oxygen has a great tendency to react with other compounds by accepting electrons from them; oxygen is a powerful oxidizing agent. Iron readily donates electrons to other compounds; it is a strong reducing agent. When iron rusts, the iron becomes oxidized and the oxygen becomes reduced.

Electron Carriers

In many cases, the two partners of an oxidation/reduction reaction bond to each other, such as iron rusting when exposed to oxygen. But in other cases, the electrons are transferred to a compound that can in turn pass them on to another compound, reducing it. Such a compound is an **electron carrier**, and the two most important ones are **$NADPH_2$** (reduced nicotinamide adenine dinucleotide phosphate) and **$NADH_2$** (reduced nicotinamide adenine dinucleotide) (**TABLE 5.1**). When $NADPH_2$ donates

TABLE 5.1. Molecules that Carry Electrons and Reducing Power

ATP carries energy to reactions. During the reaction, it loses one or two phosphate groups and becomes either ADP or AMP. The phosphate groups may be released into the cytoplasm or be attached to one of the compounds in the reaction.

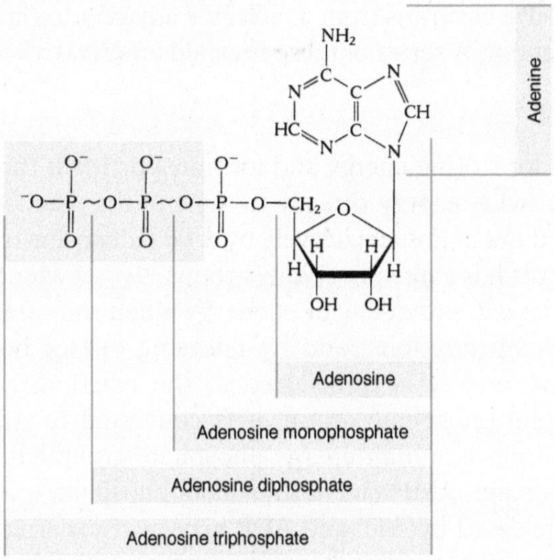

FIGURE T5.1a.

Adenosine triphosphate. Breaking off the last phosphate to produce ADP results in a more stable set of electron orbitals, and energy is given off. The same is true of removing the second phosphate but not the third.

$NADH_2$ and $NADPH_2$ both carry two electrons to materials that need to be reduced. As they transfer electrons to a compound, they can no longer hold their two hydrogens, and they are converted to NAD^+ or $NADP^+$. $NADH_2$ is produced by respiration; $NADPH_2$ is produced by photosynthesis.

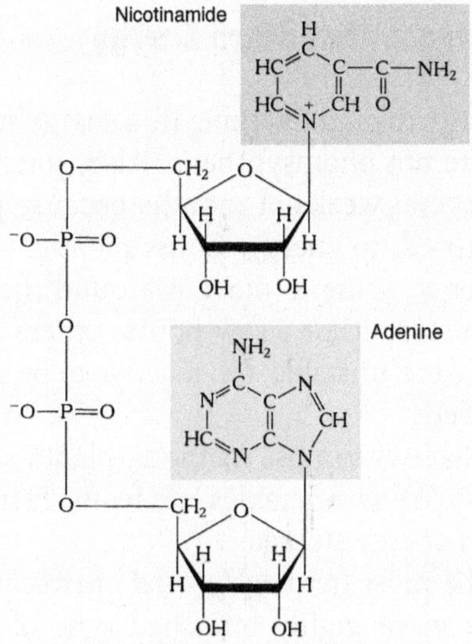

FIGURE T5.1b.

Nicotinamide adenine dinucleotide (NAD^+). The important area is the site where the positive charge is carried (purple box); this ring can pick up two additional electrons and then deposit them elsewhere. This entire molecule is the "container" for carrying two electrons around a cell. The adenine (blue box) is far removed from the site where the electrons are carried, but if the adenine is altered, NAD^+ cannot enter into the proper reactions.

its electrons to another compound, it becomes oxidized and can no longer hold on to its protons; it becomes **$NADP^+$**. This can move around the cell until it enters another reaction that will load it with electrons again, converting it back to $NADPH_2$ and ready to carry the electrons somewhere.

Other electron carriers are not so mobile; they are components of membranes and typically they receive electrons from a molecule adjacent to themselves and pass them onto another neighbor. A series of these is called an **electron transport chain**.

Energy Carriers

Sugars are useful for storing energy and for transporting it through phloem, but they do not actually provide energy directly to most of the reactions that require energy. That role is carried out almost exclusively by **ATP** (adenosine triphosphate) (Table 5.1). This small molecule has three phosphate groups, two of which are attached by **high-energy bonds** that can provide a lot of energy when they are broken or rearranged. ATP can enter many reactions, and by releasing one or both of the high-energy phosphate groups, energy is transferred to the reaction, allowing it to proceed. If ATP donates one phosphate group, it is converted to ADP (adenosine diphosphate), which still has one high-energy phosphate group left. Once that is donated, the molecule becomes AMP (adenosine monophosphate), a low-energy molecule. The phosphates released by ATP and ADP may be transferred onto one of the other molecules in the reaction, making it more energetic, or the phosphates may be released free into the cytoplasm. Many molecules must receive a phosphate group like this to become reactive; in this case the molecule has been phosphorylated.

It is difficult to overstate the importance of ATP: almost every metabolic pathway in every organism relies on the energy of ATP at one or more steps. If ATP suddenly ceased to exist, all life would stop instantaneously. ATP is synthesized from ADP and phosphate. Only two processes are especially important in producing ATP: photosynthesis and respiration.

Molecules and Organs that Store Energy

Leaves capture light energy during daytime; this energy must then be moved to other parts of the plant that are not photosynthetic. Also, some must be stored, at least for night, and usually for several weeks or months because plants don't reproduce until older, and need some stored up energy. Consequently, plants need various types of molecules to handle energy: some to move it around the cell, some to move it from organ to organ, some to store it for a few hours, others to store it for months. ATP, $NADH_2$, and $NADPH_2$ are too unstable, too reactive to be stored or moved out of a cell; other molecules are needed.

We are already familiar with most methods plants use for long-term storage of energy: we eat them every day as vegetables and fruits. Plants use a variety of chemicals and organs for long-term energy storage.

Amylose (starch) is the most frequently used chemical in plants, but some plants also use amylopectin, a more highly branched type of starch molecule present in medium-grain rice and waxy corn (it is less abundant in long-grain rice and russet potatoes). Both amylose and amylopectin are stored in amyloplasts (starch grains). Sugar cane and sugar beets are two of the very few cases in which plants covert glucose to sucrose rather than starch for long-term storage. The sucrose is accumulated in the central vacuole, and it causes a large build up of pressure, but the plants tolerate it. Unlike us animals, plants only rarely store lipids for their own later use; the oils in avocados and olives are located in fruits that entice animals to eat them and distribute their seeds. Once a plant deposits material into a fruit, it cannot be reused by the plant. Seeds like peanuts, sunflowers, and coconuts contain oils that will be used by the embryo when the seed germinates. Proteins are abundant in seeds of legumes like beans and peas, but this is a means of storing amino acids for the embryo; plants almost never respire proteins for energy the way we animals do after eating a meal with too much meat.

A variety of organs are used for long-term energy storage in plants. All seeds store at least a small amount of energy for the embryo, and many seeds store enough to allow the embryo to grow for several weeks even if there is no sunlight for photosynthesis. Coconuts store very large amounts, and the world's largest seed, the sea coconut (*Lodoicea maldivica*), weighs several pounds, most of it being high-energy compounds. In monocot seeds, most of the energy is stored as endosperm, a tissue that surrounds the embryo; a few dicots also have seeds with abundant endosperm. Most dicot seeds store energy within the embryo itself, very often in the cotyledons (the two halves of a peanut or bean), but cactus seeds use the hypocotyl (the region where the root meets the stem).

It is worth thinking about seeds a moment longer. The evolution of plants to produce seeds was a momentous occasion for animals as well as plants. Before that, there were no concentrated masses of starches or oils or proteins in the plant world. Animals could eat leaves, stems, and roots, or some insects could suck phloem sap, but no animal could feast on a fruit or a cone full of seeds. The evolution of seeds, and later of edible fruits, gave animals an entirely new type of food, one that provided much more energy to herbivores, and allowed a more active life.

In perennial, woody plants, especially deciduous ones that shed their leaves in autumn, parenchyma cells in the wood and phloem of both shoots and roots typically accumulate starch during the summer, then use it throughout winter and spring while the plant is leafless. Even when the wood contains its maximum amount of starch, there is almost always so little that few animals bother with eating wood, especially since they would have to chew through wood fibers to get to it. Phloem of woody plants also is too lacking in calories for us to eat, but rabbits, deer, and many other animals do eat bark when deep snow has covered all other possible food.

Many perennial plants go through further change than shedding their leaves in autumn: they allow all of the above-ground portion of the shoot to die. During the following spring, buds on subterranean shoots sprout and grow into a whole new shoot system. The subterranean shoots may be tubers, bulbs, rhizomes, corms, and so on, or there may be just a tiny bit of stem on top of an enlarged root, as in carrots.

Many perennial plants have no specialized organs for long-term storage. Ferns store starch in most of the cells of their cortex and pith, but they are rarely tuberous or bulb-like. Agaves, yuccas, and other succulent monocots have large, fleshy leaves that store mostly water, but they are so voluminous that if each leaf cell stores just a little starch, the total amount is great.

Photosynthesis

Capturing Light Energy

Light is part of an electromagnetic spectrum that can be thought of as being both particles called **quanta** and as waves. Some types of light (such as violet) consist of waves that have short wavelength, and each of their quanta contains more energy than does each of long wavelength light (such as red). Visible light, the portion of the

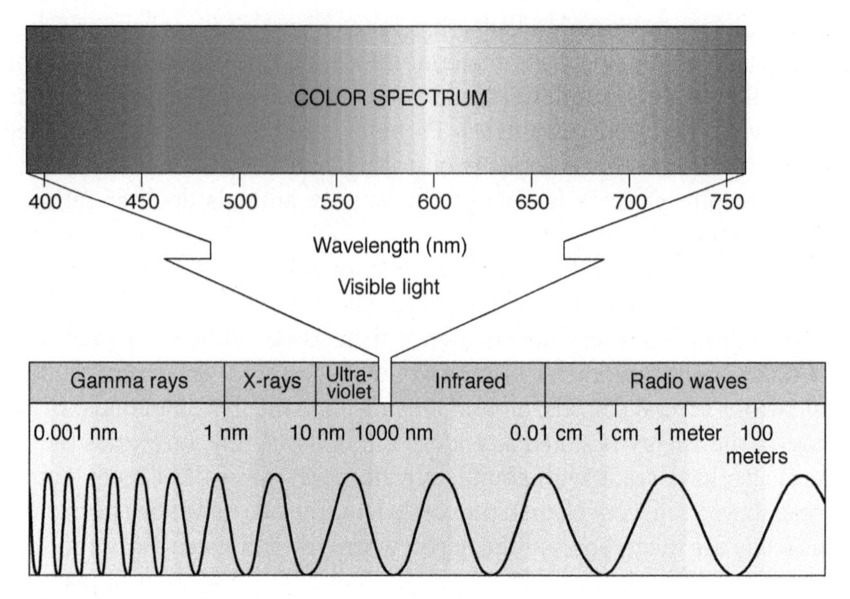

FIGURE 5.2. Visible light—the light that our eyes can see—is just a small part of the spectrum of energy. Each quantum (light particle) of gamma rays, X rays, and ultraviolet light has enough energy to damage living organisms, whereas each quantum of infrared light and radio waves has too little energy to be useful in photosynthesis.

electromagnetic spectrum we can see, consists of red, yellow, green, indigo, blue, and violet (FIGURE 5.2). Our eyes cannot detect longer wavelengths like infrared, but we can feel them on our skin. Ultraviolet wavelengths are too short to be visible to us, but they can cause sunburn and damage DNA, leading to skin cancer. Plants too can be harmed by both infrared and ultraviolet, and do not use them for photosynthesis.

A chemical that absorbs light energy is called a **pigment**. The pigments in paint, our clothes, and in our hair and skin merely absorb some light and reflect away other light; for example, red hair absorbs all colors except red. The same is true of all non-green pigments in flowers and fruits: they absorb some wavelengths, reflect the ones animals see, but do not use any of the energy that strikes them for any chemical reaction. Rhodopsin, the pigment in our eyes, absorbs light and rearranges itself, which stimulates retina cells to send a signal to the brain; but the pigment then returns to the way it was, and no chemical has been changed permanently, no energy made available for us to use. However, something completely different happens when chlorophyll *a* (plants have two types of chlorophyll, *a* and *b*) absorbs light. One of its electrons is activated and then passed to the first electron carrier of an electron transport chain in the chloroplast inner membrane. This electron and its energy can then be used to make both ATP and $NADPH_2$.

Two important concepts in photobiology are **absorption spectrum** and **action spectrum**. The wavelengths that chlorophyll *a* or any other pigment absorbs is its absorption spectrum (FIGURE 5.3). The wavelengths that provide energy to a reaction are the reaction's action spectrum. We know that chlorophyll *a* is the main pigment responsible for photosynthesis because its absorption spectrum closely matches the action spectrum of photosynthesis. You might expect that a photosynthetic pigment like chlorophyll *a* would absorb all types of light and thus capture as much energy as possible for a plant. But that does not happen: it absorbs mostly red and blue, and just a little green, and none of the really high-energy quanta of violet and ultraviolet. This is just the way evolution happened; if it were possible to have a pigment that

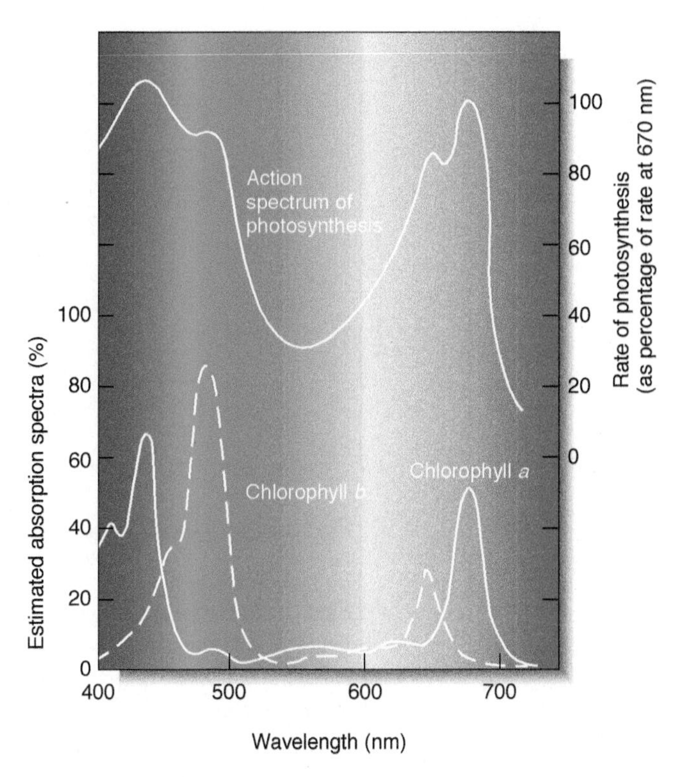

FIGURE 5.3. The absorption spectra of chlorophyll *a* and chlorophyll *b* and the action spectrum of photosynthesis. On the bottom axis is the wavelength of light, with short (blue) wavelengths to the left and long (red) ones to the right. The vertical axis of the absorption spectra is the amount of light absorbed by the pigment; for the action spectrum, it is the amount of photosynthesis carried out. Chlorophylls absorb little of the very short wavelength light at 400 nm, and little photosynthesis occurs. But light at slightly longer wavelengths, about 425 nm, is absorbed well by chlorophyll *a* and photosynthesis proceeds. Quanta with intermediate wavelengths pass right through the pigments and photosynthesis is low, but in the 650 to 680 nm range (red) considerable absorption occurs. Because the absorption spectra of chlorophyll *a* and *b* are different, more wavelengths can be efficiently harvested. If the two matched perfectly, chlorophyll *b* would be unnecessary.

could absorb and use more wavelengths, it almost certainly would have evolved by now.

An alternative to having one pigment with a very broad absorption spectrum is to use several pigments, each of which captures wavelengths the others miss. We know that this happens in photosynthesis because the absorption spectrum of chlorophyll *a* does not perfectly match the action spectrum of photosynthesis, so other pigments must be involved.

One of the extra pigments is chlorophyll *b*, which absorbs several wavelengths that chlorophyll *a* misses, then transfers that energy to chlorophyll *a*. **Carotenoid** pigments also transfer energy to chlorophyll *a*. Carotenoids and chlorophyll *b* are **accessory pigments**: they absorb light energy but cannot carry out the necessary chemical reaction. To be of use, they must transfer their energy to chlorophyll *a*.

Converting Light Energy to Chemical Energy

To understand photosynthesis, we must examine the chloroplast inner membrane. The membrane is folded into flattened vesicles called thylakoids, most of which lie side by side in stacks called grana. Grana are interconnected by sheets of inner membrane (FIGURE 5.4). Between the chloroplast outer membrane and the thylakoids is a solution called stroma. Photosynthetic pigments and electron carriers are embedded in thylakoid membranes.

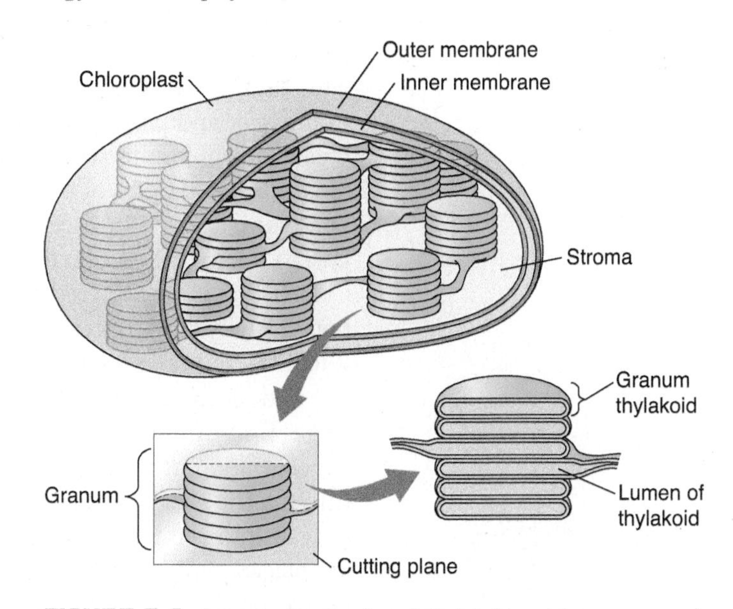

FIGURE 5.4. Grana are stacks of small thylakoid vesicles compressed together; frets are regions of thylakoid that connect one granum to another. The lumen of the thylakoid region is continuous with that of the fret region. The liquid surrounding all the thylakoids is stroma.

Chloroplasts have two electron transport chains involved in their energy metabolism, and two separate chlorophyll *a* molecules must act to completely convert light energy to chemical energy. One set of carriers and pigments is called photosystem I (PSI); the other is photosystem II. When pigments in PSI are activated by light, they pass the energy to a special chlorophyll *a* that is part of a complex called reaction center P700. This elevates an electron and passes it to the first carrier of the electron transport chain in PSI

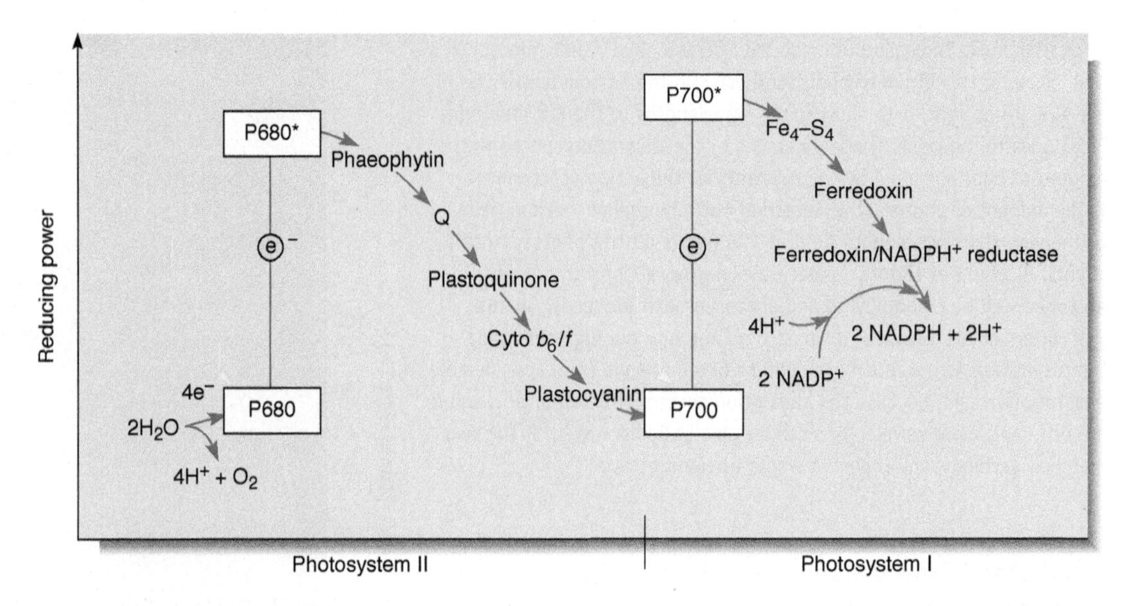

FIGURE 5.5. The two photosystems work together to transfer electrons from water to NADPH. The scale on the left indicates reducing power, the power to donate electrons to another substance. The higher a molecule is in the chart, the greater its capacity to force electrons onto another molecule. Because of its shape, this diagram is called a Z scheme.

(FIGURE 5.5). At the end of this set of carriers, the electrons are finally passed onto $NADP^+$, reducing it to $NADPH_2$. $NADPH_2$ is not part of the chloroplast membrane; instead it is free to diffuse to other parts of the chloroplast, carrying its two electrons to sites where carbon dioxide is being processed.

After chlorophyll *a* has passed an electron onto the electron transport chain, it is important that it obtain a new, low-energy electron immediately. If it does not, it will break down. The source of the electrons is the other electron transport chain, the one in PSII. When pigments in PSII capture light energy, they send it to chlorophyll *a* in reaction center P680, which sends electrons through the PSII electron transport chain to rescue the chlorophyll *a* of P700. But now the P680 chlorophyll *a* needs an electron or it will break down. The source of electrons for it is water; a plant enzyme takes electrons from water and passes them on to chlorophyll *a*, keeping it stable and safe. The water breaks down, forming oxygen and two protons, but each plant cell has trillions of water molecules, so the loss of a water molecule is not a problem. The oxygen is a byproduct, and plants usually just let it diffuse away out through the stomata. When a stomatal pore is open, carbon dioxide diffuses into the leaf and oxygen diffuses out.

$NADPH_2$ is synthesized directly as the last step of the PSI electron transport chain, but ATP is synthesized indirectly by a process called **chemiosmotic phosphorylation** (FIGURE 5.6). The water molecules that donate their electrons break down into oxygen and protons; this occurs inside the thylakoid lumen, and the protons are trapped there. Every time the P680 chlorophyll *a* takes more electrons from more water, the concentration of protons inside the lumen becomes stronger. Also, as electrons pass through the electron transport chain of PSII, one of the steps pumps protons from the stroma into the lumen, adding to the proton concentration. Finally, when electrons are placed onto $NADP^+$, it picks up protons from the stroma to become $NADPH_2$; this occurs on the outside of the lumen, reducing the concentration of protons there. The entire process builds up a large difference in proton concentrations, and there is a strong tendency for protons to migrate from inside the lumen to the outside to equalize the concentrations. The membrane has large enzymatic channels that allow the

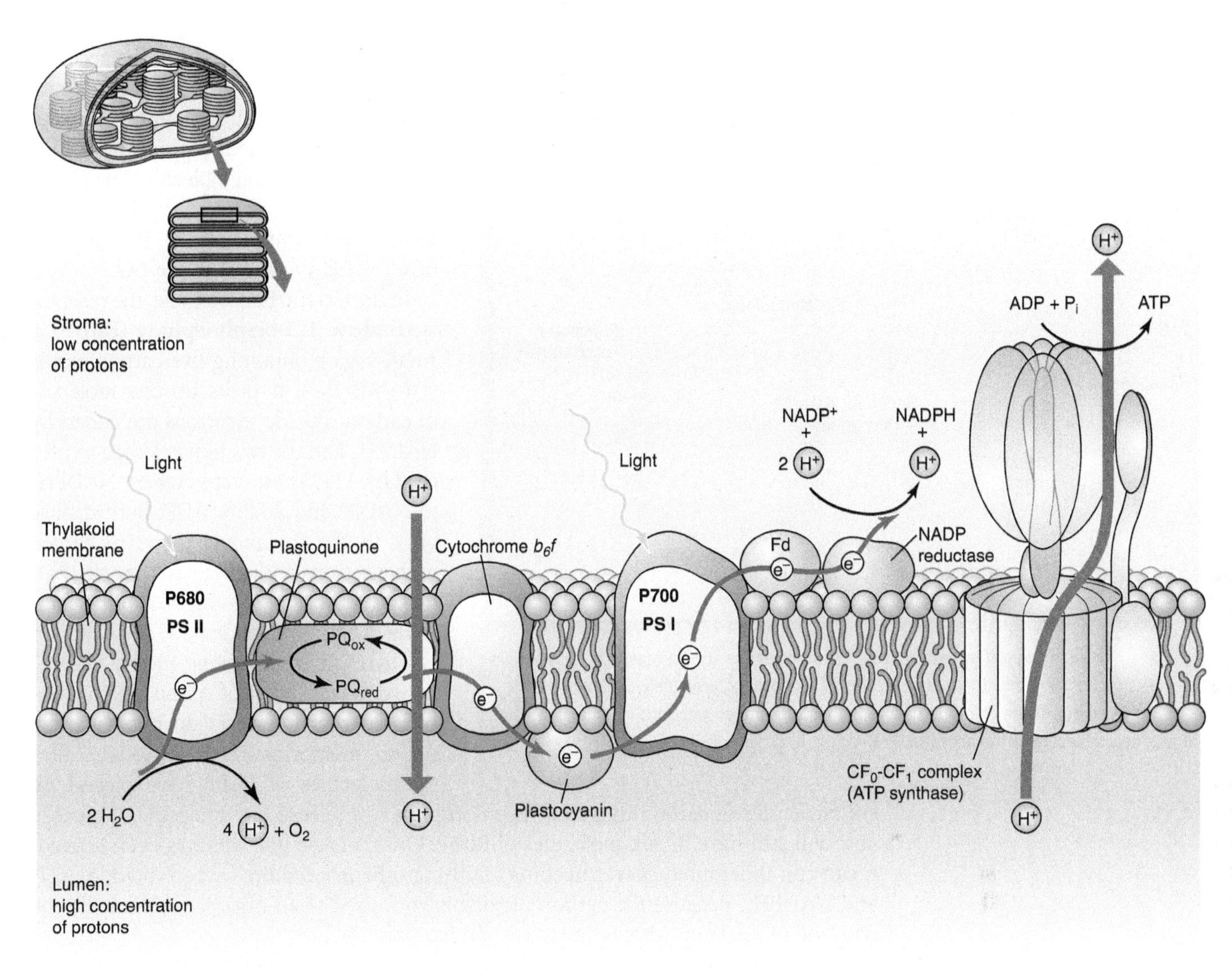

FIGURE 5.6. The water-splitting, proton-producing reactions of photosystem II take place on the lumen side of the thylakoid membrane, and various electron carriers also transport protons into the thylakoid lumen. When NADPH is formed, it picks up protons from the stroma. Passage of electrons results in a deficiency of protons in the stroma and an excess in the thylakoid lumen. Protons return to the stroma by passing through ATP synthetases, powering the phosphorylation of ADP to ATP.

excess protons to flow out, but as they pass through the enzyme, they cause it to add a phosphate to ADP, resulting in the formation of ATP. This is then free to diffuse to other parts of the chloroplast where carbon dioxide is being processed. All of these reactions that are powered directly by light are called the **light reactions**.

The ATP and $NADPH_2$ produced by the light reactions could theoretically diffuse to any part of the cell and power various reactions, but both are retained inside the chloroplast and are used almost exclusively to convert carbon dioxide to glucose.

Converting Carbon Dioxide to Sugar

The process in which carbon dioxide is converted into carbohydrate has two names; some people call it the **C_3 cycle**, others call it the **Calvin/Benson cycle**, or the stroma reactions. This type of cycle is extremely common in organisms: a small molecule (in this case, carbon dioxide) is added to a receptor that then passes through various conversions that result in the regeneration of the receptor and some sort of transformation of

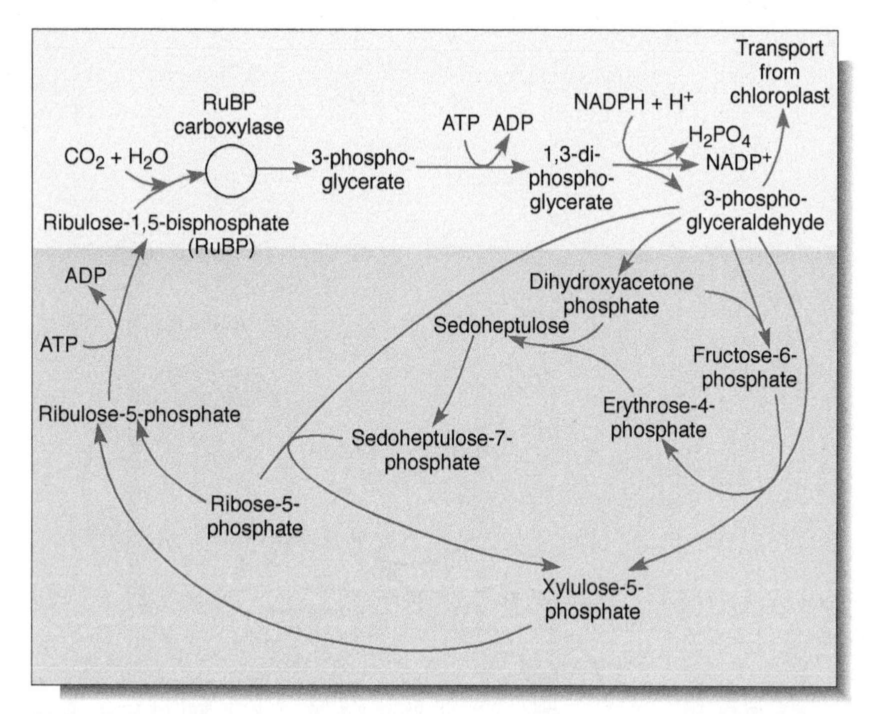

FIGURE 5.7. In the yellow area are the first steps of the stroma reactions, also known as the C_3 cycle; the product is two molecules of 3-phosphoglyceraldehyde. Some of this is transported out of the chloroplast, and the rest undergoes reactions (purple area) that form a new molecule of the acceptor, RuBP.

the small molecule. Because the receptor is regenerated each time, the cycle can, using only a small amount of receptor, operate repeatedly, processing large amounts of the small molecule. The chemical equation for photosynthesis does not indicate the complexity of the process: $6CO_2 + 6H_2O \rightarrow C_6H_{12}O_6 + 6O_2$.

In the Calvin/Benson cycle, the receptor is **ribulose 1,5-bisphosphate (RuBP)**, a small sugar containing five carbon atoms (FIGURE 5.7). It picks up one molecule of carbon dioxide, electrons are added by $NADPH_2$ and the reaction is forced to proceed by ATP. These steps convert $NADPH_2$ to $NADP^+$ and ATP to ADP; both diffuse back to the thylakoids to be recharged with electrons and energy so they can participate in the processing of another molecule of carbon dioxide.

There are many steps in the Calvin/Benson cycle, most of which are so complicated that botanists do not generally try to memorize them. Basically after six molecules of RuBP have picked up six molecules of carbon dioxide, the chloroplast can release one molecule of glucose and will still have its six molecules of RuBP. Once a cycle like this has been started, it can run indefinitely, never needing anything other than more carbon dioxide, ATP, and $NADPH_2$. Because the carbon dioxide is converted to an organic molecule by the addition of electrons, this is called carbon fixation.

The important enzyme is **RuBP carboxylase**, usually called **RUBISCO**. It is the only enzyme in the world that can carry out carbon fixation: virtually every carbon atom in every organic molecule in all organisms has passed through RUBISCO. However, this is a surprisingly poor enzyme: it has a low affinity for carbon dioxide and it occasionally makes an error by picking up oxygen. But RUBISCO originated when Earth's atmosphere had a much higher concentration of carbon dioxide and almost no oxygen, and in the hundreds of millions of years since, it has not evolved to be better adapted to the new conditions (BOX 5.1).

The glucose produced by chloroplasts is available for immediate use, but it can also be stored for use later. Typically, during daylight, chloroplasts convert the new glucose into starch and store it as a starch grain inside themselves. Later, at night when the chloroplast cannot carry out photosynthesis because there is no light, it breaks the starch back down into glucose and transports it out to the rest of the cell for use there. Some might be used immediately in respiration so that the cell has ATP during the night (all living parts of plants must respire at night). Other molecules are converted to sucrose and loaded into the phloem for transport to other parts of the plant.

Environmental Factors that Affect Photosynthesis

Many environmental factors affect photosynthesis. The quantity of carbon dioxide is critical of course, but its concentration in air is usually more or less constant from

BOX 5.1. Plants, People, and Photosynthesis

Think about yourself, the things you have done today, the food you eat, the exercise you get, the air you breathe. All these things, all aspects of your life are possible only because of the photosynthesis described in this chapter. This photosynthesis is not the only type that exists, it was not the first type to evolve, but it is the photosynthesis that made our world.

The photosynthesis described in this chapter is **oxygenic photosynthesis**. It gives off oxygen as a byproduct when it takes electrons from water. Several types of bacteria have an older type of photosynthesis, one that take electrons from sulfides and gives off pure sulfur as a byproduct. Oxygen is not involved; it is **anoxygenic photosynthesis**.

When oxygenic photosynthesis originated three and a half billion years ago, the world was very different. The atmosphere contained no breathable oxygen, no O_2; in fact there was no free oxygen anywhere: all oxygen was combined into water or carbon dioxide or other compounds. Without free oxygen, there was no aerobic respiration; most organisms produced ATP only by glycolysis and fermentation, which yielded little energy. Intense ultraviolet light scorched the Earth's surface, so there could be no life on land. All organisms were prokaryotes (bacteria and archaea).

Then, a group of bacteria, called **cyanobacteria**, evolved (**FIGURE B5.1a**). They were different from all other organisms because they had oxygenic photosynthesis. As they photosynthesized, they released O_2. For the first several millions of years, each new molecule of oxygen immediately reacted with iron: the world began to rust. Ancient rocks contain a layer of rust, rust that was only possible as O_2 finally became present. At some point, everything that could combine with oxygen had done so, and oxygen began to build up in the air. This had a beneficial impact immediately: in the presence of ultraviolet light, oxygen reacts with itself and produces ozone (O_3), a chemical that strongly blocks ultraviolet radiation. As ozone accumulated, it stopped enough ultraviolet light that Earth's surface became safe for living organisms, and life on land and in surface waters was possible. However, even now after billions of years, ultraviolet radiation is still strong enough to cause skin cancer.

With some free oxygen available, the environment could support aerobic respiration. Before this, any mutation that resulted in an enzyme for aerobic respiration would have been useless. But as soon as free oxygen was present, such mutations were advantageous, and over millions of years and through many mutations, bit by bit, enzymes for the citric acid cycle and the electron transport chain could be added to those for glycolysis, changing glycolysis from being the first step of fermentation to the first step of aerobic respiration. Soon, organisms had much more ATP available in their cells, had a more vigorous metabolism.

The origin of cyanobacteria with oxygenic photosynthesis was so momentous, we have to ask, what happened to them? Many cyanobacteria are still alive and flourishing as free-living microbes, looking much like they did billions of years ago. Almost any soil sample has some, and often a black discoloration on a moist garden wall is due to a film of cyanobacteria. Sheet-like colonies of them, large enough to be easily visible, lie on the soil in dry areas (**FIGURE B5.1b**).

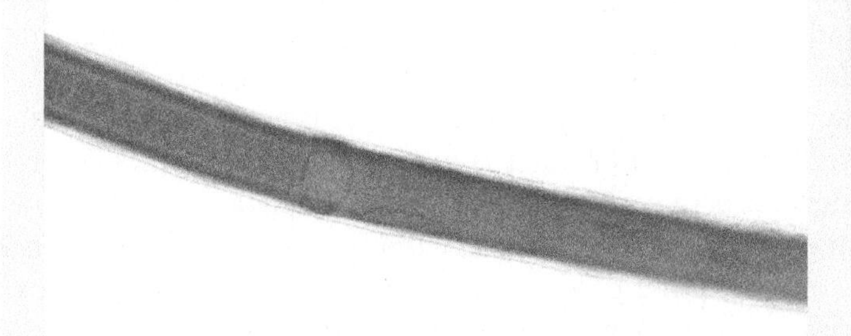

FIGURE B5.1a. This cyanobacterium has enzymes that can bind N_2 from air, force electrons onto it, reducing to a point where it can be used in organic compounds. Cyanobacteria similar to this were the ancestors of chloroplasts (×450).

FIGURE B5.1b. Think about the cyanobacteria that were the ancestors of chloroplasts. They evolved into two very different groups, both of which still successfully exist today. All the chloroplasts (as well as proplastids, amyloplasts, chromoplasts, and so on) of all plants are members of one group of descendants, and the modern cyanobacteria (pictured here) are members of the second. Cyanobacteria are abundant in soil, and even form easily visible sheets on the ground in many places. This was photographed after a rain west of Austin, Texas.

(continued)

Other cyanobacteria took an alternative road to success. They began living symbiotically inside cells of an early eukaryote. Much like symbiotic nitrogen-fixing bacteria in root nodules, and phosphate-providing fungi in mycorrhizae, these symbiotic cyanobacteria succeeded by living inside cells, providing them with photosynthetic energy while receiving nutrients and protection. Gradually, the cyanobacteria evolved to be chloroplasts: all the chloroplasts present in plants and algae today are descendants of those first endosymbiotic oxygenic cyanobacteria. And as for all the photosynthesis performed by the plants around us, it would be more accurate to say it is not performed by the plants themselves but instead by their endosymbiotic cyanobacteria. In contrast to free-living cyanobacteria that mostly inhabit soil and need to obtain all their water and mineral nutrients on their own, the endosymbiotic ones—the chloroplasts—live everywhere that plants live; they live in prairies, forests, marshes, deserts, mountains, and jungles, and always surrounded by a plant cell, and always having water and nutrients brought to them by xylem and phloem.

day to day and from place to place. Corn fields can absorb and fix carbon dioxide so effectively that they lower the carbon dioxide in the air around them on quiet days, but even slight wind brings in fresh air with more carbon dioxide. For the last century, the amount of carbon dioxide in the air has been gradually increasing; this is discussed in Chapter 11.

Light is necessary for photosynthesis but think carefully about light. Light intensity varies due to time of the day and season. It is absent at night but intense at noon when the sun is overhead and light passes directly down through the atmosphere. It is less intense in early morning and late afternoon, and photosynthesis is slower then. In winter, the sun is lower in the sky, and much of its light is scattered through the atmosphere: at latitudes far from the equator, even the brightest winter day is much dimmer than summer days, plus the days are shorter, so light is available for fewer hours per day. Light is less intense in shady areas under trees (FIGURE 5.8), in deep canyons, or beside tall buildings in cities. Less sunlight reaches plants in areas that are foggy or rainy, such as along many ocean coasts and high in mountains. Most algae and some plants are aquatic, living in lakes, streams, or oceans. Water absorbs and scatters light, and at a depth of only a few feet, light is dim even in clean, clear water. And just a shallow layer of muddy water blocks almost all light and interferes with photosynthesis.

Light also varies in quality or the relative amounts of different wavelengths it contains. All light from the sun is the same, but as light passes through a tree, chlorophyll in its leaves absorbs most of the red and blue and lets mostly just green pass through. Ferns and mosses on the forest floor thus must be adapted not only to dim light but to this altered type of light, and must have accessory pigments. Water absorbs red and violet, so aquatic plants and algae that grow deeper in water need accessory pigments that absorb green and blue.

If we look at the photosynthesis equation, we see that water is necessary as a source of electrons to keep chlorophyll stable. But water is important in other ways as well. If a habitat is too dry, plants conserve water by keeping their stomata closed; this prevents carbon dioxide from entering leaves, so photosynthesis is inhibited. This is not common in moist regions of course, but it occurs frequently in deserts, and is one of the main reasons that so few plants survive where there is little water.

C4 Plants and CAM Plants

Plants in dry regions face a problem in that they lose water whenever stomatal pores are open and allowing carbon dioxide to enter. In moist habitats, this water loss is not a problem because roots can replace it, and also the loss of water keeps leaves cool and

FIGURE 5.8. **(a)** Almost all crop plants, such as this broccoli, are cultivated in full sunlight in order to maximize the energy they have for photosynthesis. As long as they have enough water and nutrients, the more sunlight a plant has, the faster it grows. A few crops, such as coffee, are damaged by strong sunlight and are cultivated in the shade of other plants. **(b)** This shady habitat at the ground level of a dense forest of tall trees does not have enough light to cultivate crops. The plants that do occur here are adapted to low light intensity and have accessory pigments; even so, they grow slowly despite having plenty of water and a rich soil.

pulls water up through the xylem. But plants of dry regions must conserve water, and if too much water is lost relative to the amount of carbon dioxide absorbed, the plant becomes dehydrated and may die.

C_4 photosynthesis is one solution to this problem. Plants with this metabolism have a special enzyme, PEP carboxylase, that has a very high affinity for carbon dioxide molecules. Its active site binds carbon dioxide and attaches it to a small receptor molecule, PEP (phosphoenolpyruvate), and oxaloacetic acid or malic acid is produced. The acid is then transferred from the mesophyll cells into special cells that form a sheath around leaf veins (this leaf anatomy is called Kranz anatomy) (FIGURE 5.9). Inside the sheath cells, the acids release the carbon dioxide, which then can be picked up by RUBISCO and used for the ordinary Calvin/Benson cycle. The receptor molecule is transferred back to the mesophyll, and prepared to carry another carbon dioxide molecule to the vein sheath.

The benefits of adding this extra step to photosynthesis is that the initial enzyme, PEP carboxylase, has such a high affinity for carbon dioxide that, no matter how quickly carbon dioxide diffuses into the leaf, its concentration

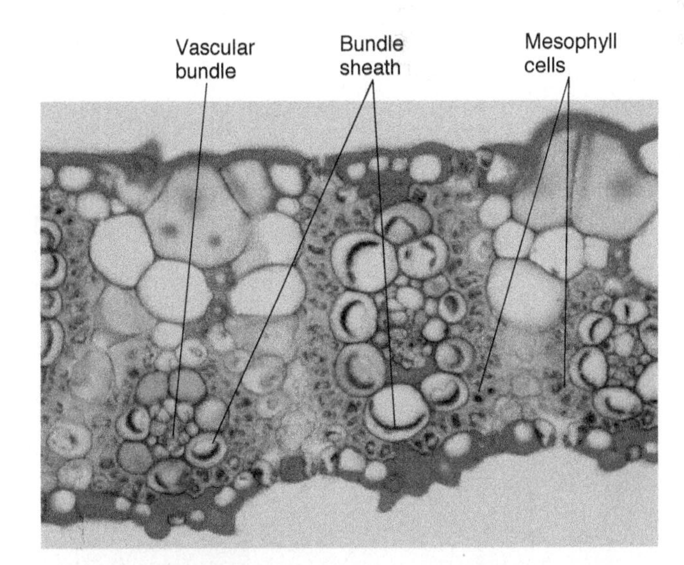

FIGURE 5.9. This transverse section of a sugar cane leaf shows Kranz anatomy; there are three vascular bundles, each surrounded by a sheath of large cells, and between one sheath and another are just two or three layers of mesophyll cells. In mesophyll cells, carbon dioxide and PEP combine into acids that are then transferred into the sheath cells, where the acids break down and the carbon dioxide is picked up by RUBISCO (×300).

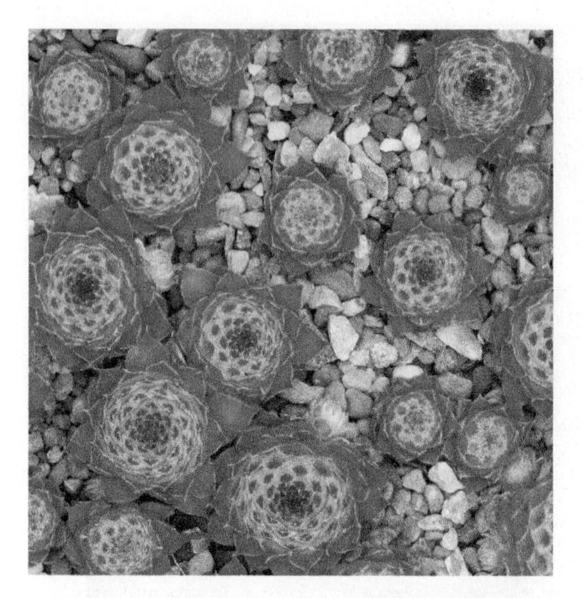

FIGURE 5.10. These are plants of *Crassula*, members of the family Crassulaceae for which CAM was named. The giant cells in their leaves have vacuoles so large they can store a considerable amount of acid molecules without the vacuole sap becoming dangerously acidic. (Image provided by Chris McCoy.)

remains almost zero. By keeping the carbon dioxide concentration low in the leaf, more carbon dioxide diffuses in rapidly. Also, the acids are collected from throughout the mesophyll and then concentrated into the small volume of the vein sheath cells. As the carbon dioxide is released, its concentration is high, which helps RUBISCO's active site to grab it and hold it long enough to catalyze the reaction with RuBP.

Crassulacean acid metabolism (CAM) is another alternative means of reducing the amount of water lost while stomata are open. CAM plants open their stomata at night, when the air is cooler and its relative humidity is higher. As carbon dioxide enters, PEP carboxylase attaches it to PEP and makes acids, just as in C_4 metabolism. But because this occurs at night, the acids cannot be used immediately. Instead, they are merely stored in the central vacuole and are not moved to any other cell. The acid concentration builds up throughout the night, then at dawn, the stomatal pores close, the acids release the carbon dioxide, and it is trapped inside the plant because the pores are closed. With sunlight present, the Calvin/Benson cycle can begin to operate, and throughout the day the trapped carbon dioxide is fixed into carbohydrate by RUBISCO.

Storing acid is dangerous to cells because it affects the protoplasm's pH if it becomes too concentrated. But CAM plants are desert succulents such as cacti and crassulas; they have thick water storage organs with giant parenchyma cells with exceptionally large central vacuoles (**FIGURE 5.10**). Even though the plants accumulate large amounts of acids during the night, that is not enough to create a concentrated solution in such large vacuoles.

An important technical detail is that the conversion of PEP to acids is not carbon fixation: no electrons are added during this process and carbon is not reduced. Reduction and fixation occur only after the acids break down and the newly released CO_2 is acted upon by RUBISCO.

Respiration

All living parts of all cells must have ATP, but the ATP produced by chloroplasts is used within the chloroplasts themselves. It is respiration that produces ATP for all other parts of the cell. Even in leaf cells during sunlight, chloroplasts export glucose, which is then respired to supply ATP in the protoplasm outside the chloroplast (**FIGURE 5.11**).

In many ways, respiration is the opposite of photosynthesis: photosynthesis uses energy to convert carbon dioxide to carbohydrate ($6CO_2 + 6H_2O \rightarrow C_6H_{12}O_6 + 6O_2$), using water as a source of electrons and giving off oxygen as a byproduct. Respiration converts carbohydrate to carbon dioxide, using oxygen as a dumping ground for the electrons that are no longer needed, and generating water as a byproduct ($C_6H_{12}O_6 + 6O_2 \rightarrow 6CO_2 + 6H_2O$) (**BOX 5.2**).

FIGURE 5.11. In the darkness after sunset, photosynthesis is not possible so all plant cells must obtain their energy by respiration rather than photosynthesis. Although moonlight may appear very bright at certain times, it does not have the proper wavelengths to power photosynthesis.

BOX 5.2. Plants, People, and Respiration

For the most part, people do not affect plant respiration very much, nor does plant respiration have too much of an impact on people. One aspect of respiration where plants and people influence each other is in food storage. Stored fruits and vegetables respire if they are alive, and if they are moist like fresh potatoes, carrots, peaches, and lettuce they respire especially rapidly. Even dry foods like beans, lentils, wheat, and rice respire, although only very slowly. Frozen and canned foods are dead and do not respire.

Respiration in dry foods is important because of the water produced. Remember that the formula for respiration is $C_6H_{12}O_6 + 6O_2 \rightarrow 6CO_2 + 6H_2O$. This water is released from the mitochondria and moistens the cell's protoplasm, causing the food to become wet and thus susceptible to attack by fungi (molds and mildews). Even if not attacked by fungi, the extra moisture may cause the cells to become more active metabolically, and seeds might even germinate while in storage. To keep wheat, corn, and similar foods dry, air is blown through the storage containers to remove any moisture produced by respiration (**FIGURE B5.2a**).

In moist stored foods, the extra moisture produced by respiration is not much of a problem. A fresh watermelon has so much water already that the little bit produced by respiration is insignificant. Instead, the problem is that respiration uses up part of their carbohydrates, the very reason that we cultivate and harvest most of them. Each week that these plants are stored, there is less carbohydrate available to us when we eat the food. A fresh potato is packed with starch, whereas a potato that has been stored for months has respired away much of its starch and is much less nutritious to us. Respiration in such foods is minimized by keeping them cool. Low temperatures slow down metabolism, and thus slow down respiratory loss of carbohydrates.

Sometimes we harm plant respiration indirectly, without even realizing it. We pollute rivers and lakes with fertilizers; this happens when we dump sewage, even cleaned up sewage, into rivers. The wastes we flush down our toilets is food to many types of bacteria, and such bacteria are able to grow rapidly and respire rapidly. With enough human fertilizer, the bacteria proliferate and use up most of the oxygen available in the water or in the mud on the bottom of a lake or river. Plant roots in those areas may then suffer from hypoxia (low oxygen) or anoxia (no oxygen), and if the roots die, the plants may die also. Of course, fish need even more oxygen than roots do, so hypoxia caused by excessive bacterial growth may cause fish to die. A current problem facing us now is that the Mississippi River carries huge amounts of human waste, as well as that from farm animals such as cattle, pigs, and chickens. In addition, it also is polluted with agricultural fertilizers that are washed out of fields by rain. As the Mississippi's water reaches the Gulf of Mexico, it fertilizes the bacteria there; the bacteria grow and use so much oxygen that no fish or algae are able to survive: our pollution causes a huge dead zone in the Gulf of Mexico (**FIGURE B5.2b**). Similar dead zones occur in other oceans where rivers drain fertilizer-enriched water from farms and cities.

FIGURE B5.2a. Grain will be stored in these two elevators for months or years, and it must be kept dry. The metal roof and walls keep rain and snow out, but just as importantly, dry air must be blown upward through the grains to remove the water produced by the respiration of the grains. Grains are never put into elevators like these unless they are already extremely dry, but even so, the very slow respiration of each kernel produces a tiny bit of water, and if it accumulates fungi can grow and spoil the grain.

FIGURE B5.2b. The dead zone in the Gulf of Mexico.

But going beyond the equations, photosynthesis and respiration are complimentary processes: photosynthesis brings energy, carbon, and electrons into the biological world from materials in the environment, and respiration distributes that energy, carbon, and electrons to other parts of the cells beyond the chloroplast's outer membrane.

There are several types of respiration, and some are complex. Numerous interconversions of compounds and many intermediates are involved, and some of these can be used as the starting material for other metabolic pathways. Portions of nucleotides, lignin, and many pigments are made in pathways that use intermediates of respiration as the initial material. Only a fraction of the glucose molecules available to plants and animals are respired all the way back to carbon dioxide and water; a large portion of them enter the respiratory pathway but when they have been converted to a valuable intermediate, they are shunted into one of numerous synthetic pathways.

Many discussions of respiration focus on energy, but another aspect is equally important: **reducing power**, the power to donate electrons to another compound. All the $NADPH_2$ produced by photosynthesis is used in the chloroplast itself. The rest of the protoplasm must obtain its reducing power, in the form of $NADH_2$, from respiration. We animals do not dedicate much of our metabolism to reducing things because most of the foods we eat are already in the proper form as amino acids, fatty acids, and so on: almost all have enough electrons to form the bonds they need. But plants often take in materials that are highly oxidized and which must be reduced—they must have electrons added—before they can become parts of organic compounds. The reduction of carbon dioxide to carbohydrate is an obvious example, but there are many others. Plants often absorb nitrates and nitrites from the soil, and these must be reduced before they can be used as the amino group in an amino acid or before they can become part of any other organic compound. Sulfur enters the plant as sulfate, and it too must be reduced before the sulfur can be used in an organic molecule. The synthesis of fatty acids requires that plants put even more electrons onto the carbons in carbohydrates, reducing them even further. Much of plant metabolism involves reductions that occur outside the chloroplasts, and respiration provides the reducing power, the mobile electron carriers, needed for that.

Three Steps of Aerobic Respiration

There are several types of respiration. We will discuss **aerobic respiration** first because it is the most common type carried out by plants and animals. It is usually considered to consist of three distinct steps. The first step is **glycolysis**, also known as the Embden-Meyerhoff pathway (FIGURE 5.12). Glycolysis starts with glucose, which may have been made in the cell by photosynthesis, or it may have been carried in from the phloem, or it may come from an amyloplast. Glycolysis has about nine steps, two that generate ATP and one that generates $NADH_2$. It also produces two molecules of pyruvate for every molecule of glucose that enters the pathway. ADP is converted to ATP by the transfer of a high-energy phosphate group from one compound onto ADP; there is no build up of a proton gradient as in photosynthesis. This is called **substrate-level phosphorylation**.

If oxygen is present, as it usually is, then the pyruvate is used in the second part of aerobic respiration, the **citric acid cycle** (also called the tricarboxylic acid cycle) (FIGURE 5.13). The pyruvate is attached to acetyl CoA, which transports it into a

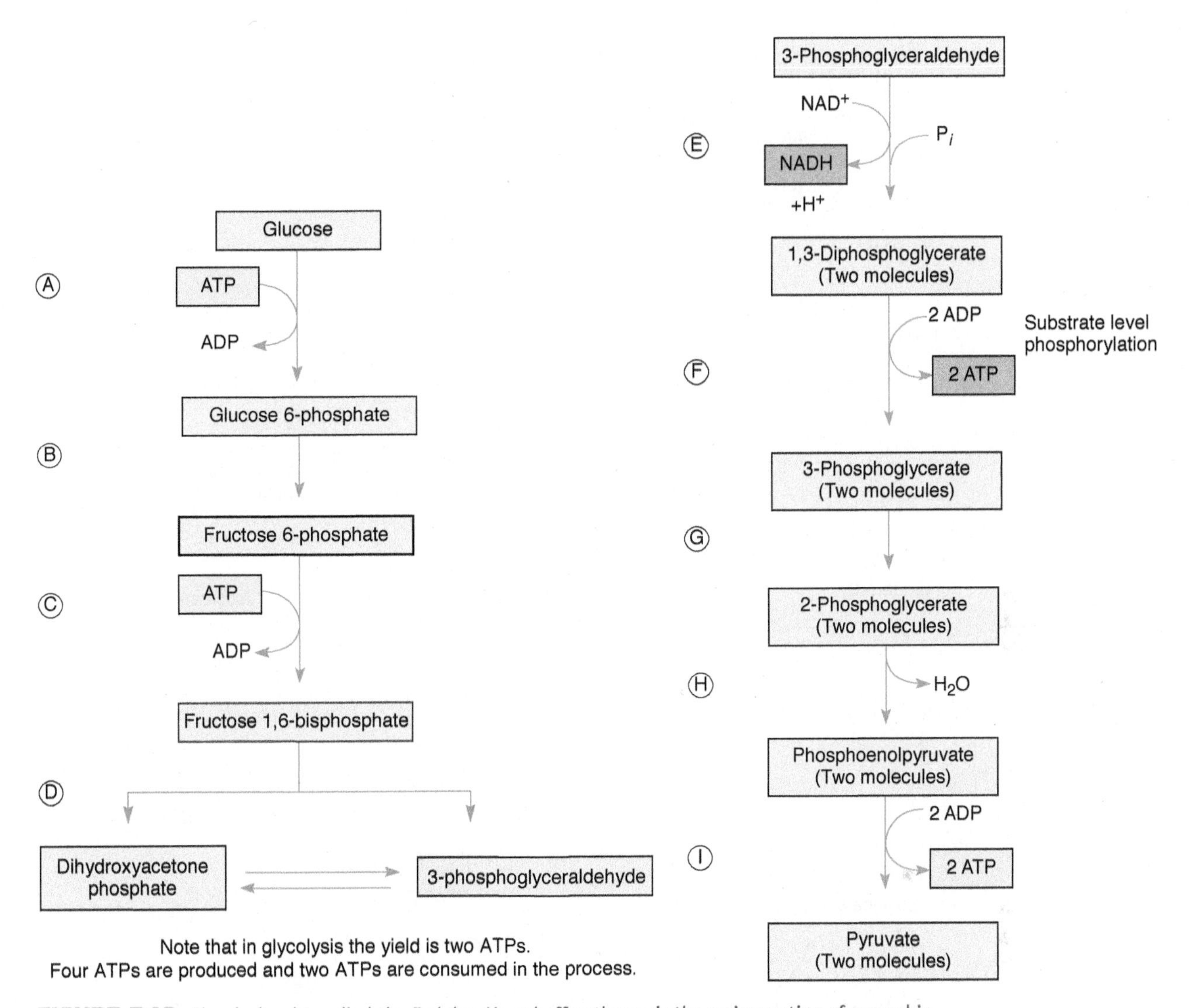

FIGURE 5.12. Glycolysis, also called the Embden-Meyerhoff pathway, is the major portion of anaerobic respiration and is also the first part of aerobic respiration. ATP is generated in steps F and I, and $NADH_2$ is made in step E.

mitochondrion. Once inside, the pyruvate (which has two carbons) is attached to oxaloacetate (four carbons) to make citrate (six carbons). At this point, we have another cycle, one in which a small molecule (pyruvate) is attached to a receptor (oxaloacetate), and then after a series of steps, the small molecule has been processed (in this case, converted entirely to carbon dioxide) and the acceptor molecule has been regenerated. A mitochondrion can convert millions of molecules of pyruvate to carbon dioxide by using a small number of oxaloacetate molecules repeatedly.

The citrate undergoes numerous reactions, a step-wise alteration of citrate until two molecules of carbon dioxide have come off, leaving a four-carbon molecule that is rearranged back into the acceptor oxaloacetate again. At one step, G, ATP is produced, and at several steps $NADH_2$ is produced.

The greatest amount of ATP production occurs in the third step of aerobic respiration, the mitochondrial electron transport chain. The $NADH_2$ diffuses to the surface of the mitochondrion's inner membranes and reduces the first electron carrier of this transport chain. $NADH_2$ does not remain bound to the carrier; it merely passes

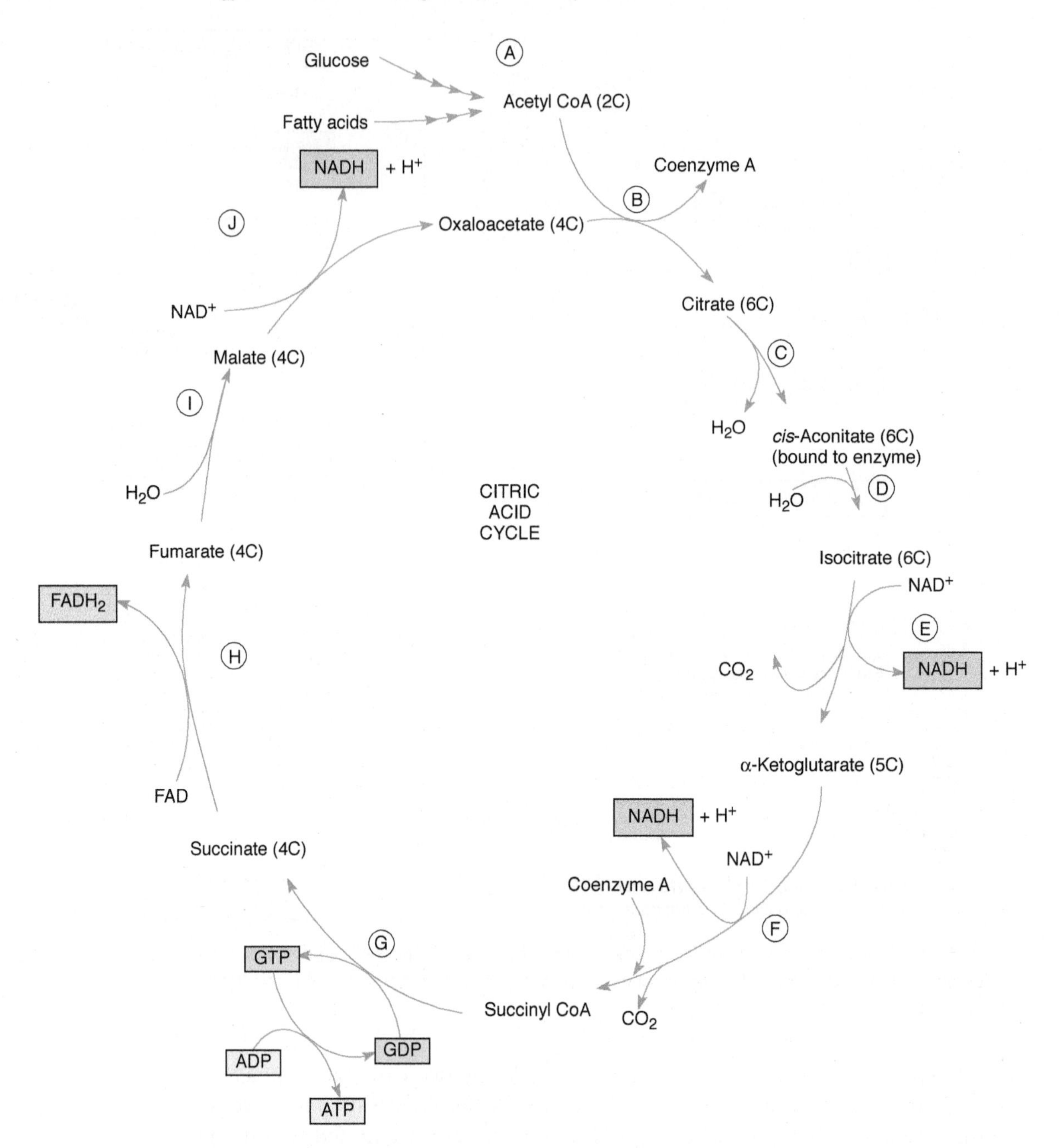

FIGURE 5.13. The citric acid cycle is an important part of aerobic respiration even though none of its steps consumes oxygen, and only one molecule of ATP is generated directly.

two electrons to it and then moves away from the membrane. Electrons are passed sequentially to the other membrane-bound carriers until finally they are deposited onto oxygen, causing it to bond with protons, resulting in the formation of a new water molecule. Oxygen is called the final electron acceptor.

Just as in the electron transport chains of chloroplasts, the mitochondrial electron carriers are located precisely in the membrane, and they cause an increased concentration of protons, a chemiosmotic gradient, in the space enclosed by the membrane (FIGURE 5.14). Just as in chloroplasts, protons flow outward through ATP synthase enzymes, and ADP is phosphorylated to ATP. ATP can then diffuse to various parts of the cell and supply energy to reactions that must be forced to go.

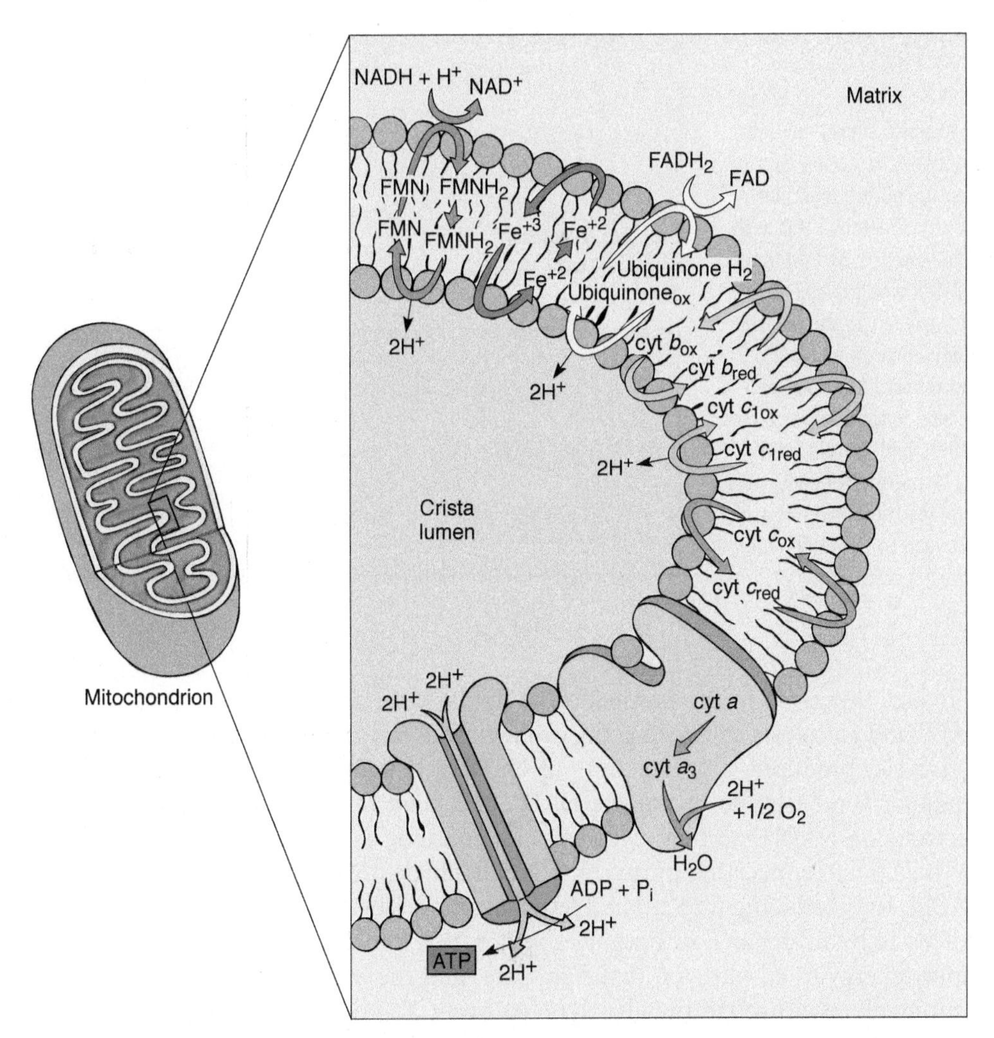

FIGURE 5.14. Only the important components of the mitochondrial electron transport chain are shown in this diagram—the steps that transport protons from the matrix to the crista lumen, establishing a proton/hydroxyl chemiosmotic gradient, just as in chloroplasts. The formation of water contributes to the gradient because protons are absorbed from the matrix but not from the cristae. FMN is flavin mononucleotide, an electron carrier similar in structure to FAD.

Respiration Without Oxygen

At certain times or places, there is not enough oxygen to support aerobic respiration. For people, a complete lack of oxygen causes us to suffocate and even die, and the same is true of plants. Floods often kill both people and plants; when plants are covered with floodwater, their roots die from not having enough oxygen to generate ATP; they drown (technically, they die of **anoxia**). But sometimes, there is some oxygen, just not quite enough. We experience this when we exercise so rapidly (fast, long running or bicycling) that our lungs and blood stream cannot supply our muscles with enough oxygen. Rather than simply stopping all activity, our muscles switch to **anaerobic respiration**, that is, respiration without oxygen. Plants do not exercise of course, but in marshy habitats with a small amount of standing water, the mud may be low in oxygen and roots of some plants can respire anaerobically.

Anaerobic respiration begins with glycolysis, just as in aerobic respiration. A problem arises because this generates $NADH_2$: this must be used up somehow so that more NAD^+ is available to keep the pathway running. Sometimes, plants need this $NADH_2$

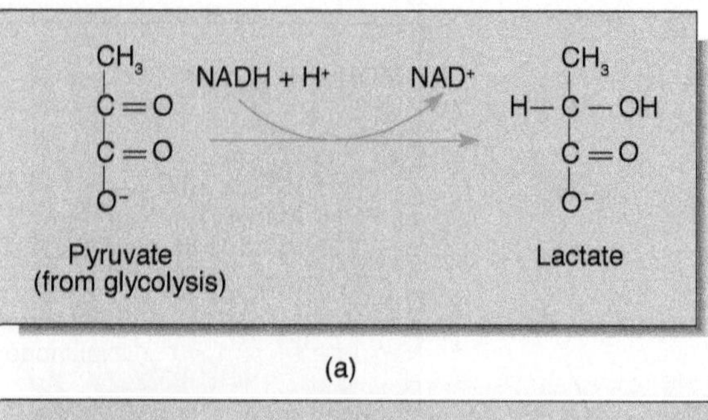

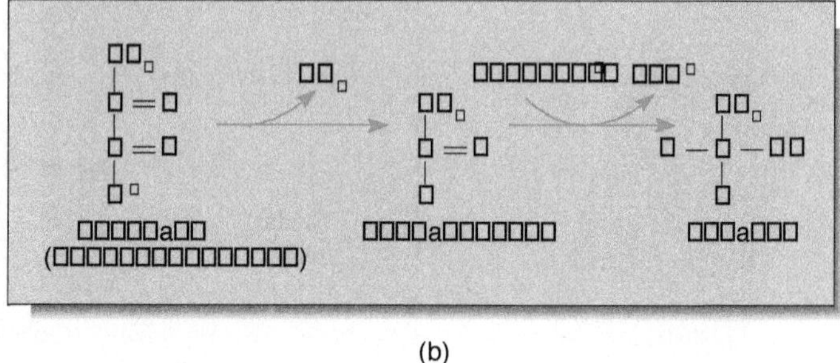

FIGURE 5.15. **(a)** When we exercise slowly enough for blood circulation to keep up, all our muscular activity is aerobic. But with rapid, intense, and prolonged activity, blood does not carry oxygen to the muscles rapidly enough and lactic acid fermentation begins. Lactate accumulation causes cramps and muscle pain. **(b)** Alcoholic fermentation involves the conversion of pyruvate to acetaldehyde before reduction by $NADH + H^+$.

to reduce things, as in the production of lipids. If so, then those reactions regenerate NAD^+ and glycolysis can keep going even without oxygen. But the next step, the citric acid cycle, produces large amounts of $NADH_2$, usually much more than any plant needs for synthesis. To regenerate NAD^+, the plant must somehow throw away the electrons on $NADH_2$ so that glycolysis can continue to run and produce ATP. Plants do this by first converting pyruvate to acetaldehyde, then transferring electrons from $NADH_2$ to it, reducing it to ethanol (ethyl alcohol) (FIGURE 5.15).

This is by no means an optimal solution to a problem. The plant needs ATP to supply energy to all parts of its metabolism, and the only way to produce ATP is by running glucose through the glycolytic pathway. To keep that running, the electrons on $NADH_2$ must be thrown away, along with the pyruvate. If oxygen were present, then the electrons on $NADH_2$ could be passed to the mitochondrial electron chain, where they would power the formation of more ATP. And, if oxygen were present, the pyruvate could enter the citric acid cycle and ultimately produce much more ATP. But instead, without oxygen, these valuable resources must be thrown away, and the result is ethyl alcohol, a chemical that can damage the plant if too much is produced. Although anaerobic respiration is not an optimal solution, in the emergency of anoxia, it does allow some plants to survive for several days or weeks. Yeast (a type of fungus) also respires glucose to ethanol if no oxygen is present; that is the basis of making wine or beer or any other alcoholic beverage.

When our muscles undergo anaerobic respiration, the muscle cells dump electrons from $NADH_2$ onto pyruvate, converting it to lactate (usually called lactic acid), which causes a burning, sore sensation. Just as in plants, our muscles are only obtaining a fraction of the energy available in glucose, but if no oxygen is available, there is no alternative.

Anaerobic respiration is also called **fermentation**, especially when done by yeasts and other microbes. Not all organisms can carry out anaerobic respiration when there is no oxygen present; they simply die. These organisms absolutely must have oxygen and are called strict aerobes. Although our muscles can undergo a bit of anaerobic exercise, our whole bodies cannot: just a few minutes without oxygen is enough to kill us. The same is true of most plants. Conversely, in many types of bacteria and other microbes, certain aspects of their metabolism are poisoned by oxygen, so they can live only in oxygen-free environments, and all their respiration must be anaerobic. These are strict anaerobes. Finally, some yeast can respire either way, depending on whether oxygen is present. If it is, they carry out aerobic respiration and produce carbon dioxide; this happens with the yeast in bread dough, and it is the carbon dioxide bubbles that cause the dough to rise. But if there is no oxygen, then the yeasts respire anaerobically and produce ethanol; why oxygen must be kept away from beer and wine as they are fermenting. Yeasts are facultatively aerobic or anaerobic.

Respiration Can Produce Many Types of Compounds

The citric acid cycle produces many intermediates, all of which can be used in the synthesis of various types of compounds. For example, one intermediate, alpha-ketoglutarate, can be converted into the amino acid glutamate, which is needed in proteins. Another, oxalic acid, is often combined with calcium; this is very insoluble so a crystal is formed, thus reducing the concentration of dissolved calcium in the protoplasm. A glucose molecule might start down the glycolytic pathway and into the citric acid cycle and then rather than being respired all the way to just carbon dioxide, bits and pieces of it are used to make other molecules.

Another type of respiration, called the pentose phosphate pathway, is also like this. Glucose molecules enter, and some are respired all the way to carbon dioxide whereas other glucose molecules might end up being converted into nucleic acids, lignin, and anthocyanin pigments.

Finally, a very small number of plants carry out a respiration that mostly just generates heat. The mitochondrial electron transport chain does not build up a proton gradient, so no ATP is generated and instead, all the energy present in the glucose is just converted to heat. This is used by skunk cabbage in springtime, letting it actually melt snow such that it can flower while nearby plants are still dormant from the cold.

Important Terms

- absorption spectrum
- accessory pigment
- action spectrum
- aerobic respiration
- anaerobic respiration
- anoxia
- anoxygenic photosynthesis
- ATP
- autotroph
- C_3 cycle
- C_4 photosynthesis
- Calvin/Benson cycle
- carbon fixation
- carotenoid
- chemiosmotic phosphorylation
- citric acid cycle
- Crassulacean acid metabolism (CAM)
- cyanobacterium
- electron carrier
- electron transport chain
- fermentation
- glucose
- glycolysis
- heterotroph
- high-energy bond
- light-dependent reactions
- NAD^+
- $NADH_2$
- $NADP^+$
- $NADPH_2$
- oxidize
- oxygenic photosynthesis
- pigment
- quantum
- reduce
- reducing power
- respiration
- ribulose-1,5-bisphosphate (RuBP)
- RuBP carboxylase (RUBISCO)
- substrate-level phosphorylation

Concepts

- All living organisms must obtain energy; there is no alternative.
- Plants are autotrophs: they obtain energy from light and carbon from carbon dioxide. Animals and fungi are heterotrophs: they obtain energy and carbon from food.
- Almost every metabolic pathway in every organism relies on ATP.
- Chlorophyll *a* differs from other pigments: the energy it absorbs can be used to synthesize new compounds.
- Photosynthesis reduces carbon dioxide and converts it into an organic molecule (glucose) inside a plant's body; this is carbon fixation.
- Respiration produces ATP for all parts—other than chloroplasts—of all cells.
- All free oxygen (O_2) in the world is produced by oxygenic photosynthesis carried out by cyanobacteria, algae, and plants.

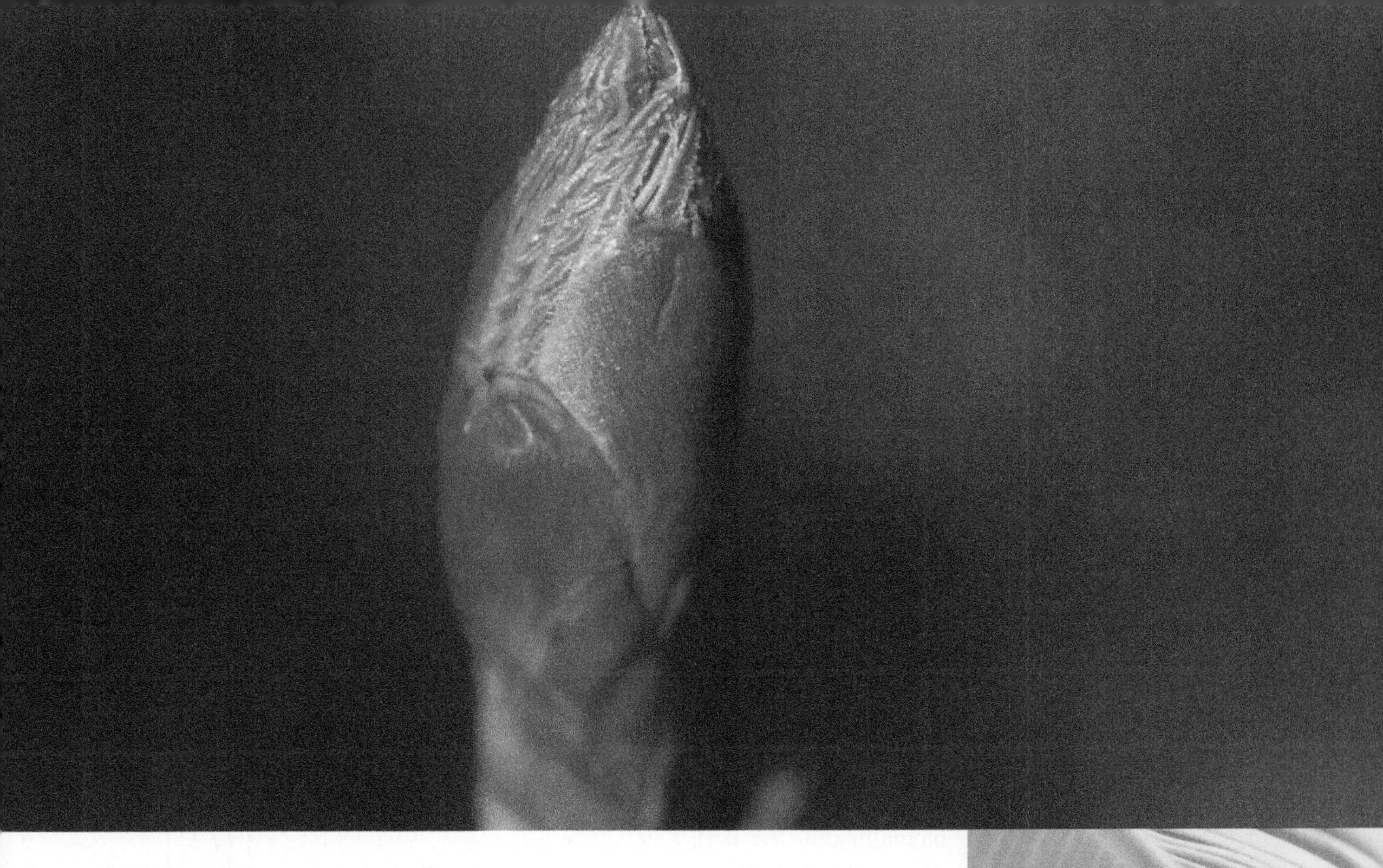

Genes, Environment, and Plant Development

This bud is opening in response to environmental signals that tell the plant it is springtime. As the bud develops into a twig and then a branch, it will communicate with various parts of the plant's own body such that the plant grows and develops correctly. All plants, just like all animals, sense and respond to important factors in both the environment and their own body.

Plants don't have brains. They don't have a nervous system, eyes, ears, taste buds, or noses. They cannot look out and comprehend the world the way we do, seeing danger, hearing the voice of a friend, smelling dinner being cooked. But just as much as we animals, plants must be aware of the world around them if they are to grow and develop properly. Also just like animals, plants must be aware of what is occurring in different parts of their own bodies: roots, stems, leaves, and flowers must coordinate their activities if a plant is to live successfully. Perceiving information about the environment or various parts of the body is only the first step. That information must then be transferred to other parts of the body and cause adaptive reactions, reactions that aid the plant's survival.

Think about the life of a plant and what it must know about its environment and itself.

- Seeds must know when to germinate and when to remain dormant.
- Once germinated, the seedling shoot needs to detect which direction is "up" and its roots need to perceive "down."
- Each newly formed cell needs to know where it is in a plant and whether it should develop into epidermis, xylem, phloem, etc.
- A plant must perceive when it is old enough to reproduce sexually.

FIGURE 6.1. This area of west Texas contains millions of plants of *Baileya radiata*; as you can see, they all bloom at almost exactly the same time. As a result, pollen from any one flower may be carried to many other flowers, and any flower might receive pollen from many other plants. Reproduction should be very successful.

- Once a plant is old enough to flower, it needs to know the proper time of year for blooming (**FIGURE 6.1**).
- After a flower has formed, it needs to know when to open, for example at sunrise or in the evening.
- Some flowers are pollinated by insects that land on them and then crawl inside; many such flowers orient themselves to be horizontal, aligned with the flight pattern of their pollinators.
- Certain axillary buds need to know when to become active, others need to be kept dormant and available for later use.
- Plants should detect if they are being attacked by animals, fungi, viruses, and so on, and then produce protective chemicals.
- Roots and shoots must coordinate with each other.
- Abscission zones need to know when to remain dormant and when to be active so that healthy leaves and immature fruits are retained but old ones are allowed to fall off.
- Newly formed embryos need to inform the plant that they exist and need not only nutrients, but also the protection of a surrounding fruit or cone (**FIGURE 6.2**).

FIGURE 6.2. Six peas are developing in this pod (fruit), but a seventh (arrow) is not. Perhaps no pollen tube brought sperm cells to it; perhaps the genes from its two parents are incompatible. For whatever cause, because the embryo is not active, it is not signaling the pod to send sugars and resources to it. Consequently, the plant will not waste resources on a seed that does not contain a viable embryo.

Numerous other aspects of the environment and of the plant itself are important; the list above is just a beginning. As you think about plants you know, keep in mind the diversity of plants and environment. Aquatic plants growing underwater in a lake have needs that differ from those of alpine plants that spend much of their life covered in snow, or parasitic plants that live partially or entirely embedded in a host plant.

Many environmental factors cannot be perceived, or even if they are, plants cannot respond in an adaptive manner. As fire sweeps through grasslands or forests, animals perceive it and flee, but plants can neither detect the fire nor respond to it. Those that already have fire-resistant bark may survive, but between the time when a fire might be perceptible and when it actually arrives there is too little time for a plant to produce extra bark. Similarly, plants neither perceive nor respond adaptively to avalanches or volcanic eruptions and so on.

This chapter examines the ways in which plants develop and behave, and how they are affected by information derived from their environment as well as from within the plant itself.

Environmental Stimuli that Affect a Plant's Growth and Development

Gravity is a constant factor that is always present in the environment (**TABLE 6.1**). It reliably indicates the direction of "up" and "down," and most roots, rhizomes, bulbs, and other subterranean organs orient their growth according to gravity. Gravity also causes plants and organs to have weight, and the heavier something is, the more strengthening tissues it needs (**FIGURE 6.3**). Tendrils of climbing plants respond to a light *touch* that does not damage cells (**FIGURE 6.4**), and all plants respond to touch that is so rough it damages cells, such as when branches rub together and form extra bark. For the latter, the true stimulus is probably cell damage, not actually touch. *Temperature* is important for plants of temperate habitats (ones that regularly have freezing temperatures in winter, in comparison to tropical habitats that almost never experience a freeze). *Light* is important as a source of energy, but it also carries information about the time of year (short days in winter, long days in summer), the direction of "up" (the brightest part of the sky), and whether a habitat is sunny (bright light) or shady (dim light). During sexual reproduction in flowers, pollen tubes carrying sperm cells are guided to egg cells by *chemicals* given off by cells near the egg.

TABLE 6.1. Prefixes for Stimuli

Stimulus	Prefix
Chemical	Chemo-
Gravity	Gravi-
Light	Photo-
Temperature	Thermo-
Touch	Thigmo-

FIGURE 6.3. Kiwi fruits grow on vines cultivated on trellises. When these fruits were still flowers, they had narrow, delicate stalks composed mostly of parenchyma. But as each fruit matured, the stalk detected the increase in weight and new strengthening cells (both collenchyma and sclerenchyma) were produced such that the stalk became strong enough to support the fruit without letting it fall before it was ripe. This is gravimorphogenesis.

FIGURE 6.4. The tendrils of queen's wreath (*Antigonon leptopus*) are slender and sensitive to touch. If they contact something, they will grow toward it, then around it (positive thigmotropism). Vining plants have numerous tendrils supporting the shoot at many places; consequently, the main stem itself does not have to be particularly strong.

Perception of Environmental Stimuli

Plants have no organs equivalent to our complex eyes and ears, or even our relatively simple taste buds. Almost all environmental stimuli are detected by cells that have such an ordinary appearance it is difficult to be certain which actually are perceptive. Light is detected by pigments; one called **phototropin** perceives the direction of light (it is sensitive to blue light) and is involved in directing organs to grow toward or away from light. A pigment called **phytochrome** detects light for measuring day length. Epidermis cells detect gentle touch, and it is believed that as the cell wall is disturbed by touch, it transfers the disturbance to microfilaments of the cell cytoskeleton. If so, we do not know how the cytoskeleton triggers further response. Biennial plants need to be exposed to cold temperature (to be **vernalized**) to know that they are passing from their first (vegetative) year to their second (reproductive) year. The temperature is detected by the shoot apical meristem itself, and the DNA of its nuclei is affected. Root caps have cells (**statocytes**) with starch grains (**statoliths**) that are so dense they sink through cytoplasm and come to rest on whichever side is lowest, indicating that that side is "down" (FIGURE 6.5).

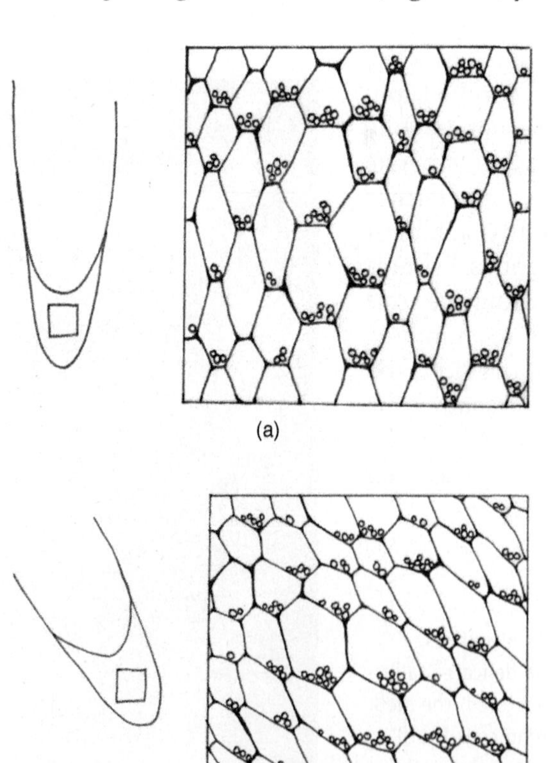

FIGURE 6.5. **(a)** Cells (statocytes) in the center of root caps contain starch grains that are especially dense: they quickly sink to the bottom of the statocyte. In taproots and other roots that grow downward, the starch grains are on the bottom of the statocyte, which is the wall farthest from the root apical meristem. The root detects this and continues to grow in that direction. **(b)** If the soil would slip such that the root tip were tilted, the starch grains would sink to the bottom of the cell, which is now the side wall; again the root would detect that and change its direction of growth until it points downward again and the starch grains are on the end wall.

Transfer of Information

In some cases, the site of perception is also the site of response. Shoot apical meristems of biennial plants perceive cold and then respond to it by producing reproductive shoots instead of continuing to make vegetative shoots. But in other cases, the responsive region is not the perceptive one, and information must be transferred. For example, after one side of a tendril touches an object, the other side of the tendril grows faster and causes the tendril to curl (see Chapter 2, Box 2.1, Figure B2.1e).

It is likely that dozens of molecules carry information from one area of a plant to another, but at this point we know of only a few plant **hormones**. These are molecules produced in very small amounts in one part of a plant, then moved to another part where they trigger a response that is out of proportion to the low quantity of hormone. This last character precludes compounds like sucrose from being considered a hormone: sucrose is produced and transported, but the response it triggers (respiration or synthesis) is in direct proportion to the amount present.

All plant hormones are involved in numerous responses; none triggers just a single, specific response. They will be introduced here then explained more fully later.

Auxin (indole acetic acid; IAA) is involved in cell elongation, apical dominance, bending growth, and suppression of leaf abscission. When applied to a cutting, it stimulates the formation of adventitious roots. Although IAA is the only known natural auxin, many artificial substances, especially 2,4-D (2,4-dichlorophenoxyacetic acid) act like auxins when applied to plants in experiments. Because plants themselves do not make these compounds, they are called **plant growth substances** rather than hormones.

There are at least two natural **cytokinins**, and like auxin, each is involved in many developmental processes, including activating dormant buds, stimulating cells to divide, promoting development of embryos and fruits, and preventing leaves from senescing. When roots are healthy and growing, they produce cytokinins that are transported up to the shoot, where they stimulate some of the dormant buds to become active. If roots are dormant, they produce little or no cytokinin and thus the shoot remains dormant also.

Abscisic acid (ABA) is mostly involved in stress responses. If a plant is heated, chilled or watered with salty solutions, its levels of abscisic acid increase rapidly and initiate many responses. It also exerts powerful control over guard cells; if plants have adequate water, then the opening and closing of stomatal pores is regulated by light, but if plants are water-stressed and need to conserve water, abscisic acid levels rise and cause stomata to close, overriding the ordinary mechanisms.

Plants contain well over one hundred kinds of **gibberellins (GA)**. Several are known to have specific hormonal activity, but others may just be metabolic intermediates as one is converted to another. Gibberellins are involved in stem elongation, seed germination, and several aspects of flowering.

Ethylene is the only hormone that is a gas. It escapes from plants through stomata and lenticels, but if certain plants are flooded, ethylene is trapped and stimulates production of aerenchyma, which increases the diffusion of oxygen. Ethylene often acts following auxin; that is, one of auxin's effects is the production of ethylene, which diffuses rapidly through a tissue or organ.

When attacked by animals and fungi, one of the first plant responses is production of **jasmonic acid**. This spreads to nearby cells and induces them to produce toxic, defensive compounds such as alkaloids. When viruses are the disease agent, plants typically respond with **salicylic acid** (the basis of aspirin), which then activates disease resistance. Recently, it has been discovered that the protein coded by a gene called *FLOWERING LOCUS T (FT)* is transported from leaves (the site of perception of day length) to shoot apical meristems (site of response) when plants are induced to flower.

The Ways Plants Respond to Environmental Information

Tropic responses involve growth that is oriented with regard to the stimulus: the organ grows toward (a positive response), away from (negative), or at an angle to the stimulus. For example, roots that grow downward are showing a positive gravitropic response, seedling shoots that grow upward (while still underground and not influenced by light) are responding with negative gravitropism, and rhizomes that grow horizontally are displaying diagravitropism. Once a seedling emerges above ground, it continues to grow upward by positive **phototropism**. To respond, an organ must change the direction in which it is growing, and this requires **differential growth**: one side of the organ grows faster than the other, and the organ curves toward the side that is growing more slowly. Because tropic responses involve growth, the plant is permanently changed.

Nastic responses involve temporary expansion or shrinkage of cells, so these responses are reversible. Also, nastic responses are not oriented with regard to the stimulus; instead they are the same in all circumstances. The closing of leaflets of sensitive plant (*Mimosa pudica*) when touched is a nastic response: the leaflets always close whether touched with an upward, downward, or sideways motion, and after a few hours, the leaflets return to their normal orientation (FIGURE 6.6).

(a)

(b)

FIGURE 6.6. **(a)** The circled region indicates one pinnately compound leaf of sensitive plant (*Mimosa pudica*) with its leaflets extended. This is the normal position of the leaflets during daylight while they are photosynthesizing. **(b)** This is the same leaf, just seconds after it was touched: its leaflets have moved quickly and folded together. The two leaf segments at the top of the circled region have stayed open; a more vigorous touch would have caused them to fold also.

(a)

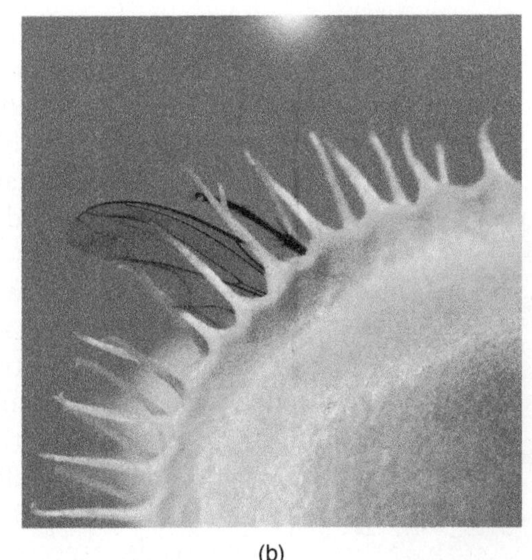

(b)

FIGURE 6.7. **(a)** This is an open leaf of a Venus flytrap (*Dionaea muscipula*). **(b)** A leaf that has closed rapidly enough to catch an insect. After a trap closes, it secretes enzymes that digest the insects proteins, then the leaf absorbs the amino acids. (a, © Jordan Tan/ShutterStock, Inc.; b, © Marie Cloke/ShutterStock, Inc.)

A **morphogenic response** is a change in the development or quality of a plant. For example, cold temperatures induce a morphogenic response when they change a biennial plant from being vegetative to floral.

We must also consider the strength of the response relative to the strength of the stimulus. In an **all-or-none response**, there is no response at all until a stimulus has acted for a certain time or reached a critical strength, then the plant gives a complete response. An example is the closing of the trap leaves of Venus flytrap (a thigmonastic response). Each leaf has six hairs, and the trap will not close until one or two have been touched strongly. Then the trap closes as rapidly as possible (FIGURE 6.7). A light touch will not cause a slow closing: that would not catch any insects. The alternative is a **dosage-dependent response**, in which a weak or brief stimulus will cause some response, but a stronger or more prolonged stimulus will give a more pronounced response. Many plants whose flowering is controlled by the length of the day show a dosage-dependent photomorphogenic response. If plants are given only one or two days that are the proper length, some will produce a few flowers, but if the plants are given many days of the proper length, all plants produce many flowers.

Examples of Plant Development

Seed Germination

Species differ greatly in the mechanisms that control the germination of their seeds. In plants that are probably most familiar to us—seeds of vegetables and flowers we plant in our gardens—seeds need only water: even if we just lay the seeds on a moist paper towel, they germinate. And many seeds in nature are similar; once they have been shed from the parent plant, they will germinate as soon as they have adequate moisture and a temperature that is not too cold. But most seeds have a dormancy mechanism that keeps them from germinating until precise conditions are met. For many desert plants, it is best if seeds germinate only after a heavy rain, when the soil will be moist enough to allow plant growth. Many of these seeds have a water-soluble germination inhibitor, and a heavy rain will dissolve that out of the seed and wash it away. A light rain will not dissolve enough to permit germination. In many species, abscisic acid is the inhibitor that must be washed out. Other seeds have tough, water-resistant seed coats that do not allow water into the seed, even when the soil is moist. After the seeds have been in the soil for several years, fungi or bacteria will finally degrade the seed coat, making it permeable and allowing germination. This is extremely haphazard, and some seeds germinate after a few years, others only after many years, but this is advantageous. It allows a plant to have some of its seeds germinate at different times, not all at once. In many species, the tough seed coat or the inhibitory chemicals are removed by an animal's digestive juices: the seeds must be eaten (but not digested) and when they are defecated by the animal, they are ready to germinate (FIGURE 6.8).

FIGURE 6.8. These are seeds of madrone (*Arbutus xalapensis*) that have been eaten and then defecated by a wild javalina in Guadalupe Mountains National Park. The javalina's digestive system removed the fruit and broke down a germination inhibitor, so now the seeds can germinate when environmental conditions are right.

Some seeds, especially small ones, need light to germinate; if they are buried too deeply in soil, they might not have enough stored carbohydrate for their seedling shoot to reach the soil surface and sunlight. But once the soil is disturbed and the seeds brought close enough to the surface for a pigment called **cryptochrome** to be activated by blue light, the seeds germinate. This occurs frequently when plowing a field or tilling a garden; turning the dirt over brings formerly buried seeds to the surface where they germinate rapidly.

Etiolation: Seedling Growth in the Dark

Shoots growing in darkness develop a special form (an **etiolated** form) that differs from the one they produce in the light. They have little or no chlorophyll, leaves remain tiny and undeveloped, internodes are long and slender (FIGURE 6.9). These are adaptive in that chlorophyll and foliage leaves cannot function in darkness, and long slender internodes will move the shoot apex to a new place—maybe somewhere sunny—faster than short wide internodes would. Basically, many resources are directed to growing as far as possible toward light. If none is reached, the plant starves to death when its stored starch is used up. But if the shoot apex does reach light, this is detected by phytochrome, which then induces de-etiolation: the plant produces chlorophyll, foliage leaves develop fully, and new internodes are shorter, stouter, and stronger. Any shoot will become etiolated if placed in dark conditions; for example if covered by a land slide. The control of plant development by light is an example of photomorphogenesis.

Seedlings that germinate underground almost always begin growth in an etiolated condition, showing the symptoms mentioned above. In addition, their cotyledons remain close together, protecting the shoot apex, and the shoot tip itself is bent

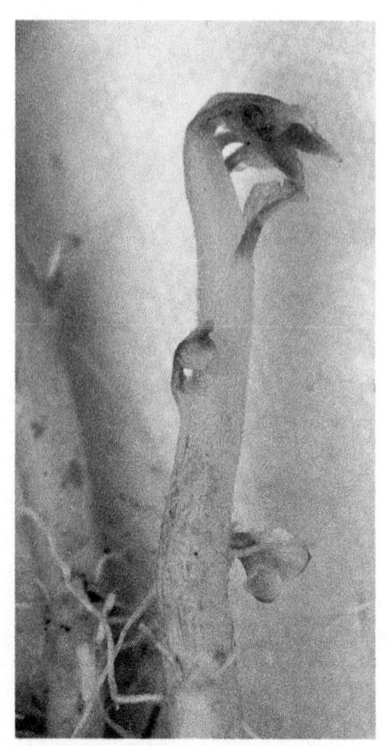

FIGURE 6.9. This shoot has grown from an axillary bud (an eye) of a potato planted about 4 inches (about 10 cm) underground. The shoot and leaves are white with no chlorophyll except the topmost leaves that had gotten close enough to the soil surface that they perceived dim light. All subterranean leaves have remained small and underdeveloped, and the internodes are very long. The shoot tip is hooked, such that the new leaves are being dragged up through the soil rather than being pushed. Numerous slender, white adventitious roots are present. This is photomorphogenesis.

FIGURE 6.10. These pea seedlings were germinated in darkness, so even after they emerged above soil, they did not encounter any light that would trigger them to change from their subterranean growth to their above-ground growth. The stems are extremely long, leaves are unexpanded, and no chlorophyll has formed. The containers in Figures 6.10 and 6.11 are the same size, which gives you an idea of the extreme size difference between seedlings that develop in darkness versus those in sunlight.

FIGURE 6.11. These pea seedlings were planted at exactly the same time as those in Figure 6.10, but their pots were placed in sunlight. As soon as the seedlings grew to the soil surface, they detected sunlight and switched from the subterranean type of growth: their stems have not elongated as much as those in the dark (Figure 6.10), their leaves have started to expand, and chlorophyll is being synthesized.

downward (called an apical hook) so that as the seedling shoot extends upward, the cotyledons and shoot tip are dragged backward through the soil rather than being pushed forward through it (being dragged backward is safer for the apex; **FIGURES 6.10** and **6.11**). Once the seedling reaches the light, the cotyledons spread apart, the apical hook straightens out, lifting the shoot apex into its more typical position directly above the stem it is producing.

BOX 6.1. Gardening Is a Good Way to Learn About Plant Development

If you have gardened even a little, or if you have a potted plant on your window sill, you probably know a lot about plant development already.

Phototropism and window sills. Indoor plants placed on window sills invariably undergo positive phototropism, growing toward the bright window rather than the darker room. That sometimes seems puzzling, because a room can look bright to us, but they rarely are; they only look that way because when we come indoors our pupils immediately dilate and admit more light to our retinas. We are fooled into thinking the room is bright, but plants are not tricked. Their phototropin detects that the window is brighter than the room, it initiates the redistribution of auxin in the stem and differential growth proceeds. With

(continued)

some plants, the bending is noticeable within just a day or two, and we must constantly turn the plant if we don't want it to be lopsided.

Hormones and rooting. It is easy to make cuttings of most house and garden plants. Just cut off the youngest part of a branch, place it in water and before long adventitious roots appear at the cut end. While any branch is growing, its apex produces auxin, which is transported downward to the roots, keeping them active. When the shoot is cut to make a cutting, the auxin accumulates at the cut end, it can go no further. Once its concentration is high enough, it induces cells to begin dividing and form root primordia (**FIGURE B6.1a**). Some plants have enough auxin naturally, or their cells are so sensitive that the cuttings root easily. But other plants do not root well on their own, the cut end remains inactive or even rots; most of these will root if the freshly cut ends are treated with a commercial rooting compound, all of which contain a high concentration of auxin.

Pruning and pinching back. Pruning plants is a method for overcoming apical dominance. The natural shape of many trees and shrubs is too tall and spare for some garden spaces, so we prune them by cutting off the distal portion of certain branches. This removes the shoot apex and the auxin it produces, so some of the repressed axillary buds become active: cutting one branch back will result in the growth of two or more branches, and the tree or shrub remains compact, more highly branched, and with more leaves and flowers in a small space (see Figure 6.15). Many herbs and small perennials also have rather spindly growth, but if we cut off their tops, the remaining shoot bases branch profusely and then later the plant produces many more flowers than it would have otherwise. Because herbs have such soft stems, we can actually remove stem tips with our own fingernails; this process is called pinching back.

FIGURE B6.1a. When a shoot tip is cut off a plant, auxin accumulates near the cut and stimulates vascular cells to differentiate into adventitious roots.

Dormancy, bulbs and depth. Nurseries and garden centers usually give very specific information about the depth at which to plant particular bulbs such as lilies, daffodils, and alliums. Some are near the soil surface, others are several inches deep, some must be planted 18 inches or more. This often is related to the size of the bulb and the severity of the winters in their native areas, but it will also affect their ability to detect sunlight and use phototropism to guide the upward growth of shoots and flower buds when the bulbs become active. Bulbs that naturally occur deep in the soil mostly use negative gravitropism to guide their shoots upward; shallower bulbs can use positive phototropism.

Most bulbs have a very pronounced seasonality. At the proper season, probably determined by day length and phytochrome in most, the bulb becomes dormant and usually allows all its foliage to die. Most have no abscission zone; the leaves just wither but remain attached to the bulb. Often, the roots die as well. The plant is completely subterranean at this time; it is resistant to cold in many species, and resistant to heat and dryness in desert bulbs. Although appearing completely inactive, cold-climate bulbs usually have a special metabolism in the winter that prepares them to flower in spring. Many of us must buy bulbs of crocus and daffodil in autumn and store them in a refrigerator for several months before planting them, otherwise they will not bloom. The stimulus for becoming active again varies. For some that bloom in early spring, warming soil is the signal for activity; cold temperatures have removed whatever had caused them to become dormant earlier. Others are guided by day length; all bulbs of oxblood lilies (*Rhodophiala bifida*) in a certain area bloom within 2 weeks of each other at the end of August: all plants respond at almost precisely the same time. Rain lilies (*Zephyranthes*) bloom reliably after a summer rain (but not after a rain in spring or autumn).

Winter annuals and vernalization. Plants described as being biennials typically germinate in the spring of one year, then bloom and die the following year. Winter

FIGURE B6.1b. This is a typical growth form for a winter annual: the shoot is so short that all leaves are produced as a low rosette. This was photographed in November, and the plant had probably germinated in late summer; in late winter or early spring, it would send up a tall flowering shoot, make seeds, and then die. The cold temperatures of December and January are necessary to vernalize the plant and allow it to flower later.

annuals are similar except they germinate in late summer or early autumn of their first year and do most of their vegetative growth on warm days in late autumn and early spring (**FIGURE B6.1b**). During the middle of winter, they become vernalized and then bloom in spring. They act like biennials but their entire life span is only a few months long.

Vines and trellises. Vines need some method to climb up trunks and branches of surrounding plants. In gardens, we have to provide them with a trellis. Climbing roses have thorns that act as anchors or grappling hooks; the plants always produce these no matter what: they are not induced by the presence of a trellis or any other support, and they do not change once they have hooked something, so these are not like other structures mentioned in this chapter. Tendrils are a slight variation: plants that produce tendrils do so all the time, again needing no environmental stimulus. But when a tendril touches something, it undergoes positive thigmotropism: differential growth causes them to wrap around the support. Twining vines lack tendrils, instead the entire stem and leaves wrap (twine) around a support (**FIGURE B6.1c**). While young and growing without any contact with other objects, the shoot tip does not grow straight up. Instead it grows slightly to one side, then slightly to another and another such that if watched in video, the shoot apex is growing upward as a helix several inches in diameter. This kind of growth is **circumnutation**, and it results from differential growth that is continuously changing direction. Circumnutation increases the likelihood that as the shoot tip sweeps around, it will touch an object that can be used as a support, and when it does so, that triggers a positive thigmotropic response.

Photoperiodism and holiday plants. Christmas cacti (*Schlumbergera*, formerly known as *Zygocactus*) and poinsettias (*Euphorbia pulcherrima*) bloom at Christmas, and Easter lilies (*Lilium longiflorum*) bloom at Easter, but that is not mere coincidence. Flowering time is determined by day length in these species (Easter lilies require longer days than do the other two), and all are controlled by phytochrome and internal circadian rhythms, but their natural response does not quite correspond with our calendar. Instead, nurseries and garden centers carefully control the length of the nights starting several weeks before Christmas or Easter to ensure that plants bloom at optimum time for sales: some a bit early in the shopping season, most right at the holiday, and none at all the day after. If we grow Christmas cacti or poinsettias at home, our interior lights are bright enough to convert P_r to P_{fr} until we go to bed and turn the lights out, so the plants do not get the long nights they need if kept in a room where people live. We usually have to place them in a dark unused room, a closet or garage for several weeks to artificially give them the long nights they require.

FIGURE B6.1c. This morning glory vine is twining up through the wire fence by means of positive thigmotropism: the shoot grows toward anything it touches.

Phototropism

Seeds, bulbs, and other shallow subterranean organs grow upward by bending toward the brightest part of the soil (above them), not the darkest. Many leaves also orient their blades to face the brightest part of their habitat; this is especially important for vines growing against a dark cliff face, tree trunk, or wall (FIGURE 6.12). Some leaves and flowers are so accurate and responsive they actually turn all day long, facing the sun as it moves across the sky, then turning back toward the east at night. Even after a shoot is growing above ground, it must constantly correct its direction of growth because slight errors are caused by small differences in growth on one side or the other, or if the shoot is blown or bent away from vertical. Whenever the shoot detects that light is brighter on one side, it turns and grows in that direction.

In phototropism, light is detected by phototropin, which is responsive to blue light. This causes a redistribution of auxin so that the brighter side receives less and thus grows more slowly whereas the shaded side receives more and grows faster. This differential growth turns the shoot to point directly to the brightest part of the environment, after which auxin moves equally on both sides.

Phototropism was first studied in seedlings of oats. A tubular structure called a **coleoptile** is the first organ to emerge above soil when an oat seed germinates, and the coleoptile responds with positive phototropism. The site of perception is the coleoptile tip: if it is cut off or covered with a tiny cap, it cannot perceive unilateral light and the coleoptile does not bend (FIGURE 6.13). It was also discovered that auxin was involved by dissolving auxin in gelatin, then placing a small bit of gelatin on a decapitated coleoptile. If placed on one side, the auxin would diffuse out of the gelatin into the adjacent side of the coleoptile and make it grow faster, as if it were the shaded side of an intact coleoptile. If the gelatin and auxin were placed in the center, the coleoptile grew straight upward.

FIGURE 6.12. This is a wall covered with a vine of Boston ivy (*Parthenocissus tricuspidata*); every leaf has turned photonastically so that its blade faces the light. No leaf is turned to face the wall.

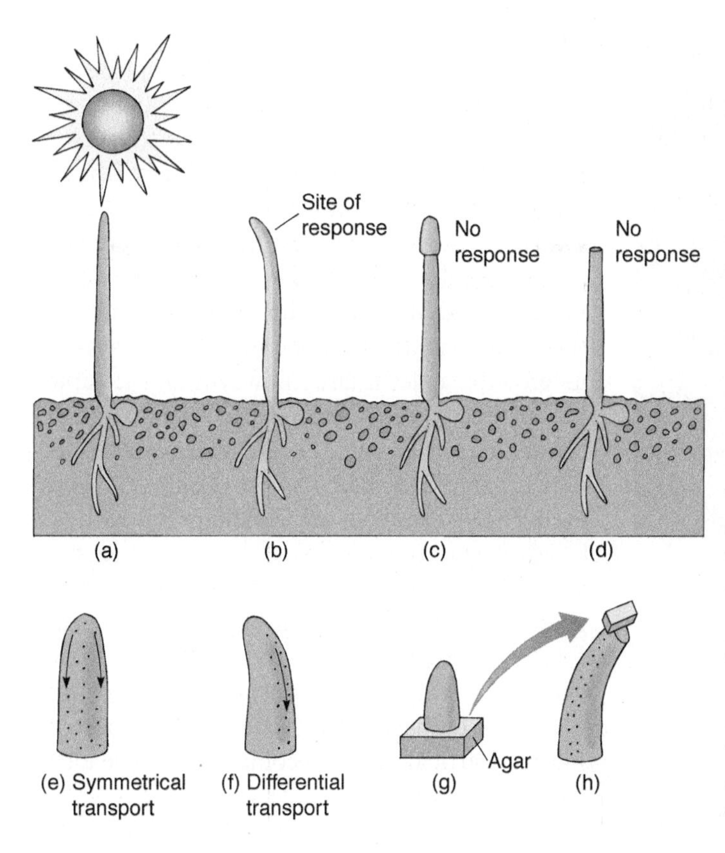

FIGURE 6.13. **(a)** When illuminated from directly above, oat seedlings grow upward. **(b)** When a young oat seedling is exposed to light from one side, its outermost sheathing leaf, the coleoptile, bends and grows toward the light. If the coleoptile apex is covered **(c)** or cut away **(d)**, no response occurs to unilateral illumination, so the tip is the site of perception. If the site of response is covered, bending occurs, so the site of response is not involved in perception at all. **(e)** In dark conditions or with overhead lighting, auxin is transported symmetrically down the coleoptile, causing equal amounts of growth everywhere. With unilateral illumination **(f)**, auxin is redistributed, with the darker side transporting more auxin than the lighted side, so the darker side grows faster, resulting in curvature. **(g)** and **(h)** Auxin can be collected in small blocks of agar or other absorptive material and then placed asymmetrically on a decapitated coleoptile; the side receiving auxin grows, but the other side does not.

Gravitropism

Many roots, rhizomes, and other subterranean organs grow so deeply in soil that there is no light at all. They orient themselves using gravitropism, being directed upward, downward or at an angle by detecting gravity. The dense starch grains (statoliths) settle to the downward side of gravity-detecting cells in root caps or in the nodes of rhizomes (see Figure 6.5). After the cells detect where the starch grains are, they cause a redistribution of auxin, just as in phototropism, and this causes differential growth. Once differential growth has caused the organ to grow in the proper direction (downward for some roots, horizontally for rhizomes and other roots), the starch grains settle to the proper side of the cell, auxin redistribution stops and the organ grows straight ahead.

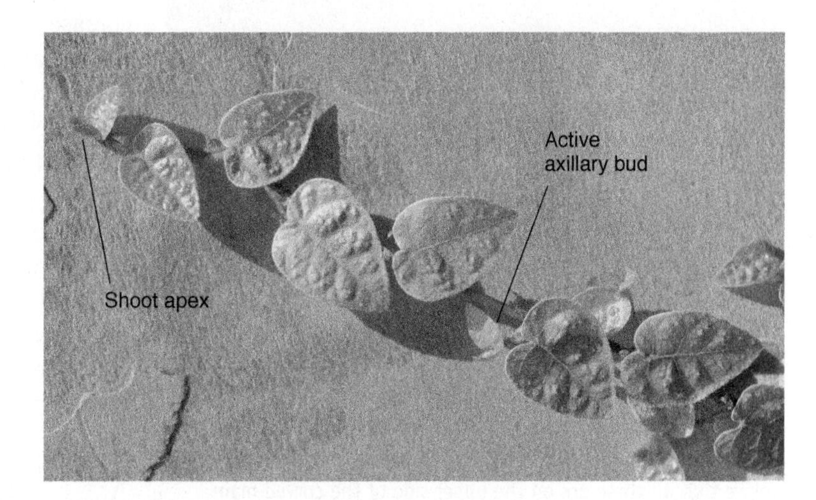

FIGURE 6.14. Apical dominance in fig ivy (*Ficus pumila*) is easy to see because the plant grows in only two planes, attached to a wall. The shoot apex on the left is dominating the axillary buds in all the newly formed leaves, but one axillary bud on the right is far enough away from the shoot apex that it has begun growing out as a branch.

Apical Dominance

Plants produce one or several buds in the axil of each leaf. Many are used to produce flowers or branches, but many remain dormant. As long as the shoot apex is healthy and growing well, it suppresses the activity of all the axillary buds it has just produced. All the uppermost buds remain quiescent. This is **apical dominance** (FIGURE 6.14).

FIGURE 6.15. This twig of viburnum was pruned 2 months before the photo was taken. The axillary buds had been completely inhibited by the shoot apical meristem at the time of pruning, but then both became active once pruning stopped the flow of auxin from the shoot apex.

The shoot apical meristem of the shoot apex produces auxin, which is transported down the stem and suppresses axillary buds. If the shoot apex is damaged, auxin production stops and some of the buds become active (FIGURE 6.15). One of the first things they do is produce auxin themselves, which keeps other nearby buds repressed. The damaged apex is usually replaced by only one or two of the uppermost axillary buds, not by large numbers of them. As long as the shoot apex is not damaged, its growth carries it farther away from the earliest buds it produced, and at some point, auxin from the distant apex is too dilute to maintain apical dominance; some buds begin growing out as branches even though the shoot apex is still healthy.

Many variations of apical dominance exist. In some plants dominance is so strong the shoot grows with no branches at all, as in many cacti (FIGURE 6.16). In other plants, apical dominance causes the branches to grow horizontally or at an angle instead of upward. If the shoot apex is damaged, then one of the uppermost branches will redirect its growth so that it is vertical and it becomes the new leader.

Leaf Senescence and Abscission

Most leaves live for only one growing season, typically being produced in spring, then dying and falling from the plant in autumn. The abscission zone is under the control of auxin produced by the leaf itself: as long as the leaf is healthy, it produces auxin that moves out of the blade and down the petiole where it represses the abscission zone. When the leaf is damaged or shaded or has become too old, its production of auxin decreases to such a low level that the abscission zone become active and releases the leaf (FIGURES 6.17 and 6.18). The same mechanism probably functions in flowers and fruits.

FIGURE 6.16. Each spine cluster on this cactus (*Acanthocereus tetragonus*) is an axillary bud. Most are inhibited by apical dominance, but three have begun to grow out as new branches. Notice that all three are on the upper side of the curved main shoot, whereas all buds on the lower side are still inhibited. Also notice that all three growing buds became active at the same time: this occurred when the shoot fell over. While the main shoot was growing straight up, its apical meristem dominated all buds, but once the shoot became top heavy and bent over, three buds at the top of the curve were released from apical dominance.

A related problem is that of keeping the leaf supplied with minerals and water as long as it is healthy. Very often all the leaves of a tree start to look old at about the same time, as if some signal has indicated that they should stop photosynthesis and instead begin breaking down proteins and salvaging the amino acids by exporting them back into the plant. This process can be delayed if the leaf is supplied artificially with cytokinin: the leaves sprayed with cytokinin remain green and active temporarily.

Photoperiodic Induction to Flower

Few plants bloom at random. If an individual plant of a species is going to have successful cross-pollination, it needs to bloom at the same time that other members of its species are blooming, and also at a time when its pollinators are present and active. Perennial plants often detect time of year so accurately that all members bloom within a month of each

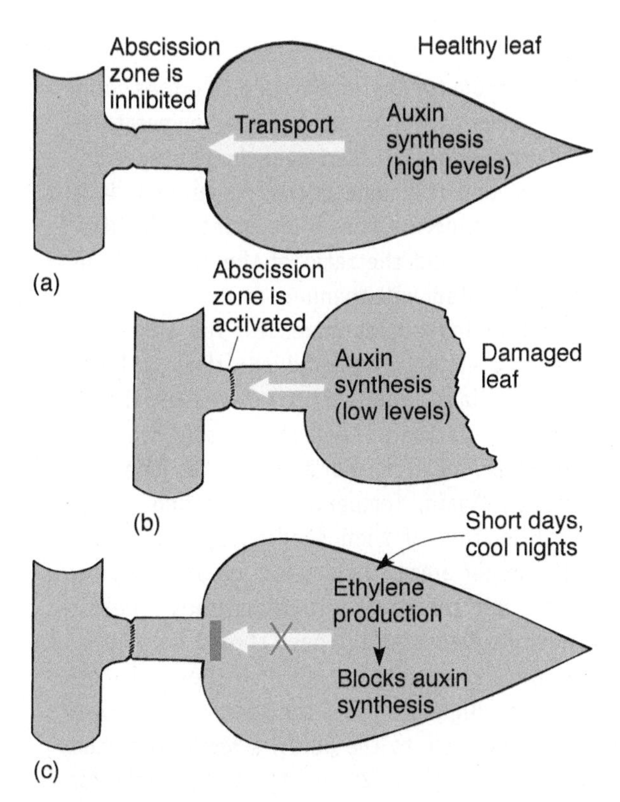

FIGURE 6.17. **(a)** A healthy leaf produces and transports enough auxin to suppress activity in the abscission zone. **(b)** A damaged leaf produces less auxin, insufficient to prevent abscission. **(c)** Autumn stimuli may cause the production of ethylene, which then suppresses auxin synthesis and transport. An alternative hypothesis postulates that autumn conditions directly suppress auxin production and ethylene is not necessary.

FIGURE 6.18. Sometimes development does not happen as it should. These leaves of big-tooth maple (*Acer grandidentatum*) should have fallen off the trees in autumn, but they are still attached in late December, with snow on the ground. Perhaps an unusually early cold night damaged the cells in the abscission zone while they were immature.

other. The conversion of an adult plant from the vegetative to the flowering condition may be the most complex of all morphogenic processes. And it appears that different mechanisms exist in different species.

Very few plants flower as seedlings; most annuals need to be at least several weeks or months old, biennials need to be in their second year, and some perennials do not flower until they are several or many years old. Plants need a mechanism for measuring their age. In certain annual species, the size of the plant appears to be the only important factor: peas and corn initiate flowers automatically after a particular number of leaves has been produced, regardless of environmental conditions. Flowering is controlled by internal mechanisms. In perhaps most species, transition to the flowering condition is triggered by **photoperiod** (or **day length**), which acts as a season indicator (**BOX 6.2**). One group of these plants, **short-day plants**, blooms when days are short (spring or fall). Another group, **long-day plants**, is induced to bloom when days are long, in summer (**TABLE 6.2**). Plants that do not respond to day length are **day-neutral plants**.

The presence or absence of light is detected by phytochrome. Phytochrome has a light-absorbing portion attached to a small protein. When phytochrome absorbs red light, the protein changes its folding, which affects many of its properties, one of which is its absorption spectrum: it now absorbs far-red light, not red light. And when this form absorbs far-red light, it refolds back to the red-absorbing form; the

BOX 6.2. Environmental Stimuli and Climate Change

Global climate change is causing our world to rapidly become warmer and wetter, but plant mechanisms for detecting and responding to environmental stimuli are changing more slowly, if at all. As we burn oil, coal, and natural gas, and as we convert forests into pasture for cattle, we increase the amount of greenhouse gasses in the atmosphere and this causes the air, soil, lakes, and oceans to become warmer (see Chapter 11 for details). As ocean temperatures rise, their surface waters evaporate faster, making the atmosphere more humid and increasing the amount of rain and snow that later fall on land. Temperatures do not increase uniformly everywhere; instead circulation patterns in the atmosphere and oceans are affected, so some areas become warmer, others cooler, some wetter, others drier.

Changing climate will have profound effects on all plants, not only on those that respond to temperatures, but also those that are controlled by day length. Increasing temperatures affect two critically important events for temperate plants: the date of the last frost in spring occurs earlier, and the time of the first frost in autumn comes later. The frost-free growing season in many areas starts earlier and ends later: plants have a longer growing season.

Plants that germinate or bud out solely based on temperatures can take advantage of this longer growing period, and many seem to be thriving. But for plants controlled by photoperiod, their critical night length does not change, they germinate or bud out at the same time in spring as they have for centuries, and go dormant at their typical time in autumn. They are not able to take advantage of the extra days of warmth in spring and autumn; instead, they are dormant when they could be photosynthesizing, growing, and reproducing. And just as bad, their respiration during dormancy is higher than before because that is controlled by environmental temperature: not only are the plants not producing sugars photosynthetically as long as they could, they are now respiring away their carbohydrates faster. They will have less reserve nutrients available when they resume growth in springtime.

Now consider the interaction of photoperiodic plants and the temperature-controlled ones. They occur together in the same habitat and compete with each other for water, minerals, room for their roots, and so on. As warm temperatures occur earlier in spring, the temperature-controlled plants get a head start over the photoperiod-controlled ones, and the same is true in autumn. It is likely that the photoperiod-controlled plants will suffer in this competition, and the ratio of the two types of plants in the ecosystem will change.

As temperature in general increases, the snow-free habitable zone in alpine areas gradually rises to higher elevations. Similarly, habitable zones are expanding northward in the Northern hemisphere, southward in the Southern. Areas near the North and South Poles are more hospitable. Again, temperature-controlled plants may benefit from this: if their seeds happen to occur in the newly warmer areas, they should be able to grow and reproduce. But the same is only partially true for photoperiod-controlled plants. These should be able to grow higher on any mountain on which they exist already: the critical night length is the same up and down the mountain. But close to the poles, a few days at the beginning of summer (June 20 or 21) have sunlight for 24 hours: there is no night at all for a few days. And at the beginning of winter (December 20 or 21), several days have no sunlight. From early winter to early summer, day length increases from 0 to 24 hours. At the equator, daylight always lasts for 12 hours and night is also 12 hours, all year long. Between these two extremes, days get slightly longer each day in lower latitudes, and much longer each day in high latitudes. For plants that need very long days to bloom (for example 17 hours, with 7 hours of night), that occurs in May or June in the northern part of the United States and southern Canada, but it occurs in March in northern Alaska and Canada. If a longer growing season would allow that species to grow that far north, its critical night length would occur too early (March), while the plant is still a seedling: it could grow in the new habitat, but not reproduce there.

It is important to remember that plants and their control mechanisms do evolve. Hundreds of different plant species differ in their critical night length, and this variation came about through evolution by natural selection. The important question is "Will these mechanisms evolve rapidly enough to allow plants to adapt to the changing climate?" We do not know the answer, but in general, such evolution is slow whereas we are causing the climate to change rapidly.

two forms are called $\mathbf{P_r}$ (red-absorbing) and $\mathbf{P_{fr}}$ (far red-absorbing). Also, P_{fr} reverts to P_r in darkness.

P_{fr} is the active form that brings about metabolic responses. After a plant is given red light, converting phytochrome to the active P_{fr} form, the phytochrome moves into the nucleus and affects the activity of many genes. It takes some time for gene control to occur, and if far-red light is given quickly enough after red light, phytochrome does not have enough time to affect cell metabolism significantly, and no effect is seen. But if the far-red comes long enough after the red, then critical changes have been initiated and far-red light can no longer cancel the red light stimulation.

TABLE 6.2. Examples of Plants that Respond to Photoperiod

Short-day plants	
	Christmas cactus
	Chrysanthemum
	Poinsettia
Long-day plants	
	Aster
	Black-eyed Susan (*Rudbeckia*)
	Carnation

A plant experiencing a short day in nature automatically receives a long night, and similarly, long days are always accompanied by short nights. Night length is actually the critical factor. A long-day plant is in reality a short-night plant. It can be placed in a growth chamber and artificially be given both long days and long nights—for example, 16 hours of light and 16 hours of dark in a 32-hour "day." If day length is the important factor, the plant should flower because it has long days; if night is the critical factor, then it should not flower because it does not have short nights. When the experiment is done, the long-day plant does not flower, indicating that night length is critical, not day length. Similar experiments have shown that short-day plants really are long-night plants: if given a 16-hour cycle (8 hours light/8 hours dark—both day and night are short), they do not flower. But if given long nights, even accompanied by artificially long days, the plants are induced to flower. In these experiments it is necessary for the light of the "days" to have some red light in it, which indicates that phytochrome is the pigment involved.

Each species has its own particular requirements for long or short nights; that is, not all "long nights" have to be the same length. Instead each species has a **critical night length**; if a short-night plant receives nights shorter than this critical length, it flowers, whereas a long-night plant must receive nights longer than its own critical night length to flower. The accuracy with which night lengths can be measured varies, but the most accurate species known is the long-day (short-night) plant henbane (*Hyoscyamus niger*); it must have nights shorter than 13 hours, 40 minutes. If the nights are even 20 minutes too long, 14 hours long for example, it does not flower.

The control of flowering and other processes such as the initiation or breaking of dormancy by day length/night length is more common at locations farther from the equator. Away from the equator, nights become progressively shorter from winter to summer, then progressively longer from summer to winter. The greater the distance from the equator, the greater the length of the longest winter night and the shorter the length of the shortest summer night. Thus, if two species are to bloom just after the beginning of May, a species in the southern United States or Mexico must have a critical night length shorter than that of a species in the northern United States or Canada (**TABLE 6.3**).

The sites of perception for night length are young leaves. It is possible to stimulate one leaf with a spotlight and induce the plant to flower, even if the apical meristem, the site of response, is not illuminated. Because the site of perception is not the site of response, a chemical messenger must be transmitted between the two. If a leaf is photoinduced and then immediately cut off the plant, no flowering occurs; if it is allowed to remain attached for several hours, the flowering stimulus is synthesized and transported out of the leaf. If the leaf is then removed, the plant still flowers. It has recently been discovered that the signal that moves from the leaf to the meristem is the protein encoded by the gene *FLOWERING LOCUS T*.

TABLE 6.3. Length of Day and Night on May 1 at a Northern and Southern Location

	Sunrise	Sunset	Day Length	Night Length
Vancouver (British Columbia)	5:51 AM	8:29 PM	14 hr 38 min	9 hr 22 min
Los Angeles (California)	6:04 AM	7:37 PM	13 hr 33 min	10 hr 22 min

* Photoperiodic plants that bloom in early May need different critical short nights depending on their latitude. If a southern species native to Los Angeles expands its range and grows farther north (perhaps due to global warming), it will bloom earlier in the springtime, perhaps so early that the plant is too young to be strong enough to flower: nights that are 10 hours and 22 minutes long occur on April 12 in Vancouver, BC.

Other plant responses are also controlled by photoperiodism. For example, in perennial species that must survive harsh winters, the short days (and long nights) of autumn are detected by phytochrome and cause shoot tips to stop producing leaves and instead convert themselves into tough, resistant dormant buds protected by bud scales.

We are continuing to discover that plant detection of light is extremely sophisticated. Numerous pigments are involved; there are several chemically distinct types of phytochrome as well as phototropin and several types of cryptochrome. Some pigments act independently of the others, but many seem to act together in a variety of light-detecting developmental programs in plants.

Endogenous Rhythms and Flowering

Plants contain **endogenous rhythms**; that is, certain aspects of their metabolism cycle repeatedly between two states, and the cycle is controlled by internal factors. The most obvious example of this is in the "sleep movements" of the leaves of plants like prayer plant (*Oxalis*; FIGURE 6.19). In the evening, leaflets drop down, and in the morning, they raise themselves to the horizontal position as motor cells increase their turgor. It is easy to assume that this is a photonastic response, but if the plants are placed in continuous darkness, the leaflet position continues to change, returning to the up position about every 24 hours. In many flowers, the production of nectar and fragrance is also controlled by an endogenous rhythm and occurs periodically even in uniform, extended dark conditions.

Endogenous rhythms are involved in numerous aspects of plant metabolism that are not easily observed, such as photosynthesis, respiration, and growth rate. The underlying mechanism that constitutes the clock is poorly understood but is known to be independent of temperature and general health of the plants. Endogenous rhythms are not affected by cooling or warming. Also, extensive studies have shown that the rhythm is truly endogenous and not controlled by an unsuspected exogenous rhythm related to Earth's rotation. Plants have been taken to the South Pole, where planetary rotation would have no effect, and have been taken into orbit. If the rhythms were actually exogenous, plants at the South Pole should lose their rhythmicity, whereas those in orbit should have a more rapid rhythm that matches the orbital period. In both cases, normal rhythm was maintained.

The underlying mechanism that constitutes the clock is poorly understood, but is known to involve a **negative feedback loop**. Imagine just two proteins, A and B, with

(a) (b)

FIGURE 6.19. **(a)** Each leaflet of the compound leaf of *Oxalis* is joined to the petiole by motor cells; here they were photographed in the morning when it was cool and the sunlight was not intense. Motor cells are turgid and leaflets are held into the sunlight. **(b)** *Oxalis* plants in full, intense sunlight. The motor cells have lost potassium and water, thus they are not turgid. Leaflets hang down, minimizing their exposure to light. Later in the afternoon, when sunlight is not so intense, or if the shadow of a tree moves across the plant, the motor cells will absorb potassium, then water, and raise the leaflets again. In addition to this response to excessively bright light, these leaflets respond to an endogenous rhythm: they fold down at sunset, then lift again at sunrise. If placed in continuous darkness, they would lower and raise their leaflets on a cycle approximately 24 hours long.

A activating B, and with B inhibiting A. As A works, B becomes more active and thus inhibits A, but as A becomes inhibited, it can no longer activate B. Consequently, B's activity lowers, which allows A's activity to increase. The system can cycle rhythmically indefinitely. In *Arabidopsis thaliana*, the negative feedback loop involves three genes rather than just two.

Many types of endogenous rhythms have a period that is not 24 hours long. Cytoplasmic streaming and the spiralling motion of elongating stem tips have periods of only a few minutes to a few hours; these are ultradian rhythms. If a period is approximately 24 hours long, it is a **circadian rhythm**, the most common kind. Some seeds have an annual rhythm of germinability: if stored in uniform conditions and periodically provided with moisture and warmth, they germinate only at times of the rhythm that correspond to springtime.

"Circadian" means that the rhythm is only approximately 24 hours long. When placed in uniform conditions, the true cycle typically differs slightly from 24 hours, being either somewhat longer or shorter. However, in nature, the rhythm is exactly 24 hours long because light is able to entrain (reset) the rhythm. The pigment responsible for detecting the light for entrainment is phytochrome. Each morning, sunrise resets the rhythm so it can never get out of synchronization with exogenous light/dark cycles.

The involvement of endogenous circadian rhythms in flowering was discovered during dark interruption experiments: a short-day (long-night) plant can be prevented from flowering by interrupting long nights with a brief (15 minutes or less) exposure of red light. This is detected by phytochrome, and the plant acts as though it has received two short nights separated by a 15-minute day. Short-day plants given a very long night—continuous darkness—have an endogenous rhythm of sensitivity to light

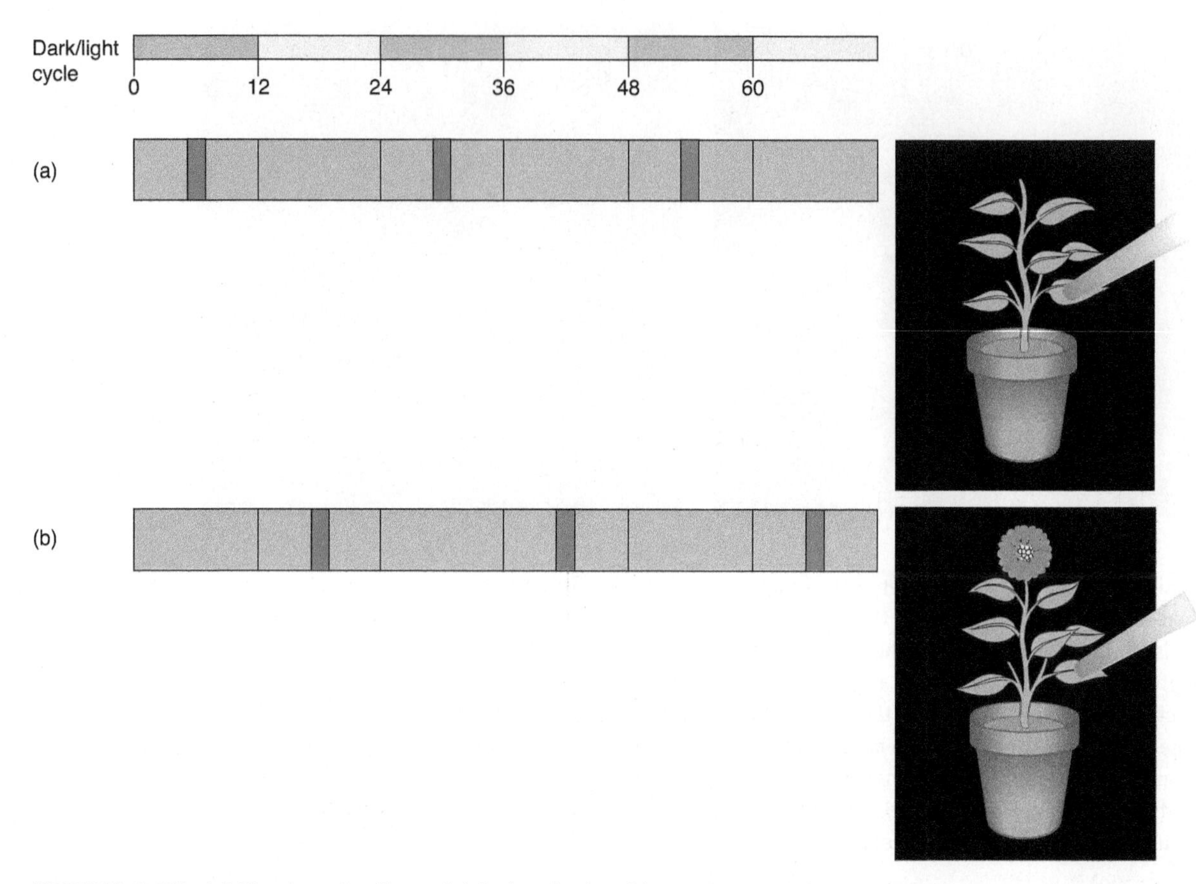

FIGURE 6.20. **(a)** If a short-day (long-night) plant is placed in continuous darkness, it can be prevented from flowering by giving it red light night breaks, but these are effective only if given at those times of the endogenous rhythm when the plant is "expecting" darkness. **(b)** If given when the rhythm is "expecting" light, the light breaks are ineffective. Whatever metabolism red light interrupts does not occur continuously in prolonged darkness, but rather periodically, controlled by the internal clock.

breaks. If the light break is given at 6 hours into the dark period, or at 30 hours (24 + 6), 54 hours (24 + 24 + 6), and so on, the light break prevents flowering (FIGURE 6.20). These times correspond to darkness in a normal environment. But if the light break is given at a time when the endogenous rhythm would be "expecting" normal daylight conditions, such as at 16, 40 (24 + 16), or 64 (24 + 24 + 16) hours after the beginning of the dark treatment, the light break does not stop flowering. Plants kept in uniform, dark conditions undergo an endogenous cyclic sensitivity and insensitivity to red light interruption of the critical night length. Just how the endogenous rhythm and the critical night length work together to stimulate flowering is not known.

TABLE 6.4. Examples of Fruits with Climacteric and Non-climacteric Ripening

Climacteric
Apple
Banana
Peach
Tomato
Non-climacteric
Cherry
Grape
Orange
Pineapple

Fruit Ripening

For many fruits, ripening is a slow and steady process, but for others, especially ones that are juicy and fleshy, ripening is slow only at first but is very rapid at the end. Rather suddenly the fruit becomes soft, its color changes, starches are converted to sugars, and flavors and aromas develop. The rapid ripening is accompanied by a rapid increase in the fruit's respiration. This is called **climacteric ripening** and it is controlled by ethylene (**TABLE 6.4**). One of the steps of fruit maturation is the production of ethylene, which stimulates many of the ripening processes. While the fruit is young, very little ethylene is present, but one of the things that ethylene stimulates is the production of more ethylene. And as more is produced, its stimulation increases and even more ethylene is produced. This is an example of a **positive feedback loop**. With climacteric fruits, ripening can be delayed by storing the fruit in a partial vacuum, which pulls ethylene out of the fruit and prevents it from building up. On the other hand, fruits can be picked while immature and hard, shipped to market or storage, and then when ripe fruits are needed, the fruits are artificially supplied with ethylene. Many of us do this at home; we place fruits that are not quite ripe in a plastic bag with an overripe banana: the banana produces large amounts of ethylene, which stimulates the other fruits to mature quickly.

Non-climacteric fruits produce little or no ethylene. Their ripening is slow and steady, and occurs only if left on the plant; if picked while immature, non-climacteric fruits will not continue to ripen.

Genes Affect Growth and Development

Environmental stimuli and internal signals from within the plant itself eventually cause a site of response to do something that it had not been doing before. Some cells grow faster or slower than before during a tropic response, whole new types of metabolism are initiated during a morphogenic response. There must be some sort of change, and that requires information that will guide the new metabolism. That information is stored in an organism's DNA, most of it in the cell nucleus, a small amount of it in plastids and mitochondria.

All cells within an individual organism, whether plant or animal, have almost identical DNA. As a zygote divides into two cells and then into four and so on throughout the growth of a plant, the DNA is duplicated (the technical term is **replicated**) as exactly as possible. The result is that all nuclei in all cells have the same DNA, the same sequence of nucleotides, and thus the same genes. Plastids and mitochondria also replicate their DNA as accurately as possible. Every single cell contains all the information needed to make all the various types of cells within a plant. Leaf chlorenchyma cells have the genes for root development, wood development, flower development, and so on. If all cells have the same information, why do cells differ in their structure and metabolism? It is because all multicellular organisms have **differential gene expression**. That means that different cells express different genes. Leaf chlorenchyma cells have the genes for root development but do not express them, whereas root cells do express root genes but not those involved in leaf morphogenesis.

The key to differential gene expression is **gene activation** and **gene repression**. The genes necessary to guide the transformation of a leaf primordium cell into a mature leaf cell must be activated whereas the genes involved in root or wood or flower development must be repressed.

Hormones and other information molecules are involved in differential gene expression, but we know few details. We do know that when red light converts P_r to P_{fr}, the phytochrome actually enters the nucleus and then somehow alters the patterns of which genes are active and which are not. Similarly, when a hormone reaches a responsive cell, it binds to a **hormone receptor**, usually located in the plasma membrane but sometimes occurring in the cytoplasm. Various things can happen after receptor/hormone binding. The receptor/hormone complex is often an enzyme that then starts a reaction even as it remains in the membrane. Or the complex may release from the membrane and enter the cell; some may even enter the nucleus.

A gene is a specific sequence of nucleotides that codes for the sequence of amino acids in proteins. That is true, but genes have more to them than just that. They also have regions called **promoters** that act more or less like a lock. A particular key-like molecule must fit onto the DNA of the promoter to either activate the gene or repress it. This is the ultimate effect of the environmental stimuli, the internal signals, the hormones, and so on: their combined effects result in the production of the proper DNA binding molecules (the keys) that will activate the correct genes in the correct cells at the correct times.

When a gene is activated, enzymes attach to it and transcribe its information into either messenger RNA or into microRNA. Messenger RNA leaves the nucleus, binds to ribosomes, and guides the synthesis of proteins. Many of these will be enzymes: enzymes for pigments in flower petals and fruits, enzymes for secondary walls and lignin in sclerenchyma cells, enzymes for starch formation in tubers, etc. Others may be membrane proteins or structural proteins. The microRNAs have many functions. Some actually bind to mRNA that already exists and stop it from being used any further. Other microRNAs seem to be the key molecules that bind to promoters and activate or repress other genes. As a result of all this, genes are expressed differently and the plant has numerous types of specialized cells.

Important Terms

abscisic acid (ABA)
all-or-none response
apical dominance
auxin
circadian rhythm
circumnutation
climacteric ripening
coleoptile
critical night length
cryptochrome
cytokinin
day length
day-neutral plant
differential gene expression
differential growth
dosage-dependent response
endogenous rhythm
ethylene
etiolation
gene activation
gene repression
gibberellin (GA)
hormone
hormone receptor
IAA (indole acetic acid)
jasmonic acid
long-day plant
morphogenic response
nastic response
negative feedback loop
P_{fr}
P_r
photoperiod
phototropin
phototropism
phytochrome
plant growth substance
positive feedback loop
promoter
replicate
salicylic acid
short-day plant
statocyte
statolith
tropic response
vernalization

Concepts

- The environment has a great deal of information that plants should perceive and respond to in an adaptive manner.
- The bodies of plants must be integrated: various parts need to know what other parts are doing and react appropriately.
- Information must be transferred from the site of perception to the site of response, if the two sites are not the same.
- All cells contain all the genetic information necessary to construct all cells, tissues, and organs of a plant.
- Differentiation of cells, tissues, and organs comes about through differential gene expression: certain genes are expressed in some cells, other are expressed in different cells.

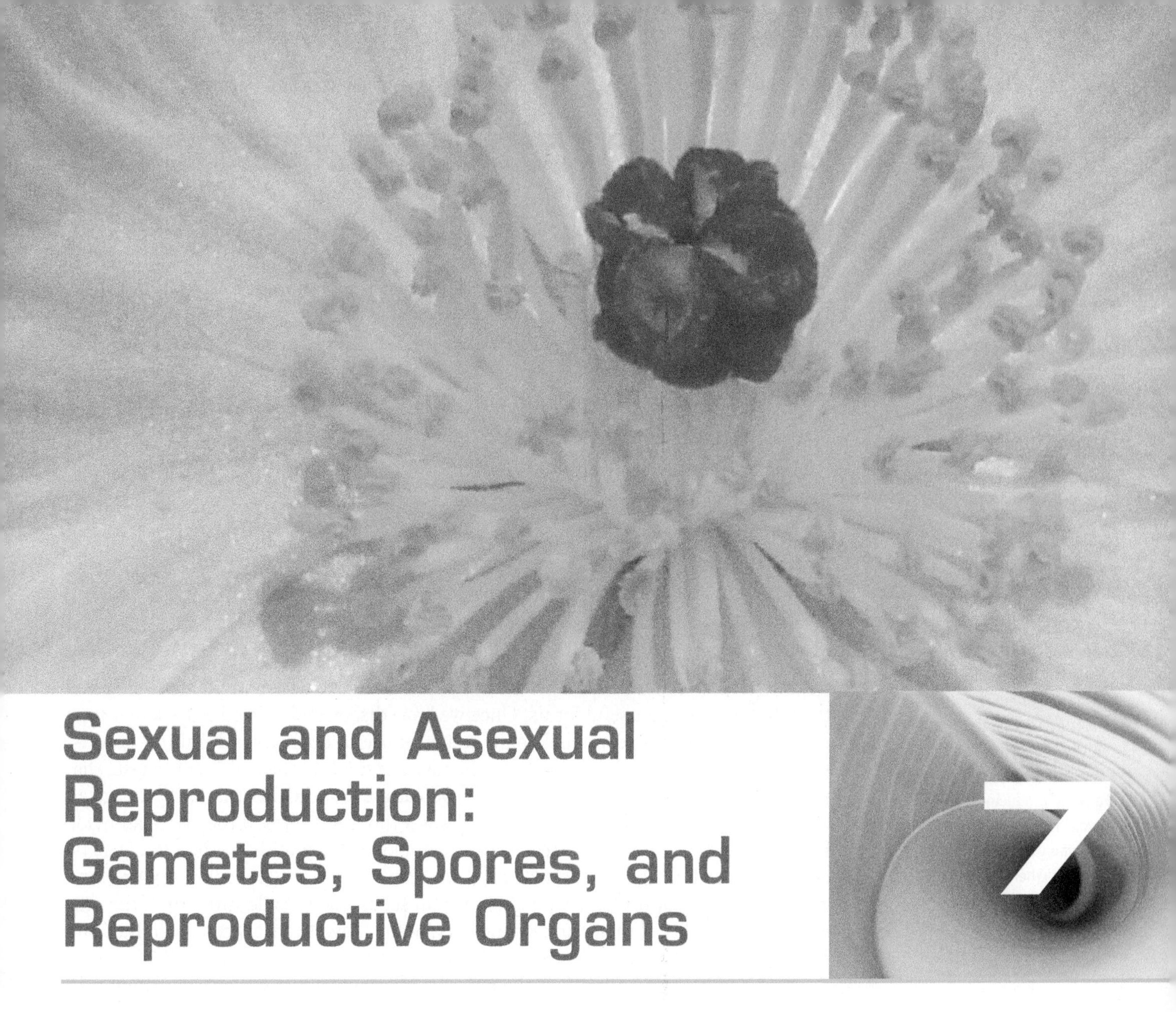

Sexual and Asexual Reproduction: Gametes, Spores, and Reproductive Organs

7

Flowers produce cells essential to sexual reproduction. The yellow stamens of this prickly poppy produce pollen grains that in turn produce sperm cells, and the red stigmas are part of a structure that leads to the production of egg cells (and later seeds and fruit). Just as in animals, sperm cells and egg cells transmit genes from parents to offspring. Nonflowering plants use cones (as in conifers) or specialized leaves (as in ferns).

Many plants live for hundreds of years, some for thousands, but none is immortal. If a species is to survive through time at least some of its members must reproduce. There are two types of reproduction that involve different processes and produce different outcomes: asexual and sexual reproduction.

During **asexual reproduction**, a single parent produces progeny that are genetically identical to itself (FIGURE 7.1). This has many advantages: because the parent has lived long enough to reproduce, it must be reasonably well-adapted to its habitat, and all its progeny will be just as adapted as it is. Also, asexual reproduction requires only the single parent, so plants growing in sparse populations where members are widely separated will be able to reproduce. Think of trees growing near treeline on a mountain: below is a forest with a high density of individuals, above are just a few widely scattered trees. With asexual reproduction, even the most isolated member of these trees can reproduce. But a significant problem is that none of the progeny are *more* adapted to the habitat than the parent is. And if the habitat changes, the parent and its progeny will no longer be adapted to it (**TABLE 7.1**).

FIGURE 7.1. The leaf tip of this *Kalanchoe* has produced an adventitious bud with stem, leaves, and roots. In most plants, cells of mature leaves never undergo cell division, but *Kalanchoe* and a few other species are unique in regularly producing plantlets on leaves. The plantlets have exactly the same genes as the parent plant, so when the plantlet falls off and becomes established as a separate plant, that is a form of asexual reproduction.

TABLE 7.1. Sexual and Asexual Reproduction

Sexual reproduction
Progeny are genetically diverse.
Some are less adapted than the parent but others are more adapted.
Progeny cannot colonize a new site as rapidly because not all progeny are adapted for it, but some can colonize different sites with characteristics not suitable for parents.
Changes in habitat may adversely affect some progeny, but others may be adapted to the new conditions.
Isolated individuals cannot reproduce.
Asexual reproduction
All progeny are identical genetically to parent and to each other.
All progeny are as adapted as parent is, but none is more adapted.
Rapid colonization of a new site is possible.
All may be adversely affected by even minor changes in the habitat.
Even isolated individuals can reproduce.

From a "plants and people" perspective, asexual reproduction in plants is often extremely valuable and useful for us. Once we have discovered or bred a particular type of plant that has an exceptional combination of useful qualities or beautiful flowers, we can be certain that all the progeny will have the same qualities if we propagate that plant asexually. Virtually all apple fruits we eat are from plants that are propagated by making cuttings, so even if an orchard has a thousand trees, if they were all made from cuttings of the original tree, they are all clones; they all have the same genetic characters; and all the apples will have the same flavor, aroma, color, texture, and other qualities. If apple trees were grown from seeds, each would differ from the others, and the quality of their fruits would vary. Many flowers too, especially orchids, are propagated by cuttings to ensure the quality of the blooms.

During **sexual reproduction**, the progeny are genetically diverse. Each parent produces **gametes** (sperm cells and egg cells) that each contain one complete set of the genes of that species. A sperm cell fuses with an egg cell resulting in the beginning of a new individual, an individual that has two full sets of genes and that differs from not only its parents but also from its siblings. What produces the diversity?

The two parents differ from each other genetically because each has different ancestors. In the many generations of these ancestors, mutations have occurred and many have accumulated, others were eliminated by natural selection (**BOX 7.1**). Also, the process that produces gametes involves a shuffling of genes (described below) such that each sperm cell and each egg cell differs from all others. Gametes are so small that many can be produced by a single plant, and many new combinations of genes can be "tested" rather inexpensively. For example, a single large tree produces thousands of flowers and millions of pollen grains, each genetically unique, yet the tree uses only a few grams (not even an ounce) of carbohydrate, protein, and minerals (**FIGURE 7.2**). In areas with oak or conifer forests, so much pollen is released into wind that sidewalks and cars become covered with the yellow dust. Similarly, thousands of egg cells can be produced inexpensively. The pollen from one plant can be blown or carried by pollinators to the flowers of hundreds or thousands of other plants, and one plant may receive pollen from numerous other individuals.

BOX 7.1. Mutations

A mutation is any change, however large or small, in DNA. If the mutation occurs in a gene, it will result in the gene having a new sequence of nucleotides. All the other copies of this same gene in other cells of the same plant, and in all the other plants of the same species, are unaffected, so there will be two versions of this gene: the original and the new version created by the mutation. Rather than giving this new version a new name, we say that both the original version and the new version are **alleles** of the gene (pronounced "ah LEEL"; the final "e" is silent). Somewhere in the plant population, a different mutation may also occur in this gene; then the gene would have three alleles. The presence of these variations is what makes individuals of a species differ slightly from each other. If there were no mutations and all genes existed in just one form, then all individuals of a species would be identical and sexual reproduction would be unnecessary.

The effect and significance of a mutation depend on many things, and some mutations are inconsequential, others beneficial, and some are life threatening. Mutations in spacer DNA may have little or no effect at all. Also, a mutation that changes just a single DNA base (an A, T, G, or C)—even if it is in a gene and creates a new allele—may not be important if it causes an inconsequential change in a part of a protein that is not critical to the protein's functioning. On the other hand, a mutation may cause the gene to code for a protein whose active site is disrupted or the protein cannot fold properly and therefore cannot function. Mutations in promoter regions or other areas where a receptor/hormone complex binds to the DNA can completely inactivate a gene or cause it to be active at the wrong time or in the wrong cell. The larger the mutation (the greater the number of DNA bases affected), the greater the probability that a critical part of the DNA is affected.

Statistically, mutations are almost always harmful. Enzymes tend to be about 300 to 400 amino acids long, and hundreds of trillions of proteins could possibly exist. Yet only a small fraction would be useful in living organisms. Any mutation that changes the structure of proteins, rRNA, or tRNA is more likely to produce a less useful than a more useful form. The majority are deleterious, the minority beneficial. Natural selection eliminates the deleterious mutations and preserves the beneficial ones.

Mutations can occur at any time in any cell, but if they happen in cells that never lead to gametes, they are called **somatic mutations**. For example, a gene in a leaf primordium cell may undergo a mutation, but because leaves are not involved in sexual reproduction, the mutation is somatic and is not passed on to the plant's progeny, regardless of whether the mutation is advantageous or disadvantageous. And because plant cells do not move from place to place, the mutated cell cannot harm other parts of the plant. When the leaf falls off in autumn, the mutated gene decomposes along with the rest of the leaf. In general, somatic mutations are not very important for most plants. In contrast, somatic mutations are a significant threat to us because our cells are mobile. For example, ultraviolet light (sun tanning, tanning beds) can mutate genes in skin cells, and even though these are somatic mutations, they can lead to skin cancer, and some of those cells can spread to other parts of our bodies and can kill us.

Mutations can be caused by a variety of agents; some are part of the organism's own metabolism, others are environmental. As mentioned in the text, the enzymes that replicate DNA are not perfect; they occasionally make a mistake, and that mistake is a mutation. Other aspects of an organism's own metabolism that damage DNA are: mitosis (chromosomes sometimes break during metaphase and anaphase), various enzymes, and oxygen (certain reactions cause forms of oxygen that are especially damaging; one benefit of antioxidants in our foods is that they control these reactive forms of oxygen).

A number of environmental agents cause mutations; they are called **mutagens** or are **mutagenic**. Many mutagenic chemicals are manmade and are increasing in our environment; examples are nitrous acid, ethidium bromide, benzene, and ethylnitrosurea. Some of the secondary metabolites that defend plants against animals are mutagenic, especially their alkaloids. Short-wave length radiation such as ultraviolet light, X rays, and gamma rays are mutagenic; they can damage DNA or even break it into pieces. Ultraviolet light occurs naturally in sunlight, but fortunately, ozone in the upper atmosphere prevents most ultraviolet light from ever reaching Earth's surface. Without the ozone layer, we would not be able to live in sunny areas.

X rays are never a serious environmental problem, but gamma rays can be. They are produced by radioactive substances, and areas such as uranium mines have higher levels of gamma radiation. A significant concern is the possibility of leakage for radioactive material from nuclear power plants, as occurred at Chernobyl in 1986 and in Japan in 2011 as well as from waste storage tanks near nuclear plants. No country on Earth, not even the United States, has yet established a safe, permanent facility to store the tons of radioactive wastes that are being produced each year. Even small amounts of this waste emit enough gamma radiation to be lethally mutagenic and carcinogenic.

FIGURE 7.2. This is only a small branch of a large tree, but it has hundreds of flowers and can produce thousands of pollen grains (each with two sperm cells) and thousands of ovules (each with one egg).

The thousands of seeds produced by a single sexually reproducing plant represent thousands of natural genetic experiments. During seed and fruit maturation, those embryos with severely mismatched genes abort and use no further resources. The tree finally produces hundreds or thousands of fruits and seeds. The total reproductive effort may be a significant drain on the tree's resources, but it produces numerous embryos, many of which are at least as genetically fit as the tree is and perhaps even more fit. In both stable environments and changing ones, sexual reproduction provides enough diversity of progeny that at least some are well adapted. The diversity itself is important for natural selection and evolution.

Think of sexual reproduction in humans: the children produced by a particular couple are variable, not identical to each other or to either parent. Some may have a particularly advantageous combination of genes and be more healthy or intelligent or athletic or creative than either parent. Others may have combinations of genes that result in congenital problems. Most children are more or less the same as the parents.

Sexual reproduction has disadvantages though. It requires two parents, so individuals living in sparse populations may not be able to reproduce. Think about the trees on the mountain again. Those in the dense population will have no trouble sharing sperm cells and egg cells, but the most isolated members might not receive pollen from other trees, and its pollen might not reach another plant. They may survive, but they will not reproduce sexually.

Growth, development, and sexual reproduction require two types of cell division. One (mitosis) creates duplicate copies of a cell's genes, and this is used as an individual grows from a zygote to an adult and until the end of its life. The other (meiosis) is used to shuffle the genes when sex cells are made and to ensure that each gamete gets one—and only one—complete set.

Duplication and Reduction Division

Cell division is only one portion of the life of a cell. When the cell is not actually dividing, its nucleus is responding to information from the environment and other parts of the plant, activating some genes, turning others off and, in general, controlling the cell's metabolism (FIGURE 7.3). If the cell is part of a meristem or other growing tissue, it will also need to prepare itself for the next round of division of its nucleus and cytoplasm. On the other hand, if the cell is part of a mature organ, then it will probably never divide again. Take a look at any large tree: it may have no dividing cells whatsoever except at the tips of twigs and roots and in the vascular and cork cambia. The entire life of a cell, from one division to the next, is called the **cell cycle**, and it consists of a growth phase and a division phase.

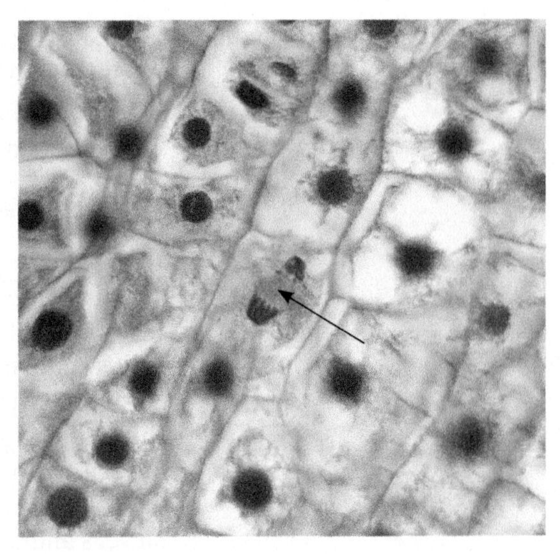

FIGURE 7.3. Of the dozen or so cells present here, only one has a dividing nucleus (arrow). Each of the nuclei were formed by nuclear divisions. Because this is a meristematic tissue, all nuclei present here probably would have divided within the next few days. These nuclei are dividing by duplication division (mitosis), so all have almost identical alleles.

Growth Phase of the Cell Cycle

In the 1800s, it was assumed that between divisions cells were "resting," so the growth phase was called the resting phase, or interphase. But the cell is active during interphase, with three distinct phases: G1, S, and G2.

In G1 (or gap 1), the first stage after division, the cell recovers from division and conducts most of its normal metabolism, including the synthesis of nucleotides for the next round of DNA replication. The length of G1 varies tremendously, depending on the type of cell, the plant's health and age, the temperature, and many other factors, and it causes the cell cycle to be variable also. Short cycle times of just a few hours occur in rapidly growing embryos and roots, but cell cycle times of 2 to 3 days or even months are not unusual in tissues or plants that grow slowly. After a cell stops dividing and begins to differentiate and mature, it enters a state similar to G1 and remains in it for life.

During S phase (synthesis), the genes in the nucleus are replicated. The entire complex of genes for an organism is its **genome**. All genes are attached to other genes by short pieces of linking DNA into a linear structure called a **chromosome**, and plant cells usually have between 5 and 30 different chromosomes (FIGURE 7.4). Even with 30 different chromosomes, an organism with 30,000 genes would have an average of 1,000 genes on each chromosome, resulting in long pieces of DNA that might break if unprotected. In onion, the DNA in each nucleus is 10.5 meters long (about 11 yards, as long as the space between lines on a football field) when all DNA molecules are measured. In eukaryotes, a special class of proteins called histones complexes with DNA and gives it both protection and structure. Chromosomes also have a centromere, usually located near the center of the chromosome, and each end is protected by a telomere.

During S phase, linking pieces of DNA as well as genes are replicated, and new histone molecules complex with the new DNA. Thus, entire chromosomes, not just DNA, are replicated. The duplicate DNA molecules remain attached to each other at the centromere. This could be called a "double chromosome" or a "pair of chromosomes," but instead it is also called a chromosome, and we call each half of the doubled chromosome a **chromatid**. After S phase, each chromosome has two identical chromatids; before S phase, each chromosome has just one chromatid.

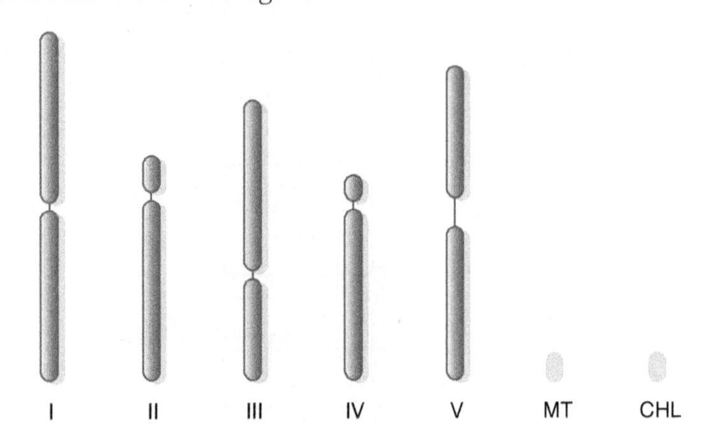

FIGURE 7.4. Each nucleus in the cells of *Arabidopsis thaliana* has two each of these five types of chromosomes, one set inherited from the paternal and one from the maternal parent. The slender area in each represents the centromere. The lines labeled mt and chl represent the DNA circles of mitochondria and plastids. The entire genome has been sequenced: we know which DNA nucleotide occurs at each site in each chromosome.

After S phase, the cell progresses into G2 (gap 2) phase, during which cells prepare for division. The actual division involves two processes: (1) division of the nucleus, called **karyokinesis**, and (2) division of the cytoplasm, called **cytokinesis**. There are two types of nuclear division: mitosis and meiosis.

Mitosis Produces Two Almost Identical Nuclei

Mitosis is **duplication division** (FIGURE 7.5). It is the method any multicellular organism uses as its body grows and the number of its cells increases. The nuclear genes are first copied, then one set of genes is separated from the other, and each is packed into its own nucleus. Because genes are linked together into just a few chromosomes, it is only necessary to make certain that one half of each doubled, large chromosome goes to one end of the cell and the other half to the other end. If that happens with each chromosome, each end of the cell automatically receives one full set of genes. Each daughter nucleus will be basically a duplicate of the original mother nucleus and a twin of the other. The enzymes that replicate DNA occasionally make errors, called **mutations**, so each nucleus is not exactly identical to the others, but they are far more similar than those produced by meiosis (described below).

Prophase. During interphase, the DNA of a chromosome is a long, extended double helix associated with histone protein. This open configuration allows enzyme complexes to find specific genes that must be read for the information they contain, but in this

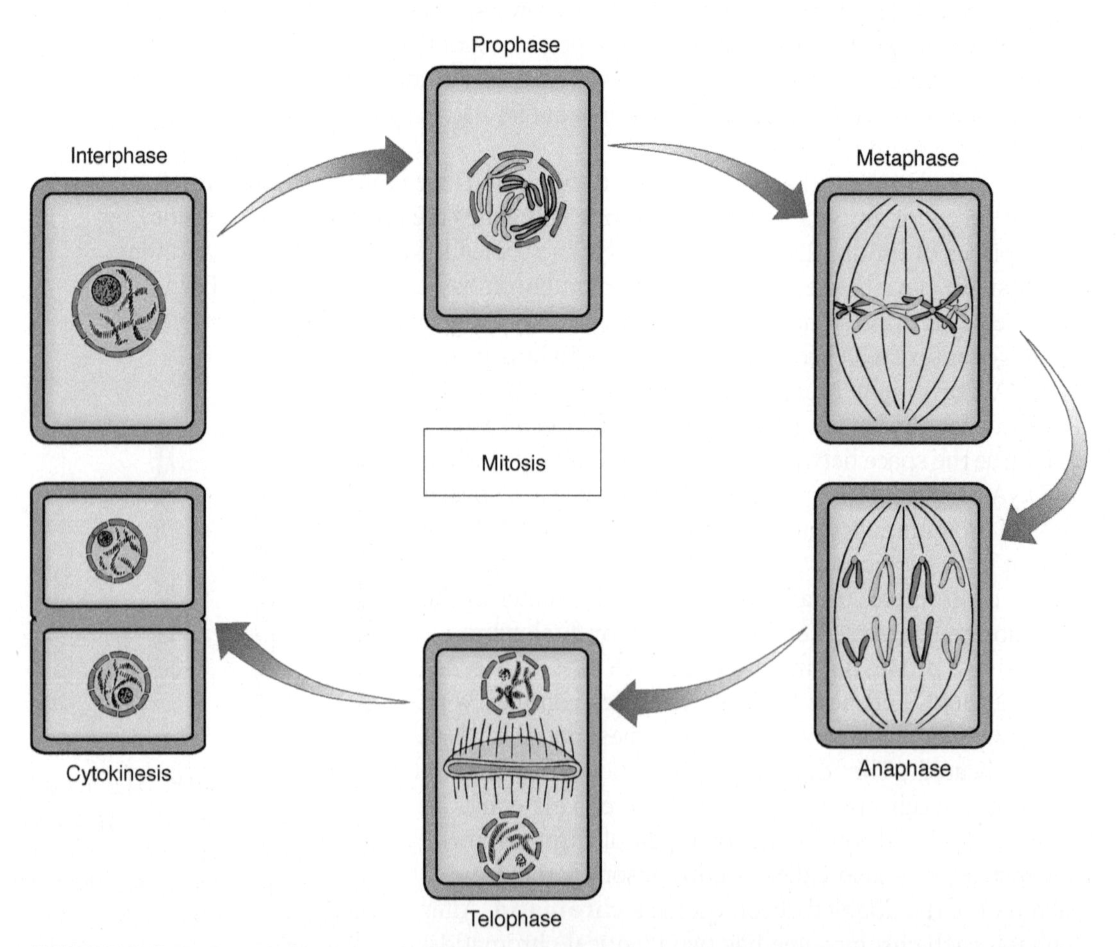

FIGURE 7.5. During mitotic nuclear division, one chromatid from each chromosome is pulled to one end of the cell, and the other chromatid of each chromosome goes to the other end. Details are given in text.

condition, a chromosome may be several centimeters (an inch or more) long, making it impossible to pull one chromatid from the other. During prophase, chromosomes condense: they coil repeatedly, becoming shorter and thicker until they are only 2 to 5 μm long. In this condensed form, we can actually see the chromosomes by light microscopy.

Also during prophase the nuclear envelope breaks into numerous vesicles, and a long set of microtubules, the **spindle**, forms in the center of the cell, extending from one side (pole) of the nuclear region to the other.

The spindle is composed of hundreds of microtubules. Some extend from one pole to the center of the cell where their ends overlap the ends of other microtubules that extend from the opposite pole. The two sets together, overlapping in the center, form a large framework. Other microtubules run from a pole to a centromere on a chromosome. The point of attachment is a kinetochore. Each centromere has two kinetochore faces: one attached to microtubules from one end of the spindle and the other face attached to the other end of the spindle.

Metaphase. Spindle microtubules gradually move chromosomes to the cell center. Their arrangement there is called a **metaphase plate**; viewed from the side, the chromosomes appear to be aligned in the very center, but viewed from the end, they are seen to be distributed throughout the central plane of the cell. At the end of metaphase, a protein-degrading enzyme frees the two chromatids of each chromosome from each other. In this step, *the number of chromosomes is doubled, but the size of each chromosome is halved*. Each chromosome is like it was in G1, before S phase.

Anaphase. During anaphase, spindle microtubules that run to the kinetochores shorten, pulling each daughter chromosome away from its twin. Long chromosomes may tangle somewhat, but microtubules exert sufficient pull to untangle them and drag them to the ends of the spindle. Because the spindle is shaped like a football, as chromosomes on each side get closer to the end, they are pulled together into a compact space.

Telophase. As chromosomes approach the ends of the spindle, fragments of nuclear envelope appear near them, connect with each other, and form complete nuclear envelopes at each end of the cell. Chromosomes uncoil, new nucleoli appear, and the spindle depolymerizes and disappears.

To summarize, after mitosis, each new nucleus contains as many chromosomes as the mother cell had, and each chromosome has one chromatid. The new nuclei are almost identical to each other and to the original nucleus that began mitosis.

A nucleus with any number of sets of chromosomes can undergo mitosis successfully: nuclei with one set or two, three, or more have no trouble at all.

Cytokinesis: Division of the Cell

Division of the protoplast is much simpler than division of the nucleus. Each daughter cell must receive some of each type of organelle, but random distribution of organelles in the mother cell usually ensures this. No matter how the cell divides, each half typically contains some mitochondria, some plastids, some ER, and so on. It is not necessary for each daughter cell to get exactly half of each. A single mitochondrion can divide, or a fragment of ER can grow until the cell has an adequate amount. The same is not true for genes: if one daughter cell is missing a gene or chromosome, the other genes cannot regenerate the missing information.

Just before prophase, a set of microtubules and actin filaments aggregates into a band running around the cell, just interior to the plasma membrane. This preprophase band identifies the plane of division, and it marks the region where the new cell wall will attach to the existing wall (FIGURE 7.6). The preprophase band is transitory: its microtubules quickly disassemble and are recycled into the mitotic spindle.

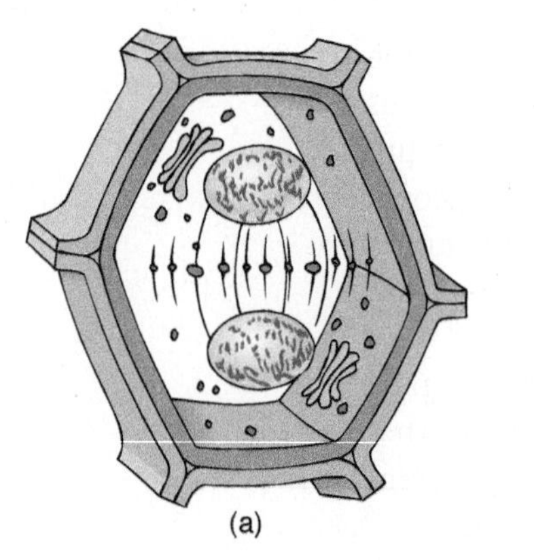

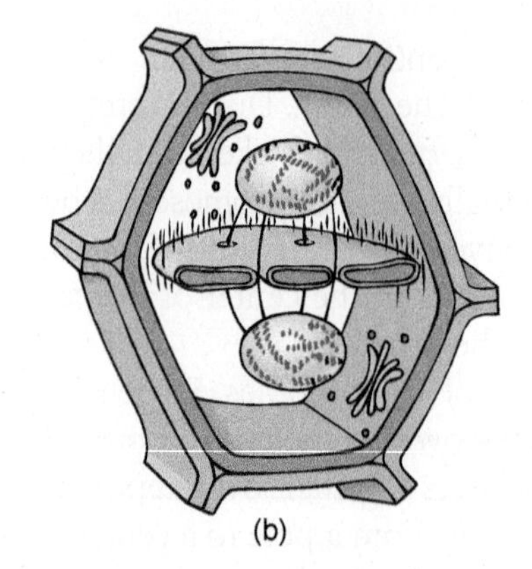

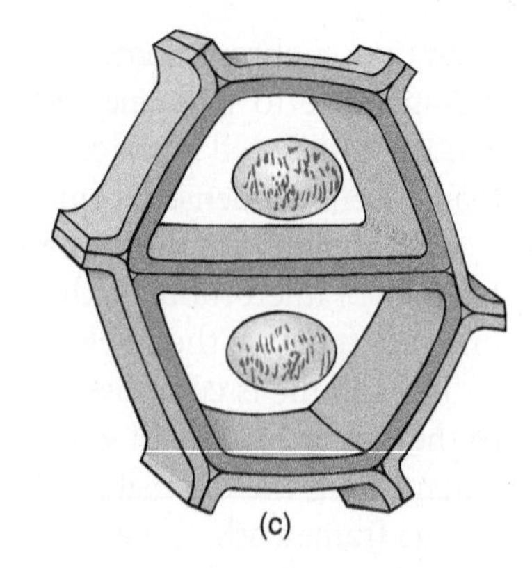

FIGURE 7.6. **(a)** Cytokinesis begins as dictyosome vesicles are trapped by phragmoplast microtubules in the space between the two new nuclei. **(b)** The small vesicles fuse into one large vesicle in which the new middle lamella and two primary walls will form. **(c)** The cell plate enlarges toward the existing cell wall as more dictyosome vesicles fuse with the edges of the cell plate vesicle. When the cell plate reaches the existing cell walls, the vesicle membrane fuses with the plasma membrane and thus becomes plasma membrane itself. The new cell plate abuts the old cell wall, and the two become glued together with hemicelluloses and pectins.

Cytokinesis in plants involves a **phragmoplast**, a set of short microtubules aligned parallel to the spindle microtubules. Actin filaments are also present. The phragmoplast forms where the metaphase plate had been, then it traps dictyosome vesicles, which fuse into a large, flat, plate-like vesicle. Inside the vesicle, two new primary walls and a middle lamella begin to form. The phragmoplast grows outward toward the walls of the original cell; simultaneously the large vesicle grows outward following the phragmoplast. The new walls extend outward along their edges. The phragmoplast, vesicle, and walls are called the **cell plate**. When the cell plate meets the mother cell's plasma membrane, the two fuse, and the vesicle membrane becomes a part of the plasma membranes of the two daughter cells. Simultaneously, the new walls meet and fuse with the wall of the mother cell, completing the division of the mother cell into two daughter cells.

Meiosis Reduces the Number of Sets of Chromosomes Within Each Nucleus

In mitosis, daughter nuclei are duplicates of the original mother nucleus. This is necessary for the growth of an organism but creates a problem when sexual reproduction occurs. Each gamete contains one complete set of chromosomes, and nuclei, cells, and organisms with one set of chromosomes in each nucleus are said to be **haploid (1n)**. The zygote has two complete sets (one from each gamete), so it is **diploid (2n)** (FIGURE 7.7). Because a zygote grows into an adult

Sperm nucleus
Egg nucleus
Fertilization
Zygote
S phase
Mitosis: growth of body

FIGURE 7.7. Sperm cells and egg cells are haploid, each having just one set of chromosomes; in this case the set contains a long chromosome and a short one. After fertilization, the zygote is diploid with two complete sets: two long chromosomes and two short ones. The zygote grows into a mature plant by mitosis; all nuclei are duplicates of the original zygote nucleus. Each is diploid, with one set each of paternal and maternal chromosomes.

plant or animal by cell divisions with mitotic nuclear divisions, all cells of the adult are diploid also.

If the adult were to produce gametes by mitosis, the gametes would be diploid as well, and the next zygote would be tetraploid with four sets. Instead, a **reduction division**, called **meiosis**, occurs somewhere. Meiosis involves two rounds of division without allowing the S phase to occur after the first division. The two divisions are called meiosis I and meiosis II, and each contains four phases similar to those of mitosis. Meiosis occurs only in the production of reproductive cells: gametes in animals and spores in plants. In angiosperms, meiosis occurs in only a few cells of flowers, and in animals it occurs in the reproductive organs (testes and ovaries). Meiosis is never used in the growth of the body of any organism.

Meiosis I. The first stage of meiosis I is prophase I (FIGURE 7.8). All events of prophase of mitosis also occur in prophase I: nucleolus and nuclear membrane break down; a spindle forms; microtubules attach to centromeres; and chromosomes condense and become visible. In addition, special interactions of chromosomes occur that are unique to prophase I; prophase I is divided into five stages:

1. In leptotene, chromosomes begin to condense and become distinguishable, although they appear indistinct.
2. During zygotene, a remarkable pairing of chromosomes occurs. There are two sets of chromosomes in every nucleus of diploid plants and animals: one from the paternal and one from the maternal gamete. During zygotene, each chromosome of one set finds and pairs with the equivalent chromosome (its **homologous chromosome**, or homolog) of the other set; this pairing is **synapsis**. The two

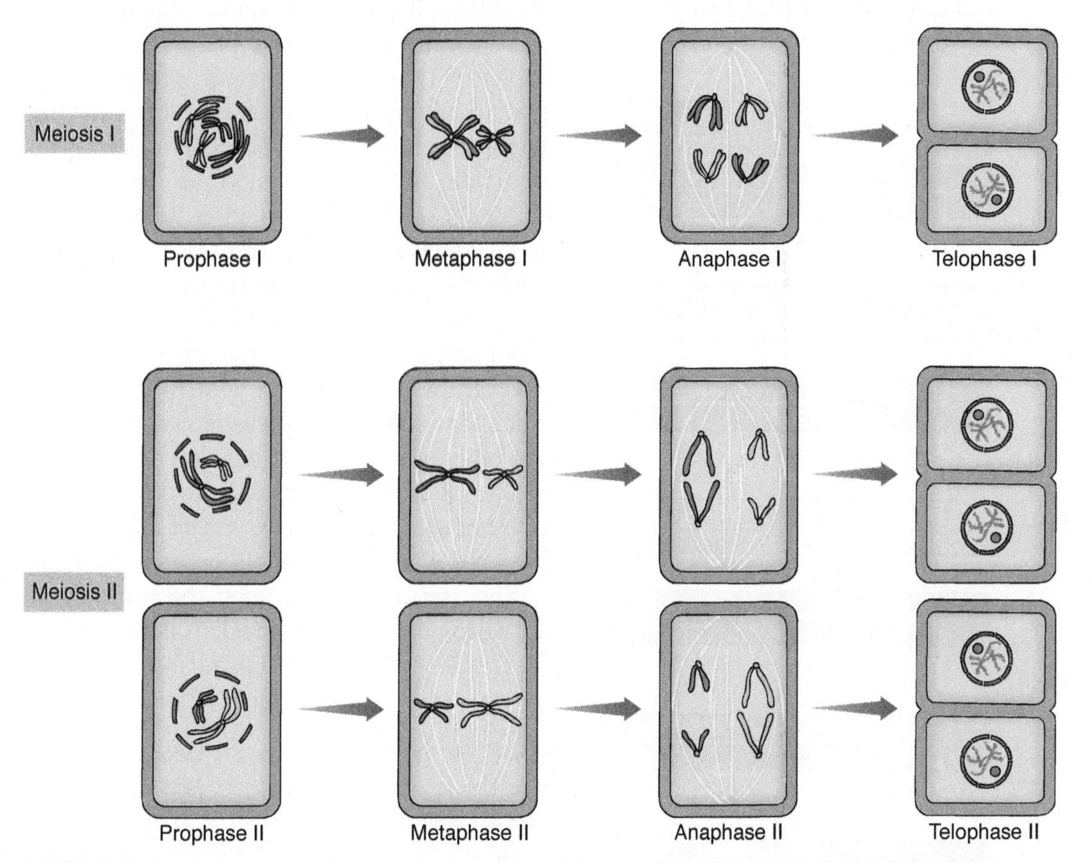

FIGURE 7.8. Meiotic nuclear division consists of two divisions without an intervening S phase. Details are given in text.

homologous chromosomes in each pair align almost perfectly from end to end. A synapsed pair of homologous chromosomes is called a bivalent.

3. Chromosomes continue to become shorter and thicker; this stage is pachytene. **Crossing-over** occurs now: in several places in each chromosome, the DNA of each homolog breaks and enzymes repair the breaks but hook the "wrong" pieces together (FIGURE 7.9). A piece of the maternal homolog is attached to the paternal homolog, and the equivalent piece of the paternal homolog is attached to the maternal one. If the maternal and paternal chromosomes are absolutely identical, nothing significant has happened, but if the genes on the paternal and maternal chromosomes are slightly different because of their separate ancestry, the new chromosomes that result from synapsis and crossing-over differ slightly from the original chromosomes. There are no new genes; instead new combinations of genes exist on each chromatid. This increases the genetic diversity of all haploid cells produced: every sperm cell or egg cell produced by any single plant or animal (yourself included) is unique. This genetic diversity is important for evolution.
4. After pachytene is diplotene. The homologous chromosomes of each bivalent move away from each other but do not separate completely because they are held together at their paired centromeres and at points (chiasmata; singular, chiasma) where they appear to be tangled together.
5. In the final stage, diakinesis, homologs continue to separate, and chiasmata are pushed to the ends of the chromosome. The homologous chromosomes become untangled and are paired only at the centromeres.

Prophase I is the most complicated stage of meiosis; the remaining stages are simple and quite similar to the stages of mitosis. During metaphase I, spindle microtubules move the tetrads to the center of the cell, forming a metaphase plate, then during anaphase I the homologous chromosomes separate completely from each other, moving to opposite ends of the spindle. Centromeres do not divide, and each chromosome continues to consist of two chromatids. This is different from the metaphase–anaphase transition of mitosis. In mitosis, chromatids are released from each other, and each chromosome divides into two chromosomes, each with just one chromatid. However, in the metaphase I–anaphase I transition, homologous chromosomes separate from each other, and each still has two chromatids. One set of chromosomes is pulled away from the other set, and two new nuclei are formed. These nuclei are now haploid because each has only one set of chromosomes; the homolog of each chromosome in one nucleus is now in the other nucleus. Telophase I is similar to telophase

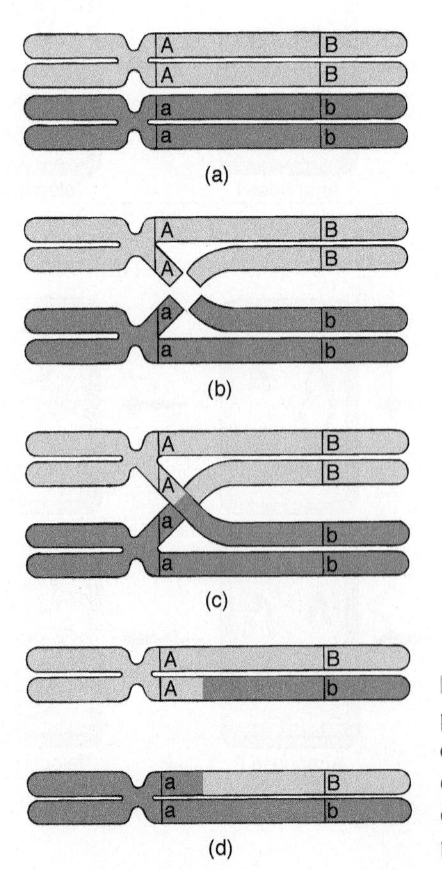

FIGURE 7.9. **(a)** During prophase I of meiosis, each paternal chromosome somehow finds and pairs with the equivalent maternal chromosome. They lie parallel to each other. **(b)** and **(c)** Breaks occur in similar sites on equivalent chromatids and repair enzymes attach maternal pieces to paternal pieces, resulting in new chromatids **(d)**.

of mitosis, but because the chromosomes still have two chromatids each, they do not need to undergo an interphase or an S phase.

Meiosis II. Prophase II is not subdivided into stages like prophase I. Metaphase II is short, and at the end of it, the centromeres divide, separating each chromosome into two chromosomes, just as in metaphase of mitosis, but different from metaphase I. Anaphase II then separates each new chromosome from its replicate, and in telophase II, new nuclei are formed. Each nucleus contains just one set of chromosomes, each with a single chromatid.

To summarize, the resulting nuclei each have a haploid set of chromosomes, each chromosome with only one chromatid. Crossing-over has resulted in new combinations of genes on each chromosome. Meiosis is difficult in haploid or other odd-ploid (3n, 5n, etc.) nuclei because pairing of odd numbers of homologous chromosomes causes irregularities during prophase I; the resulting nuclei have abnormal numbers of chromosomes.

In meiosis, nuclear division and cell division are often not directly linked. In some species, cytokinesis happens after both meiosis I and meiosis II, and four haploid cells result from each original diploid mother cell. In other species, no cytokinesis occurs after meiosis I, but a double cytokinesis occurs after meiosis II, again resulting in four haploid cells. For example, pollen grains can form either way, and meiosis results in four haploid cells. However, in many organisms, no cytokinesis occurs at all during the meiosis that leads to the formation of eggs. The final cell is tetranucleate and may remain that way, or three of the nuclei may degenerate and produce only a single uninucleate haploid egg cell by meiosis. It seems to be selectively advantageous to produce one large egg rather than four small ones.

Asexual Reproduction

Within angiosperms, numerous methods of asexual reproduction have evolved. One of the most common is fragmentation: a large spreading or vining plant grows to several meters (several yards) in length, and individual parts become self-sufficient by establishing adventitious roots (FIGURE 7.10). If middle portions of

FIGURE 7.10. This image shows three nodes of a runner (stolon) of pothos (*Epipremnum*), and one or several adventitious roots have emerged at each node. Occasionally an axillary bud will be released from apical dominance and begin to grow as a plantlet (although technically it is a branch). It produces a shoot and is supplied by the adventitious roots, so if the runner is broken or dies, the plantlet becomes a new, individual plant.

the plant die, the ends become separated and act as individuals. Certain modifications improve the efficiency of fragmentation. In many cacti, branches are poorly attached to the trunk, and the plant breaks apart easily. The parts then form roots and become independent. In some members of the saxifrage, grass, and pineapple families, plantlets are formed where flowers would be expected; these look like small bulbs and are called bulbils. Kalanchoes produce such large numbers of plantlets along their leaf margins that they can be weeds in both nature and in greenhouses (see Figure 7.1).

In willows and many thistles, adventitious shoot buds form on roots and then grow into plants. Adventitious buds may grow out even while the parent plant is still alive, and a small cluster of trees may in fact consist of just a single individual. A grove of aspens that covers several acres in Utah is a single plant.

Sexual Reproduction

Sexual reproduction in angiosperms involves flowers, which produce the necessary cells and structures. To understand flower structure, one must first understand the plant life cycle.

Life Cycle of Plants

The life cycle of animals is simple: diploid adults have sex organs that produce haploid gametes, either sperm cells or egg cells, by meiosis. Male individuals produce sperm cells and females produce eggs. One sperm and one egg are brought together forming a new single diploid cell, the fertilized egg or **zygote**, which then grows to become a new diploid individual that resembles its parents. In mammals, the egg, zygote, and embryo are retained inside the mother, whose body nourishes and protects them for several months. After that they are born.

The life cycle is more complex in plants. The plants you are familiar with—trees, shrubs, and herbs—are all just one phase of the plant life cycle, called the **sporophyte** generation, and are always diploid (FIGURES 7.11 and 7.12a). Like adult animals they have special cells (located in the flowers in angiosperms, on the underside of leaves in ferns) that undergo meiosis. In animals, meiosis results in haploid gametes, but in plants it results in haploid **spores** (FIGURE 7.12b) The difference between gametes and spores is great: gametes fuse with other gametes in a process called **syngamy** or fertilization, thereby producing the diploid zygote. A gamete that does not undergo syngamy dies (unfertilized eggs of bees are exceptional and develop into sterile workers).

Plant spores are just the opposite: they cannot undergo syngamy but each undergoes mitosis and grows into an entire new, haploid plant called a **gametophyte** (FIGURES 7.11 and 7.13). During sexual reproduction, when a sporophyte reproduces, *it does not produce a new diploid plant like itself* but rather a haploid plant. Furthermore, in all vascular plants, a haploid gametophyte does not even remotely resemble a diploid sporophyte: it is a tiny mass of cells with no roots, stems, leaves, or vascular tissues, but *it is an entire plant and it does produce gametes.* The gametes then undergo syngamy, forming a zygote that grows into a new, diploid sporophyte and the life cycle is complete.

Like animals, plants have two types of gametes: small sperm cells and large egg cells. Sperm cells are produced by one type of gametophyte and eggs by a different type, so there are male gametophytes and female gametophytes. The two types of gametophytes have grown from two types of spores: male gametophytes from microspores and female gametophytes from megaspores.

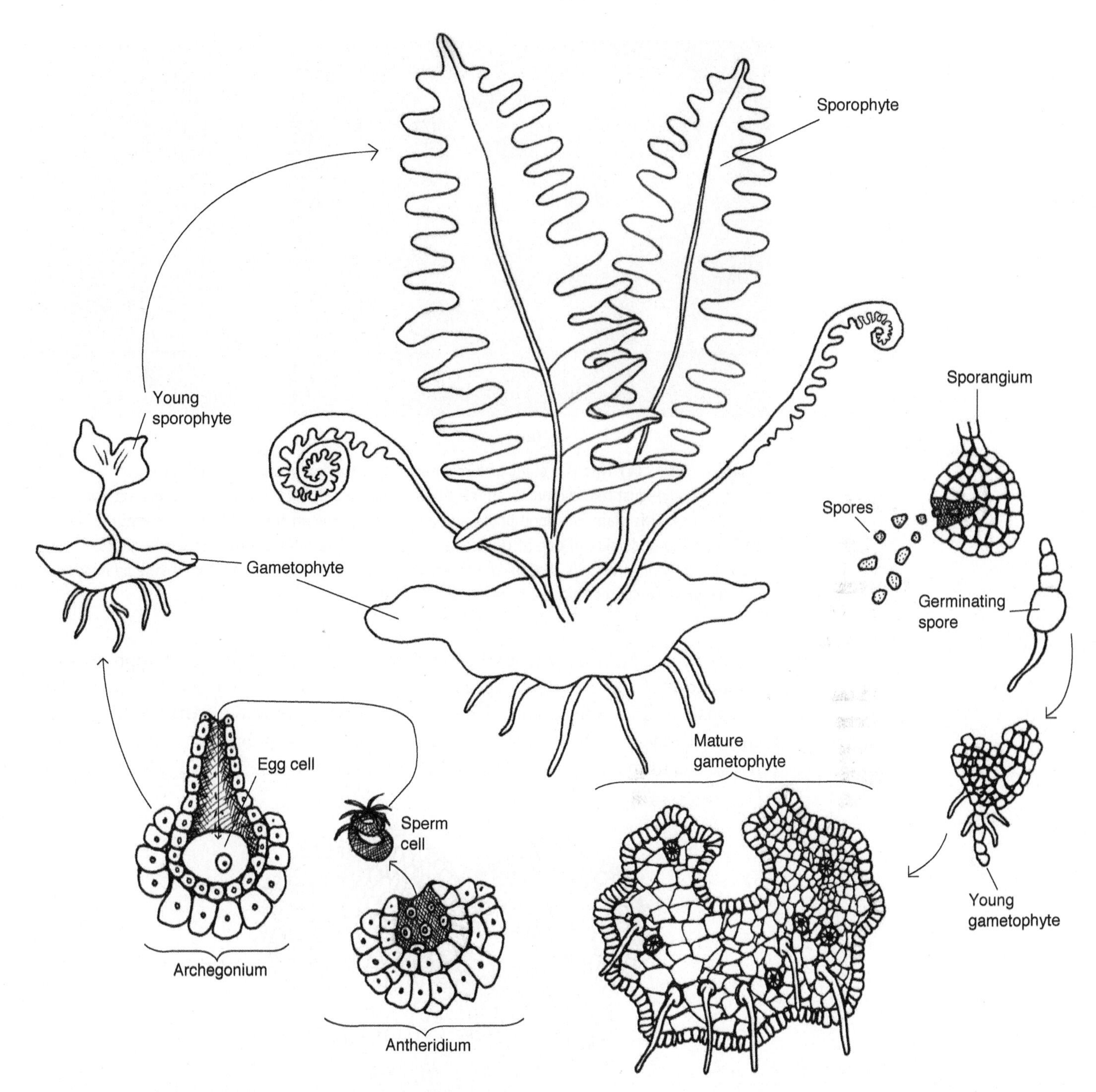

FIGURE 7.11. Diagram of the life cycle of a fern. Details are given in text. Begin with the sporophyte at the top of the diagram; its cells are diploid, just as ours are. Some cells differentiate into sporangia, and cells within each sporangium undergo meiosis and become haploid spores. When cells in our own reproductive organs undergo meiosis, they produce haploid gametes, but that never occurs in plants. The spores germinate and grow into new plants called gametophytes (at bottom of diagram). As shown here and in Figure 7.13, fern gametophytes are small and delicate, only about one cell thick, but they must live on their own: they must photosynthesize, gather water and minerals, defend themselves from fungi, and so on. The gametophytes have two types of gametangia: antheridia produce sperm cells and archegonia produce egg cells (for micrographs of actual antheridia and archegonia, see Figure B7.2h). Sperms are released from antheridia and swim to an archegonium when the gametophytes are wet with rain or dew. The fertilized egg is diploid and grows into a new sporophyte. It is important to remember that every species of ferns has two very different types of plants: diploid sporophytes that are the ferns we are familiar with and tiny haploid gametophyte ferns that most of us have never seen.

FIGURE 7.12. **(a)** This is a plant of Bird's Nest fern (*Asplenium nidus*). Every nucleus in every cell of its body—just like the nuclei in our bodies—is diploid, each has two complete sets of chromosomes. This plant is a sporophyte and will make spores on the underside of certain leaves. **(b)** This is the underside of one of the leaves in (a). The dark lines are rows of sporangia (see Figure 7.11), groups of cells that make spores. Certain cells in each sporangium divide by meiosis and thus produce spores that are haploid, not diploid.

A life cycle like this, with two generations—sporophyte and gametophyte—is an **alternation of generations**. Because gametophytes do not resemble sporophytes, it is considered an alternation of **heteromorphic generations** (**BOX 7.2**). This is a complex life cycle, with at least three distinct plants (one sporophyte and two gametophytes). No animal life cycle has anything equivalent to the haploid generation.

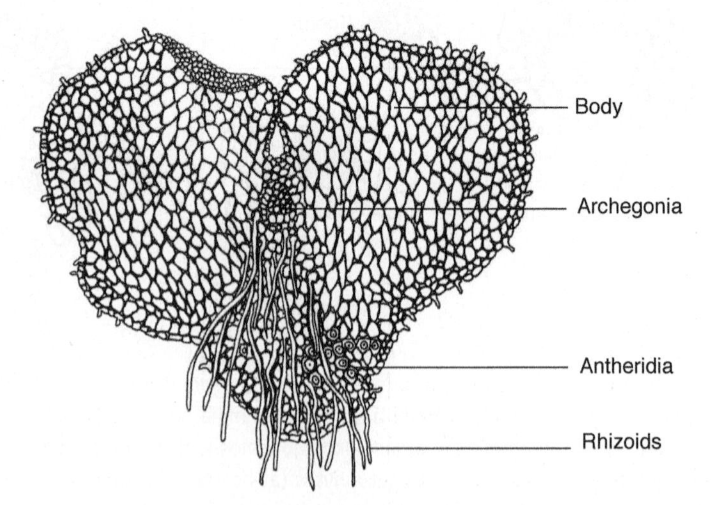

FIGURE 7.13. Fern gametophytes tend to grow into a heart shape composed of chlorophyllous parenchyma cells, with most of the body being only one cell thick. This is the underside of the gametophyte; it bears antheridia (which produce sperm cells) and archegonia (which produce one egg cell each) on the side closer to the soil.

BOX 7.2. Life Cycles in Other Plants

Each group of plants has the same life cycle as angiosperms—an alternation of heteromorphic generations—but they differ in details.

Conifers. Pines, larches, firs, and other conifers are similar to angiosperms but produce cones instead of flowers. Pollen is produced in small cones (pollen cones; **FIGURE B7.2a**) that last only a few months, and seeds are produced in larger, tougher cones (seed cones; **FIGURE B7.2b**) that last for a year or more. Pollen and seed cones occur on the same plant in many species (monoecious species) but are on separate plants in others (dioecious species).

FIGURE B7.2a. Whereas most flowers produce both pollen (with sperm cells) and ovules (with eggs), all conifers have some cones that produce only pollen, others that produce only ovules (which mature into seeds if fertilized). These are the pollen cones of lodge pole pine (*Pinus contorta*); the same plant has seed cones, but they are not visible here.

FIGURE B7.2b. These are seed cones of ponderosa pine (*Pinus ponderosa*). Two are smaller than the other three; they may not have been pollinated well enough to develop.

Ferns. Ferns never produce seeds (see Figures 7.12 and 7.13). They release all their spores; they do not retain any inside the parent body the way angiosperms and conifers retain their megaspores and female gametophytes. Instead, all spores of these nonseed plants land on the soil, germinate, and grow into small green gametophytes, haploid plants that are completely independent of the parent sporophyte fern. They are thin, flat, green sheets of chlorenchyma; they photosynthesize, gather water and minerals, and when old enough, produce sperm cells in antheridia and egg cells in archegonia. After a sperm cell fuses with an egg cell, the zygote grows into a diploid fern sporophyte with leaves, stems, and roots. *Equisetum* (horsetails; **FIGURE B7.2c**), *Lycopodium* (club-mosses [not really mosses]), and *Selaginella* (spike-mosses [not really mosses]) are similar.

Mosses. In this group of plants, it is the haploid gametophytes that are the familiar plants (**FIGURES B7.2d–f**). The green, leafy moss plants we see throughout the year are all haploid, and most are perennial, living for many years. They produce sperm cells and egg cells, and after fertilization, the zygote is retained and protected by the parental gametophyte. The moss sporophyte is small and never has leaves or roots: it consists of a foot that absorbs water and nutrients from the gametophyte, a stalk (the seta), and a sporangium (the capsule). Once the sporangium grows to

FIGURE B7.2c. The cones of horsetails (*Equisetum*, also called scouring rushes) are always located at the tip of branches and are always covered with scales that have six sides. Horsetails have only tiny leaves, and the stems are the main photosynthetic organ. The stems of horsetails are the sporophyte body and have diploid nuclei. Certain cells in the cone will undergo meiosis and make haploid spores (not visible here). The spores blow away then grow into tiny gametophytes hidden among leaf litter and soil; they are almost impossible to find. Courtesy of Chad Husby.

(continued)

FIGURE B7.2d. This is a large patch of gametophytes of the moss *Leucobryum glaucum* (sometimes called pincushion moss): all cells in these plants have haploid nuclei. Whereas gametophytes in seed plants are just the tiny pollen grains or embryo sacs inside ovules, gametophytes of mosses are the large, photosynthetic phase of the life cycle. These gametophytes have gametangia that produce sperm cells and egg cells, but they are too small to be seen.

FIGURE B7.2e. This is a patch of the moss *Polytrichum commune* (common haircap moss). The green plants are the haploid gametophytes, and they produce sperm cells and egg cells. Fertilized eggs remain attached to and receive nourishment from the gametophytes. The stalks with capsules here are the diploid sporophytes: some cells inside each capsule undergo meiosis and produce haploid spores. Moss sporophytes decompose only slowly after the spores are released, so it is common to see mosses with stalks and empty capsules.

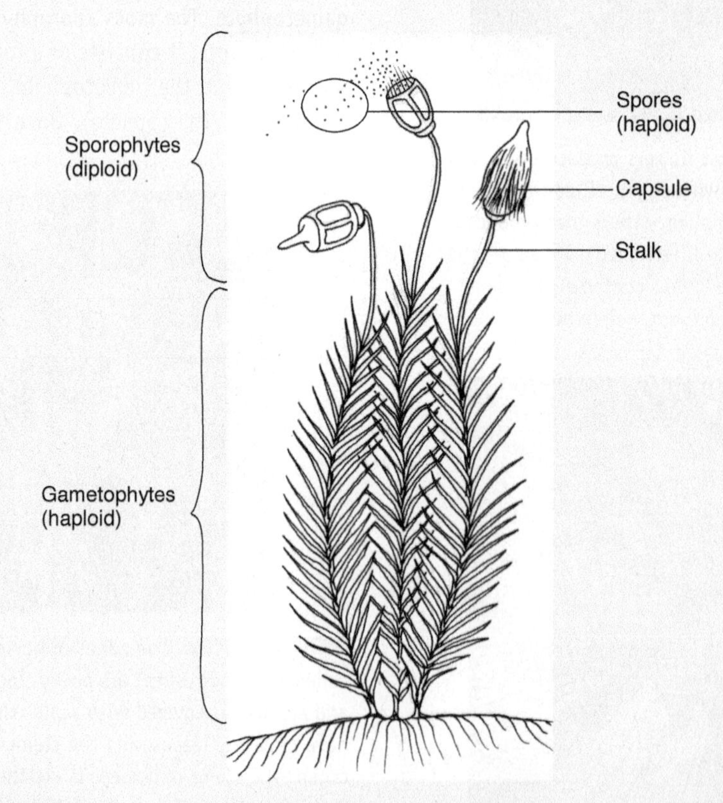

FIGURE B7.2f. This diagram shows a moss gametophyte with three leafy stems, all of which are haploid. The tip of each stem had several archegonia, and in each, one egg was fertilized and grew to be a sporophyte with a stalk and capsule. Cells of the sporophyte are diploid.

full size (just a few millimeters), the inner cells undergo meiosis and become spores. The capsule opens, releasing the spores, which will grow into new leafy green gametophyte moss plants; the capsule, seta, and foot then die.

Liverworts and hornworts. Plants are similar; the larger, green plant is the gametophyte (**FIGURES B7.2g–j**).

An alternation of heteromorphic generations occurs in every species of the plant kingdom. In ferns, both the sporophytes and gametophytes must be able to live on their own. In seed plants, the gametophytes are always nourished and protected by the sporophytes, whereas in the mosses, the sporophytes are protected and nourished by the gametophytes.

FIGURE B7.2g. This is a gametophyte of a leafy liverwort (*Fossombronia*). The "rocks" in the image are sand grains. Each leaf is less than 1 mm wide (there are 25 mm in an inch) and is only one cell thick.

FIGURE B7.2i. After a liverwort sperm cell fertilizes an egg, the zygote grows into a sporophyte with a stalk and a capsule as in mosses, but the liverwort sporophytes are very delicate and short lived; the stalk typically elongates in just an hour or so, the capsule opens and releases the spores, and then the entire sporophyte withers away in less than a day.

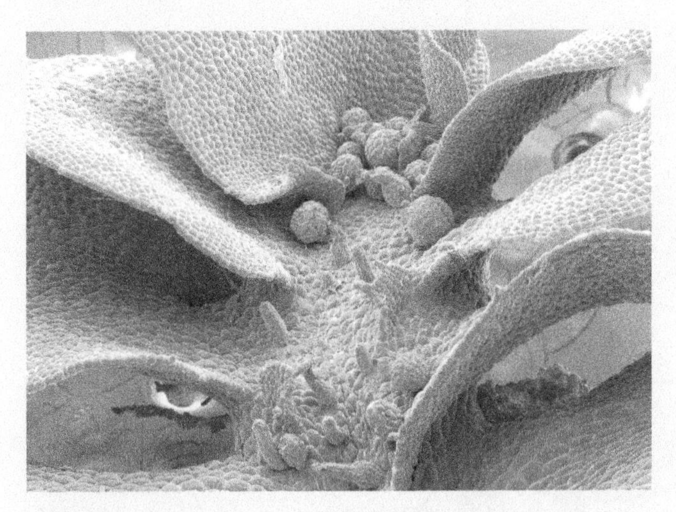

FIGURE B7.2h. This is a shoot of *Fossombronia* viewed with a scanning electron microscope. Each leaf here corresponds to a leaf in Figure B7.2g. The round spheres (antheridia) produce sperm cells and the slender tubes (archegonia) each have one egg cell at its base. Each bump on the leaves is an individual cell (×300).

FIGURE B7.2j. The flat plates of green cells are the gametophytes of the hornwort *Phaeoceros*; the small green horns are the sporophytes, each growing out of an archegonium where an egg had been fertilized. Cells in the center of the horns will undergo meiosis, develop into spores that blow away, then germinate and grow into new gametophytes.

Flower Structure

Organs of the Flower

Flowers were described briefly in Chapter 3; more details are given here. Most flowers have all four types of floral appendage (sepals, petals, stamens, and carpels) and are **complete flowers**. They typically have three, four, or five of each type; for example, lilies have three sepals, three petals, three stamens, and three carpels (FIGURE 7.14). Flowers of certain species lack one or two of the four basic floral appendages and are called **incomplete**. Many wind-pollinated flowers have no petals and are incomplete.

Sepals are the outermost of the four floral appendages. They are typically the thickest, toughest, and most waxy of the flower parts, and they protect the rest of the flower bud as it develops, keeping bacterial and fungal spores away, maintaining a high humidity inside the bud, and deterring insect feeding. Sepals may be colorful (petaloid) and help attract pollinators. All the sepals together are referred to as the calyx.

Above the sepals are **petals**, which together make up the corolla. Sepals and petals together constitute the flower's perianth. Petals are important not merely in attracting pollinators, but rather the correct pollinators. Each plant species has flowers of distinctive size, shape, color, and arrangement of petals, allowing pollinators to recognize them. Sexual reproduction results only if pollen is carried to other flowers of the same species; it cannot occur efficiently if pollen is carried to other plants indiscriminately. If a flower has a distinctive pattern and offers a good reward such as nectar or pollen, the pollinator is likely to search for and fly to other flowers with the same pattern, enhancing cross-pollination. Many flowers have pigments that absorb ultraviolet light, creating patterns only insects can see. Without light, colors cannot be seen, and night-blooming species often have white flowers lacking pigments. Their petals produce strong fragrances, and insects and bats follow the aroma to the flower.

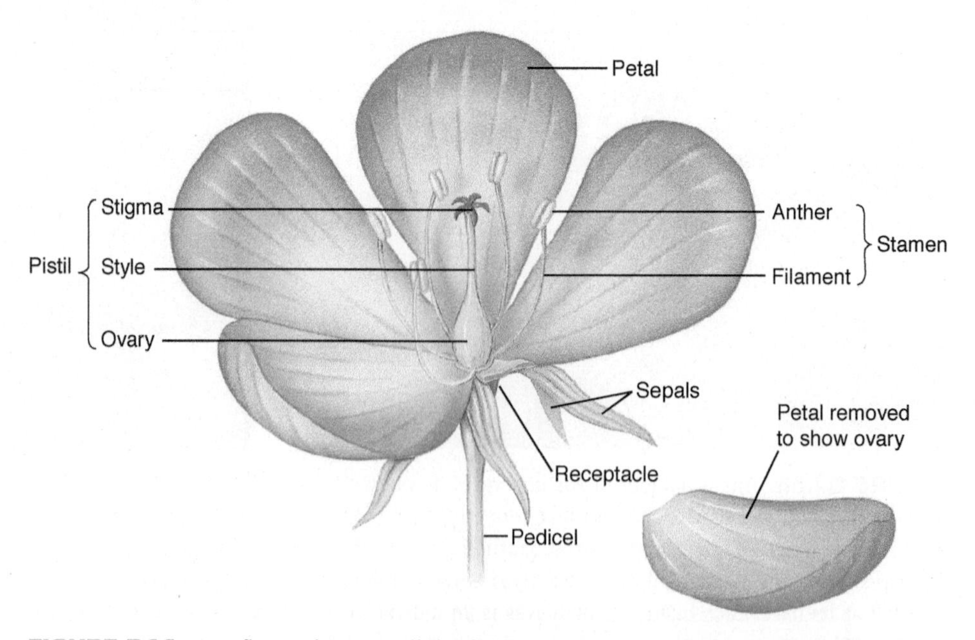

FIGURE 7.14. Most flowers have a stalk (pedicel) to which is attached a set of sepals, petals, stamens, and carpels.

Above the petals are **stamens**, known collectively as the androecium. Stamens are frequently referred to as the "male" part of the flower because they produce pollen, but technically it is just pollen that is male, not the stamen. Stamens have two parts, the filament (its stalk) and the **anther**, where pollen is actually produced. In each anther four long columns of tissue become distinct as some cells enlarge and prepare for meiosis (FIGURE 7.15). These microspore mother cells undergo meiosis, each producing four microspores. Neighboring anther cells, in a layer called the tapetum, act as nurse cells, contributing to microspore development and maturation. Microspores initially remain together in a tetrad (meiosis produces four cells), but later separate, expand to a characteristic shape, and form an especially resistant wall. They are then called **pollen**. The anthers open along a line of weakness and release the pollen (FIGURE 7.16).

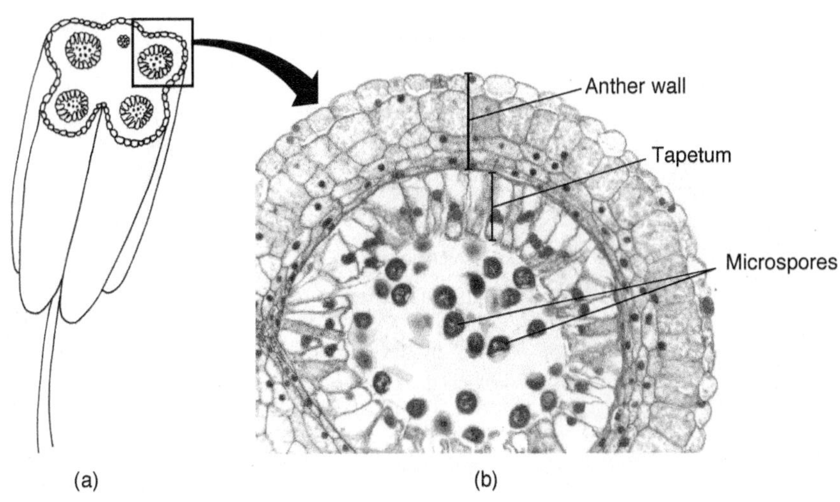

FIGURE 7.15. If an anther is cut in transverse section as indicated in **(a)**, the cells will be arranged as shown in **(b)**. This is a high magnification view of one of the four lobes of an anther. The innermost, scattered cells in the anther lumen are microspores in the process of developing into pollen grains. The columnar cells are the tapetum, and they act as nurse cells that provide important chemicals that allow the pollen to mature normally. When the pollen is mature, the anther wall must break open so that the pollen grains are released (×200).

The wall of a pollen grain is an unusually complex cell wall covered by a polymer called sporopollenin. It has one or several germination pores, weak spots where the pollen opens after it has been carried to the stigma of another flower. Sporopollenin is remarkably waterproof and resistant to almost all chemicals; it protects the pollen grain and keeps it from drying out as it is being carried by wind or animals. The pollen surface can have ridges, bumps, spines, and numerous other features so characteristic that each species has its own particular pattern. Because sporopollenin is so resistant, pollen grains fossilize well, and by examining samples of old soil, botanists can determine exactly which plants grew in an area at a particular time in the ancient past.

Carpels constitute the gynoecium, located at the highest level on the receptacle. Carpels have three main parts: (1) a **stigma** that catches pollen grains, (2) a **style** that elevates the stigma to a useful position, and (3) an **ovary** where megaspores are produced. Carpels usually are fused together into a single compound structure, frequently called a pistil. Inside the ovary are placentae (singular: placenta), regions of tissue that bear small structures called **ovules** (FIGURES 7.17 and 7.18). Ovules have a short stalk (the funiculus) that carries water and nutrients from the placenta to the ovule by means of a small vascular bundle. The center of an ovule is a mass of parenchyma called a nucellus. Around the nucellus are two thin sheets of cells (integuments) that cover almost the entire nucellus surface, leaving only a small hole (micropyle) at the top. As in anthers, some nucellus cells, usually only one in

FIGURE 7.16. The anthers of this oxblood lily (*Rhodophiala bifida*) have opened and the anther walls are curled back exposing the yellow pollen grains to pollinators. The red, three-lobed stigma is on the right.

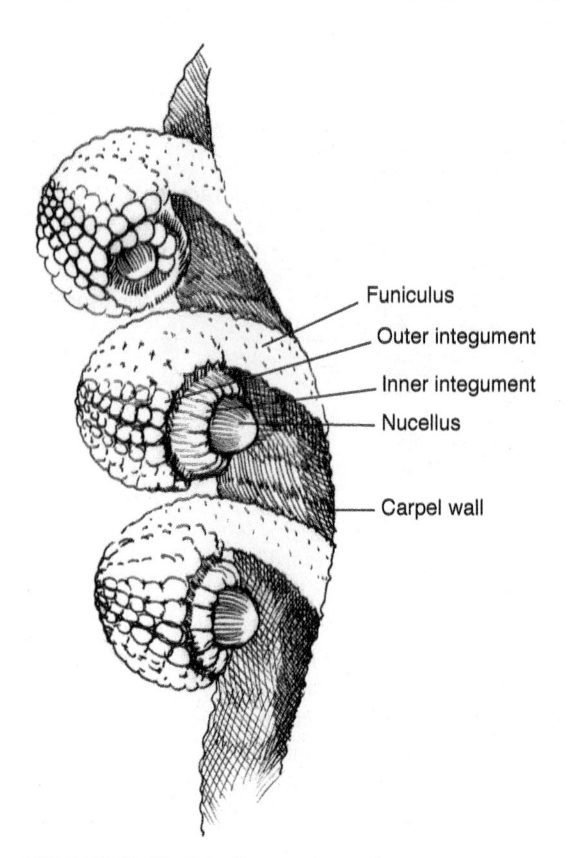

FIGURE 7.17. This diagram shows three young ovules of pea (*Pisum*) attached to the ovary wall of a carpel. The stalk of each is the funiculus, and the two integuments of each have not yet covered the nucellus. Each ovule develops into a pea seed after fertilization, and the carpel wall develops into the pea pod.

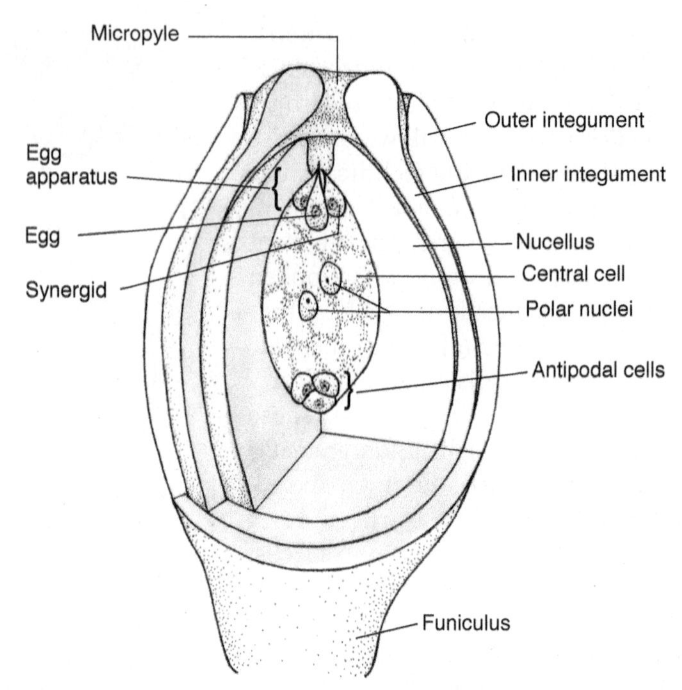

FIGURE 7.18. Inside the nucellus of an ovule, a megaspore has divided and developed into a female gametophyte. At the top is the egg cell accompanied by two synergid cells; at the bottom are three antipodal cells. A single large binucleate central cell makes up most of the gametophyte. A pollen tube grows down through the micropyle, past the two integuments, and releases two sperm cells, one of which fertilizes the egg cell. Afterward, this ovule develops into a seed, the integuments develop into the seed coat, and the fertilized egg develops into an embryo. The other sperm cell fuses with the two polar nuclei to form the endosperm nucleus. (Illustrated by Edgardo Ortiz.)

FIGURE 7.19. This is the center of a cantaloupe cut in transverse section. Each seed is attached by a white stalk (a funiculus) to one of the six placentae. All nutrients needed for the seed to develop are transported through vascular bundles up from the base of the flower, up through the placentae, and out through the stalk to the seed. The cantaloupe flesh has developed from the ovary wall at the base of the carpels.

each ovule, enlarge in preparation for meiosis; these are megaspore mother cells. After meiosis, usually three of the four megaspores degenerate, and only one survives, becoming very large by absorbing the protoplasm of the other three. Megaspores differ from microspores (pollen) because the ovule and the carpel do not open and the megaspore remains enclosed inside the carpel.

An ovule develops into a **seed** after its egg is fertilized, and the surrounding ovary develops into a **fruit** (see Figure 6.3). Each ovary might have either one or many placentae, each bearing one or many ovules; ovaries with just a single ovule develop into fruits with a single seed (such as avocado or peach). Many-seeded fruits (for example tomato, cantaloupe, squash) have numerous placentae, each with many ovules (FIGURE 7.19).

Gametophytes

Microspores develop into male gametophytes. In all angiosperms, each male gametophyte is very small and simple, consisting of at most three cells located within the original pollen cell wall. The microspore nucleus divides mitotically, producing a large vegetative cell and a small lens-shaped generative cell, which subsequently divides and forms two sperm cells. The entire male gametophyte consists of the vegetative cell and the two sperm cells. Although extremely simple and much smaller than a fern gametophyte (see Figure 7.13), the pollen grain is a full-fledged plant (FIGURE 7.20).

After a pollen grain lands on a stigma, it germinates by producing a **pollen tube** that penetrates into the loose, open tissues of the stigma (FIGURE 7.21). The pollen tube grows downward through the style toward the ovary, being protected and nourished by the style tissue. As the pollen tube grows downward, it carries the sperm cells to the ovule.

Within the ovule the megaspore develops into a female gametophyte. In one type of development, the nucleus undergoes three mitotic divisions, producing two, four, and then eight haploid nuclei all in a single, undivided cell. The nuclei migrate through the cytoplasm, pulled by microtubules, until three nuclei lie at each end and two in the center. Walls then form around the nuclei, and the large, eight-nucleate megaspore becomes a female gametophyte with seven cells, one of which is binucleate; this is called an **embryo sac** (FIGURE 7.22). The seven cells are: one large central cell with two polar nuclei, three small antipodal cells, and an egg apparatus consisting of two synergids and an egg. Like the male gametophyte, the female gametophyte is a distinct plant. As with the pollen, the female gametophyte obtains all its nourishment from the parent sporophyte.

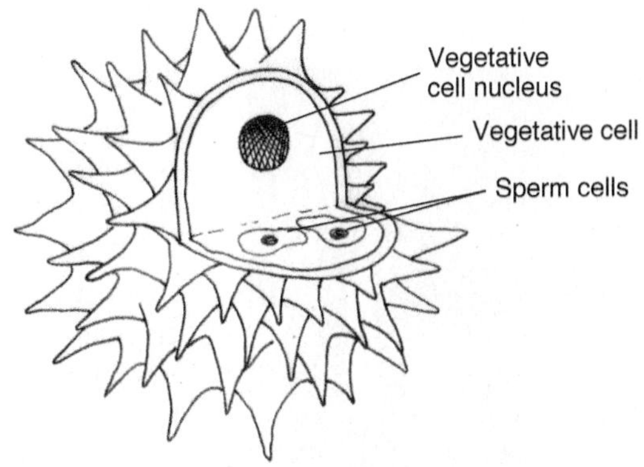

FIGURE 7.20. Each pollen grain develops into a gametophyte that contains one large cell (the vegetative cell) that fills the pollen grain; inside that cell are located two smaller sperm cells. Few cells other than sperm cells ever are located entirely within another cell.

Fertilization

Fertilization (fusion of sperm and egg) involves fusion of the protoplasts and of the nuclei. A pollen tube grows downward through the style toward the ovule then into the ovule's micropyle. The pollen tube penetrates the nucellus and reaches the egg apparatus, then enters one synergid. The pollen tube tip bursts and releases both sperm cells, one of which migrates toward the egg. As it does so, the sperm cell's plasma membrane breaks down and it loses most of its protoplasm. The sperm

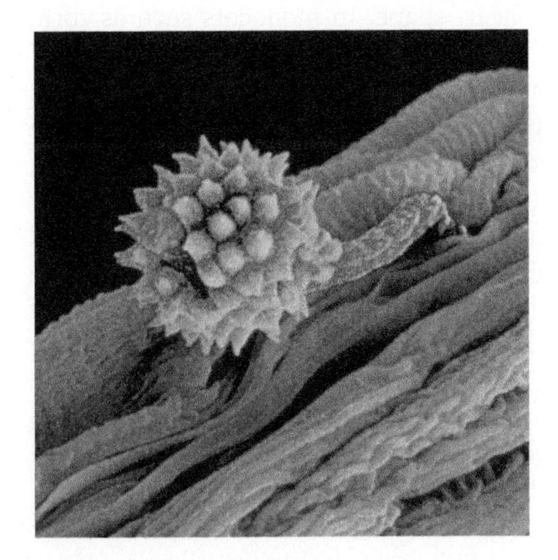

FIGURE 7.21. Once pollen grains land on a stigma, they absorb water, swell, and a pollen tube emerges from one of the germination pores on the pollen grain. The growing pollen tube is a male gametophyte, and it penetrates the stigma, then grows through the style to an ovule in the ovary. The two sperm cells are carried inside the pollen tube.

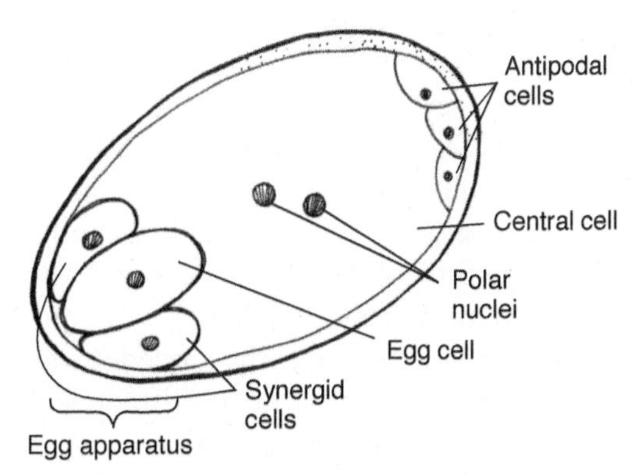

FIGURE 7.22. The female gametophyte (an embryo sac) is a large cell that contains six cells inside itself: the two synergids attract the pollen tube then pass one sperm cell nucleus on to the egg, the other sperm cell nucleus is passed to the two nuclei in the embryo sac. Although much smaller and simpler than a gametophyte of ferns (see Figure 7.13), an embryo sac is considered to be a complete plant.

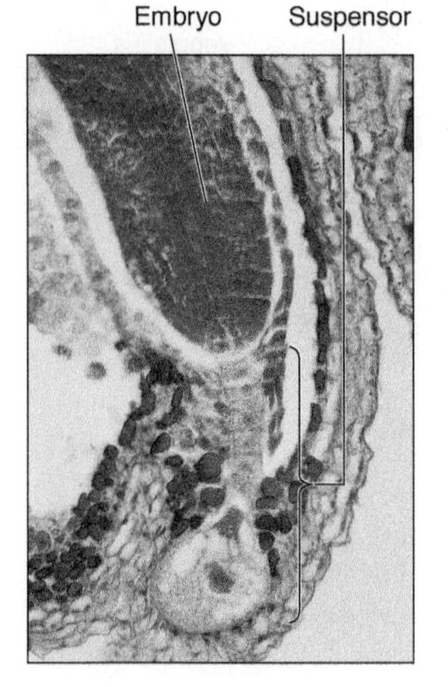

FIGURE 7.23. The suspensor of shepherd's purse (*Capsella*) has one large bulbous cell and a stalk of smaller cells. The young embryo is being pushed deep into the endosperm (×600).

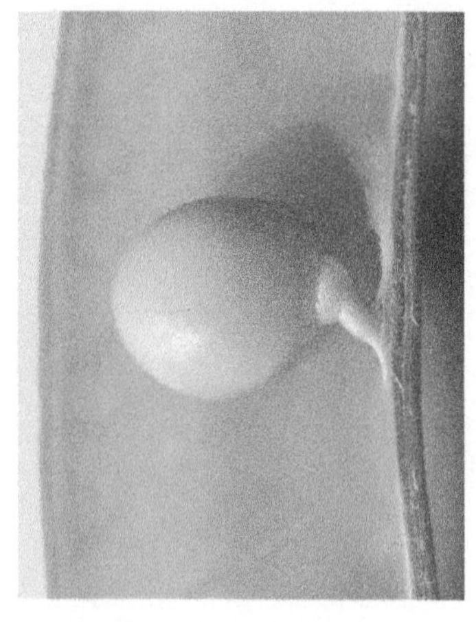

FIGURE 7.24. This pea seed has developed from a fertilized ovule similar to that depicted in Figure 7.17. Its green surface is its seed coat (developed from the integuments of the ovule); its stalk is the funiculus, which attaches to the placenta. The pea pod is the fruit, which developed from the ovary. The seed coat is thin; if we would tear it away, we would see the two cotyledons of the pea, as well as a tiny root and a stem with two immature leaves.

nucleus enters the egg, then is drawn to and fuses with the egg nucleus, establishing a diploid zygote nucleus.

Because the sperm sheds its protoplasm, it contributes only its nucleus with the set of nuclear genes. The sperm does not carry mitochondria or plastids into the egg, so the mitochondrial and plastid genes are inherited only from the ovule parent.

In angiosperms only, the second sperm nucleus released from the pollen tube migrates into the central cell. It fuses with both polar nuclei, establishing a large triploid endosperm nucleus containing three full sets of genes. Because both sperm nuclei undergo fusions—one with the egg nucleus, the other with the polar nuclei—the process is called **double fertilization**. The endosperm nucleus begins to divide very rapidly by mitosis, and the central cell enlarges enormously, usually without cell division, into a huge cell with hundreds or thousands of nuclei. Finally, nuclear division stops and walls are constructed, thus forming cells. An example of this is a coconut full of "milk." The hollow center of the coconut is one single cell, and the milk is its protoplasm. The white coconut "meat" is the region where nuclei form cells. All this tissue is called **endosperm**, and it nourishes the development of the zygote. In grains such as wheat, rice, oats, and corn, the endosperm also passes through a milk stage, but by the time of harvest, it has become cellular, starch-filled, and dry enough to be hard (but if corn is picked for corn-on-the-cob, it is collected just as endosperm is converting from multinucleate to cellular).

Embryo and Seed Development

The zygote grows into a small cluster of cells, part of which later becomes the embryo proper, and the other part becomes a short stalk-like structure, the suspensor, which pushes the embryo deep into the endosperm (FIGURE 7.23). The suspensor is usually delicate and is crushed by the later growth of the embryo; it is rarely detectable in a mature seed.

While very young, the embryo is a small sphere, its globular stage, but then two primordia grow into two **cotyledons** in dicots such as beans and peanuts. While young, the cotyledon primordia give the embryo a heart shape. In monocots such as corn, only one cotyledon primordium grows out. "Dicot" is an abbreviation of "dicotyledon," those plants whose embryos have two cotyledons. Monocots are monocotyledons, plants with only a single cotyledon on their embryos. Later the embryo becomes an elongate cylinder: a short axis is established, consisting of **radicle** (embryonic root), **epicotyl** (embryonic stem), and **hypocotyl** (the root/shoot junction). The epicotyl may bear a few small leaves, and the radicle often contains several primordia for lateral roots. Once mature, the embryo becomes quiescent and partially dehydrates, and the funiculus may break, leaving a small scar, the hilum. In green peas, the two halves of each pea are the two cotyledons, and the stalk attaching each pea to the pod (the fruit) is the funiculus (see FIGURES 7.24 and 6.2).

In most dicots, cotyledons store nutrients used during and after germination. When the seed is mature, the cotyledons are large and the endosperm may be completely used up. We are eating mostly cotyledons when we eat beans, peas, peanuts, almonds, pecans, and other seeds that easily separate into two halves. In monocots, the one cotyledon generally does not become thick and full; instead, the endosperm remains and is present in the mature seed. When eating cereals such as wheat, rice, oats, and corn we are eating almost purely endosperm. A mature seed in which endosperm is rather abundant is an albuminous seed. If endosperm is sparse or absent at maturity, the seed is exalbuminous.

The integuments that surround the nucellus expand and mature into the **seed coat** as the rest of the ovule grows. In their last stages of maturation, the integuments may become sclerenchymatous, tough, and pigmented. It is seed coat color that lets us distinguish between black, red, and pinto beans.

Fruit Development

As the ovule develops into a seed, the ovary matures into a fruit (see Figure 7.19). The stigma and style usually wither away, as do sepals, petals, and stamens, although they may persist at least temporarily. Often three layers become distinct during growth: the exocarp (also called an epicarp) is the outer layer—the skin or peel; the middle layer is the mesocarp, or flesh; and the innermost layer, the endocarp, may be tough like the stone or pit of a cherry or peach, or it may be thin. The relative thickness and fleshiness of these layers vary with fruit type, and often one or two layers are absent. The entire fruit wall, whether composed of one, two, or all three layers, is the **pericarp**.

Fruits are adaptations that result in the protection and distribution of seeds. Many different agents disperse fruits and the seeds they contain: gravity, wind, water, and animals are the most common. Fruits that are tough and full of fibers or sclereids, such as pecans, walnuts, Brazil nuts, and coconuts, offer maximum protection but are so heavy they don't travel very far and germination is difficult. More fragile fruits are better for easy germination and wide dispersal. If animals are to disperse the seeds, part of the fruit must be edible or otherwise attractive while the embryo is protected.

The term "pericarp" refers to the tissues of the fruit regardless of their origin. In most cases, this is the ovary wall, but in many species the receptacle tissues or sepal, petal, and stamen tissues may also become involved in the fruit. The terms "pericarp" and "fruit" have been applied to both types of fruit, so now the term true fruit is used to refer to fruits containing only ovary tissue, and accessory fruit (or false fruit) is used if any nonovary tissue is present. Apples develop from inferior ovaries and the bulk of the fruit is enlarged bases of sepals and petals; only the innermost part is true fruit derived from carpels. If the fruit develops from a single ovary or the fused ovaries of one flower, it is a simple fruit, the most common kind. If the separate carpels of one gynoecium fuse during development, an aggregate fruit results, such as raspberries. If during development all the individual fruits of an inflorescence fuse into one fruit, it is a multiple fruit, as in figs, mulberries, and pineapple. These are also largely accessory fruits because in addition to the ovary tissue, the inflorescence axis, bracts, and various flower parts contribute to the mature fruit.

Before moving on to the next section, think about a peach tree in full bloom. The tree has thousands of flowers, each with stamens and carpels. When it has finished producing pollen and megaspores, it has technically finished its own reproduction and now has tens of thousands of progeny in the form of male gametophytes (pollen grains) and female gametophytes (inside each ovule in each carpel). Bees carry pollen from other peach trees to the stigmas of this one, and before long it has thousands of pollen tubes growing inside its styles. You see one diploid peach tree but inside it are thousands of haploid pollen tubes (which are also peach plants) and female gametophytes (also peach plants). All the water, minerals, and photosynthates needed to keep all these thousands of plants alive are being supplied by the leaves, roots, xylem, and phloem of the diploid peach tree. We cannot say that the haploid plants are parasitizing the tree because they are essential to its production of seeds. After fertilizations have occurred, hundreds of zygotes develop into embryos, each of which is the progeny of male and female gametophytes, and the "grand progeny" of the peach tree itself. And this tree has other "grand progeny" as well, produced by its pollen that was carried to other peach trees. This tree is the ovule parent of its own seeds and the pollen parent of thousands of seeds in many other trees.

Modifications that Affect the Functioning of Flowers

Flowers are also involved in the dispersal of pollen and seeds. Because numerous mechanisms carry out these processes, numerous types of flowers and fruits exist.

Cross-pollination is the pollination of a carpel by pollen from a different individual; **self-pollination** is pollination of a carpel by pollen from the same flower or another flower on the same plant. With cross-pollination, sperm cells and egg cells from different plants unite, resulting in new combinations of genes, at least a few of which may be better adapted than either parent. But self-pollination has about the same result as asexual reproduction because all genes come from the same parent; if a plant is isolated by distance or lack of pollinators, self-pollination allows it to set seed and propagate its genes rather than lose them when the plant dies.

Self-pollination in flowers that have both stamens and carpels is prevented if anthers and styles mature at different times. In many species, anthers release pollen while stigma tissues are immature and unreceptive; exposed pollen lives only briefly, and when the stigma becomes mature there may be no living pollen left in the flower. In many species, self-fertilization is inhibited by **compatibility barriers**, chemical reactions between pollen and carpels that prevent pollen growth. If the pollen tube and carpels are closely related, chemicals on their surface match, react with each other, and prevent the pollen tube from growing.

Among incomplete flowers there is a significant difference between flowers that lack sepals or petals and those that lack stamens or carpels. The latter two organs are *essential organs* because they produce the critically important spores. Flowers that lack either or both essential organs are not only incomplete but also **imperfect flowers**. If a flower has both, it is **perfect** even though it may lack either sepals or petals or both. Sepals and petals do not produce spores and are considered *nonessential organs*.

It is necessary to consider the whole species not just individual flowers. Stamens produce pollen that results in sperm production and carpels are involved in egg production, so a species must have both types of organs. Plants that have perfect flowers satisfy this requirement. But if some flowers of the species are imperfect, having no stamens for instance, then other flowers must have stamens. A large number of combinations is possible: a species may have individuals that produce only staminate flowers and others that produce only carpellate flowers—this is dioecy, and the species (not the flower or the plant) is said to be dioecious (pronounced "dye EE cy" and "dye EE shus"). Examples of dioecious species are marijuana, dates, willows, and papaya. In dioecious species, the life cycle consists of four types of plants: (1) male gametophytes, (2) female gametophytes, (3) staminate sporophytes, and (4) carpellate sporophytes. Monoecy is the condition of having staminate flowers located on the same plant as carpellate flowers; monoecious species include cattails and corn—ears are clusters of fertilized carpellate flowers and tassels bear numerous staminate flowers (FIGURE 7.25).

FIGURE 7.25. Corn (*Zea mays*) is an example of a monoecious species: each plant has carpellate flowers (the ears) and staminate flowers (the tassels).

Most plant/animal relationships are a battle: animals try to eat plants, lay eggs in them, or do other harmful things. Plants defend themselves with poisons, spines, and sclereid barriers. Mutations that permit plants to make more effective deterrents benefit the plant but not its animal pests. But both plants and animals tend to benefit from the pollination relationship, and **coevolution** is possible: mutations that make a plant more recognizable or more convenient to its pollinators help both the plant and the animal, just as certain mutations in pollinators are beneficial to both organisms.

The shape of many flowers has been strongly affected by coevolution because a pollinator actually makes contact with it. Most flowers are radially symmetrical; that is, any longitudinal cut through the middle produces two halves that are mirror images of each other. These flowers (and stems and roots) are actinomorphic or regular (FIGURE 7.26). But all insects, birds, and bats are bilaterally symmetrical—only one longitudinal plane produces two halves that are mirror images. In many species, flowers and pollinators have coevolved in such a way that the flowers are now also bilaterally symmetrical—zygomorphic (FIGURE 7.27). When a pollinator approaches a zygomorphic flower, only one orientation is comfortable for it; any misalignment prevents the pollinator's head or body from fitting the flower's distinctive shape. As a result, as the pollinator feeds at the flower, pollen is placed on a predictable part of its body. When it visits the next flower, pollen is rubbed directly onto the stigma.

In species that are wind-pollinated, attracting pollinators is unnecessary, so mutations that prevent the formation of petals are selectively advantageous and the energy saved can be used elsewhere in the plant. Sepals are also often reduced or absent and the ovaries

FIGURE 7.26. Flowers of water lily (*Nelumbo*) have radial symmetry. If we would slice this flower from top to bottom or from left to right, or at any angle, the two halves would be mirror images of each other as long as the cut passed through the center of the flower.

(a)

(b)

FIGURE 7.27. **(a)** Flowers of Christmas cactus (*Schlumbergera*) are bilaterally symmetrical. The right and left sides are mirror image of each other, but the top and bottom are not. Most animals also have bilateral symmetry, so a bee will fit into this flower only by walking in right side up. **(b)** This is the type of view a pollinating insect has as it approaches a bilaterally symmetrical flower. The insect fits into the flower well only if it flies in right side up, as it normally would do. The flower consequently must grow such that it is oriented as the insect is. Each flower has two stamens that put pollen on the back of the insect visitor. (Courtesy of Jon Rebman, San Diego Museum of Natural History.)

TABLE 7.2. Inflorescence Types

Inflorescences with one main axis:

A raceme has a major inflorescence axis and flowers are borne on pedicels that are all approximately the same length.

Catkins are similar to racemes, differing in that the flowers are imperfect, either staminate or carpellate, and all flowers of a single inflorescence are the same, so each species must have both staminate catkins and carpellate ones. Catkins almost always contain very small wind-pollinated flowers.

A spike is similar to a raceme except that the flowers are sessile, lacking a pedicel.

A spadix (plural: spadices) is a spikelike inflorescence with imperfect flowers, but both types occur in the same inflorescence. The main inflorescence axis is thick and fleshy with minute flowers embedded in it; the entire inflorescence is subtended or enclosed by a petal-like bract called a spathe.

A panicle is a branched raceme with several flowers per branch.

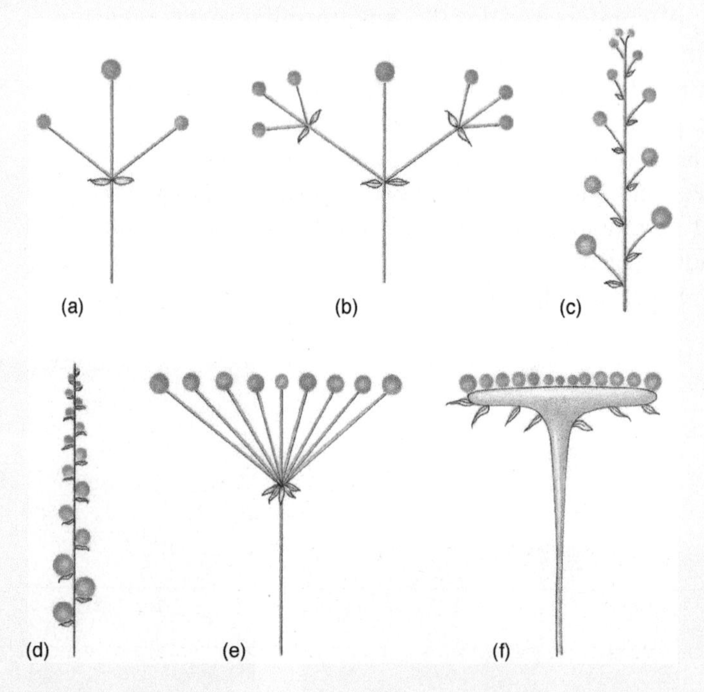

FIGURE T7.2.1. Inflorescence types: **(a)** Simple determinate inflorescence. **(b)** Compound determinate inflorescence. **(c)** Raceme. **(d)** Spike. **(e)** Umbel. **(f)** Head. Larger circles represent flowers that open earlier than those depicted as smaller circles.

Inflorescences without a dominant main axis:

In umbels, the inflorescence stalk ends in a small rounded portion from which arise numerous flowers. Their pedicels are long and arranged so that all flowers sit at the same height, forming a flat disk.

A head is similar to an umbel except that the flowers are sessile and attached to a broad expansion of the inflorescence stalk; numerous bracts may surround the inflorescence during development. Heads are almost synonymous with the aster family, sunflowers and dandelions being easily recognizable examples. In this group, the inflorescences are so compact and highly organized that they mimic single flowers; what appear to be the petals are really entire flowers, ray flowers, in which the petals are very large and fused together. The center of the inflorescence is composed of a different type of flower, disk flowers, in which the corollas are short and inconspicuous.

need no special protection, so the whole flower may be tiny. Once pollen is released to the wind, the chance of any particular pollen grain landing on a compatible stigma is small, so huge numbers of grains must be produced. Large, feathery stigmas are adaptive by increasing the area that can catch pollen grains. In general, wind-pollinated individuals produce up to several thousand small flowers; although each flower is tiny, the entire plant has a large total stigmatic surface area. Wind-pollinated species like grasses, oaks, hickories, and all conifers grow as dense populations in rangelands or forests. Within a square mile may be found thousands of plants and, more importantly, millions of stigmas.

FIGURE 7.28. Daisies, sunflowers, and asters are often mistakenly thought to be flowers, but actually they are inflorescences, clusters of many small flowers. Flowers on the edge of this cluster each have one large yellow petal, and flowers in the center of the group each have five tiny yellow petals. This is balsamroot, *Balsamorhiza*.

In some species, after all flower organ primordia are initiated at the receptacle apex (all parts are microscopically small when initiated), the primordia crowd together and the bases of the stamens, petals, and sepals fuse, creating a thick layer of protective tissues around the ovaries, which *appear* to be located below the other organs. We say either that the ovary is an **inferior ovary** or that the other parts are epigynous. Inferior ovaries also can result if receptacle tissue grows upward around the ovary. The more common arrangement, in which no fusion to the ovary occurs and the ovary is obviously above the other flower parts, is a **superior ovary** or hypogynous parts. Intermediate, partially buried ovaries are half-inferior with perigynous flower parts.

Few plants produce only a single flower; instead many flowers are produced either within a single year or over a period of many years. A grouping of flowers is an **inflorescence**, and it gives a collective visual signal to pollinators: one small flower may be overlooked, but not a hundred close together (FIGURE 7.28). Furthermore, in an inflorescence, the plant is able to accurately control the timing of the initiation, maturation, and opening of the flowers. Consequently, the plant can be in bloom and available to pollinators for several weeks even though each flower lasts only a day or two (**TABLE 7.2**).

Important Terms

1n
2n
allele
alternation of generations
anaphase
anther
asexual reproduction
carpel
cell cycle
cell plate
chromatid
chromosome
coevolution
compatibility barrier
complete flower
cotyledon
cross-pollination
crossing-over
cytokinesis
diploid
double fertilization
duplication division
embryo sac
endosperm
epicotyl
fruit
gamete
gametophyte
genome
haploid
heteromorphic generations
homologous chromosomes
hypocotyl
imperfect flower

incomplete flower
inferior ovary
inflorescence karyokinesis
meiosis
metaphase
metaphase plate
mitosis
mutagen
mutagenic
mutation
ovary
ovule
perfect flower
pericarp
petal
phragmoplast
pollen
pollen tube
prophase
radicle
reduction division
seed
seed coat
self-pollination
sepal
sexual reproduction
somatic mutation
spindle
spore
sporophyte
stamen
stigma
style
superior ovary
synapsis
syngamy
telophase
zygote

Concepts

- No individual is immortal.
- Genetic diversity is necessary for evolution by natural selection.
- Individuals with combinations of alleles that make them well adapted can maintain that combination in their progeny by reproducing asexually.
- Sexual reproduction is a means of producing many combinations of alleles and creating genetically diverse progeny that might be more well adapted than the parents, are better adapted to other habitats (allowing the species to expand its range), or better able to adapt to a changing habitat.
- All plant life cycles involve an alternation of heteromorphic generations.
- All plant species have tissues that make spores (inside anthers and ovules in angiosperms) and other tissues that make gametes (the gametophytes in all plants).

Genetics: Transferring Information from Generation to Generation

8

Think of your mother and father. Almost certainly you do not look exactly like either one of them. You may resemble your mother in certain features, your father in others. But some of your features may differ greatly from theirs: you may be taller than either of them; more or less talented at sports, music, and art; or you may have a genetic disorder that neither parent has. And you undoubtedly differ from your brothers and sisters as well (unless you are an identical twin). But you know your entire DNA and all your genes came from your parents, as did that of your siblings. Why does each family show such diversity?

Now think of going into a grocery store and examining the produce. There is no diversity in the heads of lettuce: they all look the same (FIGURE 8.1). Who could tell which bunches of celery, which carrots, and so on came from the same parent plants? Each of these plants too has a maternal parent and a paternal one, and each has siblings that grew from seeds produced by the same parent: why is there so little diversity here?

As described in Box 1 in Chapter 7, all individuals accumulate mutations as they replicate their DNA or as accidents occur in the lives of their cells. Consequently, most genes occur in various forms, called alleles, and each allele produces a protein that differs from that produced by other alleles of the same gene. The differences in proteins may affect cell metabolism and thus the general health, vigor, and characters

Our crop plants do not have their features by accident. The ancestral plants that were originally noticed by people and then domesticated had features as the result of natural selection. But after domestication, we examine fields for variations that are even more useful or more suitable and use those as the parents of the next crop; this is artificial selection. At present, we try to use parent plants that will produce seeds that grow into plants that are more resistant to disease, have more balanced nutrition, or more beautiful flowers, and so on.

FIGURE 8.1. This is a field of wheat that has been partially harvested. You can see how extremely uniform these plants are; this variety of wheat has been bred such that the plants all have the same height. Older types of wheat varied in height, and many plants were so tall and spindly they would fall over in wind or rain, damaging the crop. The harvester would have to cut the wheat close to the ground and would often take in dirt and rocks, damaging the machine. But this modern wheat all stands erect, tall enough to have enough leaves for rapid photosynthesis, tall enough for the harvester to cut it safely, but short enough to not fall over. Whereas older types of wheat used a great deal of their carbohydrate to build tall stems, this newer wheat uses more of its carbohydrate to build seeds.

of the individual organism. Some genes, such as those for hair color in people, are not essential for survival, so mutations that affect these genes are not life threatening and many alleles occur in nature. The proteins coded by other genes, however, can be so crucial that almost any mutation causes an allele to make a protein that just doesn't work: individuals with those mutations die and the allele is lost immediately. An example of this type of gene in plants is RUBISCO, the gene that codes for the enzyme critical in capturing carbon dioxide during photosynthesis. If the nucleotide sequence for this gene is examined in many individuals of the same species, no diversity at all is found. Even if the search is extended to all plant families, very little variation is found: for hundreds of millions of years, virtually every allele produced by mutations in this gene have been eliminated.

Because of the presence of alleles, individuals of a species differ from each other. Each makes gametes that carry all the same genes as any other gamete in the species, but they differ in the versions (alleles) of those genes. And because of crossing-over during meiosis, almost every gamete made by any one parent differs from every other gamete made by the same parent. The diversity of gametes is great, and the diversity of progeny is even greater.

So why are people so diverse whereas crop plants are not? Mutations occur in both, but with people, we do all we can to keep everyone alive; even when we have alleles that cause poor eyesight, allergies, or problems with our metabolism, we live and reproduce and keep those alleles available. With people, we see the results of trying to maximize genetic diversity. Crop plants and livestock could be just as diverse, but geneticists have identified many of the alleles that produce desirable characters, then they carefully select certain plants and animals to be the parents of the next crop. Here, we are seeing the results of efforts to minimize diversity.

Genetics is the scientific study of genes and their alleles, and of the characters they produce. The genes of an organism constitute its **genotype** (it should really have been named allelotype), and the characters they produce are the **phenotype**. Through genetic studies, we study how genes are passed on from one generation to another during sexual reproduction, and which alleles affect which characters.

Experiments that Consider Only One Trait: Monohybrid Crosses

Sexual reproduction between two individuals is called a **cross**, and it involves the fusion of two gametes. Because of meiosis, each gamete is haploid and each contains one complete set of genes; the zygote (the fertilized egg) has two complete sets.

Within a population, mutations produce new alleles, and the genotypes of individuals within the population differ. Of the plants that grow in an area and that can interact sexually, many may have the same allele of a particular gene and other individuals may have other alleles, other versions of the gene. Consequently, the alleles carried by a particular sperm cell may or may not be identical to the homologous alleles of the egg it fertilizes.

Traits that Have Incomplete Dominance

In a **monohybrid cross** only a single character is analyzed and studied; the inheritance of other traits is not considered. For instance, a plant with red flowers might be crossed with one that produces white flowers, and only the inheritance of that flower color trait is studied (FIGURE 8.2). Characters involving flower shape, leaf structure, and photosynthetic efficiency are also being inherited simultaneously, but in a monohybrid cross, only one is studied. This makes the analysis and understanding of the results much simpler. Once the basic principles of a particular trait are known, then its interaction with another factor (dihybrid cross) or two other factors (trihybrid cross) can be studied. Gregor Mendel (1822–1884) performed experiments with pea plants and pioneered the concept of studying only one trait at a time (BOX 8.1). His results were easy to understand and his crossbreeding experiments became the basis for modern genetics. Before Mendel, people tried to analyze many characteristics at a time and became hopelessly confused.

Consider the cross just mentioned: A plant with red flowers is bred to one with white flowers (FIGURE 8.3a). It does not matter which flower produces pollen and which produces ovules. Once the cross is made, seeds and fruit develop; the seeds are planted, and when mature, the new plants are allowed to flower so that their flower color can be examined. The flowers of all plants in this new generation are pink, resembling each parent somewhat, but not exactly like either. The parents are called the **parental generation**, the progeny of their crossbreeding are the **F_1 (first filial) generation**, and if these interbreed, their progeny are the F_2 generation.

FIGURE 8.2. Many species have a particular flower color; for example, bluebonnets are usually blue and buttercups typically are yellow (like butter). But occasionally there are mutations that change the color of flowers, and these are often extremely popular in horticulture because they are unusual. Very often, they are crossed with the ordinary plants in attempts to obtain hybrids that have intermediate colors. (© Hemera/Thinkstock.)

The molecular biology of this monohybrid, flower color cross is easy to understand. Each parent is diploid and each has two copies of the gene involved. In the red-flowered parent, both alleles produce mRNA that is translated into functional enzymes involved in synthesis of red pigment. In a white-flowered parent, both alleles are defective. It may be that each produces an mRNA that results in a protein unable to perform the necessary reaction. Or the promoter region may be mutated and can no longer interact with a chemical messenger. Whatever the cause, there is no pigment, and the flower is white.

Using genetic symbols, the red-flowered parent is *RR* and the white-flowered one is *rr*. Each parent is said to be **homozygous**, because each has two identical alleles for this gene. The pink-flowered F_1 has received an *R* allele from one parent and an *r* allele from the other, so its genotype is *Rr*; it is **heterozygous** because it has two different alleles for this gene. We use these symbols even though we have neither isolated the gene nor analyzed its nucleotide sequence; *R* and *r* are simply labels. Most genes are known only by their phenotypes and the labels given to them by geneticists. With a genotype of *Rr*, the plant produces mRNA, half of which carries the defect; thus only half the normal amount of enzyme is produced. This results in less pigment being formed, only enough to make the flower look pink, not red. Neither parental trait dominates the other, so this pair of alleles shows **incomplete dominance**: The heterozygous phenotype differs from both homozygous phenotypes.

When analyzing the possible outcomes of crosses and breeding, we must understand the types and quantities of gametes involved. All the plants we are considering are diploid and form haploid spores by meiosis. The *RR* parent has chromosomes as shown in FIGURE 8.3b, and regardless of how chromosomes separate during meiosis, all spores receive an *R* allele. In the *rr* parent, all spores receive an *r* allele. Spores are not gametes, but as they develop into gametophytes, they divide by mitosis—duplication division—so all cells of the male gametophyte (pollen) have the same alleles, as do all cells of the female gametophyte (the embryo sac inside the ovule). Because the *RR* parent produces only *R* spores, it also produces only *R* gametes, both sperms and eggs in typical bisexual flowers. Similarly, the *rr* parent produces only *r* gametes. When the two plants are interbred, an *R* gamete unites with an *r* gamete, establishing a heterozygous (*Rr*) zygote that grows into a heterozygous adult by means of mitotic cell divisions. No other outcome is possible; it does not matter which gamete is which, an *R* sperm and an *r* egg result in the same type of zygote as an *r* sperm and an *R* egg.

When the heterozygote matures and flowers, each spore mother cell produces two types of spores, not just one as is true of a homozygote. Because each cell has one *R* allele and one *r* allele, during the first meiotic division, one daughter cell receives *R* and the other receives *r*. The second division of meiosis results in two *R* spores and two *r* spores. In the anther, half the pollen grains and hence half the sperms have *R*, and the other half have *r*. In the ovules of many plants, only one megaspore survives: half the time *R* cells survive and half the time *r* cells live. In heterozygote parents, the two types of sperms and eggs are produced in equal numbers.

Crossing Heterozygous Plants with Each Other

When a plant's own pollen is used to fertilize its own eggs, the cross is a selfing. A plant can also be selfed by being crossed with another plant with exactly the same genotype. Selfing heterozygotes has interesting, instructive consequences; 50% (on average) of all sperms and eggs contain the *R* allele and 50% have *r* (FIGURE 8.4). Zygotes will not be identical genotypically: some will be *RR*, having resulted from an *R* sperm and an *R* egg, others will be *rr* (*r* sperm and *r* egg), and some will be *Rr* (either *R* sperm and *r* egg or *r* sperm and *R* egg)(see lower half of Figures 8.3a and 8.3b). Selfing a heterozygote produces three types of F_1s, some of which (*Rr*)

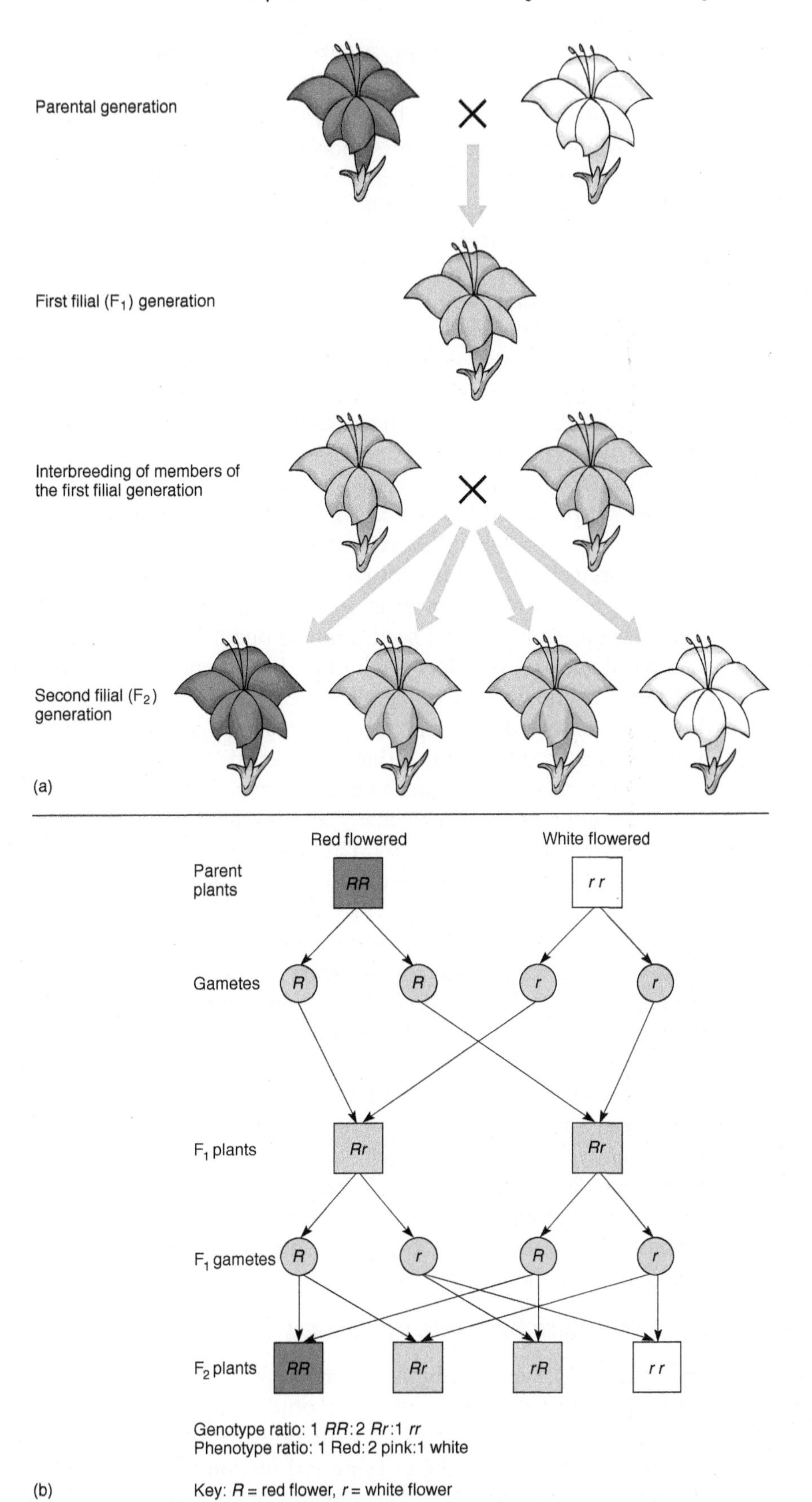

FIGURE 8.3. **(a)** A monohybrid cross analyzing the trait of flower color. Details are explained in the text. **(b)** The genotypes of parents, gametes, and progeny.

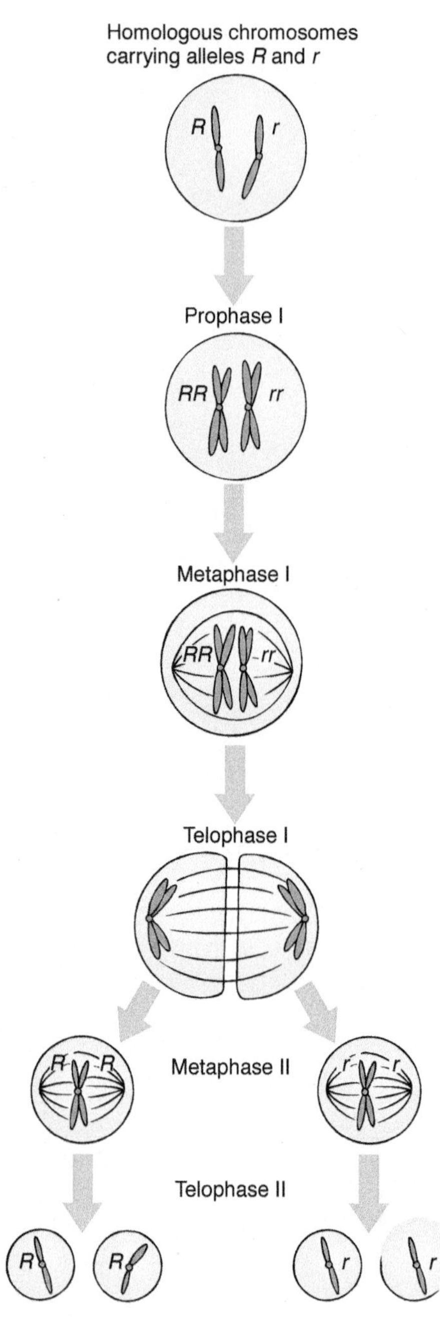

FIGURE 8.4. Production of haploid cells by meiosis in a heterozygote. Homologous chromosomes pair (synapse) during metaphase I, then the paternal chromosome is separated from the maternal chromosome during anaphase I: one type of allele is separated from the other. As sister chromatids separate after metaphase II, each daughter nucleus receives identical alleles.

resemble the parents and others (*RR* and *rr*) the grandparents. Again, to analyze the results, we must wait for the zygote to develop into an embryo, then plant the seeds and wait until the new plants are old enough to flower. Because each genotype produces a distinct phenotype, the genotype of each plant is known simply by looking at the flowers. If a large number of heterozygotes (pink-flowered plants) are selfed and large numbers of their seeds are grown, about one-fourth of the resulting plants have red flowers (*RR*), one-half pink (*Rr*), and one-fourth white (*rr*). This is an important ratio, typically represented as 1:2:1, and should be memorized immediately.

The reason for the proportion of these genotypes is explained in FIGURE 8.5. A **Punnett square** can be set up in which all types of one gamete, say, the egg, are arranged along the top of the square, and all types of the other gamete are arranged on the left side. The boxes are then filled in with the allele symbol above it and to the left. Because the gametes are produced in a ratio of 1 *R*: 1 *r*, listing them as in Figure 8.5 automatically represents their relative numbers in nature.

The Punnett square does not represent the outcome of any one cross; if a single heterozygous flower is selfed, it may produce only one or two seeds. If you plant just one seed, there is one chance in four that it will have red flowers, two chances in four pink, and one chance in four white.

The 1:2:1 ratio was one of Mendel's great discoveries. In a selfing of this type, the recovery of the parental types means that genetic material must be composed of particles, such that the *R* genetic material can be separated from the *r* genetic material in the *Rr* heterozygote. Prior to Mendel's work, it was thought that the genetic material was a fluid. Two fluids cannot mix and then separate again perfectly. The constancy of the 1:2:1 ratio made it more logical to think in terms of discrete particles, genes, that never lost their identity regardless of the crosses in which they participated.

Furthermore, the 1:2:1 ratio is most easily interpreted in terms of each individual plant having two copies (being diploid) and each sex cell having one copy (being haploid). If each plant had only one copy, pink-flowered heterozygotes would be impossible, whereas if each had three, phenotypes such as dark pink (*RRr*) and light pink (*Rrr*) should also be present. Keep in mind the state of scientific knowledge in 1865 when Mendel was working. The concept that all organisms are composed of cells with nuclei had only recently been proposed; mitosis and cell division were very poorly understood, and meiosis would not be discovered for another 23 years. The concepts of chromosomes and homologous pairs would not be well established until the twentieth century, and the existence of mRNA was not confirmed until the mid-1960s.

Traits that Have Complete Dominance

The situation in which only half as much product of an enzyme, such as the red pigment discussed above, is produced in a heterozygote is not universal. In certain species or with other traits, cytoplasmic control mechanisms may cause the enzyme to function until a specific amount of product is synthesized. The enzyme may have to work faster or longer, but the final amount of product is the same whether the plant has two functional alleles or only one. In other situations, the amount of enzyme might be monitored such that the nonmutant mRNA of the heterozygote is translated more frequently or rapidly.

In either case, the phenotype of the heterozygote is like that of the parent with two effective alleles. That trait is said to be **dominant** over the other version of the trait, which is **recessive**. An example is height. A tall plant with a *TT* genotype produces gametes that are all *T*. A short plant, genotype *tt*, produces only *t* sex cells. When the tall plant is crossed with the short one, all F_1 progeny have the *Tt* genotype, but they all have the phenotype of being tall, indistinguishable from their *TT*, homozygous dominant parent (FIGURE 8.6). We do not know what protein is produced by the *T* allele, but even with only one functional allele, enough product is made to permit normal growth. *Tt* plants are tall, and the tall character completely dominates the "dwarf" character.

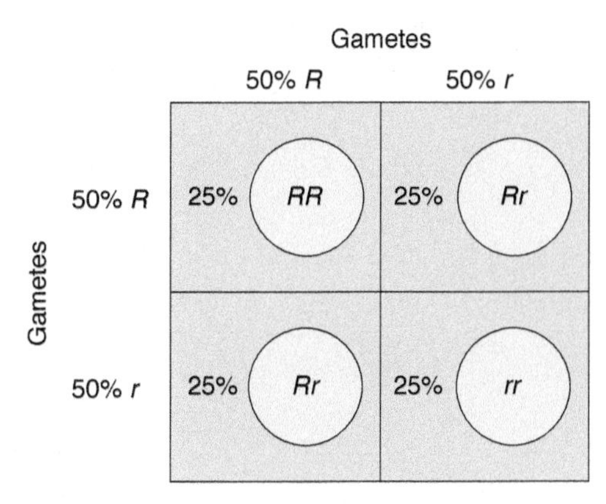

FIGURE 8.5. A Punnett square makes it easy to analyze and understand the results of a cross. Gametes from one parent are listed across the top; those of the other parent are listed on the left side.

When the heterozygotes are selfed, two types of sperm cells (*T* and *t*) and two types of eggs (*T* and *t*) are produced. The Punnett square for the cross is shown in FIGURE 8.7, and a genotype ratio of 1 *TT*: 2 *Tt*: 1 *tt* is expected. The prediction is correct, but what will the phenotype ratio be? One out of four plants will be tall due to a *TT* genotype, and two out of four will be tall due to a *Tt* genotype. Thus, three-fourths have the tall phenotype and one-fourth have the short phenotype. Whenever a cross is made and a phenotype ratio of 3:1 is seen, we should suspect that the parents are heterozygous and the character shows **complete dominance**.

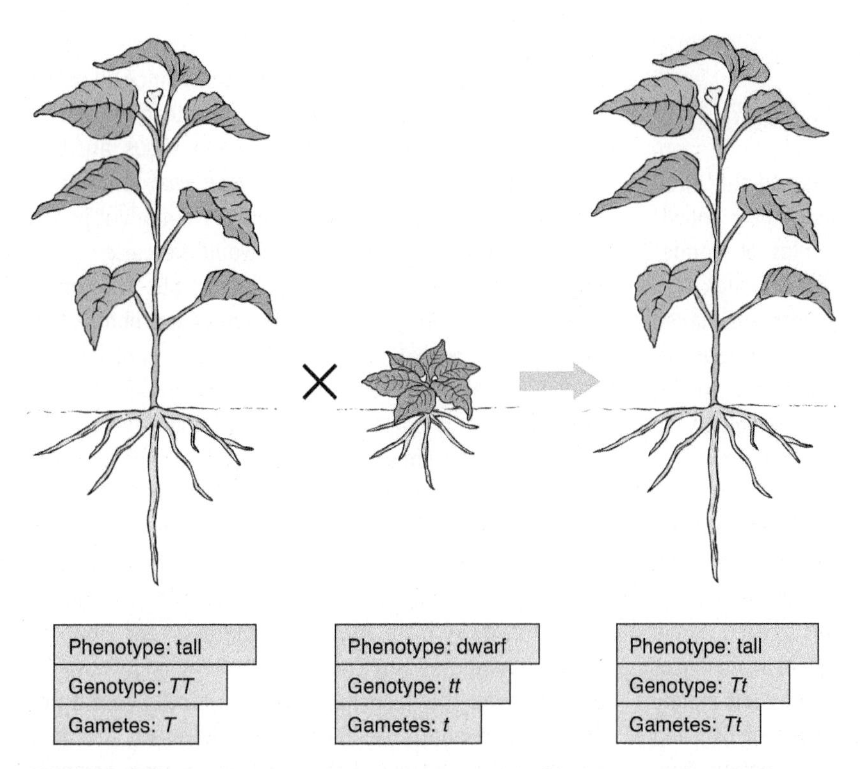

FIGURE 8.6. In this monohybrid cross, the trait "tall" shows complete dominance over the trait "dwarf." In a *Tt* nucleus, the *T* allele may be transcribed twice as much as each *T* allele in a *TT* nucleus; the resulting mRNA may be translated twice as much, the protein may work twice as long or twice as fast, or a *TT* plant may contain a level of *T* protein far above the threshold for responsiveness.

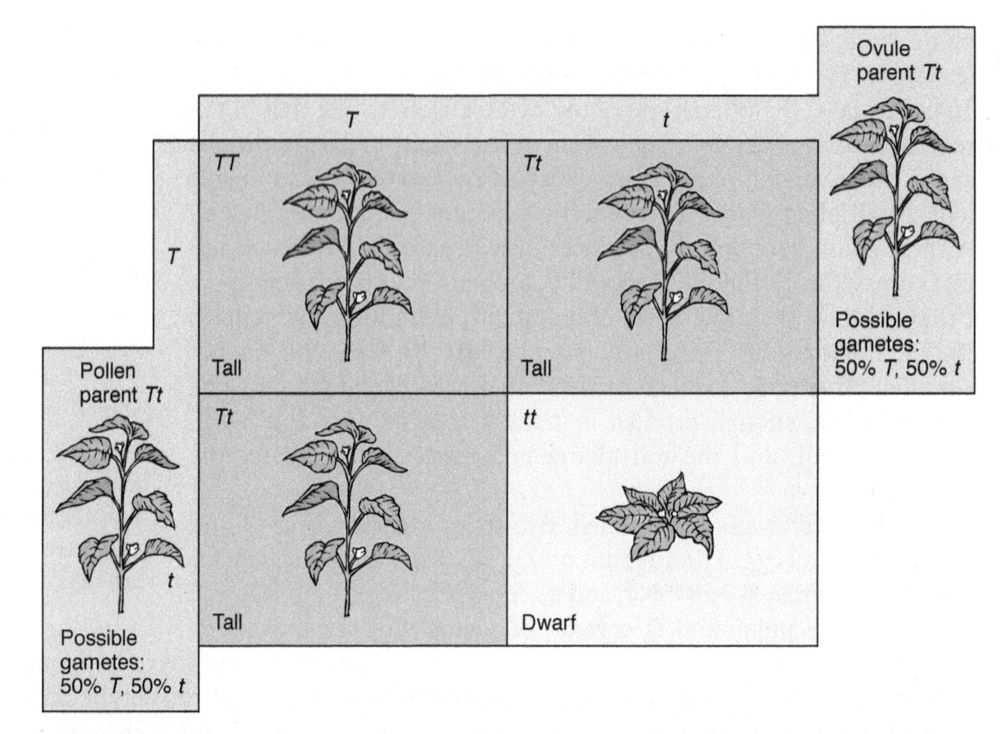

FIGURE 8.7. In setting up the Punnett square for a selfing of *Tt* plants, first establish the genotypes of the two parents. Then determine what types of gametes are produced and in what proportions, and fill in the squares with the genotypes. From the genotypes, the phenotypes in each square can be determined.

BOX 8.1. Mendel and Our Changing View of the World

It is worthwhile to think about how our ideas and concepts, how our entire view of the world develops as new knowledge becomes available. The simple principles of Mendelian genetics had profound consequences for us. Look at a Punnett square: the alleles of the male gamete are placed on one edge, those of the female are placed on another; then the spaces are filled in. Most important, the spaces are filled in equally: both gametes contribute equally to the individual being produced by sexual reproduction. This concept of equality is fairly recent.

The role of men and women, or more generally of male and female, in sexual reproduction was a mystery for centuries. People knew that both males and females are necessary, and there were many theories about the roles of each. Some believed that females were the more important contributor to reproduction: eggs are large and easily visible, not merely eggs of chickens and ostriches, but even those of fish and frogs were well known before microscopes were invented. And in us mammals, the new progeny start their development inside the mother, not the father. But other people believed that males were more important, since all important people were male: kings, popes, military leaders, etc.

Microscopes were invented in the mid-1600s, and by 1677 Antoni van Leeuwenhoek had discovered that semen contains something that moves. Your first impression might be that he had discovered sperm cells, so it would be clear: 1 sperm + 1 egg = 1 zygote. But it was not that simple. The concept of "cell" had just been proposed only a few years before, and just for cells in cork, not for other parts of plants or for any part of any animal. Also, Leeuwenhoek had previously discovered "animalcules," microscopic protozoans and animals that live in pond water. Consequently, when the moving things in semen were discovered, they were assumed to be parasitic microbes, not part of the male's body, but an infection. A problem with this interpretation was that boys before puberty were not infected, nor were very old men. How could it be that every male animal became infected? How could a parasite's eggs get into a male's testes? You might assume that people would then conclude that sperm cells were a product of the male body, like blood cells; however, at the time, people did not believe that animal bodies were composed of cells. But people did believe in spontaneous generation: that life arises repeatedly every day. Mushrooms appear overnight from nothing, and when an animal dies, maggots arise. It was believed that the world contains a mysterious force, a "vital force," that could transform itself into living creatures spontaneously. To many of the scientists at the time (all of them men), it was only natural that men should have so much vital force that it would just spontaneously turn into animalcules, and what better place than in their testes? It all seemed perfectly logical. As microscopes became better, it was realized that maggots develop from eggs of flies and mushrooms arise from fungal cells in the soil, and the concept of spontaneous generation was abandoned around the middle of the 1800s. But whereas these microscopes were good enough to see fly eggs and sperm cells in semen, they were not good enough to see details clearly, and some people thought they could see a completely formed, miniature human being (a homunculus) inside each sperm cell; their interpretation was that the sperm cell carried a human being that just needed an "incubator"—an egg—in which to develop. Such a view meant that males were the truly important entity in reproduction. But other microscopists thought they saw a homunculus inside each egg, so the only role of the sperm cell was to awaken the homunculus from dormancy; this meant that females were the truly important entity.

Between 1838 and 1842, Matthias Schleiden and Theodor Schwann proposed the cell theory, which postulated that all organisms are composed of cells and that all cells come from preexisting cells. But microscopes were not yet good enough to see certain critical details. Did sperm cells merely come close to an egg cell or did they enter it? If they only needed to come close, it must mean that the egg contained the homunculus, or at least all the information necessary to make a new individual. If the sperm entered the egg, then it must be important. It was known that moving a magnet near a wire caused the formation of electricity: could sperm cells be acting in the same way, merely giving rise to some force inside the egg without needing to do anything other than come near it? And if sperm cells did enter the egg, how many? Semen contains millions of sperm cells, the idea that only one would enter seemed unlikely. But in 1865, Mendel pointed out that his results indicated that one sperm should enter one egg, and that each would contribute equal amounts of information to the new zygote.

Unfortunately, at this same time, many other people also had been studying genetics, but not coming up with the clear results Mendel obtained. Plant breeders knew that characters of the pollen parent were important and could show up in the progeny, so males must be passing some information along, and seeds were not obtaining all their information from the egg. But the other geneticists were considering many complex features simultaneously and were not analyzing their results mathematically. It seemed that progeny were a complicated, unpredictable mix of characters from the male and female parent, a blending of the two. It was being accepted that organisms were composed of cells, but nuclei were poorly understood and were thought to be just temporary, sporadically present in some cells, and often they just broke down into pieces. We know now that the nuclei were dividing and the pieces are chromosomes, but until that late 1800s, chromosomes were thought to be the debris of disintegrating nuclei. When microscopists finally were certain that sperm cells did enter eggs, they noticed that the sperm cells seemed to dissolve (sperms shed all their protoplasm, only the sperm nucleus enters the egg). This dissolving of the sperm cells seemed to support the idea of the blending of characters, not the retention of individual genes as Mendel had proposed.

Finally, in the 1880s, several important discoveries were made. It was noticed that the number of nucleus debris fragments was always the same in a particular species, so they must be important structures, and the name chromosome was applied. Then Walther Flemming found that during the production of sex cells, certain divisions resulted in nuclei having only half the number of chromosomes as body cells: meiosis was discovered. Afterward, it was realized that the zygote had twice as many chromosomes as the gametes. It is easy to not realize how important this is: if chromosomes are real, and if they are passed from parents to progeny as structures, not as fluids, then Mendel was correct: genes are particles that retain their identity as they are inherited, not fluids that can blend. In 1902, Edmund Wilson put all the data together: one sperm cell enters one egg cell; they both carry a haploid number of chromosomes; and the zygote is diploid with *equal* contributions by both the male and female parent. All of this was consistent with Mendel's results.

And it was important in a much larger sense. Think of the social world in the 1800s. Women could not vote in much of the world, they could not own property, they could not be students in universities, and they certainly could not be professors or scientists. It was believed that women just were not capable of such things; their role in life was to make babies and then take care of them. With the realization that women's gametes have all the same chromosomes and all the same information as men's gametes, this concept of the inequality of the genders could not use genetics as a support. Mendel's genetics has therefore been a powerful force supporting the concept of human equality.

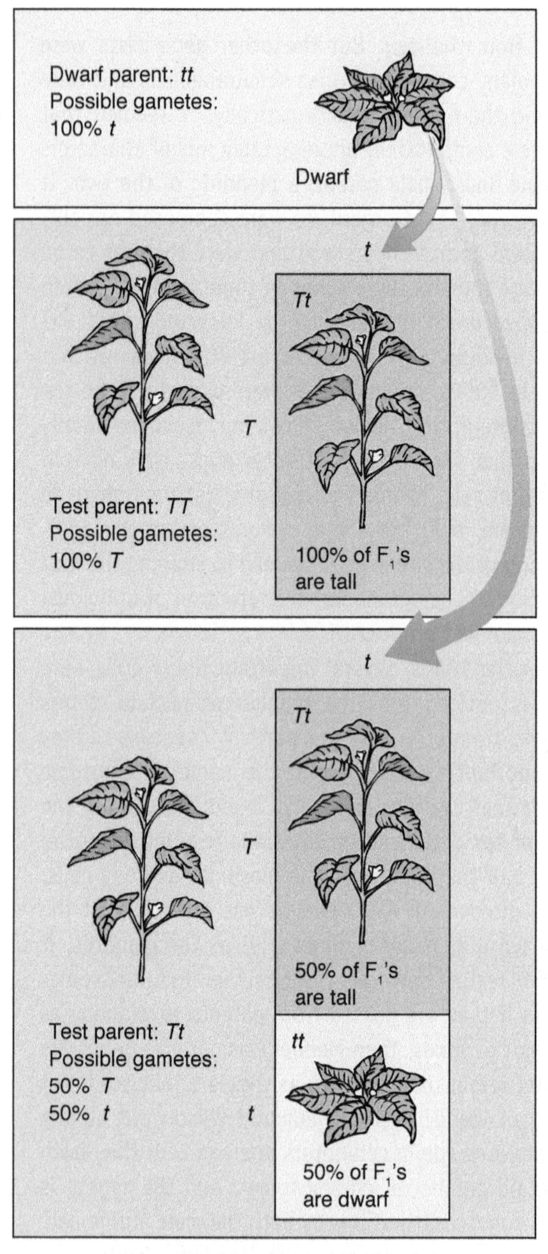

FIGURE 8.8. Testcrosses to determine if a plant with the dominant phenotype is heterozygous or homozygous. Of the data in this diagram, you would not know the genotypes of the test parents before the testcross; those data are what you are trying to discover. If you want to find only a single homozygote or heterozygote, you would need to do testcrosses on only one or a few plants. But imagine that you had done an experimental cross for a trait that you assumed would give you a 3:1 ratio; it would be necessary to do testcrosses on a large number of progeny plants with the dominant phenotype just to confirm your experiment.

Testcrosses

We can only see an organism's phenotype; to discover its genotype, we must perform crosses, and the testcross is one of the most useful. When traits with incomplete dominance are studied, the genotype of any plant is easy to determine from its phenotype. If the trait has complete dominance, it is difficult to know what the genotype of any particular plant is unless the plant shows the recessive trait. Imagine you are studying the inheritance of tallness; you have selfed some heterozygotes and planted the resulting seeds. In your greenhouse or garden, there are now hundreds of plants, approximately 75% of which are tall and 25% short. You know the short ones are *tt*, and if you need a plant with the *tt* genotype for an experiment, you know automatically to choose short plants. But if you need to experiment on plants with the *TT* genotype, how can you tell which they are? A tall plant picked at random is more likely to be a *Tt* plant, because there are twice as many of them as *TT* tall plants.

The genotype can be revealed by a **testcross**, a cross involving the plant in question and one that is homozygous recessive for the trait being studied. All gametes produced by a homozygous recessive parent carry the recessive allele, which is unable to mask the homologous allele in the resulting F_1 zygote (FIGURE 8.8). If the plant being tested is actually homozygous dominant (*TT*), 100% of the progeny will be heterozygous (*Tt*) and tall. If the tall plant is heterozygous, half its progeny in the testcross will be tall (*Tt*) and half short (*tt*).

Once the actual genotypes of plants are known, those that are homozygous dominant are gathered and planted in special areas, kept free of all natural pollinators, and allowed to breed only among themselves. Their entire progeny will be homozygous dominant and can be used in breeding experiments. The homozygous recessives can also be kept as a special line, being selfed and kept pure. Such groups are **purebred lines** and are both useful and valuable. It is not possible to maintain the heterozygotes like this, because they do not breed true; that is, their progeny are not exactly like them. By using purebred lines, seed companies cross plants whose alleles are already well known and which will produce crops that have predictable characters.

So far, we have considered only one trait, height, but for plants that are important crop or horticultural species, dozens or even hundreds of traits are cataloged. Seed companies maintain hundreds of different lines of corn, for instance, in which the genotype of many characters is known for each line. Many times, plants are collected from the wild, so nothing is known about their genotype unless it is immediately obvious from the phenotype. However, from looking at a few collected plants it is not possible to tell which characters are dominant and which are recessive; carefully controlled and recorded crosses must be made.

When testcrosses must be made on annual plants, the results are usually not known until after the plants have died. Their genotypes are then known, but the plants cannot be used for experimentation or breeding. In such cases, it usually is necessary to do both testcrosses and experimental crosses simultaneously, not knowing which plants have the correct genotype for the experiment being performed. After the testcross results are complete, the parents that had the proper genotype can be identified and the experimental crosses that involved them can be analyzed.

Experiments that Consider Two Traits Simultaneously: Dihybrid Crosses

A **dihybrid cross** is one in which two genes are studied and analyzed simultaneously, rather than just one. Every cross involves all the genes in the organism, but the terms "monohybrid" and "dihybrid" refer only to the number being analyzed.

When two genes are studied, the results of the crosses depend upon the positions of the genes on the chromosomes. If they are on different chromosomes, the alleles for one gene move independently of the alleles for the other gene during meiosis I. But when two genes are close together on the same chromosome, the alleles for one gene are chemically bound to the alleles for the other gene and move together. The situation in which the genes are on separate chromosomes is easier to understand and is explained first.

Genes on Separate Chromosomes: Independent Assortment

Consider a plant heterozygous for two traits, for instance seed coat texture and color, with a smooth seed coat (*S*) showing complete dominance over a wrinkled seed coat (*s*), and a yellow seed coat (*Y*) being dominant over a green seed coat (*y*). The plant's genotype is *SsYy*, and its phenotype is smooth yellow seeds. If we consider only the gene for color, the plant produces two types of gametes in approximately equal numbers, some carrying *Y* and some carrying *y*. Also, if we consider only texture, half carry *S* and half *s*. How do the alleles of the two genes relate to each other? As in monohybrid crosses, knowing the types of gametes that can be formed is the key to understanding the patterns of inheritance.

If the two genes are on separate chromosomes, the alleles of one gene move independently of the alleles of the other gene; this is called **independent assortment**. The chromosomes have duplicated during S phase and each has two copies of each allele. During prophase I, homologous chromosomes pair with each other. There are two *Y* alleles, two *y* alleles, two *S* alleles, and two *s* alleles at the metaphase plate (FIGURE 8.9a). During anaphase I, homologous chromosomes separate from each other, and both *Y* alleles move to one pole because they are on the two chromatids of one chromosome, still held together by its centromere. Both *y* alleles, located on the two chromatids of the homologous chromosome, move to the other pole. There is no way to predict which pole will receive which type of allele. Similarly, the *S* alleles separate from the two *s* alleles and move randomly to the poles. In some cells, the pole that receives the *Y* alleles at telophase I also receives the *S* alleles, but in other cells the *Y* and *s* alleles end up together. During meiosis II, the two chromatids of each chromosome separate from each other, resulting in four types of haploid cells in equal numbers: *SY*, *sY*, *Sy*, and *sy*. Any single microspore or megaspore mother cell produces only two types of haploid cell; the set *SY* and *sy*

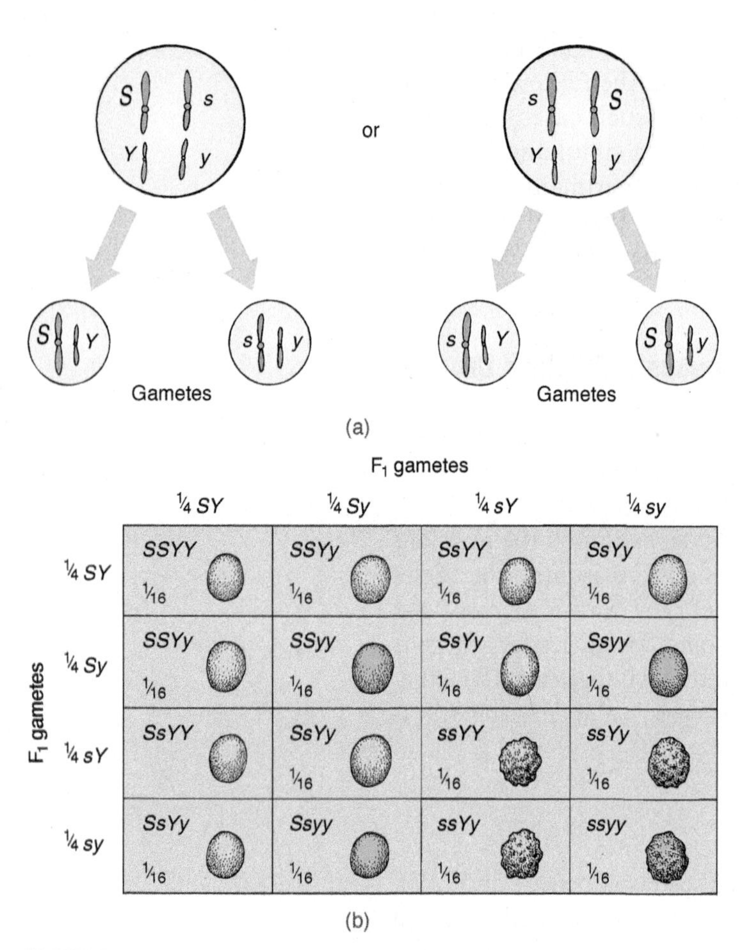

FIGURE 8.9. **(a)** During anaphase I, chromosomes move independently of each other, so one pole receives *S* and the other receives *s*; likewise, one gets *Y* and the other *y*. But in about half the cells both *S* and *Y* move to the same spindle pole by chance, whereas in other dividing cells, one spindle pole receives *S* and *y*, again by chance. **(b)** A Punnett square for a dihybrid cross is set up just like one for a monohybrid cross; establish the types and relative abundance of gametes, then fill in the squares. The table looks a little formidable, but it really consists of two 3:1 ratios intermingled.

or the set *sY* and *Sy*. All four types are produced by a single plant because some mother cells produce one set and some produce the other set.

Once the possible types of gametes are known, the Punnett square can be set up, as in **FIGURE 8.9b** (**TABLE 8.1**). Any single fertilization results from the syngamy of one sperm cell and one egg cell and produces only one of the 16 possible zygote genotypes shown in Figure 9b. All 16 types of zygote occur only if we study many fertilizations; to have the ratios come out accurately, we have to analyze hundreds of progeny. In plants, it is usually easy to obtain large numbers of fertilizations because pollen and ovules are produced in large amounts. Once pollinated, most plants produce dozens or even thousands of seeds, enough progeny that all 16 zygote genotypes occur in about the expected ratios. But in large animals such as mammals, reproduction may be infrequent and only one or two progeny are produced each year; a great deal of work is necessary to get enough progeny to verify the results of a dihybrid cross.

In a dihybrid cross involving independent assortment of two heterozygous genes, each gene showing complete dominance, a characteristic phenotype ratio occurs, just as is true of the 3:1 ratio in a monohybrid cross. The ratio is 9:3:3:1, with 9/16 of the plants having the dominant phenotype for both traits (in our example smooth yellow seed coats), 3/16 with the dominant phenotype of the first trait and the recessive phenotype of the second (smooth, green), 3/16 with the first trait recessive and the second dominant (wrinkled, yellow), and 1/16 in which the plants have the recessive phenotype of both traits (wrinkled, green) (Table 8.1). The 9:3:3:1 ratio results only if all four types of gametes are produced

TABLE 8.1. Genotype and Phenotype Ratios of a Dihybrid Cross[1]

Phenotype	Ratio	Genotype
Smooth, yellow	9/16	1/16 *SSYY*: 2/16 *SSYy*: 2/16 *SsYY*: 4/16 *SsYy* or *S*___*Y*___[2]
Smooth, green	3/16	1/16 *SSyy*: 2/16 *Ssyy* or S___*yy*
Wrinkled, yellow	3/16	2/16 *ssYy*: 1/16 *ssYY* or *ssY*___
Wrinkled, green	1/16	1/16 *ssyy* must be *ssyy*

[1] The two individuals are heterozygous for two characters showing complete dominance and independent assortment.

[2] The notation ___ indicates that either allele may be present; e.g., *S* ___ *yy* represents either *SSyy* or *Ssyy*, both of which produce a smooth green phenotype.

in equal numbers and have equal opportunity to participate in reproduction. The alleles *Y* and *y* must be independent of *S* and *s* during meiosis I; this automatically happens if they are on different chromosomes.

Notice that if only one trait is considered, it behaves as in a monohybrid cross: plants with smooth seeds outnumber those with wrinkled seeds by 3:1 and those with yellow seeds are three times more abundant than those with green seeds. Similarly, the monohybrid genotype ratios are also present—1 *SS*: 2 *Ss*: 1 *ss* and 1 *YY*: 2 *Yy*: 1 *yy*. Considering two genes simultaneously does not affect their inheritance at all.

Genes Close Together on the Same Chromosome: Linked Genes

If two genes occur close together on a chromosome, they usually do not undergo independent assortment during meiosis I; instead, the two genes are said to be **linked**. Consider a plant heterozygous for the two traits of the seed coat again, but now imagine the genes occurring close together on one chromosome. During meiosis, haploid cells that are either *S* or *s*, *Y* or *y* are formed, but a new complexity arises. Because the genes are linked, several types of heterozygote are possible: an *SsYy* individual might have the alleles *S* and *Y* on one chromosome and the alleles *s* and *y* on the homologous chromosome. But an *SsYy* individual might instead have *s* and *Y* linked together and *S* and *y* linked (FIGURE 8.10). We must consider the types of gametes that can be formed. If crossing-over is ignored, the first individual produces haploid cells with the genotypes *SY* and *sy* only, whereas the second plant would produce *sY* and *Sy* gametes only. The result of selfing is not a 9:3:3:1 ratio, but something drastically different.

The most instructive cross is a testcross using a double homozygous recessive parent: *ssyy*. If we continue to exclude crossing-over (imagine that the two genes are extremely close together), the gametes from the first type of heterozygote will be *SY* and *sy* and the gametes from the homozygous recessive parent will be all *sy*. Two types of F_1 will be produced, smooth yellow and wrinkled green, and they occur in a ratio of 1:1, as shown in Figure 8.10b. The two phenotypes are like those of the parents.

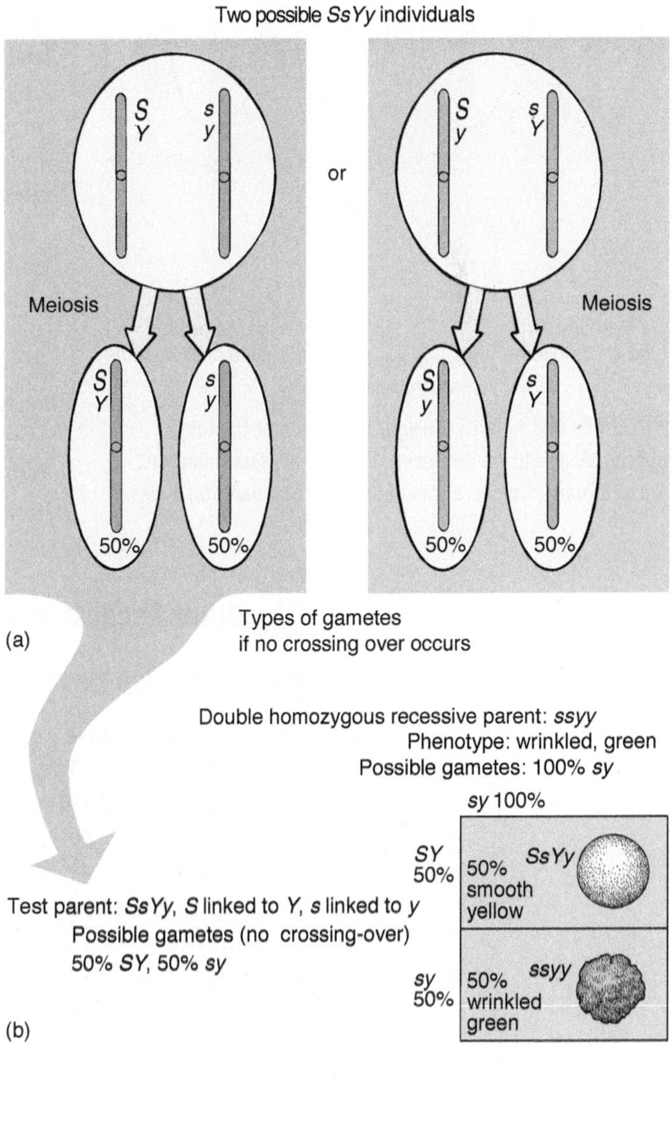

FIGURE 8.10. **(a)** Two types of *SsYy* individuals are possible, but in most instances their phenotypes are identical. Only genetic tests can distinguish which is which, based on the unique gametes produced by each. **(b)** A testcross, with the plant on the left in (a) being the test parent; try setting up the Punnett square for a testcross with the plant on the right in (a). A testcross of two closely linked genes produces results very different from those of a testcross involving nonlinked genes that assort independently. What would the Punnett square be like if *S* and *Y*, *s* and *y* were not linked?

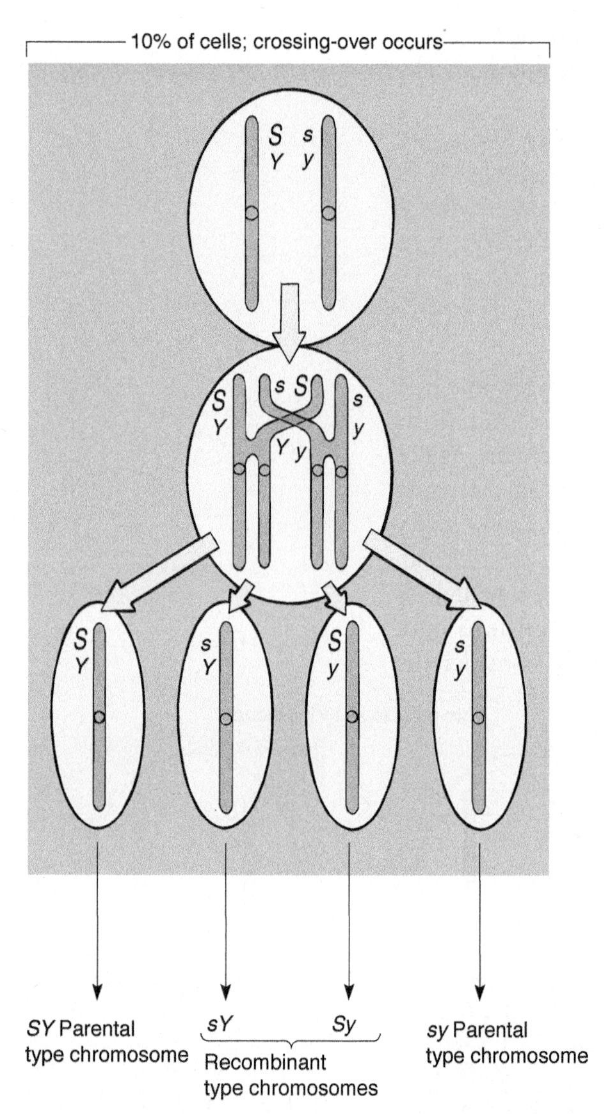

FIGURE 8.11. With linkage, four types of haploid genotypes might be produced, but not in equal numbers. Recombinant types are less abundant than parental types.

Crossing-Over

Independent assortment also occurs if two genes are located far apart on the same chromosome such that crossing-over occurs between them during prophase I (see Chapter 7). The farther apart two genes are on a chromosome, the greater the possibility that crossing-over will occur between them. Most plant chromosomes are so long that crossing-over occurs several times within each chromosome during each prophase I. Consequently, the two ends act like separate entities; if the gene for seed coat color were at one end of the chromosome and the gene for seed coat texture were at the other, they would still undergo independent assortment. However, if the two genes are close together on a chromosome, crossing-over might not occur and the two alleles might move together during meiosis I, as described below.

If the genes are not extremely close together, crossing-over may occur in a few of the spore mother cells, but not in all of them. For example, in 10% of the cells undergoing meiosis, a crossing-over might happen, so the plant would produce four types of gametes, *but not in equal numbers.* Of the gametes for the first type of heterozygote, 47.5% would be *SY*, 47.5% *sy*, 2.5% *sY*, and 2.5% *Sy*. The last two are **recombinant chromosomes** formed from a crossing-over of the homologous chromosomes and recombination of alleles. The first two are **parental type chromosomes** (**FIGURE 8.11**).

The rate of crossing-over is directly proportional to the physical spacing between the genes on a chromosome. Two genes that produce 6% recombinant F_1s are closer together than two genes that produce 10% recombinant F_1s. By analyzing as many mutant alleles as possible, we can measure "space" between them in recombination percentages, each 1% being called one **map unit** or one centimorgan; one map unit, on average, is about one million base pairs. By this means, a genetic map can be constructed.

Other Aspects of Genes and Inheritance

Each Gene Can Have Multiple Alleles

Each gene may have many alleles, not just two as in the examples discussed so far (*T* and *t*, *R* and *r*). Because mutations occur frequently, any gene may exist in many forms, called **multiple alleles**. When genes are polymorphic, having multiple alleles, numbers, such as X_1, X_2, X_3, X_4, and so on, are used rather than capital and lower-case letters (**TABLE 8.2**). Certain mutations still result in the production of a protein with the normal sequence, but most lead to altered protein structure. Some of these proteins are similar to the original protein, perhaps having similar or even identical enzymatic activity. However, many carry out the proper reaction more slowly or are not accurately controlled by regulatory mechanisms in the cytoplasm; a normal, wild-type phenotype may not be produced. With multiple alleles, the concept of dominance is

TABLE 8.2. Names of Genes

Many genes are named with a symbol or abbreviation related to their effect. The following are names of a few genes in peas (*Pisum sativum*).

Symbol	Phenotype
ch1, ch2, lum, pa, vac	Each affects chlorophyll synthesis, changing the color from green to dark bluish green
sa1, sa2, sa3	Leaves have fewer stomata than in the wild type
uni (unifoliata)	Leaflets are converted to just a single leaf

Plant scientists working with *Arabidopsis thaliana* have agreed that the wildtype form of the gene (that found in most natural populations) should be written in *ITALICIZED FULL CAPS*, whereas any mutant allele is written in *italicized lower case*; if the protein itself is discussed, it is in ordinary font. Often, genes are discovered by exposing thousands of seeds to mutagenic chemicals, then sowing the seeds and looking for any aberrant plants with interesting phenotypes. After such a procedure one plant of *A. thaliana* had many more stomata than is typical; that plant was cultivated to maintain the mutated form of the gene, and each time the plant was used in a cross, some of its progeny carried this allele. The gene was named after its mutant phenotype: *TOO MANY MOUTHS* (*TMM*) (full caps indicates this is the wildtype allele), with the newly produced allele being *too many mouths* (lower case indicates this is the new mutant allele), the protein coded by the gene is "too many mouths" (lower case indicates it is the mutant protein; lack of italics indicates it is the protein rather than the gene). Notice that an odd thing about this system is that genes are named after mutant phenotypes: the wildtype, ordinary *TMM* allele produces just the right number of stomata, but its name suggests it produces too many.

With thousands of seeds being treated, it is possible to mutate various genes that cause plants to have too many stomata, and these might be named *tmm1, tmm2, tmm3,* etc., assuming that each is caused by a different gene. People studying epidermis development might use genetic engineering techniques to modify *TMM1* into a number of new alleles, and these would then be named *tmm1-1, tmm1-2,* and so on. If the sequence of the TMM allele is known, then the equivalent genes can be searched for in other plants whose development and genomes are well known, for example snapdragon (*Antirrhinum majus*), tobacco (*Nicotiana tabacum*), and other plants commonly used in developmental studies. If the genes are found, then the species initials are used to distinguish them: *atTMM, amTMM, ntTMM.* These three genes all descended from an ancestral *TMM* gene; the forms of a single gene present in different species are called **orthologs**.

Several *Arabidopsis thaliana* genes:

Name	Phenotype
TOO MANY MOUTHS (*TMM*)	More stomata than in wildtype
FOUR LIPS (*FLP*)	Stomata occur in pairs
SHOOT APICAL MERISTEMLESS (*STM*)	Seedlings fail to make a shoot apical meristem
ROTUNDIFOLIA (*ROT*)	Leaves are short and round
CIRCADIAN CLOCK ASSOCIATED (*CCA*)	Affects endogenous rhythms
TIME FOR COFFEE (*TIC*)	Affects endogenous rhythms

more complex; one allele may produce the proper quantities of the functional protein, whereas a different allele produces more, another produces less, and a fourth produces one with altered activity. Many different phenotypes are possible.

Within a population of plants, as many types of gametes can be produced as there are different types of alleles. A heterozygous X_1X_2 plant can be crossed with a heterozygous

X_3X_4 one, resulting in progeny such as: X_1X_3, X_1X_4, X_2X_3, and X_2X_4. Four distinct types of F_1 plant are produced, none of which has the genotype of either parent.

Each Character Can Be Controlled by Multiple Genes: Quantitative Trait Loci

Individual phenotypic traits are the result of complex metabolic processes involving numerous enzymes; therefore, many separate genes may affect any single trait (**BOX 8.2**). The gene *R* was described as affecting flower color by producing an enzyme that synthesized a red pigment. But that enzyme requires the proper substrate, which is present only if it is synthesized by a different enzyme controlled by a distinct gene. If this gene is present as an ineffective allele, there will be no substrate and thus no pigment, regardless of whether the first flower color gene is present as allele *R* or allele *r*.

Most synthetic pathways involve at least four or five intermediates, and their four or five genes all affect the same trait. Traits such as leaf shape, cold-hardiness, and general vigor are much more complex than presence or absence of pigment. They are the result of dozens genes and rather than having phenotypes that are simply dominant and recessive, the trait may show a continuous gradient of values. Furthermore, some intermediates may be produced by several pathways, each with its own enzymes; a mutation in one is not necessarily severe because the alternate pathway may increase its activity and produce sufficient amounts of the intermediate. If this happens, the mutation is masked and is not reflected in the phenotype, even if it is present in the homozygous condition. It may be necessary to perform extremely complex crosses involving hundreds or thousands of progeny to determine what fraction of a particular phenotype is correlated with a particular gene. The genes or other portions of DNA associated with such traits are **quantitative trait loci**.

BOX 8.2. Whose Genes Do You Have?

You got half your chromosomes from your father and the other half from your mother. So half of the DNA you had when you were a zygote—a single cell—was your father's and half was your mother's. But beyond that? One quarter of your genes came from your paternal grandfather (your father's father), one quarter from your paternal grandmother, and so on for your maternal grandparents. Can we say that one quarter of your DNA came from each of your grandparents? That is not so easy. Of the thousands of cells in your father's testes, one in particular underwent meiosis and produced four sperm cells, one of which gave rise to you. During prophase I of that meiosis, the chromosomes your father got from his father paired (synapsed) with the ones he got from his mother (your grandmother), and they underwent crossing-over. If each homologous chromosome were to break in exactly the same place, then when they are repaired and meiosis continued, each would have exactly the same amount of DNA. But often homologous chromosomes do not break at quite the same spot; after repair, one might have a bit more DNA, the other a bit less, one might be mostly paternal, the other mostly maternal. As long as it is not too unbalanced, the chromosomes will be functional. But because of this, we cannot say that exactly one-quarter of your DNA came from each of your grandparents: some might have given you a bit more, others a bit less.

With each generation we go back, the number of ancestors doubles, and the number of genes we have received from each drops by half. By the time we get to the fifteenth generation, we have 32,768 great-great-etc.-grandparents (2^{15}), and each contributed 1/32,768 of our genes. But we think that humans only have somewhere between 30,000 and 35,000 genes. If we assume each generation is about 20 years (some people have children earlier, some later), then this fifteenth generation lived about 300 years ago, about 1700 CE, not long after the first microscopes were invented. Of course, each of us is descended from many more generations before that, but even so, we may not have actually gotten any of their DNA.

Conversely, when an intermediate is part of several metabolic pathways and is produced by only one enzyme, a mutation in that enzyme's gene affects all the pathways and alters several different traits. Multiple phenotype effects of one mutation are called **pleiotropic effects**. For example, any mutation that affects the protein portion of phytochrome affects all developmental processes controlled by phytochrome, and mutations that alter the level of pyruvate affect the citric acid cycle, amino acid synthesis, and C_4 metabolism.

Uniparental Inheritance

In the crosses described above, the alleles of both parents are transmitted equally to the progeny, a situation called **biparental inheritance**. All genes in the nucleus undergo biparental inheritance. An unexpected feature of both plants and animals is that during fertilization, the sperm cell loses most of its cytoplasm and only the sperm nucleus enters the egg. Consequently, the zygote obtains all its plastid and mitochondrion genomes from the maternal parent; this is **uniparental inheritance**, more specifically **maternal inheritance**.

Mitochondrion genetics are difficult to study for several reasons, but plastid genetics are easier if alleles affecting chlorophyll synthesis are studied. Consider a plastid whose DNA has mutated so that the plastid can no longer produce any chlorophyll; it may be red or orange because carotenoids are no longer masked by chlorophyll, or it may be white if no pigment at all is produced. If a zygote receives only plastids like these, the embryo and seedling will be red, orange, or white. Such seedlings are rare, but in large commercial nurseries that grow millions of seedlings, they appear from time to time. The plants can be kept alive artificially if they are grafted onto a normal plant that supplies them with carbohydrates. Grafted plants can then be used for genetic experiments.

When an achlorophyllous plant is crossed with a green, chlorophyll-bearing plant, the outcome depends on which plant is the **pollen parent** and which is the **ovule parent**. If the achlorophyllous parent provides the pollen, 100% of the zygotes will be green, because they receive all their plastids from the normal parent by way of the egg; all mutant plastids are destroyed during sperm differentiation and syngamy. But if the achlorophyllous plant is the ovule parent, all progeny also have mutant plastids and are achlorophyllous.

Lethal Alleles

The phenotypic result of a mutation can vary in severity from almost undetectable to **lethal**; that is, its presence can kill the plant. Genetically inherited lethal mutations are most often recessive and are fatal only if present in the homozygous condition; a heterozygous plant has enough normal protein to survive. The reason for the recessive nature of most lethals is simple: a dominant lethal would kill the plant, even in the heterozygous condition, so it would rarely be passed on to any progeny.

Polyploidy: Multiple Sets of Chromosomes and Gene Families

Whereas most animals have diploid body cells and haploid gametes, plants are much more diverse. Rarely, a spore is formed without undergoing meiosis, resulting in diploid gametophytes and diploid gametes. If these fertilize a normal haploid gamete, a triploid zygote results. Although this would be instantly lethal in almost all animals, plants tolerate it well and a triploid sporophyte develops from the zygote. However, triploids are sterile because they cannot undergo pairing of homologous chromosomes during prophase I; chromosomes come together in threes and the rest of meiosis is aberrant.

Occasionally, cells fail to undergo mitosis after S phase DNA replication, and the cell becomes tetraploid (or diploid if it happens in a gametophyte). Tetraploid cells are usually perfectly healthy; if they produce a part of the plant that initiates flowers, diploid pollen and eggs are produced, again raising the possibility of triploid zygotes if fertilized by haploid gametes. These processes can occur in virtually every conceivable combination, and plants that are 3n (three sets of chromosomes, triploid), 4n, 5n, 6n,..., up to several hundred n are common. All plants with more than two sets of chromosomes are **polyploid**. All plants with an even ploidy level can undergo meiosis and are fertile; all odd ploidy levels are sterile like triploids. Also, most polyploid plants are larger and more robust than their diploid relatives, and they produce more fruit and seeds. Very often, one of the simplest methods of increasing the crop yield of a species is to search for polyploid individuals and use them to breed the seeds that are sold to farmers.

Change in chromosome number per nucleus in plants can also come about by nondisjunction: during the second division of meiosis, the two chromatids may remain together (fail to disjoin), so one daughter cell receives both copies of the chromosome (the two chromatids) whereas the other receives none. The cell with the extra chromosome probably survives quite well and forms a functional sperm or egg cell. If this gamete, which is diploid for one chromosome, is involved in fertilization, the new zygote will be triploid for that chromosome. It received a normal, haploid chromosome set from one parent and a haploid set plus the extra chromosome from the parent in which nondisjunction occurred. In plants, this condition is tolerated often, and only by studying the karyotype can it be discovered. In animals, this is almost invariably fatal. The human congenital disease Down syndrome is caused by nondisjunction of our very small chromosome 21. Down syndrome individuals have three copies of just a few genes yet have significantly disrupted metabolism.

Plants with even ploidy levels grow and reproduce successfully, often more vigorously than diploid individuals of their species. Initially, these have complex genetics. For example, a tetraploid can have any of the following genotypes for red/white flower color: *RRRR*, *RRRr*, *RRrr*, *Rrrr*, or *rrrr*. Its gametes can be *RR*, *Rr*, or *rr*. Notice that polyploids have more copies than needed for every gene. In all diploid plants, even heterozygotes with only one functional allele are usually healthy. For nonessential genes, a complete absence of functional alleles may not be harmful, such as when *rr* produces white flower color. Consequently, the two or three extra copies of each gene in a tetraploid plant are surplus DNA. Mutations that completely incapacitate one of the four alleles of a tetraploid usually are not deleterious, as one, two, or three functional alleles remain in every nucleus. For polyploid plants, deleterious mutations tend to have little effect on phenotype, so they do not affect survival as much as they would in a haploid cell or a diploid plant. As a result, deleterious mutations are not eliminated quickly by natural selection, and the extra copies of a gene may rapidly become nonfunctional; as long as the genes are still so similar to the wildtype allele that they can be recognized as having originated as duplicates of it, they are called paralogs. Before many generations have passed, so many genes have changed that polyploids begin to act like diploids, their "extra" copies having mutated. Almost half of all flowering plant species are actually polyploids; those that now appear to be perfectly normal diploids probably underwent conversion to the polyploid condition so long ago that all their extra alleles have mutated extensively and can no longer be considered alleles of those that are still functional.

What do the extra alleles mutate into? This extra DNA is actually extremely valuable—it is the raw material for the evolution of new genes. Some may mutate into genes that produce enzymes almost identical to those being coded by the original form of the allele. The original enzyme may work best at low temperature, whereas the new form may have a higher optimum temperature. The plant can now produce two types of enzymes and function well in both warm and cool days, or the species may be able to extend its range, the new enzyme allowing it to survive in hot deserts and the original enzyme allowing it to live in its original habitat. Mutations in the promoter region instead may allow the structural portion to produce the same protein as the original gene but at a different time or place in response to a different chemical messenger. Genes that are obviously related like this are members of a **gene family**.

In other cases, mutations in the extra DNA may result in totally new genes that produce proteins whose function is not at all related to that of the original gene. Mosses, ferns, conifers, and flowering plants have evolved from green algae. Because algae have no genes for flower color, lignin synthesis, and so on, the evolution of these species has involved the evolution of whole new metabolic pathways. All of these genes had to arise by the gradual mutation of surplus alleles into nucleotide sequences that code for useful proteins.

Important Terms

biparental inheritance
complete dominance
cross
dihybrid cross
dominant allele
F_1 (first filial) generation
F_2 generation
gene family
genetics
genotype
heterozygous
homozygous
incomplete dominance
independent assortment
lethal allele
linked genes
map units
maternal (uniparental) inheritance
monohybrid cross
multiple alleles
ortholog
ovule parent
parental generation
parental type chromosomes
phenotype
pleiotropic effect
pollen parent
polyploidy
Punnett square
purebred line
quantitative trait loci
recessive allele
recombinant chromosomes
testcross

Concepts

- Many genes exist in multiple forms called alleles; other genes have only one or two alleles.
- In order to analyze the results of a cross, it is necessary to know the types of gametes that participate.
- Genes are structures (DNA sequences) that retain their individuality during mitosis and when passed from one generation to the next.
- Homologous genes do not blend together in the progeny of sexual reproduction.

9 Adapting to Changing Environments: Evolution, Diversification, and Systematics

These redwing black birds are feeding on rice left over after the field was harvested. They will digest most of the rice, but a few seeds may accidentally survive. As the blackbirds fly to a new area, they inadvertently carry the seeds (and the genes they contain) to new areas where they may germinate, grow, and then reproduce sexually with local rice plants. This is an effective means for genes to move through the environment, and mutations that occur in one area can be spread throughout a population.

DNA mutates, and it is passed from generation to generation by sexual and asexual reproduction. What are the consequences of its mutability and inheritance? The short answer is that DNA evolves, as do the phenotypes it codes for. A longer answer is found in the rest of this chapter, but the medium-length answer is that DNA guides all aspects of the individuals that make up a species, and as DNA evolves, so does the species. If a mutation produces a new allele that causes a beneficial change in phenotype such that the species becomes more adapted to its habitat, then individuals with that allele should be healthier, should grow better, and should reproduce more than others that lack the new allele. If this happens, the new allele becomes more abundant in the population, and individuals with the original versions of the gene may be crowded out.

Sexual reproduction allows old and new alleles from throughout a species to be brought together in new combinations as various individuals reproduce sexually with each other. Each progeny produced by sexual reproduction is a genetic experiment: the offspring that are most well adapted survive (as do their alleles), those that are less well adapted don't. Ideally, the mutability of DNA and the recombination during sexual reproduction should allow a species to evolve to become ever more adapted to its habitat.

But habitats change. Habitats have both nonbiological (climate and soils) and biological components (all the organisms that share the habitat). To survive for long, a species must be adapted to its existing habitat as well as continuously adapt to the changes. A plant's neighbors see it as a competitor, as food, as something to climb on or nest in, or just something that gets trampled without notice. These other organisms are evolving to exploit the plant more effectively. A plant species must evolve to maintain its defenses against its neighbors; the alternative is to go extinct.

A species may also evolve such that it diversifies into two or more species. Some individuals of a species may become adapted to one set of conditions in their habitat, other individuals adapt to other conditions. Gradually the two groups of individuals start to differ in metabolism, shape, and other aspects of their biology. For example, think of a species with small plants adapted to live in a forest. Some may become even more adapted by evolving to be more shade tolerant, by accumulating alleles that increase the efficiency of photosynthesis. Others of the same species may become more adapted by becoming epiphytic and living on branches of the trees, in brighter sunlight. They would need alleles that allow their roots to grip bark rather than penetrate soil. As the two sets of individuals continue to adapt in different ways, they would finally differ so much from each other and from their ancestors that at some point we would consider them separate species.

This evolutionary diversification causes an increase in the number of types of plants, animals, fungi, and other organisms. It is necessary to keep track of them, so we need a system to provide names for each species. That science is taxonomy. At first, names were given that just reflected the common names. But people realized that some plants resemble other plants not just by accident but because they are closely related. It was decided to invent a system of names that helps keep track of which plants are related to which. This is called a **natural system of nomenclature**, and it was soon applied to all organisms.

Population Genetics: A Group of Interbreeding Individuals that Share Alleles

A **population** is the set of individuals of a species that live in a particular area at the same time and that can interbreed with each other (FIGURE 9.1). **Population genetics** deals with the abundance of different alleles within a population and the manner in which the abundance of a particular allele increases, decreases, or remains the same with time.

The Gene Pool of a Population

The total number of alleles in all the sex cells of all individuals of a population constitutes the **gene pool** of the population. Imagine that gene *A* has four alleles: *A1*, *A2*, *A3*, and *A4*. The alleles are probably not present in equal numbers; for instance, 60% of all gametes may be *A1*, 20% *A2*, 15% *A3*, and 5% *A4*. Will this ratio be the same for the population next year or in the next generation? If only sexual reproduction is considered, the ratio remains constant over time; even if an increase or decrease occurs in the total number of individuals or gametes, the ratios do not change. Sexual reproduction alone does not change the gene pool of a population.

FIGURE. 9.1 This forest of conifers is a population of individual plants that interbreed with each other. Pollen blows from each to many of the others, carrying sperm cells with a full set of genes. Squirrels and birds do not eat all the seeds they carry, so the seeds too contribute to moving genes through the habitat. Trees in the center of the forest will be more successful at reproduction; those on the edges where there are few individuals will not receive as much pollen and much of their own pollen will just land on the ground. Other consequences of living in a dense population is that fire and disease spread quickly.

Factors that Cause the Gene Pool to Change

Mutations occur continually, and bring new alleles into existence. At first the new allele will be just one single copy, and if it is a somatic mutation, it doesn't matter. But if it occurs in a cell that will ultimately produce sex cells, it may be passed on to a zygote and become part of the next generation of individuals.

Genetic drift occurs due to events that an organism cannot adapt to, such as volcanic eruptions, exceptional droughts, unusually cold winters, and so on. When a volcano erupts a large region is destroyed by molten lava, hot ash, and poisonous gasses. All organisms, along with all their alleles, are eliminated. If by chance the volcano's destruction had included the single individual that had just obtained *A5* by mutation, then *A5* would no longer exist. If the population in the volcanic area has the same gene frequencies as the general population, the alleles are eliminated in the same proportions as they exist generally, and no change in allele frequency occurs. If the volcanic area had an unusually high number of a particular allele, for instance *A3*, then *A3* is affected more than the other alleles, so the allele frequency of the gene pool of the survivors is altered. However this change of allele frequency does not make the species any more able to survive another volcanic eruption.

Genetic drift is more common in populations that contain a small number of individuals rather than in large populations. Infrequent floods, hailstorms, droughts, or other chance events can damage or kill plants in a small area; if that area contains most of the population, then the gene pool of the survivors may be quite different from that of the original population. But if the species is widespread, then a flood or hailstorm in one restricted area will have no effect at all on any of the members outside the area, and there will be little effect (little or no genetic drift) in the species. Our northern forests have so many millions of individual trees of firs, spruces, maples, and other trees that those species experience little genetic drift. But small islands surrounded by oceans have so few individual plants of any species that genetic drift is frequent in those small populations (**BOX 9.1**).

Genetic drift affects individual alleles differently. Once an allele is formed, *A5* for instance, its numbers may increase as it is used in sexual reproduction; after a few years there may be ten individuals with the *A5* allele. The individuals are closely related and probably located close together because most seeds do not travel very far. A local disturbance might eliminate all ten: an avalanche, a herd of grazing animals, or the construction of a highway. This is one of the real dangers of habitat destruction by people: we will probably not cause the extinction of widespread species, but we often do completely destroy species that consist of just a few individuals in a small area. And attempts to rescue plants from habitats that are

BOX 9.1. Biological Preserves and Genetic Drift

Genetic drift, adaptive radiation in small populations, and artificial selection are important factors in conserving endangered species. With only a small number of individuals maintained in zoos, botanical gardens, biological preserves, and national parks, the gene pools of the protected plants and animals are so small that rapid fluctuation and drift can cause them to evolve rapidly. An endangered species protected in captivity may be lost because characters evolve that make it adapted to highly artificial, humanly maintained conditions. Even if it still looks like the ancestral species that was placed into the zoo, it differs in its resistance to infections, parasites, predators, and its ability to understand and perform mating interactions. Most zoologists now believe that habitat preservation is the only true means of maintaining an endangered animal species. Zoos cannot accommodate the numbers of individuals necessary and cannot provide realistic habitats that act as agents of natural rather than artificial selection. Many of the progeny born in zoos would be poorly adapted for survival in the wild; if they are kept alive by antibiotics and special diets, the presence of nonbeneficial alleles is maintained in the population. Natural populations have a tremendous death rate among juvenile animals, a factor necessary for natural selection; with this eliminated in zoos, the gene pool of the captive population rapidly diverges from that of the natural population, being greatly enriched in deleterious mutations. The zoo population may become incapable of surviving without human care. The same is true of plant species maintained in botanical gardens, where often the most unusual varieties are given special attention, increasing the presence of exotic alleles that would be selectively disadvantageous in natural conditions. Artificial and natural selection result in very different gene pools. This is not a condemnation of zoos or botanical gardens: they have provided an essential service in saving many species from extinction, which at present would be the only alternative for them. But it should be remembered that a species in the wild consists not only of its individuals but of its genetic diversity interacting with its environment.

In order to provide truly safe sanctuary for endangered species, each park or habitat preserve must encompass a population large enough to be stable against genetic drift. Even Yellowstone National Park is not large enough to maintain a genetically stable population of grizzly bears, although the buffalo seem to be doing well. Many individuals of a plant species may survive in a small area, such that 1 Km^2 (about one-quarter square mile) may hold a large enough number for their genetic stability on a theoretical basis. But we must maintain the pollinators and seed dispersers as well: the habitat must be diverse enough to contain realistic selection pressures, including pathogens, herbivores, and so on. If these are eliminated owing to insufficient habitat preservation, the plant species experiences unnatural selection. Recent forest fires are a good example. Fire is a natural part of many ecosystems and is an agent of natural selection for many features; if fires are suppressed, nonresistant individuals survive and the allele frequency of the population is altered.

FIGURE 9.2. Pansies are a type of violet (*Viola*) that has been bred by botanists to have large, showy flowers; this flower is about 5 cm (2 inches) in diameter. See Figure 9.3 for natural violets. Because people have decided which plants will be bred and which will be discarded, this is artificial selection.

FIGURE 9.3. Natural violet flowers, which have not been subjected to artificial selection, are less than 1 cm (about one-third of an inch) in diameter. Both this violet and the pansy in Figure 9.2 are members of the genus *Viola*.

about to be destroyed usually only manage to save several dozen to several hundred plants of any species: the rescued plants constitute a small population subject to genetic drift.

In **artificial selection**, humans purposefully change the allele frequency of a gene pool. Plant breeders continually examine both wild populations and fields of cultivated plants, searching for individuals with desirable qualities such as resistance to disease, increased protein content in seeds, and ability to survive with less water or fertilizer. Such plants are collected and used in breeding programs to produce seed for future crops. Consider just wheat, rice, and corn: almost the entire world populations of these three species consist of cultivated plants; very few of the natural ancestors still exist in the wild. The gene pool for each is made up almost entirely of alleles that have been artificially selected for thousands of years. Artificial selection is also used to produce ornamental plants that flower more abundantly or for a longer time, to alter their flower color and size, and to make the plants hardy in regions where they otherwise could not grow (**FIGURES 9.2** and **9.3**). The trees cultivated for lumber and paper are also subjected to artificial selection.

Natural selection is the most significant factor causing gene pool changes because it affects all species, both natural and cultivated ones. It is usually described as survival of the fittest: those individuals most adapted to an environment survive whereas those less adapted do not. Two conditions must be met before natural selection can occur:

1. The population must produce more offspring than can possibly grow and survive to maturity in that habitat. This condition is almost always valid for plants anywhere. Most plants produce hundreds of seeds, and even in species with wind-dispersed seeds most seeds do not travel far. Consequently, the ground can be covered with hundreds of seedlings crowded so closely together there simply is not enough room for all of them.

Besides limited resources, the number of individuals that can survive in a particular habitat is affected by predators, pathogens, and competitors. Animals not only eat plants but may also lay eggs in them, bore into tree trunks for nesting sites, walk on them, and rob nectar without carrying out pollination. Pathogenic fungi and bacteria are similarly harmful. Competitors are other organisms that use the same resources the plant needs to survive. When root systems grow together, the two plants compete for the same water and nutrients. If two species are pollinated by the same species of insect or bird, they must compete for the attention of the pollinators. All these activities limit the number of offspring that can grow and survive in an area.

2. The second condition necessary for natural selection is that the progeny must differ from each other in their types of alleles. If all individuals of a species are equally susceptible to a pathogenic fungus, no increase in fitness occurs as the result of a fungal attack. Even if some survive, they are identical genetically to those that died, so no change occurs in allele frequency; natural selection has not occurred. But if some members have an allele that gives them increased resistance, those plants should fare much better during an outbreak of fungus than those lacking the allele. If the fungus is so virulent that it often kills the plants it attacks, the allele frequency of the population is changed radically after infection. Natural selection operates even if the fungus only weakens plants but does not kill them outright; the weakened plants should produce fewer seeds than do the resistant, healthy plants.

Natural selection is the differential survival of organisms that have different phenotypes. It does not cause mutations, it only acts on preexisting alleles. The presence of a fungus does not cause plants to become resistant; if none had been resistant before, all individuals would be adversely affected. The advantageous allele must exist first. But if an allele for resistance does exist, natural selection can cause the *population* to become resistant by the preferential survival of resistant individuals, even though it cannot cause an *individual* to become resistant. A population or species evolves, but an individual does not.

Any factor that causes one plant to produce more progeny than other plants is a selective factor. If an allele causes chloroplasts to photosynthesize more efficiently, plants with that allele produce carbohydrates more rapidly than plants that lack the allele; the former plants grow faster and produce more seeds, at least half of which carry the advantageous allele.

An allele that could improve fitness may never have the opportunity to help the species. For example, a mutation that improves cold hardiness may be eliminated from the gene pool if the plant carrying the new allele is killed by fungus or drought. However, if the new allele for cold hardiness does survive, it may be able to improve the species. If cold winters are common, this allele greatly improves fitness, and its frequency may increase rapidly. If cold winters are infrequent, they do not exert a strong selection pressure, the allele does not improve fitness very much, and its frequency may remain low for years until a harsh winter does occur.

Natural selection usually changes a population very slowly. Most populations are relatively well adapted to their habitat or they would not exist. Very few mutations produce a new phenotype so superior that it immediately outcompetes all other members of the population. At the extremes, there are many species of seedless plants (lycopods, *Equisetum*, ferns) that have persisted for tens of millions of years without diverging into new species (FIGURE 9.4). In contrast, very rapid speciation is occurring in a group of asters in Hawaii: shortly after Kauai formed 5.2 million years ago, an aster seed arrived, thrived, reproduced, and spread rapidly in the unpopulated island. Its descendants have diversified into three distinct genera, each

FIGURE. 9.4 The ancestors of the horsetails (*Equisetum*) originated hundreds of millions of years ago; the oldest fossils of this group are as much as 360 million years old. At present this group is not diversifying very much; few if any new species are being formed anywhere. However, several species of *Equisetum* are adapted well enough to dominate certain landscapes. This *E. hyemale* covers thousands of square miles in Alaska; the Chugash foothills are shown here. The yellow flowers belong to dandelions growing among the *Equisetum*.

with many species, and on average, a new species arises in this group once every 500,000 years (**FIGURE 9.5**).

Evolutionary changes that result in the loss of a structure or metabolism can come about quickly. If a feature becomes selectively disadvantageous, mutations that disrupt its development become selectively advantageous. Because disruptive mutations outnumber constructive mutations, loss can occur relatively rapidly. For example, the ancestors of cacti lived in a habitat that became progressively drier; as the habitat changed, large thin leaves that had been advantageous because they carried out photosynthesis now became disadvantageous because they lost too much water. Mutations that disrupted formation of leaves were

FIGURE 9.5. This plant is silver sword (*Argyroxiphium*), a Hawaiian member of the aster family. After the first seed or seeds of asters became established in the newly formed Hawaiian Islands millions of years ago, their descendants diversified rapidly and have produced many new species that occur naturally only in Hawaii. (© Steven Maltby/ShutterStock, Inc.)

FIGURE 9.6. The earliest members of the cactus family had large, thin leaves, and in the line of cactus evolution that resulted in the genus *Pereskia*, plants still produce ordinary foliage leaves. Most pereskias live in moist areas; their large leaves are beneficial because they photosynthesize well.

FIGURE 9.7. In the line of evolution that resulted in prickly pear cacti (*Opuntia*), mutations occurred that prevent the formation of a leaf blade. These are fully mature leaves; they will never become longer or wider, and if the area where the cactus is living experiences a drought (which usually happens every year), the leaves will fall off such that the cactus has a smaller surface area and loses less water.

advantageous, and cacti lost their leaves in perhaps as little as 10 million years, whereas the evolutionary formation of leaves in seed plants had required over 200 million years (FIGURES 9.6 and 9.7).

Natural selection does not include *purpose*, *intention*, *planning*, or *voluntary decision making*. Whenever we say that "plants do something in order to...," we are suggesting that plants can plan their activities and have purpose, which is not true. It is not correct to say that certain plants have disease-resistance genes *in order to* protect themselves against fungi. The resistance of a population of plants to a fungus is nothing more than the result of the preferential survival of the plants' ancestors because they had an allele for resistance whereas their competitors did not. Although plants that have this allele in the presence of the fungus have a selective advantage, *the plants do not have the allele in order to protect themselves*. Similarly, plants do not produce nectar in order to attract pollinators; those plants in the past that did produce nectar happened to be pollinated more often than others that did not, so the alleles for nectar production increased in the population. At present those plants that secrete nectar are visited by pollinators, but there is no purpose, intent, or planning by the plant. We humans might eat food in order to get energy, but we do not digest the food in order to obtain its energy. Instead, our autonomic nervous system and cell metabolism have automatic responses to the presence of food in the small intestine that cause the secretion of digestive enzymes, the absorption of monomers, their distribution through our blood, and their respiration by cytoplasm and mitochondria. We have no voluntary control over this. (This may seem to be just a trivial problem of wording, because everyone knows what we mean, but statements should be accurate, not sloppy. If we are not meticulous in how we express our ideas, we will not be meticulous in how we think.)

Speciation

It is not possible to give an exact definition of **species** that is always valid, but generally, two organisms are considered to be members of distinct species if they do not produce fertile offspring when crossed. If two plants freely interbreed in nature, they are members of the same species; if they do not interbreed, even when manually cross-pollinated, they are separate species. Most species are composed of many populations distributed across a large geographical area, so some individuals of the species cannot interbreed simply because they are located too far from each other.

As natural selection operates on the various populations for many generations, the frequencies of various alleles and the phenotypes of the population change. At some point, so much change has occurred that a particular population must be considered a new species, distinct from the species that existed at the beginning. Natural selection has caused a new species to evolve, a process called **speciation**. Speciation occurs in two fundamentally different ways: (1) **phyletic speciation**, in which the entire species gradually becomes so changed that it must be considered a new species, and (2) **divergent speciation**, in which some populations of the species evolve into a new, second species while other populations either continue relatively unchanged as the original, parental species or evolve into a new, third species.

Phyletic Speciation

The critical feature of phyletic speciation is that as new beneficial alleles arise and are selected for, they become spread throughout the entire population. The movement of alleles from the area where they arise by mutation is called **gene flow**, and it occurs in many ways, such as pollen transfer, seed dispersal, and vegetative propagation. Every pollen grain contains two sperm cells, each of which carries one full set of genes, so all alleles of a plant are present in its pollen grains. If a new allele is carried by wind-distributed pollen, such as that of ragweed, grasses, and conifers, it can move to very distant plants. If the pollen grain's sperm cells fertilize an egg, a new seed is formed whose embryo contains the new allele. Both birds and insects tend to spend most of their time in a small area, so even though birds can fly long distances, they usually spread pollen through a smaller area. Nevertheless, allele movement and gene flow do occur.

Many species have long-distance dispersal mechanisms for their fruits and seeds (FIGURE 9.8). These can be carried by wind, floods, and stream flow, or if they are spiny or gummy, they can stick to the fur or feathers of animals, or pass through their digestive tract. Migratory animals disperse seeds over long distances.

FIGURE 9.8. Each of these dandelion parachutes is a fruit that contains one seed. Wind will carry each seed away from the parent plant, distributing the plant's alleles over a wide area. If a seed germinates and grows into a new plant that is able to reproduce, then the parent plant's alleles will be spread even farther.

Divergent Speciation

If gene flow does not keep the species homogeneous throughout its entire range, divergent speciation may occur. If alleles that arise in one part of the range cannot reach individuals in another part, the two regions are *reproductively isolated*. Two fundamental causes of reproductive isolation are abiological and biological reproductive barriers.

Any physical, nonliving feature that prevents two populations from interbreeding is an **abiological reproductive barrier**. Mountain ranges are frequently reproductive barriers because pollinators do not fly across entire mountain ranges while feeding or gathering pollen. Seeds may occasionally be carried across mountains by birds or mammals, but probably too rarely to be significant. Rivers are often good barriers for small animals, but they rarely prevent plant gene flow by means of seed dispersal. Deserts and oceans are effective barriers.

Any biological phenomenon that prevents successful gene flow is a **biological reproductive barrier**. Differences in flower color, shape, or fragrance are effective barriers if the species is pollinated by a discriminating pollinator. A mutation that inhibits pigment synthesis and causes a plant that normally has colored flowers to have white flowers might prevent the pollinator from recognizing the flower. The flower is not visited and gene flow no longer occurs between the mutant plant and the rest of the species even though the individuals grow together. Timing of flowering can be important: if a mutation causes some flowers to open in the evening, they probably will not interbreed with those that open in the morning. Differences in flowering date are critical because of the brief viability of pollen once it leaves the anthers.

Evolutionary changes in pollinators can also act as reproductive barriers for plants. If a plant population covers a large area, some parts of the range probably have characteristics different from those of other parts, such as elevation, temperature, and humidity. These variations may be important to pollinators and seed distributors if not to the plant. If so, the animals may diverge evolutionarily into distinct species, each limited to a small part of the plant's range. Little or no flow of the plant's genes occurs owing to the restricted movements of pollinators and seed dispersers.

If a plant species occupies an extensive, heterogeneous range, mutations will gradually arise that are particularly adaptive for specific regions of the range. When these new alleles arrive at that part of the range by gene flow, their frequency there is increased by natural selection. When they arrive at other parts of the range, their frequency remains low or they are eliminated there if they are selectively disadvantageous for conditions in these sites. Even with active gene flow and interbreeding, different subpopulations of the plant species emerge, each adapted to its own particular portion of the total range.

Adaptive Radiation

Adaptive radiation is a special case of divergent evolution in which a species rapidly diverges into many new species over an extremely short time, just a few million years. This usually occurs when the species enters a new habitat where little or no competition or environmental stress exists. Good examples are the colonization of newly formed oceanic islands such as the Hawaiian or Galapagos Islands. After being formed by volcanic activity, they are initially devoid of all plant and animal life, but eventually a seed arrives, carried by a bird, wind, or ocean currents. Once the seed germinates and begins to grow, it is free of danger from herbivores, fungi, bacteria, or competition from other plants. It must be relatively adapted to the soil, rainfall pattern, and heat/cold fluctuations, but otherwise its life is remarkably free of dangers. If this plant is self-fertile or if it reproduces well vegetatively, it successfully colonizes the area.

Initially, all offspring greatly resemble the first, **founder** individual(s), because the initial gene pool is extremely small; if just one seed is the founder, the original gene pool consists of its two sets of alleles. This homogeneity may last only briefly because, with the lack of competition, pathogens, and predators, fewer forces act as selective agents. Consequently, new alleles are not eliminated as rapidly as would occur in the parental population on the mainland. While the population is small, it is more subject to chance events, so genetic drift occurs and the gene pool changes rapidly and erratically. The island population soon becomes heterogeneous.

Adaptive radiation also occurs in mainland populations if the environment changes suddenly and eliminates the dominant species of a region. With the absence of these species, competition changes and other species that had not been able to compete well before can now occupy the new areas. While few in number, they undergo genetic drift, rapidly producing many unusual genotypes. Within a short time, many new species are recognizable, and adaptive radiation has occurred. Genetic drift is always a significant concern when establishing nature preserves: if the protected area only contains a small population of a threatened species, drift may cause it to quickly evolve so much that it differs significantly from the original species (see Box 9.1).

Convergent Evolution

If two unrelated species occupy the same or similar habitats, natural selection may favor the same phenotypes in each. The two may evolve to the point that they resemble each other strongly and are said to have undergone **convergent evolution**. The most striking example is the evolutionary convergence of cacti and euphorbias (FIGURES 9.9 and 9.10). Cacti evolved from leafy trees in the Americas (see Figure 9.6); as deserts formed, mutations that prevented leaf formation were advantageous because they reduced transpiration. Other selectively advantageous mutations increased water storage capacity (succulent trunk) and defenses against water-seeking animals (spines). In Africa, the formation of deserts also favored a similar phenotype, and the euphorbia family became adapted by these means. As a result, the succulent euphorbias are remarkably similar to the succulent cacti, even though the ancestral euphorbias are quite distinct from the ancestral cacti. Two groups never converge to the point of producing the same species; only the phenotypes converge, not the genotypes. For example, cactus spines are modified leaves whereas euphorbia spines are modified shoots.

FIGURE 9.9. Like many cacti, this *Weingartia* is adapted to grow in deserts that have little water. During the evolution of its ancestors, natural selection resulted in the loss of leaves, the evolution of a wide cortex capable of storing water, and spines that defend the plant from thirsty animals. Cacti originated and evolved in the Americas. Cactus spines are modified bud scales.

FIGURE 9.10. Like many cacti, this *Euphorbia meloformis* is adapted to grow in deserts. The ancestors of this plant were completely different from the ancestors of cacti, so this species resembles cacti due to convergent evolution; it is not closely related to cacti at all. This group evolved in Africa. *Euphorbia* spines are modified branches. (Plant cultivated by Bob Barth.)

Systematics

Starting about 400 million years ago, a long series of mutations and natural selection began to change a group of green algae into the plants that are alive today. During the intervening eons, that evolutionary line has progressed and diversified, branching into more and more lines of evolution. Thousands of these have become extinct and are known only by fossils; thousands of other evolutionary lines are represented by the approximately 280,000 species of living plants. The goal of modern plant **systematics** is to understand each of these evolutionary lines and to have a system of names that reflects their relationships accurately. Taxonomists have the goals of (1) developing a natural system of classification in which closely related organisms are classified together, and (2) assigning names on the basis of evolutionary relationships.

Levels of Taxonomic Categories

Because organisms have varying degrees of relatedness, a natural classification system reflects this in its numerous levels. The most fundamental level of classification is the species, which ideally and theoretically is a set of individuals closely related by descent from a common ancestor. (The word "species" is both singular and plural.) Members of a species can interbreed with each other successfully but cannot interbreed with individuals of any other species.

Closely related species are grouped together into **genera** (singular: genus; **TABLE 9.1**). Deciding whether several species are closely related enough to be placed together in the same genus is difficult. No objective criterion exists; the decision is subjective and often the cause of arguments. Some taxonomists, generally referred to as "lumpers," believe that even relatively distantly related species should be grouped together in large genera. Other taxonomists, called "splitters," prefer to have many small genera, each containing only a few species that are extremely closely related. For example, some taxonomists believe that cranberries and blueberries are so similar that they should go into the same genus, *Vaccinium*; others think that cranberries are distinct enough that segregating them into their own genus, *Oxycoccus*, more accurately reflects evolutionary reality. Both groups agree that the two sets of species are closely related, they just have different opinions as to how much evolution has

TABLE 9.1 The Taxonomic Categories (Examples)*

	Basal Angiosperms	Monocot	Eudicot
Kingdom	Plantae	Plantae	Plantae
Division	Magnoliophyta	Magnoliophyta	Magnoliophyta
Class		Liliopsida	Magnoliopsida
Order	Nymphaeales	Asparagales	Fabales
Family	Nymphaeaceae	Amaryllidaceae	Fabaceae
Genus	*Nymphaea*	*Hymenocallis*	*Lupinus*
Species	*Nymphaea odorata*	*Hymenocallis caribaea*	*Lupinus texensis*
Common name	Fragrant water lily	Spider lily	Texas bluebonnet

* The scientific name of a species is always a **binomial**, consisting of the genus and species names. The species epithet is often descriptive: *Lupinus texensis* occurs in Texas; other common species epithets are "longifolia," "grandiflora," "vulgare" (common), "acaulis" (rosette, without a stem), etc.

occurred since the time of the most recent common ancestor. The critical concern is that the genera are natural, that all the species included in the genus are related to each other by a common ancestor, and that all descendants of that common ancestor are in the same genus. Such a group is **monophyletic**. In an unnatural, **polyphyletic group**, members have evolved from different ancestors and may resemble each other only as a result of convergent evolution. Taxonomists always try to avoid grouping species into a polyphyletic genus.

The level above genus is **family**, each family being composed of one, several, or many genera. Most families are well defined, with widespread agreement as to which species and genera belong in a particular family. As examples, consider how easy it is to recognize the following families: cacti, orchids, daisies, palms, and grasses.

The levels above family are **order**, **class**, **division**, and **kingdom**. You might expect universal agreement at the level of kingdom, but not everyone agrees about which organisms should be called plants and classified in Kingdom Plantae. Some believe that green algae should be included because they are almost identical biochemically to vascular plants. Others believe that, despite the biochemical similarity, green algae should be excluded because they are so different morphologically and anatomically.

The **scientific name** of a species is its genus and species designations used together and either underlined or italicized. For example, tomato is *Lycopersicon esculentum*. Note that the species name is not "esculentum" but *Lycopersicon esculentum*. Esculentum is the **species epithet**, the word that distinguishes this species only from the other species of the genus *Lycopersicon*. We cannot refer to tomatoes as "esculentum" because that species epithet is used for many different species in different genera: for example, buckwheat (*Fagopyrum esculentum*) and taro (*Colocasia esculenta*). The word "**taxon**" (plural: taxa) is used to refer to any of the above groups in a general way. For example, some systematists study smaller taxa such as species and genera, others are concerned with higher taxa such as orders and divisions.

Cladistics

Cladistics is the study of evolutionary relationships (**phylogeny**) that centers on examining the similarity of one species to others (**BOX 9.2**). A species evolves into two species as different populations accumulate distinct alleles; even when enough divergence has occurred to create separate species, the two still resemble each other strongly. As they continue to evolve, each acquires its own mutations, and because they cannot interbreed, they cannot share the new alleles. Thus, they differ from each other more as time passes. Distantly related plants have been on separate lines of evolution for millions of years, time enough for so many mutations to accumulate that they resemble each other only slightly. For example, pinto beans (*Phaseolus vulgaris*) and lima beans (*Phaseolus lunatus*) are so closely related they differ very little and are placed in the same genus; but the ancestors of peas (*Pisum sativum*; see Figure 6.2) diverged from the ancestors of beans so long ago that more differences have accumulated and peas and beans are easy to tell apart and peas are in a different genus. However, peas and beans are still so similar they are classified into the same family (the legume family, Fabaceae).

Studies of evolutionary relationships (phylogenetic studies) are complicated by the fact that plants can resemble each other for two distinct reasons: (1) they have descended from a common ancestor, or (2) they have undergone convergent evolution. Features similar to each other because they have descended from a common ancestral feature are **homologous features** (**synapomorphies; singular: synapomorphy**). For instance, almost all members of the legume family are easily recognizable because

BOX 9.2. Genealogy Versus Clades: Your Family History Is the Opposite of a Clade

It is easy to confuse cladistics with genealogy. When you try to trace your family tree, you consider your parents, their parents, the next set of parents, and so on into your past. You are trying to locate your past relatives, but the real point of a genealogy is to map out all the people who might have contributed alleles to your own genome. You got all your alleles from your two parents, but those alleles came from your four grandparents, your eight great grandparents, and your 4096 great, great, great, great, great, great, great, great, great, great grandparents, and so on. At each generation, you are considering the potential input of new alleles, and also you double the number of ancestors you are considering. Your family tree becomes broader as you go farther back.

A cladogram is just the opposite. We are interested in one single set of ancestors or founder individuals, and we are interested in how their progeny change with time. The **ancestral group** will be genetically isolated from all other groups (remember, only members of the same species can interbreed), so their only source of new alleles is mutations within the descendants. A cladogram attempts to map out the origins of new alleles or of important new combinations of alleles that lead to detectable changes in phenotype.

A genealogy maps the genetic convergence of characters from many ancestors into one descendant over a few generations. A phylogeny maps the evolutionary divergence of characters and progeny over thousands or millions of generations.

of their fruits are pods that split lengthwise in two and contain a row of seeds (see Figure 6.2). All members have these structures because all have inherited their flower and fruit genes from a common ancestor. The pods are homologous in all species of legumes. Homologous features are the ones critically important for making taxonomic comparisons and the only ones that can be used to conclude that species are related.

The second cause of resemblance, convergent evolution, results when two distinct evolutionary lines of plants respond to similar environments and selection pressures. Under these conditions, natural selection may favor mutations in each line that result in similar phenotypes. Features like this are **analogous features** (**homoplasies; singular: homoplasy**) and should never be used to conclude that plants are closely related. A striking example is the convergent evolution of cacti and the succulent euphorbias discussed earlier in this chapter. Determining whether a similarity is due to homology (common ancestry) or analogy (convergent evolution) can be extremely difficult.

A corollary to the assumption that similar plants are closely related is the assumption that dissimilar plants must not be closely related. Studying lack of similarity can also be difficult because in some cases a small genetic change results in dramatic phenotypic changes. Mutations that affect hormones result in large changes of the phenotype between two closely related species. Also, mutations that affect early stages of development such as the embryo or bud meristems can cause closely related species to look deceptively dissimilar.

Taxonomists study virtually every aspect of plants using a wide variety of tools. Simple observation of major parts is still important, but scanning electron microscopy to study hairs, stomata, cuticle, and waxes is also used. Internal structure is studied with both light and electron microscopy. Various aspects of metabolism are important, ranging from the types of pigments in flowers to the presence of CAM and C_4 metabolism and specialized defensive, antipredator compounds.

DNA sequencing is a new tool for analyzing evolutionary relationships. Two plants are considered separate species only if they differ in significant, heritable ways; therefore, the sequence of nucleotides in the DNA of each species must differ

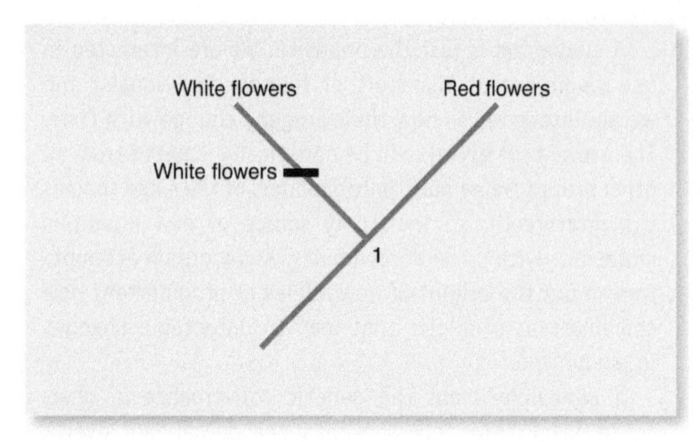

FIGURE 9.11. A simple cladogram showing the relationships between an ancestral group (at node 1) and its two descendants. The ancestral condition was to have red flowers, but we have to know that from some other source of information; we could not determine that just by looking at the two descendants. The ancestral group diverged into two, and now one group has white flowers (an apomorphy) and the other still has red flowers. Be careful to note that this is treating only a very limited amount of information. It does not indicate if we are dealing with two species and their ancestral species, or whether we are comparing two genera or two families.

from that of all other species. When plant phenotypes are studied, the actual objective is to determine differences in genotype. In the past, this could be done only by examining phenotypic features. Now DNA can be examined directly and mutations can be identified, even if they do not cause any detectable change in phenotype.

Phylogenetic Trees

A **phylogenetic tree** (also called a **cladogram**) is a diagram that shows evolutionary patterns by means of a series of branches (**FIGURE 9.11**). Each point at which a cladogram branches is called a node, and it represents the divergence of one taxon into two. All branches that extend from any particular point represent the descendants of the original group, and the ancestor at that point is their **common ancestor**. Any ancestor (any node) and all the branches that lead from it constitute a **clade**. Because each node represents a taxon dividing into two, it also represents some detectable change that creates a new group: after the divergence, one of the taxa differs from the other. For example, a plant in a species with red flowers might undergo a mutation that gives it white flowers. As long as the two can still interbreed, this is not especially significant. But the change in flower color may cause different pollinators to visit it, and it might become reproductively isolated from its red-flowered ancestors. If the white-flowered individuals thrive and become numerous—but cannot interbreed with the red-flowered ones—this is a new species, and the divergence should be represented as a node with two branches coming from it. In this case, white flowers are a derived condition (an **apomorphy**) and red flowers are the ancestral condition (**TABLE 9.2**).

Both populations might continue to evolve and diverge, giving rise to a phylogeny like that shown in **FIGURE 9.12**. Node 1 here represents the same taxon as node 1 in Figure 9.11, but now with more time, both its lineages have diverged: at node 5, some that had continued to have red flowers have diverged into two groups, one continuing to have red flowers, the other mutating to have white flowers; white flowers have now originated twice. Synthesis of pigments requires many steps and many enzymes, and mutations in any of those genes could cause the same phenotype. The two sets of white flowers almost certainly have different genotypes. This is homoplasy (convergent evolution). The original white-flowered lineage has diverged twice: after node 2, one group has trichomes on its leaves, the other continues to have smooth leaves. At node 3, dwarfism originates in one line; at node 4 one group mutates to have narrow petals.

Notice that we can speak of many clades in this phylogenetic tree: all the species depicted, plus their common ancestor at node 1, form a clade. Species E and F plus node 5 are a smaller clade, as are the four species (A to D) that have descended from node 2. It is especially important to notice what is not a clade: there are now five species with white flowers (A, B, C, D, E), but we cannot classify E with the others; they do not form a clade. The *most recent common ancestor* of all white-flowered species is node 1, and a natural, monophyletic clade includes the ancestor and all its descendants. If we would try to classify A–E as a group but leave out F, we would be creating a group that has two separate origins (a polyphyletic group).

TABLE 9.2. Terms Used in Cladistic Studies

The diagrams in Figures 9.14 and 9.15 are cladograms, branching trees that show the evolutionary lines and relationships of the existing species or genera being studied. The forking points in cladograms represent times when one group evolved into two distinct groups, usually with a new feature being present in one of the forks, the ancestral feature being present in the other fork. Many, perhaps most, studies of evolution now use cladograms, and there is a set of important terms that describe characters.

Plesiomorphies are characters that were present in a group's ancestors. For example, some cacti still have leaves like their ancestors had; the presence of leaves is a plesiomorphy in cacti.

Symplesiomorphies are ancestral characters shared by two or more groups. These are often called shared primitive traits. Symplesiomorphies are not usually helpful in analyzing evolutionary relationships. Most dicots have simple leaves, and the ancestor of flowering plants probably did too. The fact that many dicots now still have this symplesiomorphy does not help us understand the relationships among them.

Apomorphies are derived characters, new characters that were not present in the ancestors.

Synapomorphies are shared derived characters, that is, they are derived characters that occur in two or more modern groups because those groups are closely related. They share a recent ancestor that had this character, and they have inherited it. Synapomorphies are important characters we search for. The pod-like fruits of peas, beans, and other legumes are synapomorphies that indicate that all members in this family are closely related and form a natural group.

Autoapomorphies are unique derived characters. If a derived character occurs in only one group, it indicates that the group should be considered as a unit, but it does not help us understand what it is related to.

Homoplasies are the result of convergent evolution. They are characters that appear to be the same in two or more groups, but in fact they did not evolve from the same ancestral character. Plants called spurges in the genus *Euphorbia* resemble cacti but are not closely related to them. The similarities that fool us when we examine cacti and spurges are homoplasies. Trying to distinguish homoplasies from synapomorphies is always a concern in evolutionary studies. If we accidentally classify two unrelated groups together because we mistake a homoplasy for a synapomorphy, we have created an unnatural, polyphyletic group.

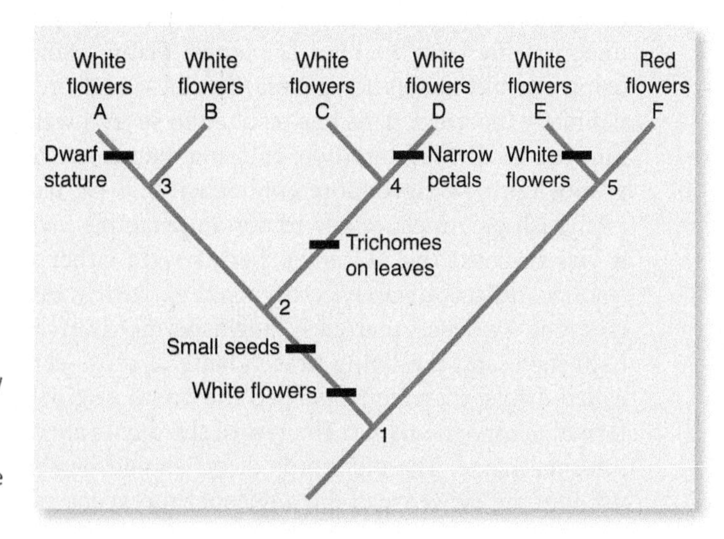

FIGURE 9.12. A more complex cladogram with more taxa and characters. We have data for the following characters: flowers (white or red), seeds (large or small), leaves (with or without trichomes), plant height (tall or dwarf), and petals (wide or narrow). In all taxa, the fruits are berries. Imagine that the ancestral group had red flowers, large seeds, leaves without trichomes, tall shoots, and broad petals (we probably would not know this beforehand, the cladogram is helping us determine what the ancestors were like). See the text for details.

Think about what a systematist is faced with. There are six species, some with red flowers, some with trichomes on their leaves, some with a dwarf stature, some with narrow petals. In constructing a phylogenetic tree, we search for shared derived characters, synapomorphies. Species C and D both have trichomes on their leaves but no other species does, so this is a synapomorphy that shows these two are closely related. Clade A–D share the synapomorphy of having small seeds. Species A is the only one that is dwarf: no other species has this feature, so it does not help us understand the phylogeny; it is said to be uninformative. All species still produce fruits that are berries, this has not changed anywhere; this shared ancestral condition (a **symplesiomorphy**) also does not help us in arranging these species. Any character that occurs in all species or in only a single species is never helpful.

Given these six species and these characters, a systematist could create any number of phylogenetic trees. Between node 1 and species A we could imagine many steps in which flower color changed from white to blue to yellow and then back to white, with the blue and yellow forms having gone extinct. And this might have happened: we can see the six living species and their characters, but we cannot see any of the ancestors (they are all long since dead), and there may have been other descendants that died out and thus are missing from our phylogenetic tree. But given the information we do have, we use a principle called **parsimony**: we prefer the simplest possible hypothesis; we do not make a hypothesis any more complicated than it needs to be. We move on to a more complex hypothesis only if data indicate that the simple hypothesis is not accurate.

In constructing phylogenetic trees, many more data are used than those shown in Figure 9.12. For living species, we can obtain DNA sequences in which each nucleotide is a data point, so we might have dozens or hundreds of differences when comparing six species. But when a computer program uses the data to construct a phylogenetic tree, it almost always comes up with several that are equally simple (**equally parsimonious**) but which have the taxa arranged differently. When this happens, the various trees are analyzed to see if any clade appears repeatedly; if so, we have confidence that that particular clade might be accurate. For the other taxa, we need more information and they become the focus of more intense study.

Phylogenetic Trees and Taxonomic Categories

How are names or taxonomic levels assigned to a phylogenetic tree? The only taxonomic unit with an objective definition is species: individuals that can interbreed. But there is no such objective definition for genus, family, etc. It has been proposed that every node and every clade in phylogenetic trees be named, but that would result in an unacceptably large number of names. Many taxonomists have decided to continue using the old names for genera, families, and orders unless they are shown to be definitely incorrect. This has resulted in several well-known families being combined into one or divided into two. Informal names are temporarily being used for certain groups until we have more confidence that we truly understand their evolutionary relationships. An especially important example involves angiosperms. Until recently, it was believed that all angiosperms were either monocots or dicots and that the "monocot/dicot" divergence occurred extremely early (FIGURE 9.13). But now there is strong evidence that early angiosperms diverged into several clades before that happened, and the living descendants of those early-diverging clades are no longer called dicots; they are just called the **basal angiosperms** and have not been given a formal taxonomic name. The rest of the dicots are called **eudicots**, and this too is an informal name. The old word "dicot" should be abandoned but many botanists still use it in the old sense of "all angiosperms except monocots."

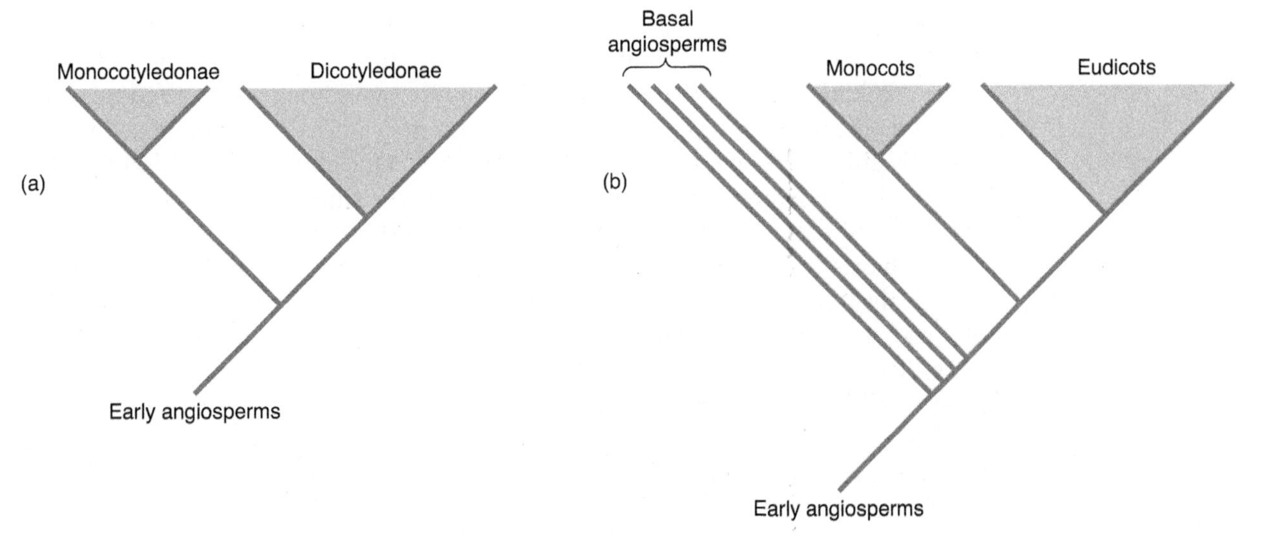

FIGURE 9.13. **(a)** A cladogram that represents the historical view that all angiosperms were either monocots or dicots. These had the formal taxonomic rank of class: Monocotyledonae and Dicotyledonae in some books, Liliopsida and Magnoliopsida in others (Table 9.1). **(b)** This very simplified cladogram summarizes a current hypothesis of evolutionary relationships within angiosperms. The "basal angiosperms" diverged from early angiosperms before the monocot/dicot split. All basal angiosperms had been known as dicots, as had all eudicots. But that concept of dicot would be a paraphyletic group, it would leave out some descendants (the monocots) of the early angiosperms. We do not use paraphyletic clades, so we need a new name for the basal angiosperms (they are usually just called basal angiosperms) and for the dicots to the right of monocots (they are now called eudicots).

The Major Lines of Evolution

All organisms are grouped into three domains: Bacteria, Archaea, and Eukarya (FIGURE 9.14). The most significant event in evolution was the origin of life itself, probably about 3.5 billion years ago. The first organisms must have been simple, consisting of a cell membrane, protoplasm, and some means of inheritance; they almost certainly had no distinctive nucleus or other membrane-bounded organelles. Such organisms, either living or extinct, are prokaryotes. This line of evolution diversified into numerous clades, and even though many have gone extinct, there are still thousands of living species of bacteria and archaeans. Another significant step was the evolution of oxygenic photosynthesis based on chlorophyll *a* (see Chapter 5). We do not know exactly which species of bacterium this occurred in, but their descendants are known as **cyanobacteria** (see Figure B5.1b) a clade within Bacteria, and one was the ancestor of chloroplasts.

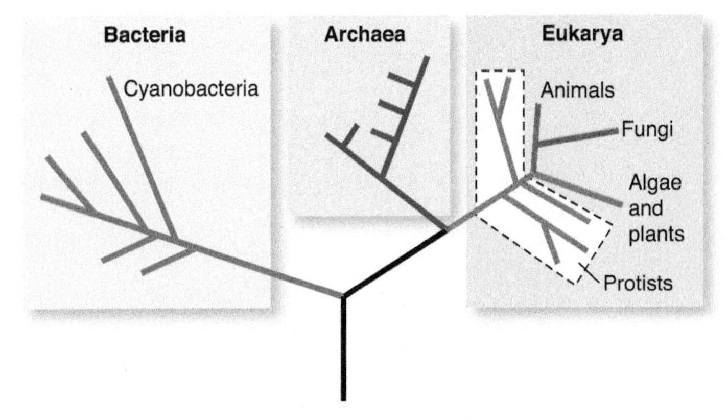

FIGURE 9.14. This cladogram representing our latest hypotheses about the interrelationships of the three domains: Bacteria, Archaea, and Eukarya.

The next major evolutionary event was the conversion of a prokaryote into a eukaryote, having a membrane-bounded nucleus. This must have been an extremely gradual procedure because several living species still have characteristics intermediate between prokaryotes and eukaryotes. A significant aspect was the origin of mitochondria; a bacterium capable of aerobic respiration began living inside the protoplasm of an early eukaryote whose own capacity for aerobic respiration was less sophisticated. Both organisms would have benefited from their association, so this was a symbiosis, not a parasitism, and because one lived inside the other, it was an **endosymbiosis**. This is the hypothesis of the endosymbiotic origin of mitochondria.

Once eukaryotes evolved to the level of having mitochondria, endoplasmic reticulum, and a true nucleus, numerous evolutionary lines emerged. Those that are extremely simple are often referred to as "protists." One clade contained organisms that would later diverge into animals and fungi. Surprisingly, fungi are more closely related to animals than they are to plants. The other early clade established an endosymbiosis with cyanobacteria, which then evolved into chloroplasts, producing the first algal cells. The early algae continued to diversify, and about 400 million years ago, some became adapted to living on land, establishing the clade of true plants, Kingdom Plantae. From these early pioneers, the major evolutionary lines that diverged were simple plants with neither seeds nor vascular tissues (mosses, liverworts, and hornworts), plants that do not produce seeds but do have xylem and phloem (ferns and similar plants) and the seed-bearing vascular plants (gymnosperms and angiosperms; FIGURE 9.15).

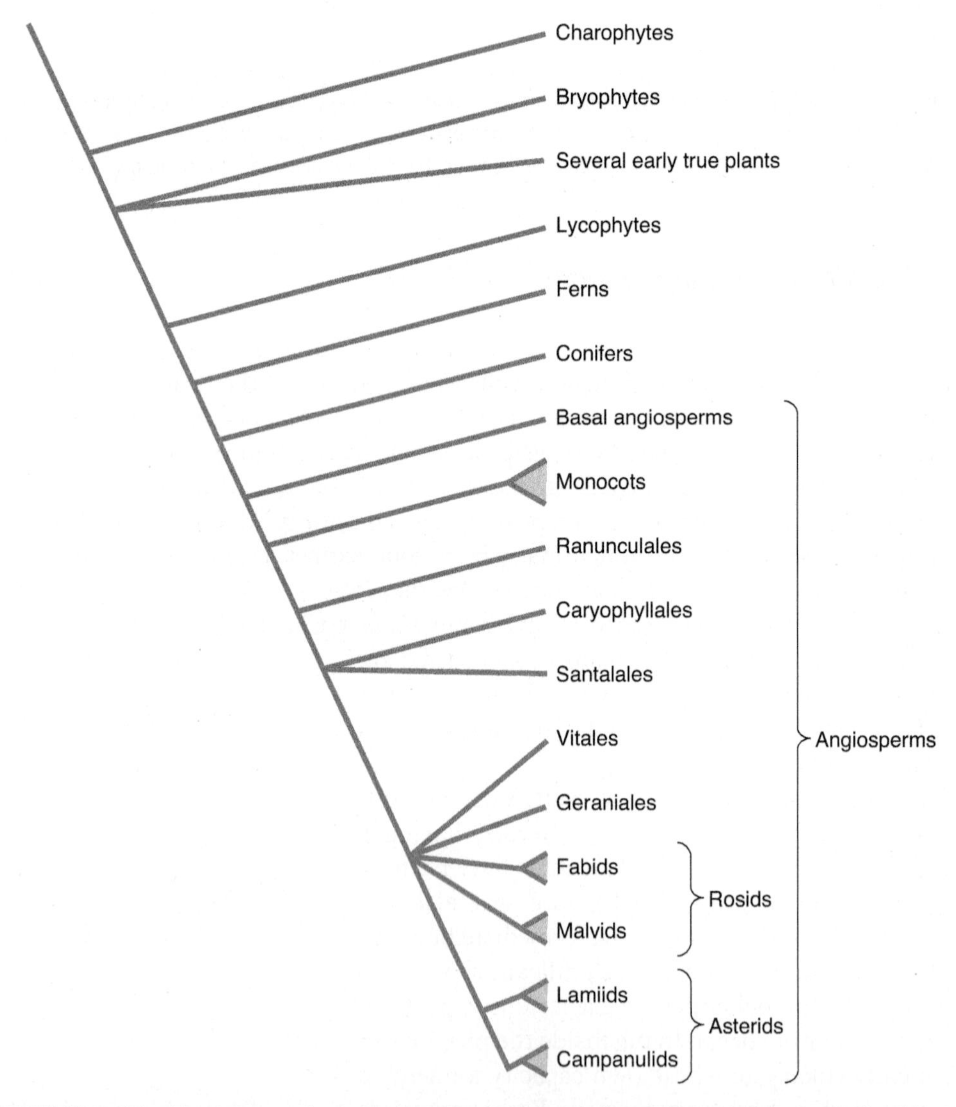

FIGURE 9.15. This phylogenetic tree represents our theories about the evolutionary relationships of plants and their nearest relative, the charophyte algae. Details are given in the text.

The Major Lines of Plant Evolution

Charophytes: The Algal Relatives of Plants

The closest relatives to true plants are a group of algae called Charophyta (FIGURE 9.16). These live in freshwater streams and are large enough to see easily with the naked eye. Their cell walls become encrusted with calcium carbonate, which makes them feel rough, giving them their common name of stoneworts.

Early True Plants

The criterion for saying that an organism is a plant rather than an alga is that plants make gametes and spores using multicellular organs, and only the innermost cells become reproductive cells. The outer cells form a sterile jacket, a layer of cells that protects the inner reproductive cells. In algae, all cells become reproductive, there is no protective layer. The earliest fossils of true plants are simple cylindrical stems with an epidermis, cuticle, stomata, cortex, and a slender vascular bundle. They had no leaves or roots (FIGURE 9.17). Because they had a cuticle and stomata, we know that parts of them were in the air, not in water or mud. Several fossils are known as *Cooksonia*, *Rhynia*, *Aglaophyton*, and *Horneophyton*. All are now extinct, but their descendants are the plants that are alive today.

Bryophytes

Mosses, liverworts, and hornworts used to be classified together as Division Bryophyta, but now each is placed in its own separate clade, and we use "bryophyte" as an informal term for all of them. Before we talk about bryophytes it is best to recall that in all other plants except bryophytes, the large plants we see are diploid sporophytes (see Figure 7.11 about plant life cycles). The haploid gametophytes are just the pollen grains and embryo sacs. But in bryophytes, the green photosynthetic bodies we are familiar with are the gametophytes. For some reason, the early plants gave rise to two extremely different types of plants: some with dominant gametophytes (the bryophytes) and others with dominant sporophytes (all the rest of the plants). At the time this divergent evolution occurred, the ancestral early plants were still simple,

FIGURE 9.16. This is an individual of the alga *Chara*. It grows in freshwater streams and during summer can become extremely abundant. It will survive brief drying if the surrounding mud stays moist, but unlike true plants, it has no cuticle and cannot effectively retain water for long.

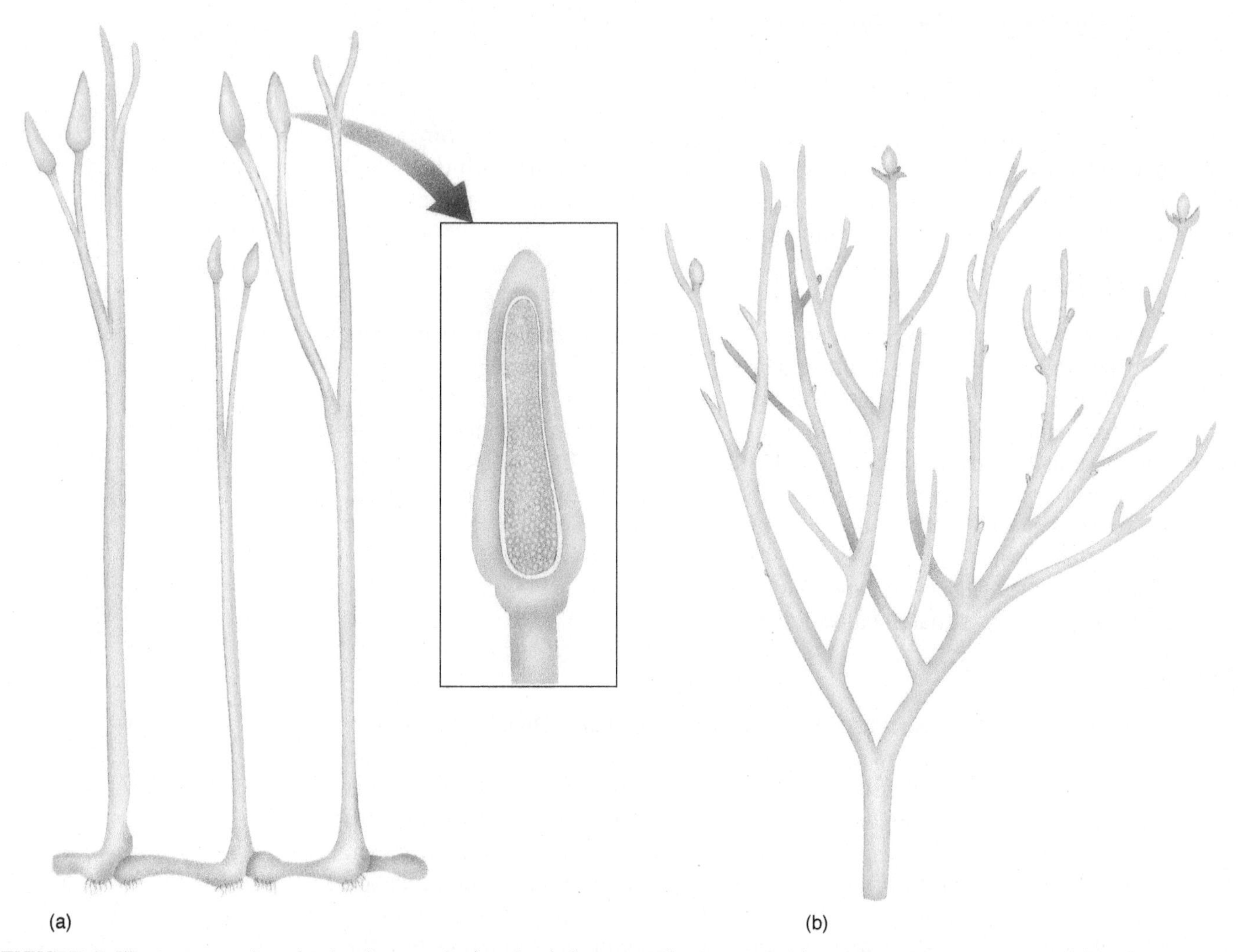

FIGURE 9.17. Reconstruction of *Aglaophyton major* (previously known as *Rhynia major*) **(a)** and *Rhynia gwynne-vaughanii* **(b)**. They strongly resemble *Cooksonia*, and we know that they definitely had rhizomes, upright stems, and rhizoids. Inset shows the sporangium cut away, revealing spores.

and apparently natural selection caused one group to have complex gametophytes whereas in the other it produced complex sporophytes.

Mosses are very common in almost all habitats so you should be able to find some easily and examine them with a hand lens (**FIGURE 9.18**; see Figures B7.2d, e, and f). Each has a stem and leaves, but none ever has a root. We use the term "leaves" for mosses even though the evolutionary origin of these gametophyte leaves is completely distinct from the origins of leaves on sporophytes in other plants. These two types of leaves are an example of convergent evolution, and if we tried to say that a particular moss and a certain seed plant are closely related because their leaves are similar, we would be making the mistake of using analogous features (homoplasies).

There are many families and genera of mosses, and they are extremely diverse in structure and physiology. Even though most people associate mosses with shady, cool, damp habitats, many are adapted to live in hot, full-sun desert areas. In numerous species, moss plants consist of tightly packed vertical shoots; in other species they are more open, loose masses of horizontal shoots. Moss plants are typically small, less than a centimeter tall (less than half an inch), but individuals of the common hair cap moss (*Polytrichum*) are several centimeters tall (3 or 4 inches)(see Figure B7.2e). In some moss species, a single plant will produce both sperm cells and egg cells, in other species these are produced on separate male and female plants. A few of the larger mosses have types of conducting tissues that somewhat resemble xylem and phloem.

Liverworts tend to be smaller and more delicate than mosses, and generally are not as common or as easy to find (see Figures B7.2g, h, and i). Your best chances of locating liverworts is to search stream banks for the large bodies of *Marchantia* (FIGURE 9.19) or *Conocephalum*, or look on streamside trees with smooth bark for tiny leafy liverworts that look like dark spider webs (FIGURE 9.20). These represent the two very different types of liverwort bodies: many liverworts have leafy stems and resemble mosses; others have a flat, ribbon-like body that completely lacks any leaflike structure (such a body is called a thallus).

The body of a hornwort is just a small disk of photosynthetic parenchyma, usually much smaller than 0.5 cm in diameter and just a few cells thick (see Figure B7.2j). They are never found in hot, dry areas, but during cool, rainy weather of spring, their spores germinate and grow quickly to a body capable of sexual reproduction. Hornwort sporophytes are long, slender horn-shaped cylinders; rather than having a stalk, the sporophytes grow by means of a basal meristem, so they gradually become longer and produce spores for many weeks.

Bryophytes, especially mosses, are important to us by helping in soil formation. Mosses secrete acids that accelerate the breakdown of rock, and tufts of mosses trap moving soil particles and hold them in place, creating good places for spores and small seeds to germinate. Mosses that grow along streams stabilize the soil and prevent erosion. Sphagnum moss thrives in calm water, and grows so extensively that it forms bogs that are the homes for many species of aquatic plants and animals, especially water birds. Sphagnum moss is harvested from old bogs and sold as peat moss, which is used by gardeners to lighten and acidify soils.

Lycophytes

Lycophytes and ferns are called vascular cryptogams: "vascular" because they have xylem and phloem, and "cryptogam" because they never make seeds. All modern lycophytes are small green herbs with leafy stems; they grow flat on the ground in some species, upright in others (FIGURES 9.21–9.23). Sporangia occur along the stems in many species, but are grouped at the shoot tip in a small cone-like structure in other species. Lycophytes are rather widespread and you may see them on nature walks in many areas. Shaded, damp areas often have the upright members of *Lycopodium*; sunny, dry semi-desert areas have the prostrate, low-growing species of *Selaginella*. It is easy to see a *Selaginella* and assume it is a large coarse moss (Figures 9.22 and 9.23).

FIGURE 9.18. Like many mosses, this *Thuidium* grows as a loose, open mass of horizontal branches. Each branch has many stems and leaves. All these plants are haploid; each nucleus has only a single set of genes.

FIGURE 9.19. Plants of *Marchantia* have some of the largest bodies of any liverwort. They have no leaflike structures; their bodies are ribbons that periodically branch in two. The cup-like structures produce groups of cells (called gemmae) that can grow into new plants when knocked out of the cup by rain. These plants are haploid gametophytes.

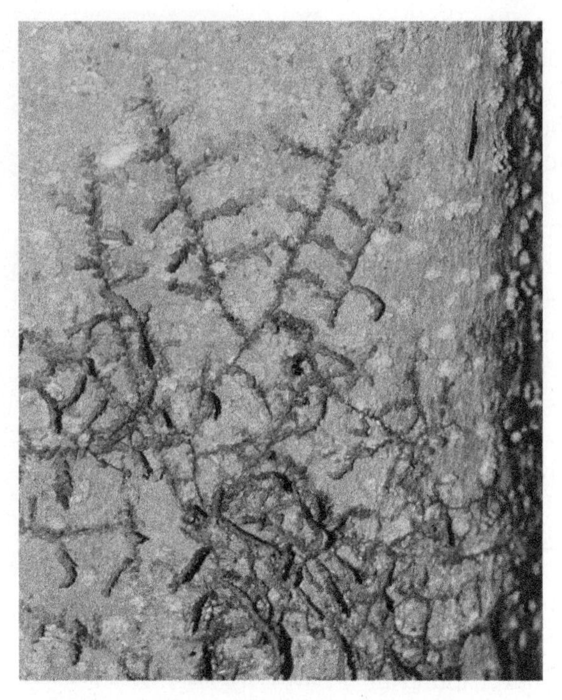

FIGURE 9.20. This is one plant of the leafy liverwort *Frullania*, growing on the smooth bark of a tree. A single spore probably germinated at the center of the photograph, then it produced several branches that each grow radially outward.

FIGURE 9.21. This is a plant of *Lycopodiella cernua*, a species in the family Lycopodiaceae. Like all lycophytes, it has stems, leaves, xylem, and phloem and, thus, resembles the plants we see every day. However, careful study of the anatomy and fossils of lycophytes have shown that these leaves had a fundamentally different evolutionary origin than did the leaves of ferns, conifers, and flowering plants. Consequently we classify the lycophyte clade as being distinct from that of other vascular plants. The cones in this plant contain sporangia: some cells in the sporangia undergo meiosis to produce haploid spores. (Courtesy of Chad Husby.)

FIGURE 9.22. The low, moss-like plant growing on otherwise bare stone in the semi-desert region of Fort Davis, Texas, is *Selaginella wrightii*. Most people would mistake it for a moss. Cacti and desert grasses are growing with the *Selaginella*.

Lycophytes are common today, but they never dominate a landscape. However, about 350 million years ago, many lycophytes were large woody trees that formed extensive forests covering thousands of square kilometers of land. In fact, much of the coal mined and burned today consists of the long-dead bodies of ancient lycophyte trees. For some reason, all the large, wood lycophytes became extinct, and only small herbaceous ones exist today.

FIGURE 9.23. A close-up view of *Selaginella wrightii* shows its moss-like nature, but these plants have both xylem and phloem and their leaves have stomata. These plants are diploid sporophytes.

Ferns

Ferns are also vascular cryptogams and resemble lycophytes (FIGURE 9.24; see also Figure 1.11 and Chapter 11). But fossil evidence and many technical features lead us to believe that ferns are a different clade than lycophytes. One critical difference is that leaves of ferns and seed plants had a different evolutionary origin than did leaves of lycophytes.

The large, familiar fern plant is the diploid sporophyte; the fern life cycle was described in Box 7.2 and Figure 7.11. In ferns, both sporophytes and gametophytes must live on their own: each must carry out photosynthesis, gather minerals and water, and protect themselves from herbivores and pests. Ferns are so abundant in most places, and they reproduce so well that if you look carefully in cool, moist times of the year, you have a good chance of finding gametophytes in nature.

Modern ferns have remarkable diversity of growth forms. All are herbs, none ever has a vascular cambium, wood, or bark, but some grow to be tall trees (the so called tree ferns; FIGURE 9.25). Many ferns grow as small rosettes of tightly clustered leaves attached to short stems whereas others are vines and some are tiny aquatic plants that live only floating on water. The aquatic fern *Salvinia molesta* grows so rapidly that it clogs streams, rivers, and canals and is considered a "noxious weed."

Because they are cryptogams, ferns never make seeds or fruits, so they are not a good source of food the way that many angiosperms such as wheat and apples are. But young fern leaves (fiddleheads; FIGURE 9.26) are often eaten as a vegetable in many areas. Fiddleheads of bracken fern may cause stomach cancer, but those of ostrich fern appear to be safe.

Equisetum (called both scouring rush and horsetail) are now classified with ferns (see Figures 1.12 and 9.4). Their stems are hollow and made of joints that can be pulled apart easily, so they are very recognizable. Some have a single upright stem, others are highly branched, and all have a fringe of tiny leaves at the top of each joint. Many have a cone at the stem tip (see Figure B7.2c); others produce cones only on special branches that barely emerge from underground. The jointed stems also occurred in

FIGURE 9.24. *Astrolepis cochisensis* is one of many ferns adapted to live in desert regions. This is in its dry state in November, in West Texas near Marfa. After a rain, the leaves quickly uncurl, turn green, and carry out photosynthesis. If cultivated in "typical" fern conditions of low light and high moisture, it would quickly die.

FIGURE 9.25. The trunks of tree ferns are strong but not because they have wood (no fern ever produces wood); instead, they have masses of fibers near the stem surface, and adventitious roots grow downward through the outer cortex, strengthening the stem. (Courtesy of Chad Husby.)

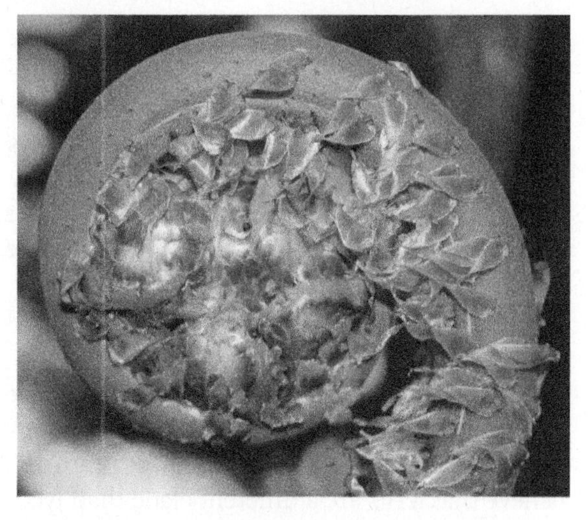

FIGURE 9.26. Young leaves of ferns develop in a coil, then uncurl as they expand to be a mature leaf. Leaves in this stage are called fiddleheads (*Cyathea arborea*).

FIGURE 9.27. *Cycas revoluta* is a cycad commonly used for landscaping in warmer areas. They have palmately compound leaves at the top of a single, unbranched trunk (the trunk has a small amount of very soft wood). Although branches almost never form on the side of the trunk, new branches often arise at the very base of the trunk, producing a dense cluster of stems. Some plants produce pollen cones at their shoot tip, others produces seed cones.

ancient, early species of this group, and rocks as much as 360 million years old contain fossils that look very similar to the plants that are alive today.

Seed Plants: Gymnosperms

Plants referred to as gymnosperms are the conifers (pines, larches, firs, cedars, junipers, and others; see Figure 9.1), cycads (sago palms; **FIGURE 9.27**), *Ginkgo biloba* (the maidenhair tree; **FIGURE 9.28**), and the gnetophytes (three genera of unusual, uncommon plants). "Gymnosperm" is an old term that we now use only informally; you will often see it in books about plant identification and roadside flowers. All gymnosperms have vascular tissues, wood and bark (secondary xylem and phloem), and seeds. We no longer use the term "gymnosperm" formally because we now realize these groups are separate clades.

The greatest number of gymnosperms are conifers and we will focus on them here. Conifers are large, perennial trees; none is either an annual or an herb. Their leaves are needles or scales that live for many years; only larches, bald-cypress, and dawn redwood lose their leaves each autumn (**FIGURE 9.29**). Conifer wood is softwood because it has no fibers. Conifers produce pollen cones and seed cones (see Figures 7.30 and 7.31). The ability to make seeds is a major difference between seed plants (**spermatophytes**) and nonseed plants (**cryptogams**): in cryptogams, all spores are released and all must live on their own. In seed plants, the megaspore is retained and develops into a female gametophyte while still on the parent plant. The female gametophyte

FIGURE 9.28. *Ginkgo* trees have very characteristic fan-shaped leaves, and they turn beautiful yellow in autumn. Ginkgoes are often used as landscape trees in cities because they are beautiful and withstand air pollution. *Ginkgo biloba* is a dioecious species: some plants produce pollen, others produce seeds.

(a)

(b)

FIGURE 9.29. Bald-cypress (*Taxodium distichum*) is one of the few conifers that loses its leaves in autumn and then produces a whole new set in springtime. **(a)** All the leafless trees in this swamp are bald-cypresses in November. Unlike most other conifers, bald-cypresses prefer to grow in streams or the shallow edges of lakes. This was photographed at Chicot State Park, Louisiana. **(b)** This is a young branch of bald-cypress producing new sets of leaves; a few of last year's leaves are still present.

makes one or several eggs, which are fertilized and which grow into an embryo all while still being retained on the parent plant. An embryo and surrounding tissues make a seed.

"Gymnosperm" means "naked seeds." When eggs are ready to be fertilized, the tiny, immature seed cone expands slightly, its scales separate a bit, and pollen grains fall right at the base of the scale next to the ovules. If we look closely, we can see the ovule without doing any dissecting: the ovule is naked. This never happens in angiosperms: the carpels do not open before fertilization, the ovules cannot be seen without dissection, so the only way pollen grains can get the sperm cells to the egg cells is by having a pollen tube that grows through the stigma and style. "Angiosperm" means "clothed seeds."

Conifers are extremely valuable commercially as sources of lumber, fiber for paper, and chemicals and fuel for wood-burning stoves. Esthetically, conifers populate many of the extensive forests where we like to hike, bicycle, kayak, camp, and just sit and enjoy natural beauty. Conifer forests are home to thousands of species of plants, animals, and fungi. Many ornamental plants used for landscaping are conifers.

Seed Plants: Angiosperms

The angiosperms (Magnoliophyta) are the flowering plants. At present, this clade contains more families, genera, and species than any similar group of plants. It also has the greatest diversity of metabolism, morphology, ecology, and reproduction. Flowering plants range from being annuals to long-lived perennials, from tiny herbs to giant trees, and extensive rhizomatous clones. It is in this group that we find bulbs, corms, vines, epiphytes, parasites, and many other growth forms. Almost all our food plants are angiosperms, and many plant-derived medicines and drugs come from them as well.

FIGURE 9.30. The very first set of angiosperms diverged into two clades: one now contains only *Amborella trichopoda*; the other contains all the rest of the living angiosperms. This does not mean that *A. trichopoda* is a primitive plant; its clade could have diverged into numerous species, genera, and families in the past, and many new features undoubtedly evolved, but *A. trichopoda* is the single surviving species, all others became extinct. (Courtesy of Chad Husby/Montgomery Botanical Center.)

Many features distinguish flowering plants from other plants, but one significant one is that both stamens and carpels occur together in most flowers (and in the early, ancestral flowers). This does not happen in most other groups, which keep microspore-producing structures separate from megaspore-producing ones.

Early angiosperms diversified into several groups. The first are referred to as the "basal angiosperms" and include water lilies and a species called *Amborella trichopoda*. *Amborella* is now believed to be the most basal living flowering plant (FIGURE 9.30), which means that its ancestors diverged into two groups, one of which gave rise to *Amborella*, the other produced all the other flowering plants. After the basal angiosperms, a few other clades arose, including those of magnolias and black pepper; the next diversification produced two groups that have since evolved to contain many diverse plants: the monocots and the eudicots. These two contain almost all the genera and species of flowering plants.

Monocots

Monocots are believed to have arisen from early angiosperms about 80 to 100 million, perhaps even 120 million, years ago. All monocots lack ordinary secondary growth and wood; the early angiosperms were woody plants, so evolution of the monocots must have involved a loss of the genes necessary to produce a vascular cambium. Almost all monocots have three of each flower part: three sepals, three petals, three stamens, and three carpels. Sepals and petals look so similar in most monocots they are called *tepals*. Monocot tepals rarely fuse to each other to form a tube; instead they remain separate from each other. Early monocots diverged into a series of clades whose extant members are classified into about ten orders.

The order Alismatales contains aquatic monocots such as *Sagittaria* (arrowhead) and many aquarium plants such as *Hydrocharis*, *Najas,* and *Hydrilla*. These plants are most often found in swamps and marshes, partly or entirely submerged, and some (known as "seagrasses") grow completely submerged in seawater. The submerged monocots have no transpiration, and most no longer produce stomata. Air chambers make the plants buoyant, so mutations that prevent the formation of lignified fibers are selectively advantageous; the plants do not waste carbon and energy by producing fibers that are not needed for support. Such plants tend to be thin and delicate, having very little sclerenchyma and almost no xylem. Some members of Alismataceae have part of their bodies emergent above the water's surface, and these plants are less highly modified, often having large leaves, considerable amounts of fiber, and a thick cuticle on the leaf epidermis.

Family Araceae (aroids) contains numerous familiar houseplants: *Philodendron* (250 species), *Anthurium* (500 species), and *Arisaema* (jack-in-the-pulpit; 100 species). *Dieffenbachia* (dumb cane) and 1800 other species also belong here. This family is characterized by the evolution of a distinctive inflorescence: tiny flowers, either unisexual or perfect, embedded in a thick stalk, the spadix. Staminate flowers are located near the top of the spadix, carpellate flowers near the base. The spadix is surrounded by a large bract, the spathe (FIGURE 9.31).

Liliales is a large order with 11 highly derived families and 1300 species. One of the largest families, Liliaceae, contains so many ornamental plants, mostly bulbs that people are familiar with, and think of them as the "typical" monocots. Liliaceae contains *Lilium* (lilies), *Tulipa* (tulips), *Calochortus* (mariposa lily), and *Fritillaria* (FIGURE 9.32).

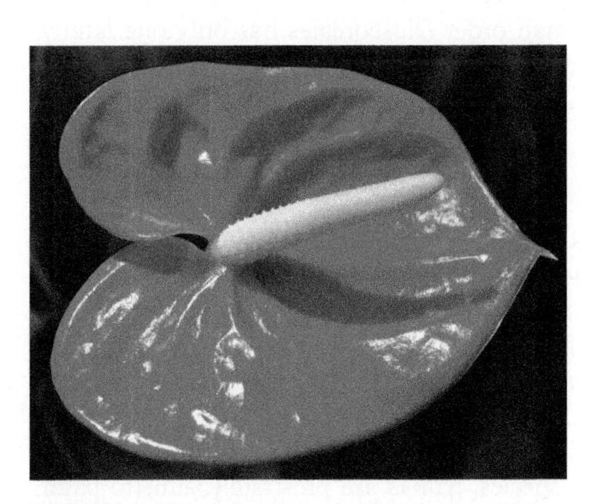

FIGURE 9.31. *Anthurium* and other members of its family have a characteristic inflorescence, consisting of a large leaf-like spathe and a columnar fleshy spadix that contains numerous unisexual flowers. The flowers are almost completely embedded in the spadix; often only the anthers or stigmas are visible.

FIGURE 9.32. Easter lilies (*Lilium longiflorum*) have representative lily flowers: they have three sepals and three petals that are indistinguishable, so we say the flower has six tepals. (© Workmans Photos/ShutterStock, Inc.)

FIGURE 9.33. You might recognize the bases of these shoots as being similar to the vegetable *Asparagus* because they actually are *Asparagus*. The *Asparagus* we eat is harvested just after the shoots have emerged above ground from a rhizome; the shoots must be harvested many times during spring so that they are not allowed to become tall and fibrous. At some point, harvesting is stopped and the shoots are allowed to grow, produce leaves, and photosynthesize enough to keep the rhizome healthy so that more shoots can be harvested the following year. White *Asparagus* is produced by harvesting the shoots before they have emerged from the darkness of soil. Fruits are present on these plants.

FIGURE 9.34. The complex sizes, shapes, fragrances, and colors of orchids make them some of our favorite cultivated plants. (© Barbara Burns/Dreamstime.com.)

Colchicaceae contains *Gloriosa* (flame lily) and *Colchicum*, the source of the alkaloid colchicine, which prevents microtubule polymerization.

Asparagales is a large clade with many families, species, and types of biology: by examining this clade, we get a sense of evolution as diversification. There are several morphological and DNA synapomorphies (shared derived characters) that unite the Asparagales. Most form nectaries by the partial fusion of their carpel bases; they do not fuse all the way up to the style, and the open areas secrete nectar. Beyond this, the group is extremely diverse in morphology, ranging from small, delicate bulbs like *Hyacinth*, chives, and onion (*Allium*), to vining epiphytes such as many orchids. Bulbs, rhizomes (such as *Asparagus*, FIGURE 9.33), and corms are common. Agaves and yuccas have giant, fibrous perennial leaves (see Figure 2.22); *Iris* has flattened sword-shaped leaves; and many members such as *Narcissus* (daffodils) have delicate leaves that live only during spring and summer. Several members of Agavaceae (*Yucca brevifolia*, Joshua tree) and Ruscaceae (*Dracaena*, dragon tree) have anomalous vascular cambia and grow to be large, highly branched trees. Asparagales has diversified biochemically as well, with different groups having distinctive chemical compounds, such as the sulfur compounds that make onion and garlic so pungent. Flowers are large, colorful, and showy for the most part, and in Orchidaceae, numerous types of insect pollination have evolved (FIGURE 9.34).

The small order Dioscoreales has only one family, Dioscoreaceae, and is mentioned because it has a familiar, important food crop: yams. These starchy "tubers" are produced by several species of *Dioscorea* and are a major source of carbohydrates for many people of tropical areas.

Arecales contains the palms, in family Arecaceae (an old name for the family is Palmae). There are about 3500 species of palms, and all are easily recognizable by their solitary trunk, which varies from only 1.0 cm (less than half an inch) in some, up to 100 cm (about 40 inches) in others. Leaves of palms always occur only near the shoot apex, never distributed along the length of the stem. In a few species, trunks are prostrate (palmetto palm, *Sabal*) or vines (*Daemonorops*). All species have simple leaves that, once fully expanded, are torn by wind into either a pinnate pattern (feather palms) or palmate one (fan palms). Coconuts (*Cocos*) and dates (*Phoenix*) are two types of palm fruit, but many others exist. Palm flowers are seldom seen because they are usually tiny, about 5 mm across (about one quarter inch), and are formed only high up in the tree.

Poales contains the grass family Poaceae as well as several other familiar groups such as cattails (*Typha*), bromeliads, and sedges. Poaceae contains about 8000 species and are much more than just the plants in the lawn. They also include most foods, such as wheat (*Triticum*), barley (*Hordeum*), oats (*Avena*), rye (*Secale*), corn (*Zea*), rice (*Oryza*), and sugar cane (*Saccharum*) as well as bamboo (the subfamily Bambusoideae). Humans began farming grass crops as early as 10,000 years ago, and about 50% of all calories consumed by people come from grass seeds. Also, our main sources of meat—cattle, chickens, pigs and sheep—are raised on grassland and fed corn. Closely related to grasses are sedges (Cyperaceae) and rushes (Juncaceae).

FIGURE 9.35. This is a "tank bromeliad"; they live as epiphytes attached to tree branches in rain forests, and their leaves overlap so tightly that they make a tank that catches rain water and holds it. The bases of the leaves gradually absorb the water: this plant obtains water through its leaves and uses its roots only to attach itself to a tree branch. Flower buds develop in the tank, while covered in water. Most bromeliads do not do this: pineapples and their relatives live in soil with ordinary leaves and roots, and bromeliads like Spanish moss and ball moss are epiphytic but absorb water from fog and mist rather than having tanks.

Other members of Poales are cattails (Typhaceae) and bromeliads (Bromeliaceae). Cattails grow in ponds and marshy areas. Each plant spreads by thick, horizontal subterranean rhizomes, with some axillary buds growing out as more rhizomes, other axillary buds growing upward and producing the familiar green leaves: what appears to be a cattail plant is really just a single branch. Their characteristic reproductive structures consist of a tip with many staminate flowers and thousands of carpellate flowers, although by the time we see them, the staminate flowers have usually withered and the carpellate flowers have matured into tiny fruits.

Bromeliaceae contains some of the most beautiful tropical epiphytes, their large, brightly colored inflorescences being easily visible even in thick jungle vegetation (FIGURE 9.35). Epiphytic species extend as far northward as the subtropics, Spanish moss and ball moss (*Tillandsia*) occurring from Florida to Texas. Other species are terrestrial and usually xerophytic, such as pineapples (*Ananas*). The xerophytic terrestrial forms are believed to be similar to the ancestral condition, and their capacity to withstand drought may have made it easier for bromeliads to adapt to life as epiphytes. Because bromeliads occur only in the Americas, they probably evolved after South America separated from Africa about 80 million years ago. Had they evolved earlier, numerous species should occur in the African rain forests and coastal deserts.

Zingiberales contains some of the most familiar of all houseplants: *Maranta*, *Calathea*, canna lilies (*Canna*), and gingers (*Zingiber*, *Hedychium*), as well as some that are best known in the warmer southern states—banana (*Musa*) and bird-of-paradise (*Strelitzia*) (FIGURE 9.36).

FIGURE 9.36. This ornamental ginger, cultivated for its beautiful flowers and leaves, is related to the ginger species we use to give flavor to ginger ale, ginger snaps, and other foods. (© mmattner/ShutterStock, Inc.)

Eudicots

The eudicots constitute a much larger group than the monocots and have diversified into numerous clades, some of which contain many families, others having fewer members. Virtually every type of organ,

FIGURE 9.37. Buttercups are familiar plants, but their flowers are unusual in having many carpels that are not fused together. (© motorolka/ShutterStock, Inc.)

tissue, and metabolism has several or many variations, so they are more difficult to characterize than are the monocots. One especially distinctive feature is that their pollen grains have three germination pores. Current phylogenetic trees of eudicots (Figure 9.15) have a large number of orders, too many to be covered here. Those that have special importance and which contain plants you might know are presented below.

Ranunculales contains families that diverged very early from the ancestors of the rest of the eudicots, so they lack many of the features that evolved later in other groups. For example, flowers of Ranunculaceae have little fusion of parts: each flower usually has many stamens and carpels, all of which remain separate of the others; examples are buttercups (*Ranunculus*; **FIGURE 9.37**), windflower (*Anemone*), and *Clematis*. The poppy family, Papaveraceae, is well known for its numerous ornamental species such as *Argemone* (prickly poppy), *Eschscholzia* (California poppy), and *Papaver* (poppies). *Papaver somniferum* is opium poppy, the source of a milky latex harvested for opium; morphine, a strong pain killer, is extracted from opium, as is codeine.

The Caryophyllales also arose early in the evolution of eudicots. Examples are cacti (Cactaceae), iceplant (Aizoaceae), portulaca (Portulacaceae), bougainvillea, four-o'clocks (Nyctaginaceae), spinach, beets, Russian thistle (tumbleweed) (Amaranthaceae), and carnations and chickweed (Caryophyllaceae; **FIGURE 9.38**). The core Caryophyllales share many derived characters, but one is especially important. Whereas other flowering plants produce anthocyanin pigments, almost all Caryophyllales instead produce a group of water-soluble pigments called betalains. The dark red color of beets is due to betalains, and betalains provide the color of flowers and fruits in Caryophyllales. The buckwheats (Polygonaceae), pokeweeds (Plumbaginaceae), two groups of carnivorous plants, Droseraceae (sundews and Venus's flytrap), and Nepenthaceae (pitcher plants) are also in the Caryophyllales.

FIGURE 9.38. The brilliant color of bougainvilleas results from betalain pigments, which are found in almost all members of the Caryophyllales but not in any other group. The presence of betalains unites this group; it is a synapomorphy. (© cycreation/ShutterStock, Inc.)

Santalales is a small order of highly modified plants, most of which are parasitic. The sandalwood family (Santalaceae) contains the large tree *Santalum* from which sandalwood incense is obtained. It appears to be an ordinary tree, but its roots make fine, almost imperceptible connections to roots of surrounding plants and parasitize them. Also in this family are common mistletoes, *Viscum* and *Phoradendron*, which are used as decorations at Christmas (**FIGURE 9.39**). These mistletoes have chlorophyll and are photosynthetic, so they are just hemiparasitic, but some members are holoparasitic, having no chlorophyll at all.

The remaining eudicots are members of two very large, very diverse clades, the rosids and asterids (see Figure 9.15). The rosid clade (named for the rose order Rosales) consists of many families that, taken as a whole, are so diverse with respect to vegetative body, flowers, chemistry, and ecology that it is difficult to see they are all related. But some share enough characters with others to indicate a relationship, and those of the second group share different features with a third group and so on until a larger picture of phylogenetic relationships can be distinguished.

The rosids consist of several small orders and two large groups. One of the small orders has few members but great economic significance. Vitales contains Vitaceae, the grape family. *Vitis* is the genus of grapes which give us juice, raisins, and table grapes. *Vitis vinifera* is the wine grape, and despite the enormous variety of wines, virtually all are made from varieties of this one species. The other small order, Geraniales, contains the geranium family, Geraniaceae. By a taxonomic quirk, the geraniums we grow at home are in the genus *Pelargonium*, not in *Geranium*.

The two large clades of rosids are the fabids and the malvids. These two clades together contain more than one hundred families and it is difficult to give any universal characters. Rosids are especially interesting in that none of them has any of the highly relictual features found in many early eudicots.

Although this clade is named for the order Rosales, roses should not be considered typical, they are just one group of many. Because the rosids consist of 14 large orders with over 50,000 species, only a few can be mentioned here. Five of the orders contain almost 75% of the species: Fabales (legumes) 19,000 species; Myrtales (*Eucalyptus*, evening primrose) 9000; Malpighiales (poinsettia) 1600; Rosales (roses, elms, marijuana) 6300; and Sapindales (maples, horse chestnuts, creosote bush, and the species whose resins are valued as frankincense—*Boswellia*—and myrrh—*Commiphora*) 5800. Members of this subclass include roses, of course, and legumes (peas, beans, peanuts, *Mimosa*, redbud, and clover in the family Fabaceae [its old name was Leguminosae]), *Fuchsia*, evening primrose (Onagraceae), dogwood (Cornaceae), the spurges that look like cacti and often have an extremely poisonous milky latex (Euphorbiaceae; see Figure 9.10), maples (Aceraceae), dill, celery, carrot, parsley, and hemlock (Apiaceae [its old name was Umbelliferae]).

The rose family is important not only in an evolutionary sense but also economically. Rosaceae contains numerous ornamental genera, including *Rosa* (roses), *Crategus* (hawthorn), *Spiraea*, *Cotoneaster*, *Pyracantha*, *Photinia*, *Potentilla*, *Chaenomeles* (flowering quince), and *Sorbus* (mountain ash). The family also provides most of the fruits that can be grown in temperate climates: *Malus* (apple), *Pyrus* (almond, apricot, cherry, nectarine, peach, plum, prune), *Eriobotrya* (loquat), *Fragaria* (strawberry), and *Rubus* (blackberry, loganberry, and raspberry; **FIGURE 9.40**).

The legume family (Fabaceae) is also a large family having bodies that vary from herbaceous annuals to woody shrubs, vines, and long-lived trees. It is important economically as a source of many foods, drugs, dyes, woods, and many of its species are the dominant plants in arid areas and deserts. Numerous species have root nodules with symbiotic associations with nitrogen-fixing bacteria, so they can grow in poor soils yet still produce protein-rich seeds. Legumes such as beans, lentils, and peanuts are a critically important source of proteins for people in arid areas.

FIGURE 9.39. Mistletoes that we see at Christmas are widespread parasitic plants that are common on many trees. The mistletoe sold in stores is *Phoradendron flavescens* in North America (in Europe, *Viscum album* is harvested instead). These both have green leaves and carry out some photosynthesis, but other mistletoes have almost no leaves, very little chlorophyll, and are much more dependent on their host plant. Shown here is *Phoradendron tomentosum*.

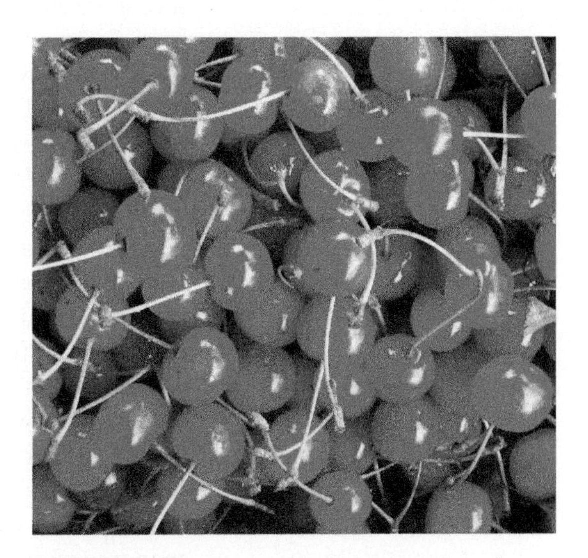

FIGURE 9.40. The rose family (Rosaceae) produces some of our most popular fruits. These cherries were bred by artificial selection from the wild cherry *Prunus avium*. In addition to fruit, the trees are loved for their flowers. (Courtesy of Chris McCoy.)

FIGURE 9.41. Flowers of moon vine (*Ipomoea alba*) are sympetalous; this means that the edges of each petal are united with the edges of the adjacent petals, forming a tubular structure. The length and width of the tube controls which animals can reach the nectar (located at the base of the tube) and pollinate the flower. This helps ensure that pollen will be carried only from one moon vine flower to another, not to just any random flower nearby.

The most derived large clade of eudicots is the asterid clade, which contains plants such as sunflower, periwinkle, petunia, and morning glory. Asterids, being a sister clade of rosids, originated perhaps as recently as 60 million years ago, and even its most basal members were much more highly derived than plants in the early eudicots. The majority of asterids can be easily distinguished from other angiosperms on the basis of three features: (1) their petals are fused together into a tube (sympetalous flowers; FIGURE 9.41); (2) they always have just a few stamens, not more than the number of petal lobes; and (3) stamens alternate with petals. Asterids exploit very specialized pollinators that recognize complex floral patterns, and such plants could not evolve before derived, sophisticated insects appeared.

Many chemical differences exist between this group and all others. Certain families produce numerous, very potent chemicals that deter animals or kill them outright. Apiaceae (celery, dill, fennel) have secretory canals containing oils and resins, and several other toxic compounds. The poison Socrates was given to kill himself came from *Conicum* (hemlock) in the Apiaceae.

Asterids have the greatest number of species (about 60,000) but they are grouped into two small orders (Cornales [dogwoods and hydrangeas] and Ericales [azaleas, blueberries, tea]) and two groups of orders, lamiids and campanulids. One family (sunflowers, daisies; Asteraceae [the old name was Compositae]) contains fully one-third of all the species and is the largest family of eudicots. Examples of asterids are milkweeds (Asclepiadaceae), potato, tomato, red peppers, eggplant, tobacco, deadly nightshade, petunia (Solanaceae), morning glory (Convolvulaceae), thyme, mints, lavender (Lamiaceae), and Asteraceae (sunflowers, dandelions, lettuce, *Chrysanthemum*, ragweed, and thistle).

Many asterids are extremely important medicinally: Apocynaceae (oleander family) contains periwinkle, *Vinca*, from which are extracted vinblastine and vincristine, two of our most potent anticancer drugs. Rubiaceae contains, in addition to coffee (*Coffea*), *Cinchona*, from which we derive the antimalaria drug quinine. Heart disease is treated with cardiac glycosides extracted from *Digitalis* (Scrophulariaceae, the snapdragon family); these compounds make heart muscle beat more slowly and strongly, increasing output of blood from the heart and improving circulation.

The order Lamiales (families of olive, *Penstemon*, and snapdragon; gesneriads, acanths, and trumpet-creeper) is quite derived in its floral characters; the flowers tend to be bilaterally symmetrical and of sizes and shapes that permit only certain

FIGURE 9.42. Members of the very large family Asteraceae have tiny flowers that develop close together in an inflorescence that resembles an individual flower. Each petal here really is a petal, but each one belongs to a different flower. The yellow "stars" in the center of the inflorescence are the stigmas of several individual flowers. After being pollinated, each flower will make a separate fruit, which most people call "sunflower seeds"; we crack open the shell (the fruit) to get the true seed.

insects to enter and effect pollination. The floral bilateral symmetry further forces the insect to enter in one particular orientation (see Figure 7.27). Pollinators capable of doing this tend to be sophisticated and able to recognize different flowers easily. They learn which flowers provide the greatest rewards, then search for other flowers of identical shape, color pattern, and fragrance, thus providing efficient pollination for the plant species.

The giant family Asteraceae (asters, daisies, sunflowers) contains 1100 genera and 20,000 species distributed worldwide in almost all habitats except dense, dark forests (FIGURE 9.42). They range from important food plants (*Lactuca*; lettuce) to ornamentals to weeds. The characteristic daisy/sunflower type of inflorescence makes them instantly recognizable. Members of Asteraceae have a wide range of unique chemical defenses against herbivores: sesquiterpene lactones, monoterpenes, terpenoids, and latex canals that contain polyacetylene resins. The presence of these chemicals makes composites extremely resistant to animals that eat plants or lay eggs in them; it also causes them to be irritating to human skin, resulting in numerous cases of contact dermatitis. The Asteraceae is a young family, perhaps no more than 36 million years old.

Important Terms

abiological reproductive barrier
adaptive radiation
analogous features
ancestral group
apomorphy
artificial selection
basal angiosperms
binomial system of nomenclature
biological reproductive barrier
clade
cladistics
cladogram
class
common ancestor
convergent evolution
cryptogam
cyanobacterium
divergent speciation
division
endosymbiosis
equally parsimonious
eudicots
family
founder
gene flow
gene pool
genetic drift
genus
homologous features
homoplasy
kingdom
monophyletic group
natural selection

natural system of classification
order
parsimony
phyletic speciation
phylogenetic tree
phylogeny
polyphyletic group
population
population genetics
scientific name
speciation
species
species epithet
spermatophyte
symplesiomorphy
synapomorphy
systematics
taxon

Concepts

- As DNA evolves, so do the organisms that contain it.
- A species may evolve and become more adapted to its habitat, or remain adapted as the habitat changes, or it may diversify into new species.
- The alleles present in a population (a group of interbreeding individuals) are the gene pool.
- Natural selection is the most significant factor causing gene pools to change.
- Natural selection requires that (1) a population produces more progeny than can possibly survive in the habitat, and (2) the progeny must differ from each other genetically.
- Threatened and endangered species have small populations, which are subject to genetic drift.

Plants, People, and the Biosphere

II

Plants and people share the world with animals, fungi, and all other organisms. Almost every aspect of our human biology, and most aspects of plant biology, influence all other creatures, and they influence us. People have set aside a large area of habitat as national parks and preserves to protect animals and the plants they depend on, as well as all the other organisms in those areas. Photographed in the Chisos Mountains in Big Bend National Park.

Plant Biogeography: The Distribution of Plants on Drifting, Changing Continents

This is a salt marsh on the southwestern coast of Louisiana. The occurrence of these plants here is not an accident; instead it is an aspect of biogeography resulting from many factors: the low flat coastal plain devoid of hills or mountains in an area frequently swept by strong storms; occasional freezing winters that kill tropical plants; lack of solid ground that would allow large grazing animals to live here, so much salt in the water that no trees can grow, and so on.

Earth is a changing, dynamic place, and because it is our home, we should think about the various changes that occur. Daily changes are familiar and have profound effects on the biology of all organisms. Every 24 hours, there are changes of light and darkness, warmth and coolness. Seasonal changes of temperature, precipitation, flooding, drought, and so on are also familiar. Changes such as wildfires, earthquakes, landslides, and volcanic eruptions do not occur on a regular cycle but happen often enough to be familiar to us and to have impacts on all forms of life.

On longer and larger scales are changes that are more major but not familiar to us. The continents move across the surface of Earth, changing their shapes, changing the shape of the ocean basins, and affecting the patterns in which water and air circulate. Mountains build where none were before, and others erode away; rivers change their courses. At times in the past, Earth has been much hotter than now, at other times it was almost completely frozen over. Eons of humid climate have existed, as have eons of dryness. As these changes occurred, some plants, animals, fungi, and other forms of life were able to adapt. Some individuals had beneficial alleles and even though some of Earth's changes caused many individuals to perish, those that were adapted to the altered conditions survived. Evolution by natural selection kept Earth populated with a rich diversity of life.

On at least two occasions, catastrophic change occurred and survival may have depended mostly on luck. A time period called the Cretaceous Period (**TABLE 10.1**) was brought to an end 65.5 million years ago when a giant asteroid struck Earth and killed most living creatures (including the dinosaurs); this is called the Cretaceous-Tertiary extinction event. Before that, 251 million years ago, one or several catastrophes caused the Permian-Triassic extinction event, during which the great majority of species were eliminated. Not just the majority of individuals, the majority of entire species. Up to 94% of all marine species and 70% of all land vertebrates. Of all the life forms that had evolved up to that point, most were eliminated. But in the time since, the survivors of those extinctions have produced all the plants and animals around us today, including ourselves. Life is resilient and dynamic.

TABLE 10.1. The Geological Time Periods

Millions of years ago	
Paleozoic Era	
540	Cambrian Period
490	Ordovician Period
450	Silurian Period
415	Devonian Period
360	Carboniferous Period
300	Permian Period
Mesozoic Era	
250	Triassic Period
200	Jurassic Period
145	Cretaceous Period
Cenozoic Era	
65.5	Paleogene Period
23	Neogene Period
2.58	Quaternary Period
2.58	Pleistocene Epoch
0.01	Holocene Epoch (started 10,000 years ago; this is the epoch in which we live)

Biogeography, the distribution of plants, animals, and other forms of life on Earth, is heavily influenced by random events, events that change the size, shape, positions, topography, and climate of the land and water that life inhabits. Mutations continually generate genetic diversity, and natural selection causes organisms to become more adapted to their environments. Millions of years ago, there would have been no way to predict where the continents and oceans would be now, and no way to predict which types of animals and plants would exist now. And of course we still cannot predict such things for the future. But we can predict that as long as some life exists, and as long as some individuals survive major climate and geological changes, the survivors will diversify and repopulate the Earth.

Of the many factors that influence biogeography, several are particularly important. First is Earth's rotation on a tilted axis, which causes seasons to change during a yearly cycle and which drives massive flows of air in the atmosphere and water in the oceans. Second is the position of Earth's land masses, the continents; when they lie in warm moist areas they are covered with life, but when they drift to inhospitable regions, such as when Antarctica drifted to the South Pole, they become barren. A third important factor is the origin, about 470 million years ago, of true plants and their subsequent evolutionary diversification as the ground beneath them drifted and the climate around them changed. The origin of true plants is significant because most live on land, and they are ultimately the food for all terrestrial animals. Before the evolution of true plants, no animal ventured more than a few feet onto dry land, all remained near the safety of oceans or rivers. Land was barren and lifeless prior to the origin of plants.

World Climate

Earth's climate is affected by many factors, especially its tilted axis of rotation and the presence of our atmosphere and oceans.

Effects of Earth's Tilt

Imagine a planet whose axis of rotation is exactly vertical, that is, perpendicular to the plane of its orbit. If we lived at the equator, the sun would rise exactly in the east every day, pass directly overhead, and set in the west. We would have no change during the year, no seasons. At all sites not on the equator, 40 degrees N for instance (the latitude of New York City), the sun would never be overhead; every day it would rise in the southeast, pass low in the sky, and set in the southwest. The equator would be hot all the time; all other regions would receive only oblique lighting and would be much cooler. If there were no atmosphere or oceans, heat could not be transferred from the equator to the poles, and there would be a tremendous difference in temperature between those regions.

Earth's axis of rotation is not vertical; it is tilted 23.5 degrees away from perpendicular. At summer solstice, June 21 or 22, the North Pole points as directly toward the sun as possible, and in the Northern Hemisphere, the sun has reached its highest point in the sky. The sun is overhead at noon for people located at 23.5 degrees N latitude, the Tropic of Cancer, which runs just south of California, Texas, and Florida (FIGURE 10.1). The days are their longest and summer has officially begun. Plants in the northern hemisphere are receiving the maximum amount of light for photosynthesis, and for plants that respond to photoperiod, days are long and nights are short. North of the Arctic Circle, the sun is visible even at midnight.

As Earth continues its orbit, the axis of rotation points less toward the sun, which appears to rise and set more to the south. By September 23, autumnal equinox, Earth has made one-quarter orbit, the sun is directly over the equator, and days are exactly 12 hours long: autumn begins. After another 3 months, winter solstice, December 21 or 22, the South Pole points as directly as possible toward the sun; the sun is directly over the Tropic of Capricorn, 23.5 degrees S latitude. Summer begins in the southern hemisphere, but in the north, winter begins and days are at their shortest. By March 21, the vernal equinox, the Sun is over the equator again.

The Air and Oceans Distribute Heat and Moisture

Our atmosphere and oceans, being fluids, develop massive flows when heated in one area and cooled in another. They distribute heat from the tropics (the zone between the Tropic of Cancer and the Tropic of Capricorn) to the temperate regions and all the way to the poles.

Much of the tropical zone is occupied by the Pacific, Atlantic, and Indian Oceans. Throughout the year, the water and air receive solar heat, causing tremendous amounts of evaporation; as the moist air warms, it expands and rises high into the atmosphere. Once high enough, the air cools, water vapor condenses into rain and falls in torrential storms, producing tropical rainforests in Central America, northern South America, central Africa, and Southeast Asia.

After rising, the air is pushed northward and southward by the continued rising of more tropical air below it. By the time it reaches about 30 degrees N or S latitude (the horse latitudes) it has cooled, contracted, and sinks. The sinking compresses the air, warming it and increasing its capacity to absorb moisture. Land areas below this descending air contain the world's hot, dry desert biomes.

Once back at Earth's surface, part of the air spreads toward the equator and part flows toward the poles. Earth's rotation causes the air moving toward the equator to be deflected westward as a northeast trade wind in the northern hemisphere. Air spreading toward the poles from the horse latitudes is deflected eastward and blows as a prevailing westerly (winds are named for their sources: a westerly wind comes from the west).

The area of descending dry air is farther north when the sun is near the Tropic of Cancer (northern hemisphere summer) and farther south when the sun is near the Tropic of Capricorn. The United States, during its winter, is located entirely within

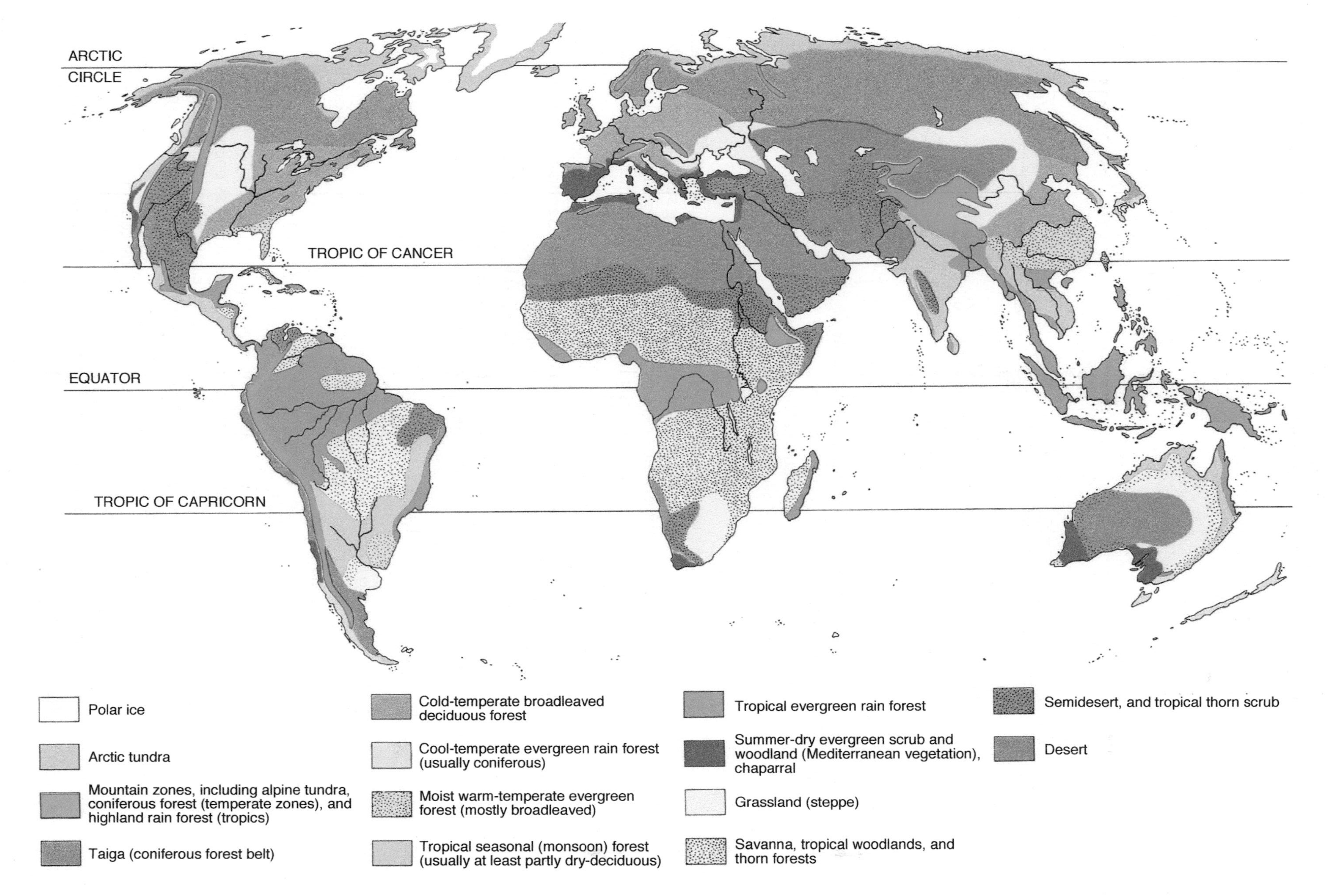

FIGURE 10.1. Most land on Earth is distributed rather far north at present, and the tropical zone between the Tropics of Cancer and Capricorn is mostly water. The greatest amount of solar energy falls on water, causing great evaporation and humidifying the atmosphere but leading to relatively little temperature change. If more land were in the tropics, the soil would heat more, which in turn would heat the air, but the oceans would stay cool and evaporate little moisture to the atmosphere; all regions would receive much less rain.

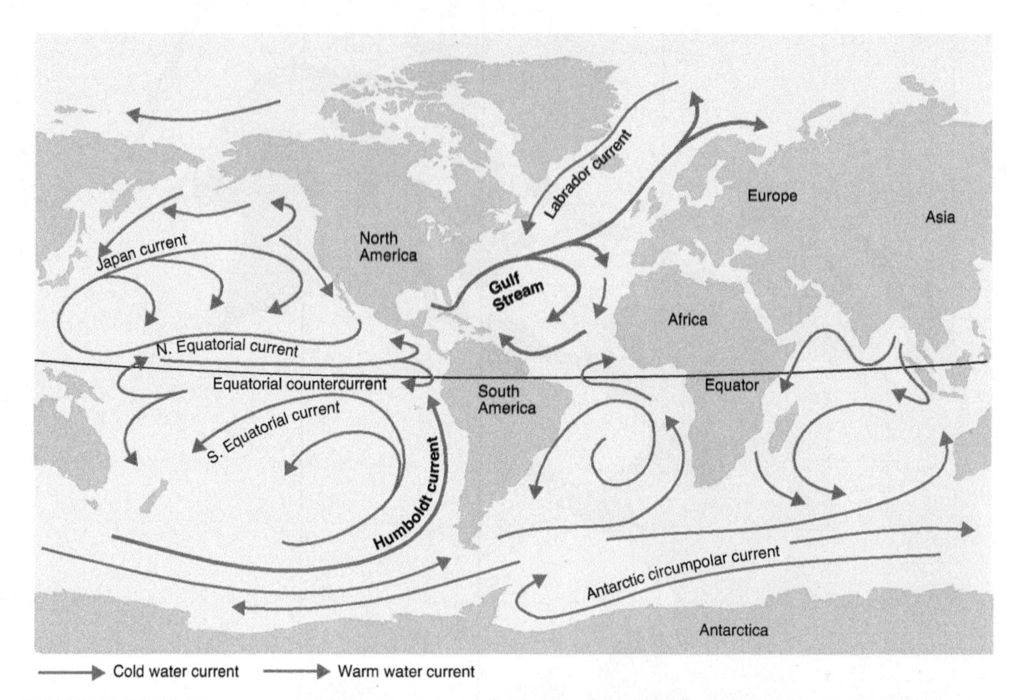

FIGURE 10.2. Ocean currents distribute heat from near the Equator toward the poles, and also affect humidity in the air, and thus rainfall patterns.

the influence of the prevailing westerlies; winter weather comes from the Pacific and Arctic Oceans, moving eastward across the continent. During summer in the United States, the northeast trade winds have moved northward far enough to influence the states along the Gulf of Mexico; northeast trade winds bring summer storms westward from the Atlantic onto the east coast and gulf, supplying summer rains.

Air circulation patterns drive water in the Pacific and Atlantic ocean basins in four giant circular currents, clockwise in the northern hemisphere and counterclockwise south of the equator (FIGURE 10.2). Ocean currents distribute heat from the tropics to the poles, lessening the temperature differences that would otherwise exist. As warm tropical water moves to higher latitudes, large amounts of water evaporate into the temperate prevailing westerlies, giving them more humidity than they would have if the oceans did not circulate.

As trade winds blow across the tropics from east to west, they push surface waters into equatorial currents. During the weeks that a particular mass of water flows along the equator as part of the Atlantic Equatorial currents, it absorbs huge amounts of energy and warms significantly. At the western side of the ocean basin, it is deflected northward by the tip of Brazil, then part is deflected by Florida and enters the Gulf of Mexico. The water's warmth permits high evaporation into the air and keeps much of the gulf coast humid. The rest of the current moves along the east coast as the Gulf Stream; because this is a latitude dominated by westerly winds, the warm current does not keep the land as warm and wet as one would expect. Near New Jersey and New York, the current turns eastward toward Europe; at the turning point, cold polar water moves south along the east coast of Canada and the northeastern United States.

The Pacific Ocean has a similar pattern with westerly equatorial currents (ocean currents are named for their direction of flow, the opposite of the names for winds). The Philippines and Indonesia act as a barricade, deflecting water north and south.

Much northern water heads northward then eastward toward Canada and Washington. Once at the west coast of North America, some water turns northward as the Alaska current, but the bulk turns south as the California current. The westerly winds absorb huge amounts of moisture as they blow for thousands of miles across the warm Pacific waters, and even though much falls as rain over the ocean, a large amount remains in the air and keeps the coasts wet.

The climate phenomenon known as **El Niño** shows us that weather and climate patterns can be unstable. At irregular intervals of between 3 and 7 years, warm water moves eastward along the equator and comes to rest in the eastern Pacific Ocean, near South America. Because this water is warm, much of it evaporates into the air and causes tremendous rain storms along the coasts of Peru and Chile, which normally are dry. Part of the moist air moves north, providing extra rainfall across the southern United States, but drier weather in the Pacific Northwest. El Niño conditions usually last for only a few months, then weather returns to normal. If something should happen that the pool of warm water would remain in the eastern Pacific rather than dissipating, weather all along the Americas would be dramatically and instantly altered. This would undoubtedly cause major shifts in the types of plants and animals that could survive where they do: our southwestern deserts would certainly become too wet for cacti, agaves, and yuccas, and the conifers of the Pacific Northwest would probably not tolerate the prolonged dry conditions. Certain rivers would flood every winter while others would have less water. Agricultural regions would be devastated.

Effects of Land and Mountains

The size and surface features of a land mass influence the weather it receives. Larger islands such as Japan, Madagascar, the Philippines, Cuba, and the Hawaiian Islands have mountains that force air to rise as it blows across them. Air cools as it rises and rain forms (FIGURE 10.3). Low-lying, small islands like the Florida Keys are too flat to affect the air; these **desert islands** are extremely dry, often with no fresh water streams or lakes.

Continents also cause air to rise, cool, and drop precipitation. If the topography is fairly flat, air rises gradually and rains are distributed over an extensive area.

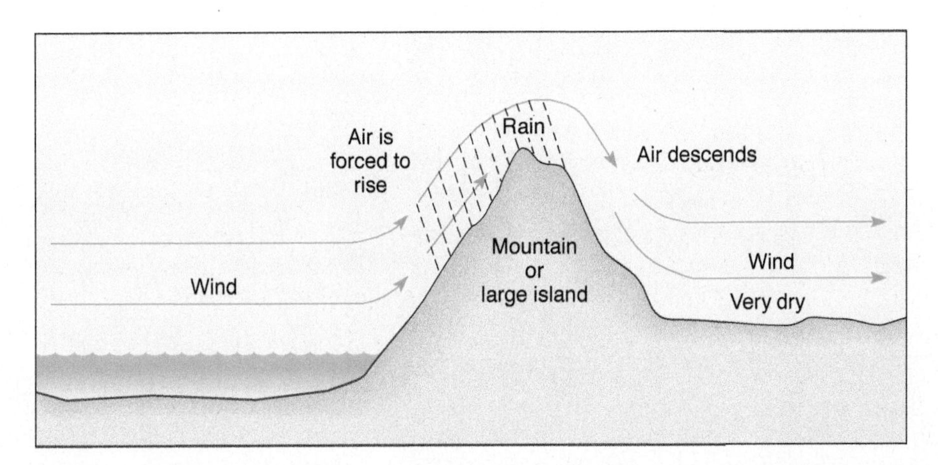

FIGURE 10.3. When a mountain or other large land mass forces air to rise, the air cools below its dew point and rain or snow forms. On the lee side, descent of the air compresses and heats it, raising the air's ability to hold water. Rather than bringing rain to the area, air may actually dry out the soil.

FIGURE 10.4. The Olympic and Cascade Mountains in the Pacific Northwest near Seattle cause moist winds from the Pacific Ocean to rise and cool. Rain, snow, and fog form and keep the western side (the ocean-facing side) moist, creating conditions for a dense forest with many conifers, ferns, mosses, and epiphytic lichens. Because this area is so far north, freezing winters occur, so this is a temperate rainforest, not a tropical one.

Summer storms from the Atlantic and Gulf of Mexico move through the eastern United States, the Mississippi Valley, and the plains states; the land rises so gradually that rainfall covers half a continent, and this area supports rich agricultural lands. No southern coastal mountain range blocks this movement otherwise the central and eastern United States would be very dry.

If topography is mountainous, as on the United States' west coast, rain is dropped in a narrow area (FIGURE 10.4). The prevailing westerlies bring moist air from the Pacific Ocean onto land over California, Oregon, Washington, Mexico, and western Canada. Air is forced to rise immediately by coastal mountain ranges, strong rains fall on the western slopes, supporting dense forests, many of which are heavily exploited for lumber and paper. But after the air crosses the summit, it descends, warms, and rains cease, forming a **rain shadow**: the eastern slopes are much drier than the western ones (FIGURE 10.5). Land east of the Rockies is grassland that is too dry for farming but is used instead for grazing cattle.

The size of a land mass also affects its temperature fluctuation. On a large island, moist air is always rising over the mountains, and clouds are frequent. Temperatures have a narrow range. Along the coasts of a continent, cloudy, rainy weather is common and temperature fluctuation is mild. Farther inland, air is dry and clear; lack of clouds exposes the land to full daytime sunshine, and heating is rapid and extreme. At night, clear skies allow the land to radiate infrared energy into space, with none reflected by clouds; cooling is also rapid and extreme.

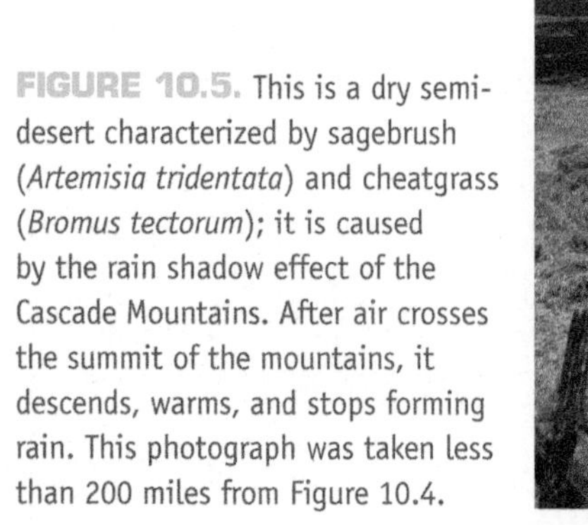

FIGURE 10.5. This is a dry semi-desert characterized by sagebrush (*Artemisia tridentata*) and cheatgrass (*Bromus tectorum*); it is caused by the rain shadow effect of the Cascade Mountains. After air crosses the summit of the mountains, it descends, warms, and stops forming rain. This photograph was taken less than 200 miles from Figure 10.4.

Continental Drift

The Earth's surface is not static. Molten rock rises from inside the Earth and emerges through volcanoes and fissures. Land-based volcanoes such as Mount St. Helens and Mt. Vesuvius are familiar, but the greatest number occur under water on the ocean floors, in regions called **mid-ocean ridges**. The solid parts of Earth's crust are broken into giant pieces called **tectonic plates**, and mid-ocean ridges are the boundaries between adjacent plates. As molten rock rises between the plates, it pushes on them, causing them to slide across the molten rock that makes up the bulk of our planet's interior. The movement is **continental drift**, and because of it, the continents and oceans that we know today have not always existed, and they will continue to change in the future (**BOX 10.1**).

BOX 10.1. Continental Drift and People

This chapter talks a lot about land, air, water, and plants but does not mention people. We humans simply did not exist during the events described in the text: as a species, we *Homo sapiens* are too new. By the time we originated, virtually everything in this chapter had happened and all continents were in their present positions.

People are members of the family **Hominidae**, which are known both as hominids and as great apes. Our family has several genera of extinct hominids, such as *Australopithecus*, but only four genera that still have living species: *Pongo* (orangutans), *Gorilla* (gorillas), *Homo* (humans), and *Pan* (chimpanzees). The most recent common ancestor of these four lived about 14 million years ago, when orangutans became distinct. Gorillas diverged about 8 million years ago, and then our line of evolution separated from that of chimpanzees about 5 million years ago. After this, our line diversified into various species, for example *Homo erectus*, *H. antecessor*, and *H. neanderthalensis*. Our species gradually became distinct and the oldest fossils that are considered to be modern humans, *H. sapiens*, are only 200,000 years old.

A new species or family has a better chance of surviving if its members disperse throughout a wide geographic area. Most plant families arose so early that their seeds had been blown or carried by animals throughout Pangaea or Gondwana, so as those supercontinents broke apart, each carried members of many plant families. But by the time our family Hominidae originated, not only had Pangaea broken apart, but Gondwana too had fragmented into the widely separated continents of Africa, South America, Antarctica, Australia, and India. Hominidae arose in Africa (that is the only area that contains fossils of our ancestors and relatives), and *Homo* itself arose in southwest Africa. Fortunately for us, Africa had drifted slightly north, touching Eurasia and dividing the Tethys Sea into the Mediterranean on the west and the Indian Ocean on the east: there was and still is a land bridge connecting the two continents.

The dispersal of animals that must walk is more difficult than that of plants, whose seeds might be carried for miles by wind, by being caught in a bird's feathers (or gut), or attached to an animal's fur. But members of *Homo* did become dispersed, first throughout Africa and then to the other continents. We know of at least three distinct migrations. Members of *Homo erectus* left Africa about 1.8 million years ago, those of *H. antecessor* left about 800,000 years ago, and those of *H. heidelbergensis* about 600,000 years ago. Our particular ancestors, *H. sapiens*, began their dispersal from the region of Angola and Namibia about 70,000 years ago and spread through Eurasia and Southeast Asia in as little as 30,000 years. In many areas, our ancestors encountered the descendants of the other species of *Homo*, but for one reason or another, we are the only ones who survived. For example, *H. sapiens* and *H. neanderthalensis* coexisted in Europe for a while, but by 30,000 years ago, all the Neanderthals were extinct.

Dispersal to the Americas was more difficult. The closest point between the Americas and any other continent is the Bering Strait, which at present is 53 miles wide, 53 miles of cold, stormy open ocean. We are not certain how people migrated to the Americas, but we do know that those who originally populated the Americas are genetically similar to people in Central Asia, and that the earliest signs of human presence in the Americas dates to about 14,500 years ago. One theory is that the Bering Strait was dry land at that

(continued)

time because Earth was in a glacial period. Even today, we are technically in an ice age, it is just that we are in a temporary warm period called an interglacial. But 15,000 years ago, Earth was in an especially cold period called a glacial, and there were so many glaciers on mountains and ice fields across Canada, the northern United States, and Europe that sea levels were much lower. So low that a land bridge connected Russia and Alaska. The continental shelves were exposed, and it would have been possible to walk from China or Russia to Alaska, south along all of North America, Central America, and even to the southern parts of South America. There are objections that this is a long way to simply walk; it would have been more difficult considering that these people would have been foraging for food. Once they had entered an area in which they could live, there would have been little incentive to wander into new areas that had different plants, animals, climate, and so on. People such as modern Eskimos know how to survive in a cold, snowy habitat but would not be able to find food in California or Mexico. An alternative theory is that people did follow this route along the western coasts of the Americas, but in canoes or boats rather than on foot. People adapted to living along a coast know how to build boats, how to fish, and find fresh water; traveling and exploring long distances in coastal canoes might have been more feasible and faster.

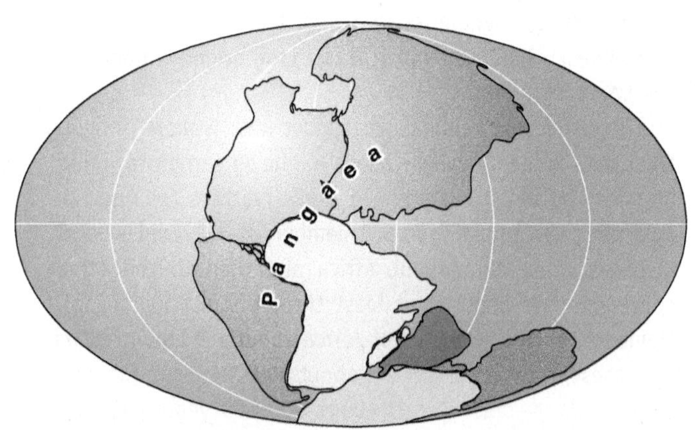

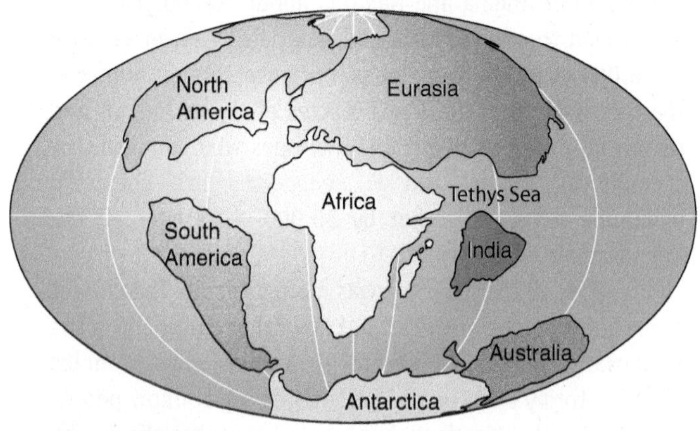

FIGURE 10.6. The continents have changed not only position, but size, shape, and profile as certain areas rise into mountain chains and other areas are covered by ocean when the sea level rises because of polar ice cap melting. Oceanic circulation changes as continents deflect or channel current movement, and this affects both transfer of moisture to the atmosphere and precipitation patterns.

Past Positions of the World's Continents

During the Cambrian Period (542 to 488 million years ago), all life was still aquatic and nothing lived on land. Several separate continents were distributed in a vast ocean in the Southern Hemisphere (FIGURE 10.6). Continental drift pushed western Europe against eastern Europe and Asia, causing the formation of the Ural Mountains. This ancient land mass was farther south than at present, located in a warmer climate. In the Southern Hemisphere was a giant continent called **Gondwana**, composed of South America, Africa, India, Australia, and Antarctica.

The Middle and Late Paleozoic Eras (425 to 250 million years ago) was the time life was beginning to move onto land and the earliest true plants, the rhyniophytes, were evolving. During this time, all continents drifted together. First, North America collided with Eurasia, forming the Appalachian Mountains, which subsequently have almost entirely eroded away in the intervening 400 million years. This new continent is known as **Laurasia**. Next, Gondwana moved north and ran into Laurasia, pushing up the Alps. Virtually all land on the entire globe was located together as one supercontinent, **Pangaea**. The Laurasian portion was still in the tropics, and much of Gondwana was in the southern temperate and polar regions.

Pangaea existed for millions of years, through the end of the Paleozoic Era, and had a diverse climate. The Laurasian portion was warm, moist, and tropical, whereas southern Gondwana was frozen and heavily glaciated. Large regions were swampy lowlands. Initially,

Pangaea must have been relatively flat, but the collisions of continents caused the formation of extensive mountain ranges in its center: the Appalachians and the Urals, as just mentioned. These must have caused rain shadows in the interior of Pangaea, and its enormous physical dimensions would have caused the central regions to be dry anyway. It was in the diverse conditions of Pangaea that most major groups of land plants arose. The earliest fossil spores of land plants are found in rocks 470 million years old, and less than 100 million years later, there were extensive forests of the earliest trees, giant lycophyte plants called *Lepidodendron* and *Sigillaria*. Mixed with these were plants called seed ferns (pteridosperms): their leaves resembled those of ferns, but they bore seeds. Some of the seed ferns were the ancestors of modern seed plants.

The Paleozoic Era ended and the Mesozoic Era began with a catastrophe called the Permian-Triassic Extinction Event, or just the Great Dying. As many as 95% of all marine species and up to 70% of all land vertebrates became extinct. We are not certain what caused the mass extinction, and it may actually have been three separate catastrophes occurring close in time. But suddenly, almost all the diversity of life ended and Earth had to be repopulated by the few survivors: the time of that repopulation is the Mesozoic Era.

In the early Triassic Period, the first portion of the Mesozoic Era, the climate worldwide warmed and became more arid. There is no evidence of glaciation anywhere; although polar regions remained cool, away from the poles, the climate was warm and equable. Many new plant groups evolved at this time: cycads and their relatives, ginkgos, and the early conifers. Ferns were already abundant, but they increased in number. The most famous creatures of the Triassic and Jurassic Periods were the dinosaurs. Many of the first major groups of plants started to disappear as they either went extinct or evolved and gave rise to new species, genera, and families.

In the Jurassic Period, Pangaea began to break up as the North American segment moved northwestward. As it separated from Eurasia, the north Atlantic Ocean formed, producing a maritime influence on the two new coasts. Eurasia also began moving northward, separating from Gondwana. This resulted in the formation of a waterway, called the **Tethys Sea**, between the northern and southern continents. This area had alternated between marshy swampland and a shallow coastal sea, but with continental separation, the Tethys Sea became deep, and oceanic circulation between southern and northern continents occurred on a large scale. The northward movement of the North American and Eurasian continents brought them out of the tropics and into the temperate zones, their northern regions extending into the north polar zones.

The breakup of Pangaea came at a critical time in plant evolution, just as the angiosperms originated. The earliest fossils of flowering plants are about 135 million years old, and of course undiscovered ones may be older. If the angiosperms had originated later, they might have been restricted to just one of the resulting fragments of Pangaea, but all continents do have flowering plants. Because the breakup occurred as some of the ancestors of modern families originated, certain flowering plant families are especially abundant on certain continents. For example, African deserts would be ideal for many cacti, but none occur there. This is interpreted to mean that the cactus family originated only after South America had separated from Africa and the southern portion of the Atlantic Ocean was wide enough to prevent seeds from the newly evolved cacti from being carried by birds to Africa. Fortunately for people and human survival, the grass family (Poaceae) and legume family (Fabaceae) originated earlier and just in time such that grasses and legumes occur on all continents. This is important to us because most of our food crops come from these two families: wheat, rice, corn, barley, sugar cane, oats, beans, peas, peanuts, lentils, and many others. Early humans originated in Africa where both of these plant families are plentiful. When our ancestors later migrated throughout Europe and Asia, they always encountered food plants of those two families. If the grasses and legumes had originated later and on some continent other than Africa, human survival—or at least human diet—might have turned out differently.

At about the same time, the Antarctic-Australian segment separated from the rest of Gondwana, and later Australia broke away and moved northward into the south temperate zone. India separated from Africa and moved rapidly over a long distance. At the time of separation, India was located at the edge of the south temperate zone (about 30 degrees S latitude) and had a flora adapted to such a climate. But continental drift caused it to migrate rapidly, in just a few tens of millions of years, into the tropics, across the equator, and into the north temperate zone. The rapid fluctuation of climate in less than 70 million years caused massive extinctions. Once India arrived at the Eurasian continent, the two collided, forming the Himalayan Mountains. As the two continents neared each other, plants could invade the foreign territory, some migrating northward onto the mainland, others spreading south from Eurasia onto India. This was to the detriment of India; its flora and fauna had not had time to become well adapted to the conditions of the north temperate zone, whereas the invaders from the mainland were well adapted.

In the Cretaceous Period, starting about 110 million years ago, South America separated from Africa, forming the southern Atlantic Ocean and allowing more humid oceanic air to reach west Africa. Until the breakup of Pangaea, Africa had the misfortune of being the dry center of the supercontinent. Its climate was so arid and severe that it had few species of plants or animals. With the formation of the Atlantic Ocean, Africa's climate became much more humid and conducive to plant growth, but even today the continent as a whole has many fewer indigenous species than one would expect if it had had a mild climate for a longer time.

The latest major event in continental drift was (is) the collision of South America and North America. The two are coming together, and as always happens when continents collide, a mountain range forms. This collision started so recently, about 5 to 13 million years ago, that the mountains are short and are mostly still submerged below sea level. Only their tops protrude and are known as Central America and the Caribbean Islands. The formation of Central America created a continuous land bridge between the two continents, allowing plant species to be interchanged.

Present Position of the World's Continents

The present position of the large continent of Antarctica at the South Pole affects Earth's entire climate. It allows huge amounts of fresh water to accumulate there as snow fields and glacial ice, lowering sea levels, and exposing more land surface. It also increases ocean salinity by trapping fresh water for thousands of years. If Antarctica were farther north, its ice would melt causing all ocean levels to rise, and seawater to be less salty.

If Central America did not exist, water could circulate between the Pacific and Atlantic Oceans, resulting in new oceanic circulation patterns that would affect heat distribution to the poles.

Eurasia is in the north temperate region and receives large amounts of moist, mild air, but it is such a large land mass that its center is a rain shadow caused by various mountain ranges. The United States would be more moist if its highest mountains were in the east or if it were located farther south where the northeast trade winds could bring moisture in across the low Appalachians and Adirondacks.

The World Biomes at Present

Earth's land surface is covered almost entirely by **biomes**, extensive groupings of many ecosystems characterized by the distinctive aspects of the dominant plants (see Figure 10.1). For example, some of the biomes of North America are the temperate deciduous forests, subalpine and montane coniferous forests, grasslands, and deserts. Plant life is sparse (rarely absent) only in the harshest deserts (the Atacama in Chile and Peru, the Sahara in Africa, the Gobi in China; FIGURE 10.7) and in the land regions covered permanently by ice (most of Antarctica and the tops of high mountains). In all other areas, the rock and soil are at least temporarily free of ice and have some liquid water; plants, algae, and cyanobacteria carry out photosynthesis and support food webs of consumers and decomposers.

Biomes vary from extremely simple, as in tundra (FIGURE 10.8), to more complex grasslands, forests, and tropical rainforests (FIGURE 10.9). Biome complexity is strongly influenced by climate and soil. A particular type of biome, such as grassland or temperate deciduous forest, may occur in various regions of Earth because the same set of climatic and soil factors occurs in various regions. At all sites, the physiognomy, the appearance, of a biome is similar, but often the actual species differ considerably from one area to another. For example, temperate grasslands are easily recognizable in the central plains of the United States, the steppes of Russia, the pampas of Argentina, and the veldt of Africa; all are dominated by grasses and large mammals and are devoid of trees except along rivers (FIGURE 10.10). Species that become adapted to the grassland niche of one continent do not occur on other continents because there are no birds, mammals, winds, or other means of carrying seeds such long distances. Despite the lack of gene flow between widely separated units of a biome, the physiognomic similarity persists as a result of convergent and parallel evolution: climate, soil, and other habitat factors in each area select for similar phenotypes.

FIGURE 10.7. The Pacific Ocean west of Peru and Chile is extremely cold, and consequently little moisture evaporates into the winds that blow eastward onto land. As the winds come ashore, they rise and cool but are too dry to form rain or snow until they have risen high up in the Andes. In most areas, the lower elevations virtually never receive rain, only an occasional fog. The result is the Atacama Desert, one of the driest areas on Earth; virtually no life exists in many parts of the Atacama.

FIGURE 10.8. Plants in the alpine tundra biome face short, cold growing seasons, and often snow is never gone from the shaded areas below cliffs. Soils are thin and ultraviolet light is intense.

If portions of a biome are not too widely separated, they will actually have some species in common.

As a result of continental drift, the United States and Eurasia are currently located almost exclusively in the north temperate zone. In contrast, the south temperate zone is occupied mostly by oceans with the only land being New Zealand, and the southern parts of South America, Africa, and Australia. The region between the Tropics of Cancer and Capricorn have large parts of South and Central America, Africa, and Southeast Asia as well as extensive regions of ocean. Fortunately, most of Earth's land has now drifted into regions that provide both suitable temperatures and adequate moisture for plants, animals, and fungi to thrive. Only Antarctica is more or less devoid of life.

FIGURE 10.9. Tropical rainforests never experience freezing conditions or dry periods that last longer than a few days. There are many ways of being adapted to such mild conditions, so tropical rainforests have high diversity, with hundreds or thousands of species living side by side in a small area. (© Matt Tilghman/ShutterStock, Inc.)

Moist Temperate Biomes

Moist temperate biomes receive moderate to abundant rainfall and experience freezing temperatures in winter. The northwest coast of the United States is formed by a series of mountain ranges that force westerly winds from the Pacific Ocean to rise as soon as they come ashore. Rain on the western side of the Olympic Mountains in Washington is often above 300 cm/yr (120 in/yr). Rains are reliably present through autumn, winter, and spring, with only a brief period of summer dryness (**BOX 10.2**). Winters are mild but light freezing temperatures occur from year to year; summers are warm but not hot. These conditions extend from northern California to Alaska, but their eastern edge stops at the summit of the Coastal Range, about 200 km (about 120 miles) from the coast.

FIGURE 10.10. Much of the Midwestern United States and Canada was covered in temperate grassland. Trees are rare except along rivers, and the area is much flatter than regions to the west and east. This is in eastern Wyoming where there is just enough rainfall to cultivate wheat and other cereal grasses.

BOX 10.2. Wetlands

Wetlands such as swamps, marshes, and bogs are areas in which the soil is saturated with water either permanently or for a long enough period that only specially adapted plants survive. The water may be fresh or marine, and it may be flowing or standing, but the important thing is that the soil or mud has very little oxygen much of the time (**FIGURES B10.2a** and **B10.2b**). Roots of

FIGURE B10.2a. The Everglades National Park in Florida has been called a "sea of grass," but the plant this name refers to is actually a sedge (*Cladium jamaicense;* sawgrass). Sawgrass dominates all regions of the Everglades that are less than 60 cm (less than 2 feet) above sea level, but even the slightest rises in elevation are covered with shrubs and trees instead. This ecosystem has been badly disturbed by people. Fires were routinely suppressed until it was realized that fire is necessary to burn off the old foliage and release the locked up minerals, thereby fertilizing the soil and enhancing plant growth.

FIGURE B10.2b. This is a mangrove wetland along the coast of Texas. Almost all the plants in both the foreground and the background are black mangroves (*Avicennia germinans*) growing in shallow saltwater less than one meter (three feet) deep. These sturdy plants withstand hurricane-force winds and their dense branching, leaves, and roots break up ocean waves, protecting the coastline in the background. Unlike many other wetlands, mangrove wetlands have low species diversity because few other plants tolerate seawater.

(continued)

wetland plants must be adapted to these low-oxygen (hypoxic) conditions by some mechanism, and often they have internal air canals that permit oxygen to diffuse downward from stems to roots.

Wetland habitats are extremely rich in the number of species they contain, not only of plants but also birds, fish, crustaceans, insects, and so on. Familiar examples of herbaceous plants are cattails, waterlilies, sedges, irises, and marsh marigolds, and among trees are willows, cottonwoods, salt cedar, bald cypress, and mangroves. Some wetland plants, such as *Hydrilla*, *Myriophyllum*, and *Ranunculus aquatilis*, grow entirely submerged except for their flowers.

Wetlands occur in all the various biomes of the world, anywhere that water saturates the soil for a prolonged period. There are tropical and temperate wetlands, as well as alpine and low-lying ones. Deserts have small wetlands wherever aquifers produce a spring or seep. When rivers flood, their waters spread out, slow down and drop silt, creating flat bottomlands that are also wetlands. Before flood control was introduced, the Mississippi River, along with the Ohio, the Missouri, and all other rivers in the central United States were accompanied by bottomlands on either side that supported rich forests and extensive habitat for animals. As rivers approach an ocean, their flow slows and silt is deposited, building up a delta if the ocean is not too deep at that point. The entire southern portion of Louisiana was built up in this way, and much of southern Florida consists of the wetland known as the Everglades.

Wetlands have always been tempting targets for development. They tend to be flat and to have the water needed for people to drink and use. Canals have been dug to drain wetlands, which have then been cleared for farming, ranching, or to build cities, factories, or airports. For some reason, wetlands were thought of as waste lands, good for nothing at best, or for breeding mosquitoes at worst. But now we realize that wetlands have tremendous biological diversity and productivity. They are important not only for resident birds but also migrating waterfowl that link widely separated ecosystems. Wetlands prevent flooding of rivers and erosion of coastlines and they purify the water that flows through them. Because of the low oxygen levels, dead plants and animals that settle to the bottom of wetlands decompose only slowly, so wetlands are important for keeping carbon dioxide out of the atmosphere.

Bottomlands along rivers have been cleared and farmed, and cities have been built where the river provides a means of transportation. But bottomland areas are then prone to flooding, and people have subsequently built dikes and levees along the river to protect the bottomlands in their area. But this only channelizes the floodwater, it does not eliminate it, and the extra water flows downriver until it encounters the first area that has not been diked. Every year we see news stories of people filling sandbags to temporarily protect their towns. A better solution would be to remove existing levees upstream and restore some of the bottomland to a more natural condition that would provide both flood control and wildlife habitat.

Coastal wetlands too have been drained, cleared of natural vegetation, and then built up (**FIGURE B10.2c**). Extensive canals were constructed in southern Florida to create dry land for airports, suburbs, and citrus orchards, and this has disrupted the flow of water to the Everglades. This vast wetland has been gradually drying up since the canals were built upstream of it. Most of southern Louisiana too has had a network of canals built, which drains away water and allows the land to be used for cultivating rice and sugar cane. All the lower portions of the Mississippi River have been lined with levees to prevent flooding, but a consequence is that all the silt that used to be spread across southern Louisiana is now dumped directly out into the Gulf of Mexico where currents take it gradually eastward. Without new layers of silt being deposited on southern Louisiana, the entire coastline is gradually eroding away with every storm and hurricane, and Louisiana is rapidly losing land. In the past, the vast wetlands protected New Orleans and other cities from storm surges caused by hurricanes, but as land is lost, the cities are more vulnerable.

At one time it was popular to offset wetland destruction by creating new wetlands elsewhere. Fortunately, it did not take long to realize that simply flooding an area of land does not actually create a wetland, at least not one that has biological diversity and richness. Even planting an artificial wetland with several species of dominant plants does not recreate the thousands of species of plants, animals, and microbes that occur in natural wetlands.

Recently, public pressure has slowed the destruction of wetlands, but still several tens of thousands of acres are being lost every year. Many regions have programs to fill in canals and restore natural water flow such that wetlands can restore themselves.

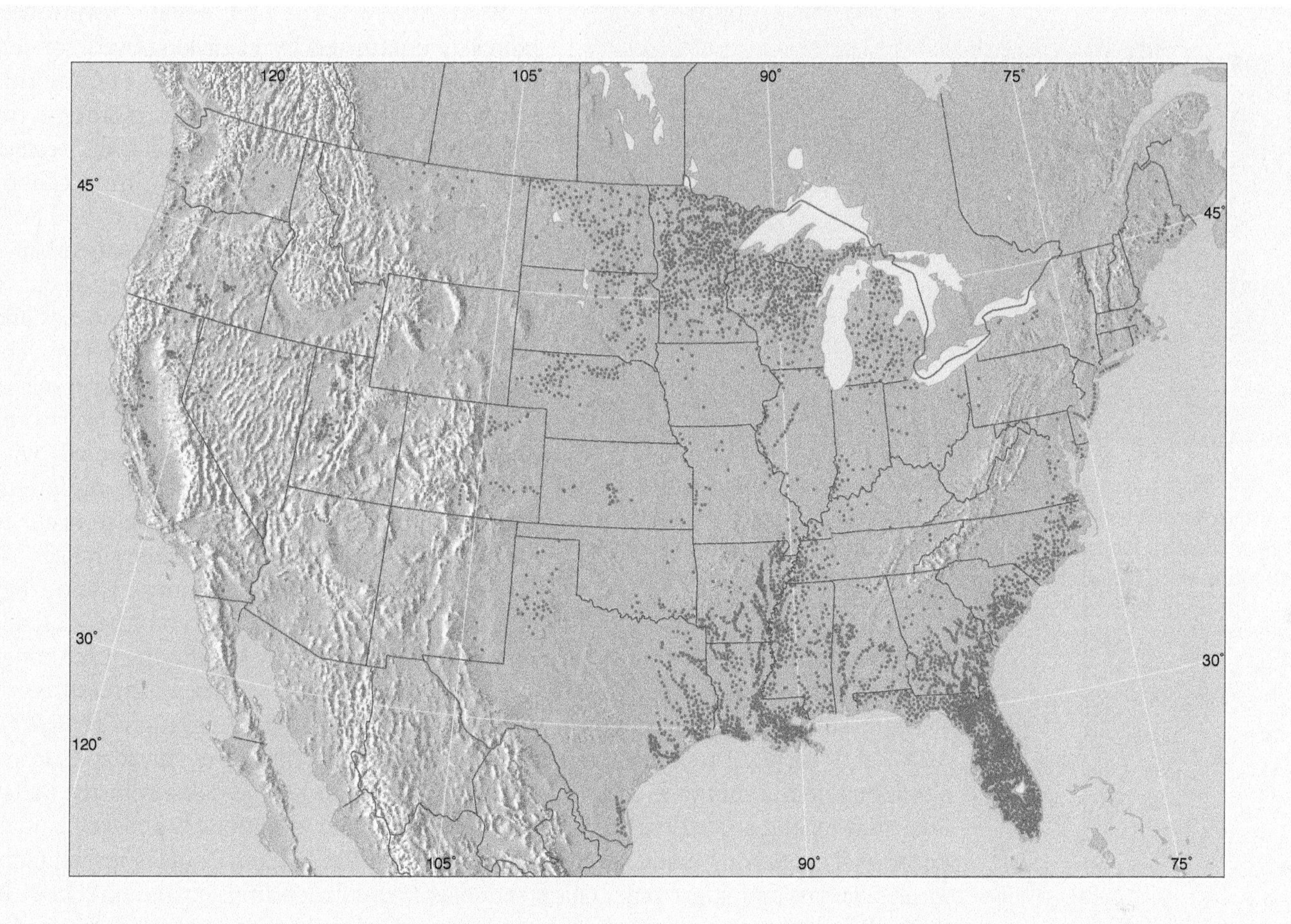

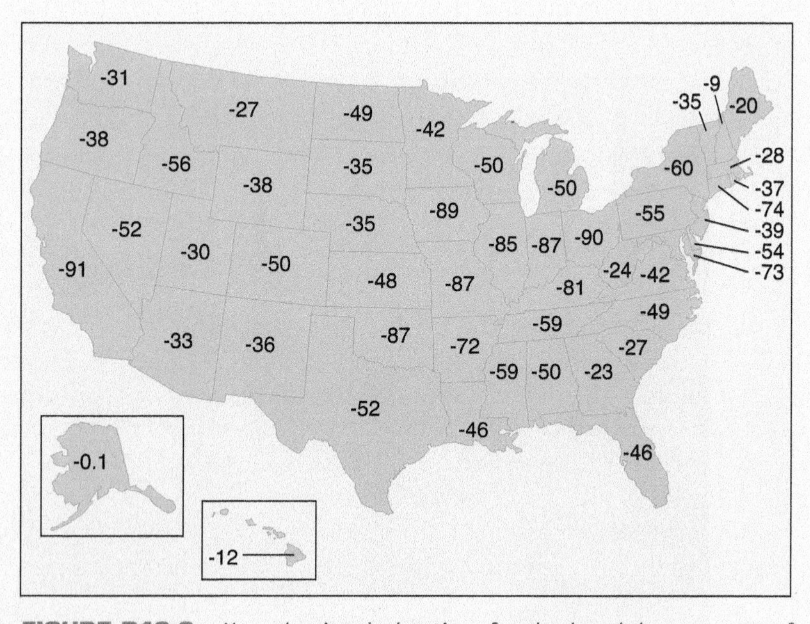

FIGURE B10.2c. Maps showing the location of wetlands and the percentage of wetlands that have been lost in each state.

FIGURE 10.11. Montane forests occur at lower elevations on mountains. In western North America, montane forests often contain ponderosa pines (*Pinus ponderosa*), the tree shown here. The low shrubs are sagebrush (*Artemisia tridentata*), which indicates that this is a dry area with cold winters.

Plant life in the **temperate rainforest** biome is dominated by giant long-lived conifers (see Figure 10.4). Coastal redwoods of California are up to 100 m tall (about 325 feet, more than the length of a football field), and Douglas fir, western hemlock, and western red cedar form a canopy that reaches upward 60 to 70 m (197 to 230 feet). Forests can attain old age with little disturbance other than being harvested for lumber or cut to make room for towns and cities. Hurricanes and tornadoes do not occur, and fires are rare. The stability of the forest and the daily fog result in a rich growth of epiphytic mosses, liverworts, and ferns, and the ground is covered with shrubs, herbs, and ferns. Temperate rainforests, containing totally different species, also occur in southwest Chile, which has the same climate.

As the westerly winds continue inland, they are drier and shed less rain on the Cascade Mountains and the Sierra Nevada, and even less on the Rocky Mountains. Each range feels the rain shadow effect of the preceding mountains. **Montane forests** occur at the bases of these mountains, **subalpine forests** at higher elevations (FIGURES 10.11 and 10.12). The first inland range in California is the Sierra Nevada; its lower elevations hold a montane forest of ponderosa pine, and some oaks from the valley floor may extend some distance up the slopes. At higher elevations is a mixed conifer forest, one containing many species of conifers: ponderosa pine, Douglas fir, white fir, incense cedar, and sugar pine. Giant sequoias (*Sequoiadendron giganteum*) occur in the California coastal subalpine forests; they are the largest organisms in the world, much more massive than whales, being 80 m (262 feet) tall, up to 10 m (33 feet) in

FIGURE 10.12. Subalpine forests occur in mountains at elevations above those of the montane forests. The two types of forest intergrade; there is not a sharp boundary between the two. This is a spruce/fir (*Picea/Abies*) forest in Idaho.

FIGURE 10.13. This is a photograph of Aspen Vista in Santa Fe, New Mexico. Quaking aspens are a striking feature of subalpine forests in the Rocky Mountains. They are broadleaf (dicot) trees in forests dominated by conifers, and typically only become established if the conifers have been damaged by fire, logging, landslides, or some other disturbance. Seedlings of spruce and fir can grow up through the shade of the aspens, and after several hundred years the conifers will crowd out the aspens, returning the area to a conifer forest. In the mean time, disturbances elsewhere will allow new groves of aspens to become established.

circumference, and weighing over 400 metric tons (440 tons; a compact car weighs about 1 ton) when dry. Giant sequoias have been greatly harmed by lumbering and are currently found in only 75 groves. At higher elevations are subalpine forests of lodgepole pine, whitebark pine, and mountain hemlock. Some of the most famous residents are the quaking aspens (*Populus tremuloides*), trees with white bark and with leaves that flutter in the slightest breeze and turn brilliant yellow in autumn (FIGURE 10.13). Along the higher elevations, subalpine forests form open stands and then give way to alpine meadows at their upper boundaries.

The Rocky Mountains are the most massive mountain range in the United States and Canada, extending in a broad band from Alaska south through Canada, Idaho, Montana, Wyoming, Utah, Colorado, Arizona, New Mexico, and Mexico. Being the third range from the coast, it is driest, receiving as little as 40 cm (about 16 inches) of rain per year; the southern Rockies are especially dry and warm. The subalpine forest contains Engelmann spruce over the entire length of the range, but in the north other trees are alpine larch and whitebark pine; in the south, varieties of bristlecone pine are part of the subalpine forest. The montane forests typically contain Douglas fir in their higher elevations, ponderosa pine in lower ones. The montane forests on the Rocky Mountains are frequently subjected to fire, as often as every 5 years under natural conditions. Ponderosa pine is well adapted to fire and survives it easily. Other plants are killed, and fires create open grassy areas around the pines.

In drier montane and subalpine forests, soil is shallow, rocky, and very well drained; it tends to be acidic. Water stress in summer is not uncommon, due both to sparse rain and to rapid runoff through porous soils on steep slopes. Much of the available moisture comes as the melting of winter snow.

In addition to causing air to rise and release precipitation, thereby affecting rainfall across the entire continent, the western mountains are important for many other

FIGURE 10.14. Many people find mountains and their biomes refreshing, exciting to some, calming to others. Many of our national parks are in western mountain biomes.

reasons. They are so tall that many have snow fields that last into the summer, thus storing winter snowfall and releasing water into rivers throughout the year. The forests are habitats to thousands of species of all sorts of organisms. They are also the area where we have established most of our national parks and preserves, indicating how important they are to our spiritual well-being. The western mountains certainly provide innumerable sites for outdoor recreation (FIGURE 10.14).

In the eastern United States, the Adirondacks and Appalachians are tall enough to support montane and subalpine forests. In the Adirondacks, a spruce/fir subalpine forest extends down to 760 m (about 2500 feet) and contains balsam fir. In the warmer southern climate, the lower limit for subalpine forest is higher: 990 m (3250 feet) in the Appalachians and 1400 m (4600 feet) in the Smoky Mountains. Above these elevations are stands of red and black spruce and Fraser fir. Below the subalpine conifer forests are montane forests composed of hardwoods (broadleaf, angiosperm trees), not conifers; these are described in the next section.

A **temperate deciduous** forest biome is produced by a climate that has cold winters and warm but not hot summers and relatively high precipitation in all seasons (FIGURE 10.15). Whereas the drier montane and subalpine forests of the Rocky Mountains have streams with running water restricted mostly to springtime when snowpack is melting, the eastern temperate deciduous forest has streams that flow year round (FIGURE 10.16). Much of the precipitation in the northeastern region is derived from summer weather systems that move northward out of the central Atlantic Ocean or from winter storms that blow southward from Arctic seas.

Temperate deciduous forest in the United States occupies lower, warmer regions. Dominant trees vary geographically, but tall, broadleaf deciduous trees such as maple and oak are frequent everywhere, maples being more common in the north and oak in the south. Intermixed and forming a subcanopy are dogwood, hop hornbeam, and blue beech. The shrub layer is sparse because of heavy shade provided by broadleaf dominants, but witch hazel, spicebush, and gooseberry are common. The ground layer is covered with herbs that thrive during spring just before trees leaf out. Such a spring sunny period does not occur in evergreen forests.

The foliage of these angiosperm trees contains fewer defensive chemicals than do the needles and scale leaves of a conifer. There may be 3 to 4 metric tons (3.3 to 4.4 tons)

FIGURE 10.15. We can tell that this photograph of the Blue Ridge Parkway in North Carolina is of a deciduous forest because it was taken in spring just as the trees were putting out their new leaves, so they are easily distinguished from the conifers, which are evergreen. During winter and early spring when the broadleaf trees are leafless, sunlight reaches the forest floor, allowing photosynthesis by bryophytes and ferns.

FIGURE 10.16. Mountains in the eastern United States have streams that flow all year. Some of the water is from snowmelt, some from summer rainstorms, other water is released from aquifers. Waterfalls such as this provide mist and spray that humidify large areas, creating a habitat in which ferns, lycophytes, mosses, and liverworts thrive.

of broadleaf foliage per hectare (a square 100 m [328 feet] on each side; city blocks vary in size, but most are close to 1 hectare) in a temperate deciduous forest, and as much as 5% is consumed by herbivores; the rest is abscised in autumn and decays quickly in the humid conditions produced by frequent rain. A thick layer of litter does not accumulate.

Forests in these areas have a rich diversity of plant species, which in turn provides considerable habitat diversity for animals and fungi. Much animal life is located above or below ground, not at its surface. Most birds make their nests on branches or in holes in tree trunks. Small mammals may burrow, but many are arboreal. As in the coniferous forest, strong climatic differences exist between summer and winter, accented by the deciduous nature of the broadleaf forest. Autumn leaf fall ends the feeding period for aerial insect leaf-eaters but initiates the season for decomposers and insect herbivores of the litter zone. Many birds migrate south to wintering grounds, but other species may immigrate from farther north. Disturbance is not common; fires are rare and hurricanes that enter the regions usually lose destructive power quickly and change into widespread weaker storms.

The southeastern **evergreen forest** biome occurs at the southern edge of the oak/pine component of the temperate deciduous forest, along the northern portions of the Gulf States, the top of Florida, and the coast of the Carolinas. This biome shows the powerful effect that disturbance and soil type exert. Its climate is similar to that of the inland region except that winters are warmer; frosts occur but the ground does not freeze. Precipitation is higher than in inland temperate deciduous forest areas, but the soil is so sandy and porous that rainwater percolates downward rapidly and runs off into streams. Shortly after a rain, the soil is dry. The region has a definite dry aspect to it.

In addition to rapid drainage, the southeastern evergreen forest biome is shaped by frequent fires. Lightning initiates fires that burn rapidly through the litter of fallen pine needles that decompose only slowly. As a result of fire disturbance and soil conditions, the forest consists almost purely of fire-adapted longleaf pine with occasional oaks. The fact that fire rather than climate is the primary biome determinant has been proven by fire prevention: repeatedly suppressing fire lets broadleaf seedlings survive and quickly changes the forest from pines to oaks, hickories, beeches, and evergreen magnolias.

Dry Temperate Biomes

The entire central plains of North America, extending from the Texas coast well into Canada is—or more accurately, was—**grassland**, often referred to as prairie (see Figure 10.10). This part of North America has no mountains, being too far from the various continental collision zones; it is remarkably flat, with at most low, rolling hills. It is characterized climatically as drier than the forests discussed so far, being located in rain shadows of the prevailing westerlies but out of reach of many Atlantic weather systems. Rain is only about 85 cm/yr (almost 3 feet). Seasons vary from bitterly cold winters, especially in the north, to very hot summers. Climate and vegetation factors have produced some of the richest soil in existence anywhere. Rainfall is sufficient to promote reasonably rapid weathering of rock, but it is not so great as to leach away valuable elements. Grasses produce abundant foliage that, rather than abscising and falling to the soil, is eaten by herbivores, often large mammals such as cattle or, in the past, buffalo. Vegetable matter is returned to the soil as manure that decomposes rapidly, enriching the soil and increasing its water-holding capacity. Because of the lack of trees, except along rivers, no treetop habitats are available for animals like squirrels. All small mammals burrow and birds nest on the ground.

Because the soil is so rich, almost all the grasslands have been converted to farms; virtually nothing remains of the original biome (FIGURE 10.17). Efforts are being

FIGURE 10.17. The Midwestern grasslands area has deep, rich soil and is flat enough to be farmed. Since the westward migration of settlers in the 1800s, most of the Midwestern grassland biome has been plowed and converted to farms. Almost no natural habitat remains in this region. (© AbleStock.)

FIGURE 10.18. Many plants in the chaparral in California are drought-adapted short trees and large shrubs with resinous leaves. Leaves catch fire easily, not only fallen leaves on the soil but also leaves still on plants. This encourages fires that kill nonadapted plants whereas chaparral-adapted plants survive and resprout. If fires would be constantly suppressed, the natural vegetation would be outcompeted and eliminated by plants that are not fire-adapted.

made to reestablish grassland prairie by removing all domesticated plants and seeding in grasses and other species known to have occurred originally, such as composites, mints, and legumes.

Extensive grasslands occur between the Cascade Mountains and Rocky Mountains, the result of the Cascade's rain shadow. These extend from northern Washington south through Oregon and Nevada. Much of them remains, but they are grazed; large parts have been cleared and used for irrigated farms.

A **woodland** is similar to a forest except that trees are widely spaced and do not form a closed canopy. If grass grows between the trees, the woodland is known as a **savanna** instead. **Shrublands** are similar except that trees are replaced by shrubs. Woodlands and shrublands often occur as the transition between a moist forest and a dry grassland or between grassland and desert. Soils often have a high clay content.

The **chaparral** in California is a well-known shrubland (FIGURE 10.18). Its climate consists of a rainy, mild winter followed by a dry, hot summer; drought occurs every year. Much of the California chaparral is dominated by short shrubs, 1 to 3 m tall (about 3 to 9 feet), but at higher elevations manzanita, buckthorn, and scrub oak occur. Many plants have dimorphic root systems: some roots spread extensively just below the soil surface, but a taproot system reaches great depths. The two together allow plants to gather water from the deep, constant water table and from brief rains that penetrate only a few centimeters (an inch or two) into the soil.

Fires occur frequently in California chaparral and are always in the news because of the houses they destroy (FIGURE 10.19). Low rainfall allows dry leaf and twig litter to accumulate without decomposing, and dead shrubs persist upright as dry sticks. An area typically burns every 30 to 40 years, with fires most frequent in summer. Winter rains cause flooding, erosion, and mudslides after a fire because no vegetation remains to hold soil in place. Although the shrubs and trees are fire-adapted and resprout quickly, the main new growth is by annual and perennial herbs. These are present before the fire as seeds, bulbs, rhizomes, or other protected structures, and after the burn they grow vigorously, free of shading by charred shrubs (FIGURE 10.20). Release of minerals from the ash also enriches the soil. A few years after a fire, larger shrubs dominate the biome again.

Farther east, drier climates and higher elevations result in pinyon/juniper woodland instead of chaparral shrubland. Rainfall is only 25 to 50 cm (10 to 20 inches), and soils

FIGURE 10.19. Many of California's cities are located in chaparral areas, and numerous people build homes in shrubland that is highly flammable. Fires inevitably sweep through these areas every summer, and not only are houses destroyed but firefighters are often killed. (© Peter Weber/ShutterStock, Inc.)

FIGURE 10.20. Only one year after a fire, this oak tree is resprouting from buds located at or slightly below soil level. Other chaparral plants have seeds that germinate immediately after a fire.

are rocky, shallow, and infertile. The vegetation is a savanna of pinyon pine—small, slow-growing trees with short needles—and juniper trees that may have the stature of large shrubs (FIGURE 10.21). In Arizona and New Mexico, oaks may be important. Trees are widely spaced and between them is a grassy vegetation; the species are bunch grasses that grow in clumps, not the mat-forming species of the central plains. Between the bunch grasses is bare soil. Sagebrush and bitterbush occur in the northern parts of the biome.

FIGURE 10.21. Pinyon pines and junipers dominate this area, and the soil between the trees is covered with bunch grass (grass that grows as clumps rather than as turf-like lawn grass). Small oak trees are present but inconspicuous. This is a pinyon/juniper grassland in New Mexico.

FIGURE 10.22. Deserts are some of the most variable biomes. Certain deserts are extremely dry and have less vegetation than shown here; others receive more precipitation and have even more shrubs and grass than this area of the Anza-Borrego Desert in southern California. After rains, deserts often have numerous short-lived small wildflowers whereas during their driest periods, shrubs may be leafless and herbs may all be dead.

The driest regions of temperate areas are occupied by **deserts**, where rainfall is less than 25 cm/yr (10 inches; FIGURE 10.22). Deserts are either cold or hot, based on their winter temperatures. A hot desert has warm winter temperatures. In the United States, this climate occurs in rocky, mountainous areas of southern California, Arizona, New Mexico, and west Texas, but in other parts of the world it may occur on flat, sandy plains, as in the Sahara or the Atacama. Three separate and highly distinct deserts actually occur in the southwestern United States and northern Mexico: the Chihuahuan Desert in west Texas, New Mexico, and Mexico, the Sonoran Desert in Arizona and northwestern Mexico, and the Mojave Desert in southeastern California, southern Nevada, and northwestern Arizona.

Deserts soils are rocky and thin; what little soil occurs may blow away, leaving nothing but pebbles. Brief, intense thundershowers wash soil out of mountains and deposit it in large alluvial fans at valley entrances; fans often have the deepest soil and their own distinct vegetation. The most abundant plants in our hot deserts are creosote bush, bur sage, agaves, yuccas, and prickly pear (FIGURE 10.23). Most perennial plants have one or several defenses against herbivores—chemicals and spines being the most common. Joshua tree, an arborescent lily, grows in the Mojave desert, and numerous other needle-leaf yuccas and agaves are abundant; cacti are ubiquitous. Deserts are highly patchy ecosystems, and slight variations in soil type, drainage, elevation, or covering vegetation can cause abrupt changes in vegetation. In valleys or mountains, slopes that face the equator intercept light almost perpendicularly and are much warmer and drier than those that face away from the equator and are lighted obliquely. The hotter side has the richer xerophyte vegetation.

FIGURE 10.23. Most people think of cacti when they think of deserts, and cacti certainly are common in all deserts in the Americas, but they do not naturally occur on any other continent. Also, many other families found in deserts almost as often as cacti are grasses, lilies, agaves, bromeliads, legumes, and mustards. This is saguaro (pronounced either sa HUA ro or sa GUA ro; *Carnegiea gigantea*) in Arizona's Sonoran Desert.

The biome located above the highest point at which trees survive on a mountain, the treeline, is **alpine tundra** (see Figure 10.8). In the equatorial region where temperatures are generally warm, it is necessary to go up to elevations as high as 4500 m (14,763 feet, almost 3 miles) before we encounter areas that cannot support tree growth, but in the cooler regions at 40 degrees N (about the middle of the United States), elevations as low as 3500 m (11,483 feet, a little over 2 miles) are too severe for trees. Alpine tundra is cold much of the year, with a short growing season limited by a late snow melt in spring and early snowfall in autumn. Soils are thin and have undergone little chemical weathering. Summer days can be surprisingly warm and clear, and many plants flourish in the brief summer. Nights are generally cold in all seasons, and severe, violent weather can occur at any time. The dominant forms of plant life are grasses, sedges, and herbs such as saxifrages, buttercups, and composites. Dwarf plants growing with densely packed stems and leaves are common and are known as cushion plants. Much of the alpine tundra land occurs as flat meadows with shallow alpine marshes.

It is not known for certain why trees do not grow above the tree line. Those near the highest elevation are short and typically have branches mostly on the side away from the wind. At slightly higher elevation, trees are extremely misshapen and gnarled, known by the German word *krummholz*, and the forest is called an elfin forest. It is believed that blowing snow and ice abrade the trees' surfaces, causing desiccation and death.

In North America, most alpine tundra biomes occur on the tall mountains in the west, but Mt. Washington in New Hampshire and Whiteface Mountain in New York are high enough to have regions of alpine tundra.

Polar Biomes

Polar regions contain few significantly dry areas. Precipitation, usually snow, may be low, but evaporation is also low, so moisture persists. In the extreme northern latitudes, the ground freezes to great depths during winter, and summer is too cool and brief to melt anything more than the top few centimeters. Below this, soil is permanently frozen and is known as **permafrost**.

Like alpine tundra, the **arctic tundra** biome has a short growing season of 3 months or less, and temperatures are cool, averaging only about 10°C (50°F; FIGURE 10.24). Freezing temperatures can occur on any day of the year. Arctic soils have a high clay content and are poor in nitrogen because nitrogen-fixing microbes are sparse. Permafrost prevents drainage when the soil surface melts in summer, so soils are waterlogged and marshy. Bogs, ponds, and shallow lakes are common.

Arctic tundra vegetation contains even more grasses and sedges than does alpine tundra, as well as many more mosses and lichens (FIGURE 10.25). Almost nothing is taller than 20 cm (about 8 inches), even the dwarf willows and birches that occur. Many plants have underground storage tubers, bulbs, or succulent roots; more than 80% of a plant's biomass may be underground even during summer.

FIGURE 10.24. Arctic tundra tends to have many pools during summer because melted snow cannot seep into the ground due to the presence of permafrost just a foot or two below the surface. Many plants must have adaptations for a waterlogged habitat. (Courtesy of U.S. Fish & Wildlife Service)

Because so much of Earth's land occurs near the Arctic Circle, arctic tundra is extensive, covering all of northern North America and Eurasia. Conversely, the only land near the Antarctic Circle is Antarctica, which is too cold to support any plant life other than two species of vascular plants and a few species of mosses.

Just south of arctic tundra is a broad band of forest, the **boreal coniferous forest** (FIGURE 10.26). Boreal means northern just as austral means southern (Australia). The boreal coniferous forest occurs completely across Alaska and Canada and throughout northern Eurasia in Scandinavia and Russia. The Russian name for this biome is **taiga**, a term frequently used in the West. The boreal forest is an ancient biome, and its formation was strongly influenced by the evolutionary diversification of the conifers just as the North American and Eurasian plates were breaking away from Pangaea and their northern

(a)

(b)

FIGURE 10.25. **(a)** Because so much of the Arctic tundra is wet with permafrost just below the surface, plants with deep roots are at a disadvantage. Lichens, however, either lie completely on the surface of the soil or penetrate it just a fraction of a centimeter and are able to thrive. In many areas, the soil of Arctic tundra is covered by lichens rather than by plants. This is *Stereocaulon tomentosum*. **(b)** Close-up view of *Stereocaulon tomentosum*.

parts were leaving the tropical zone and entering the north temperate, subarctic zone. In the last several million years, it has been the site of huge ice sheets and glaciers during the ice ages.

Boreal forest is almost exclusively coniferous. Conifers appear to be adapted to this climate because they are evergreen and capable of photosynthesizing immediately whenever a sunny day occurs. Deciduous angiosperms would be limited to the short growing season, less than 4 months long. Conifers in the boreal forest have drooping branches that shed snow easily; without this architecture, snow loads could easily break off limbs. Although there are many species of conifers, the boreal forest is not rich in diversity; often only two or three species completely dominate thousands of square miles. Black spruce and white cedar may dominate western areas, while white spruce and balsam fir cover much of the eastern area (FIGURE 10.27). Shrubs are not abundant but include blueberries, blackberries, and gooseberries. Herbs are also sparse. Not much disturbance occurs in the boreal forest; fire may occur in the south and insect plagues in northern parts. This biome does contain many large mammals such as moose, caribou, deer, grizzly bear, and timber wolves. The boreal forest is continuous with and grades into subalpine and montane forests that occur farther south. Many species occur in both biomes.

FIGURE 10.26. The boreal forest is extensive and it often extends as far as the eye can see. Boreal forests are dominated by just a few species of conifers at very high population densities, and wind-pollination is effective. Numerous broadleaf trees and shrubs occur here but are sparse; even though this was photographed in June, several of the broadleaf trees in the foreground had not yet put out their new leaves.

Tropical Biomes

Most tropical biomes are characterized by a lack of freezing temperatures. On high mountains in the tropics, cool or cold nights and winters occur, but only at very high elevations is frost encountered. Under conditions of high rainfall, tropical rainforests develop, but the drier areas contain tropical grasslands and savannas.

FIGURE 10.27. The northern edge of the boreal forest is dominated by black spruce (*Picea mariana*), which tolerates extreme cold climates and wet soils. In such conditions, trees grow extremely slowly and add only 1 or 2 cm (less than an inch) of height each year, and just a fraction of a centimeter in diameter. These small trees are probably more than 100 years old.

Tropical rainforests occur close to the equator (Figure 10.9). The Amazon Rainforest is probably the most famous, but extensive tropical rainforests occur in Central America, sub-Saharan Africa, and Southeast Asia. In the United States, Hawaii, Puerto Rico, and Guam have extensive rainforests. Precipitation is high, typically over 200 cm/yr and often as much as 1000 cm/yr: that is 10 m—over 30 feet—of rain every year. Rains typically occur every day; the morning may be cool and fresh, but clouds develop rapidly and rain almost invariably falls by noon. After a rain, there are large clouds in a clear sky and bright sunlight; the temperature quickly rises and the relative humidity of the air is close to 100%.

High temperatures and moisture cause much more rapid soil transformation here than in other biomes. Many minerals are leached from the soil, leaving behind just a matrix of aluminum and iron oxides. Humus decays rapidly, and there is little development of soil horizons. An extensive system of roots and mycorrhizae catch and recycle minerals as litter decays. Almost all available essential elements exist in the organisms, not in the soil.

The dominant trees are angiosperms, virtually no conifers occur naturally in tropical rainforests. The canopy is 30 to 40 m above ground level, but numerous large trees emerge above all others. A subcanopy may occur at about 10 to 25 m. Some trees have massive, gigantic trunks, but most are slender. In an undisturbed area where the canopy has remained intact for years, ground vegetation is minimal and it is easy to walk through the forest. Localized disturbances happen when a large tree dies and falls, creating a gap in the canopy. Suddenly, light is available on the forest floor, and herbs and shrubs proliferate. However, a tree soon grows up and fills the canopy, blocking light.

Virtually all small shrubs and herbs occur as epiphytes, located high in the canopy, nearer the light (FIGURE 10.28). Orchids, bromeliads, aroids, and cacti are common. Numerous vines are anchored in the soil, but their stems grow to the canopy, then branch and leaf out profusely. The leaves and roots of a plant can be separated by a narrow stem 50 m long: water that enters root tips travels up to 164 feet before escaping from a leaf or flower, and sugars produced by leaves must travel the same distance downward before they can be used by the roots.

Tropical rainforest is synonymous with species diversity. A single hectare may have well over 40 species of trees, often up to 100, and a single tree may harbor thousands of species of insects, fungi, and epiphytic plants. With such diversity, each hectare may have only one or two individuals of a particular species, especially of trees. No dominants occur. With such low population density for most species, wind pollination is not successful. All plants must be pollinated by animals, and being noticed by a pollinator can be difficult; many subcanopy species have brilliantly colored flowers that are easy to see in the dim light, and scents tend to be so strong that insects and birds find the flowers quickly. Large trees can undergo a massive flowering that no pollinator could possibly ignore.

Those areas of the tropics with lower rainfall develop as **tropical savannas** (a few trees) or **tropical grasslands** (no trees). The best known savannas are in Africa, but they also occur in Brazil (the cerrados), Venezuela (the llanos), and Australia (FIGURE 10.29).

FIGURE 10.28. Many of the small herbs in tropical rainforests live as epiphytes on large trees, such as this one in Guatemala. By living on top of the higher branches of a tree, the herbs have more sunlight than if they lived on the ground. These are tank bromeliads: the bases of their leaves overlap, forming a tank that catches and absorbs rain water; their roots serve only to attach the plant to the tree. (Courtesy of T. Sultan Quedesnsley.)

FIGURE 10.29. This is an unusual type of tropical grassland called "paramo." It occurs at high elevations on mountains, above the treeline but below the area of permanent snow. Although located in the topics, the elevation is so high that temperatures are always cool or cold. Most of the vegetation of this tropical grassland is indeed grass (*Calamagrotis effusa* here), but a few other herbs (*Espeletia* is visible in foreground) and some shrubs occur. (Courtesy of Oscar M. Vargash.)

Under natural conditions, undisturbed by humans, the vegetation consists mostly of bunch grasses up to 1.5 to 2 m tall. Most of the trees are open, flat-topped, widely scattered legumes. The South American savannas do not have many large grazers, but those of Africa are famous—zebras, wildebeests, and giraffes. An unexpected but very important grazer is the termite; termites are abundant and build giant nests of soil particles and plant debris. They bring in large amounts of plant material, digest it, and add their fecal material to the termite mound. Colonies are so abundant and the mounds so large that termites are a major link in nutrient cycling.

Important Terms

alpine tundra
arctic tundra
biogeography
biome
boreal coniferous forest
chaparral
continental drift
deciduous forest
desert
desert island
El Niño
evergreen forest
Gondwana
grassland
Hominidae
Laurasia
mid-ocean ridge
montane
Pangaea
permafrost
rainforest
rain shadow
savanna
shrubland
subalpine forest
taiga
tectonic plate
temperate
Tethys Sea
tropical
tropical rainforest
tundra
woodland

Concepts

- Conditions on Earth's surface change each day, each year, and on long periods of millions of years.
- Catastrophes occur irregularly, and on several occasions, a large percentage of all living organisms have been killed.
- Ever since life arose approximately 3.5 billion years ago, mutations have generated diversity, and natural selection has allowed some living organisms to survive every change of conditions. They also allowed organisms that survived major extinction events to diversify and gradually repopulate the Earth with a multitude of new taxa that were adapted to the new conditions.
- Major factors that influence the distribution of plant life on Earth are: the tilted axis of rotation of Earth; the position of land masses relative to climatic zones; and the origin of plants at a time when continents were close to each other or actually merged as Pangaea.
- Air and ocean currents distribute heat from the equatorial region to the poles; air carries water vapor from oceans to land, where it falls as precipitation.
- Continents drift across Earth's surface, changing their own shape as well as that of the ocean basins.
- Biomes are extensive groupings of many ecosystems characterized by the dominant plants.

Climate Change: The Roles of People, Plants, and Carbon Dioxide

11

This chapter is about balance, equilibrium, and the things that happen when a system is not in balance. Light from the Sun carries energy to Earth, causing it to become warmer and to radiate energy away, out into space. If the amount of energy coming to Earth equals the amount it sends out, the Earth is in equilibrium and its temperature is constant. If something would cause Earth to retain more of the Sun's energy, the Earth would become hotter and radiate even more heat away until it reached equilibrium again. Conversely, if less of the Sun's energy reached Earth, the planet would cool down until it radiated less energy and again came into balance. Depending on where the energy balance is reached, Earth might be frozen and covered with glaciers, or it might be pleasant and life sustaining as it is now, or it might be roasting hot and devoid of life. Mars, Earth, and Venus are all more or less the same distance from the Sun, yet Mars is cold, Earth is warm, and Venus is scorching.

For each planet, it is the atmosphere that is the key. Earth's atmosphere is like the glass in a greenhouse: it allows sunlight to come in, but it prevents some heat from escaping. As a result, Earth's temperature is warmer than it would be if we had no atmosphere at all. The two main gases in our atmosphere, nitrogen (N_2) and oxygen (O_2) play no role in this, only a few rare components act as **greenhouse gases**: water vapor,

This refinery converts oil pumped out of the Gulf of Mexico into many products, including gasoline and diesel for our cars and trucks. As petroleum products are burned, they add carbon dioxide to our atmosphere and contribute to global climate change. It is easy to demonize a factory such as this as being a cause of pollution, but this refinery exists only because we buy the company's products: it is our own cars and trucks that burn the fuel. And even though this complex covers a huge area, it is surrounded by marshes that appear healthy and unpolluted (the image at the beginning of Chapter 10 was photographed just a few miles from this refinery).

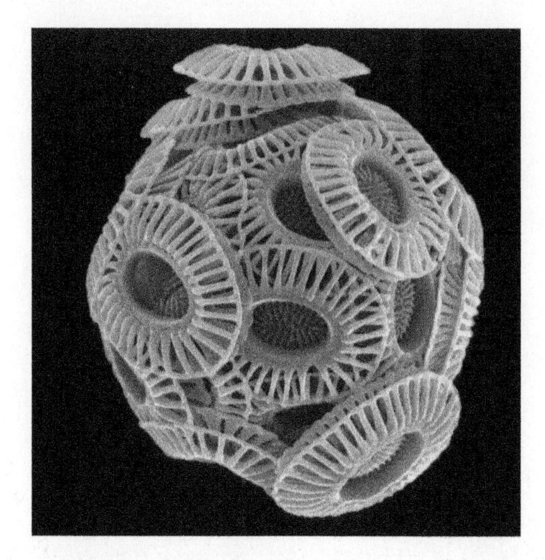

FIGURE 11.1. These calcium carbonate scales cover the bodies of unicellular coccolithophorid algae. As each alga makes a new scale, carbon is locked into the scale, which later falls to the bottom of the ocean when the alga eventually dies and its protoplasm decomposes. The carbon in these scales will remain out of the atmosphere for millions of years. The blue color was added to the image by an artist; these algae are green when alive and white when viewed by a scanning electron microscope. (© Steve Gschmeissner/Photo Researchers, Inc.)

carbon dioxide (CO_2), and methane (CH_4). Two of these, carbon dioxide and methane, are carbon compounds, and Earth's surface contains many carbon compounds, most of which have no greenhouse effect at all. For example, limestone (calcium carbonate), coal, oil, lignite, peat, wood, and shells of mollusks (clams and other bivalves; calcium carbonate) occur in the soil and do not affect Earth's temperature.

Our current concern about global climate change is with the equilibrium of these various forms of carbon: if the inert forms are converted to carbon dioxide or methane, their concentration in the air will increase, and Earth's temperature will rise. If carbon dioxide is removed from the air by photosynthesis or by animals building shells, then the atmospheric levels of carbon dioxide will fall and Earth's temperature will cool. An additional complexity is that volcanic eruptions carry carbon dioxide from deep inside Earth and deposit it in the air: the total amount of carbon at the Earth's surface is increasing.

Since the beginning of the Industrial Age in the middle 1800s, people have been burning coal, oil, and natural gas, converting neutral forms of carbon in the soil to harmful forms in the air, causing Earth's surface to become hotter. Natural changes have occurred in the past, long before humans had even evolved. After a type of algae, called **coccolithophorids** (**FIGURE 11.1**), with calcium carbonate shells evolved and became plentiful, they removed large amounts of carbon dioxide from the air: they built their shells, then when they died the shells sank to the frigid bottoms of the oceans where decomposition was almost impossible. Mollusks also use calcium carbonate for their shells. Later, the evolution of woody plants allowed photosynthesis to take carbon dioxide out of the air and lock it up as long-lasting wood, much of which became coal. The day-to-day lives of algae, mollusks, and trees caused the atmospheric concentration of carbon dioxide to drop considerably. At other times, massive volcanic eruptions on a worldwide scale released so much carbon dioxide that atmospheric concentrations of it reached levels 400 times greater than it is now. In each of these pre-human cases, the changes in carbon dioxide caused the Earth's energy input to be out of balance with its output, and global temperature and climate changed dramatically, but life always survived. The important question for us now is not "Are we humans causing global climate change?" but rather "How much climate change are we causing, and will humans survive it?" This chapter examines the most important aspects of these processes.

The Greenhouse Effect

Causes of the Greenhouse Effect

The greenhouse effect is a simple mechanism that is easy to understand. Ordinary glass like that in greenhouses or cars is transparent to visible light, so the visible light of sunshine passes through it, carrying all its energy. Once inside the greenhouse, the light is absorbed by colored objects: red flowers absorb all wavelengths except red and the energy they contain. Similarly, blue flowers, orange clay pots, and dark flooring each absorb most of the energy that strikes them and each becomes warmer. Any warm object radiates energy away from itself as infrared radiation. We cannot see infrared but we can sense it when a warm object is near our body.

The objects radiate infrared energy back, but glass is partially opaque to infrared, so it absorbs some of the energy rather than allowing it to pass outside

(FIGURE 11.2). The glass itself warms up as does the air and all objects in the greenhouse. Energy enters easily and part of it is trapped there. This is the **greenhouse effect**. The same process occurs when a car is parked in the sun with its windows up. The temperature does not increase forever; at some point the objects and the glass itself become so warm that their outward radiation equals that of the inward radiation. And in a real greenhouse, cold air outside cools the glass and carries away some of the accumulating energy.

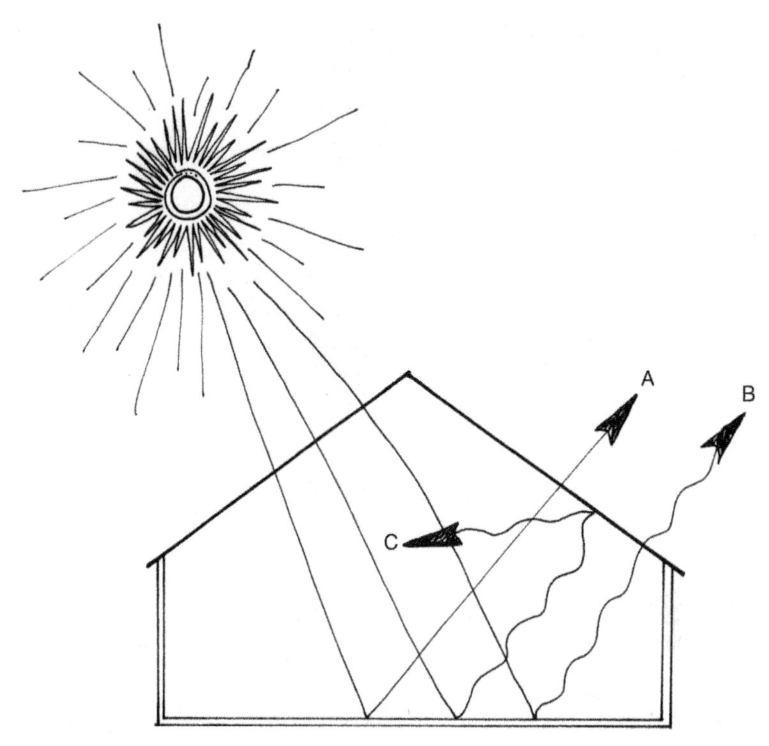

FIGURE 11.2. Diagram of the greenhouse effect. **(a)** Reflected visible light. **(b)** Infrared light that passes through the glass. **(c)** Infrared light trapped by the glass.

Carbon Dioxide and Methane as Greenhouse Gases

Our atmosphere acts like greenhouse glass. Visible light from the Sun passes right through it, then strikes the land, water, plants, animals, rocks, everything, all of which absorb some of the energy, warm up, and then radiate infrared outward. Again, just like glass, Earth's atmosphere lets some of the infrared energy pass through, but it traps part of the energy and the air itself is warmed by it. The total effect is that Earth's surface is warmer than it would be if the air had no greenhouse gases. In fact, if our atmosphere had no greenhouse gases at all, Earth's surface would be about 33°C (59°F) cooler than it is now.

Earth's temperature will not climb forever. As our atmosphere warms, it too begins to give off infrared radiation into space, thus preventing Earth from warming any more. The planet Venus has an atmosphere composed almost purely of carbon dioxide, and it traps almost all infrared radiation. Visible sunlight passes through its atmosphere, warms the surface rocks of Venus (it has no life and no liquid water), which radiate infrared energy back out. But Venus's atmosphere is so opaque to infrared that almost all the heat is trapped, and Venus's surface temperature is 460°C (860°F) before its balance is reached. Rather than lakes of water, Venus has pools of molten lead. Mars is just the opposite; even though its atmosphere is mostly carbon dioxide, the planet's low gravity and lack of a magnetic field allow it to hold on to only a thin, shallow atmosphere. There is so little carbon dioxide that energy enters and leaves easily, and the planet is cold.

Earth's atmosphere has only a tiny amount of carbon dioxide, about 387 ppm (parts per million; of one million molecules of air, 387 are carbon dioxide molecules), but the atmosphere is thicker and denser than that of Mars. Consequently, our atmosphere has properties intermediate between those of Venus and Mars: our atmosphere traps more infrared energy than does the thin atmosphere of Mars but not so much as the dense, carbon dioxide rich atmosphere of Venus. The concentration and amount of carbon dioxide are critically important.

Methane is a second important greenhouse gas. Each molecule of methane traps eight times as much energy as does a molecule of carbon dioxide, but fortunately, methane is much less abundant in our air, only about 1800 ppb (parts per billion). Also, methane reacts with oxygen and gradually breaks down into carbon dioxide;

on average, each molecule of methane only survives about 7 years once it is in the atmosphere. Unfortunately each molecule of methane breaks down into four molecules of carbon dioxide, each of which stays in the atmosphere for thousands of years.

Other Greenhouse Gases

Two other greenhouse gases are water vapor and ozone, but we have little influence on them. Water vapor accounts for 36% to 72% of the greenhouse effect on Earth, changing with seasons of cold, dry air versus those with warm, humid air. Although we spray water on lawns and crops, overall, people have little impact on water vapor concentration in the atmosphere; evaporation from oceans is the dominant source.

Ozone has a complex role in our atmosphere. It contributes only a small amount, between 3% and 7%, of greenhouse warming, but more importantly it blocks ultraviolet light. There is an ozone layer located high in the atmosphere, and it is critically necessary to our survival: without it, so much ultraviolet light would reach Earth's surface we could not live here. Human-produced chemicals (chlorofluorocarbons) used as coolants in refrigerators and air conditioners had been leaking into the atmosphere and destroying the ozone, but since the 1990s, their use has been banned and the ozone layer is recovering. Although it warms our planet slightly, we must have it to survive.

Human Activities that Produce Greenhouse Gases

Several human activities produce carbon dioxide and add it to our atmosphere (**BOX 11.1**). One fundamental activity is just our own metabolism: we eat food that consists of organic material and respire it to carbon dioxide and water. Every time we exhale, we add more carbon dioxide to the atmosphere. And the same is true for all animals, including our pets and our livestock, the animals we keep for meat, milk, wool, and so on. Even plants respire, and at night all plants give off carbon dioxide, even stored crops such as wheat, rice, corn, and potatoes. But this type of metabolically produced carbon dioxide has occurred as long as animals and other aerobic organisms have existed.

The concerns about human activities that add carbon dioxide to the air center on our burning of fossil fuel and the deforestation we do. **Fossil fuels** are things such as coal and petroleum, fuels that are the remnants of ancient plants and animals. Over millions of years, these have been changed from being the bodies of dead organisms into hydrocarbon remains, and when they are burned, they are converted mostly to carbon dioxide and water. Few of us have any day-to-day contact with petroleum, but we do use its refined products all the time: gasoline, diesel, and heating oil. Other fuels, such as firewood (used by hundreds of millions of people in developing nations), natural gas, charcoal, and lignite also are hydrocarbons that produce carbon dioxide when burned. It is easy to think that we have insignificant impact because we drive only a little, and maybe even bicycle or walk much of the time, but most of the electricity we use is generated by burning coal or natural gas, and all our food, clothing, and other necessities are delivered by trucks that burn gasoline or diesel. Those few people who own electric cars add carbon dioxide to our atmosphere whenever they recharge their batteries, unless their electricity comes from a nuclear, hydroelectric, solar, or wind turbine power plant.

Every year thousands of square kilometers of forests burn, releasing tons of carbon dioxide into the atmosphere. If caused by lightning, forest fires are a natural process, but many fires are started accidentally or by arson, and then the added carbon dioxide is **anthropogenic**, caused by people. Also, a type of logging called **clear cutting**

BOX 11.1. Zero Population Growth

There are so many people on Earth (6.8 billion in 2010) and each of us consumes so many resources and produces so much waste that we are dramatically altering the biosphere. Forests are cut to supply lumber for our houses and paper for books and magazines. Land is cleared for towns, farms, airports, highways, and parking lots. Fossil fuels are burned and rivers are dammed to generate electricity. Tons of radioactive waste are stored at many sites. Human waste travels through sewers to rivers. Already, the ever increasing number of human beings has crowded out other organisms, reducing their populations and driving many to extinction. We already have caused more extinctions than any other organism. Only asteroids or massive volcanic eruptions have been more destructive than we people. Of course, some people cause greater harm than others; each person in highly developed countries consumes and pollutes more than does an individual in a poorer, less-developed country. It is definitely possible for us to minimize the harm we do by consuming less, driving less, recycling more, and so on. But even if we simplify our lives as much as possible, still each individual does consume, does pollute, does occupy space and resources that are no longer available to members of other species. Human populations simply cannot continue growing forever. At some point we will reach the end of some necessary resource. We too have limiting factors as were discussed in Chapter 4. It may be that we finally run out of space for growing more food, or there are no more energy resources to tap, or finally our pollution has made the air and water unhealthful.

When we reach such a point, it does not mean that humans will go extinct; it is just that population growth will stop. Several scenarios are possible. Assume that at some point in the future there is just no more land that can be brought into cultivation and genetic engineering has made crop plants as productive as possible. As our population approaches that point, food will become scarce and more valuable. In some regions there will be famine and starvation (this already occurs periodically in Africa) or there may be war to obtain food or land. The human population will either stop growing or it may actually decrease. But all the farmland will still exist and we will continue to feed some number of people. Another scenario is that we run out of energy: we will have burned all the coal and natural gas, we will have cut down all trees that can supply firewood, and so on. Wind, nuclear, and hydroelectric power resources will have been exploited to the maximum. With all fossil fuels and firewood trees gone, our total supply of energy will fall greatly. This may cause a significant decrease in population: there will not be fuel to grow as much food as before, or to transport it, to heat homes, or to run factories. Economies will falter and countries will become poorer, with less capacity to maintain essential services such as hospitals, water purification, and sanitation. Disease could increase, large numbers of people would die and the total population would decrease. Another scenario is global climate change caused by our release of greenhouse gases. This could alter so many things that many people, perhaps most, would die within a few years.

An alternative to our continual population growth and the problems it will bring is **zero population growth**. This is the concept that the number of children born would be just the number necessary to offset the number of people who die. Most couples would have only two children, a very few might have three, and some people would have none. It takes many years to achieve zero population growth because in almost every society there are many more young children than there are old people. Even after all couples limit themselves to just two children, during the next few years the number of young people reaching reproductive age will exceed the number of people who are dying. But at some point, the population will stabilize and stop increasing. Many people believe that rather than just stopping human population growth, we should actually reduce the population to a lower level and then keep the number of individuals constant at that reduced level.

Another factor in the reality of Earth's finite resources is the amount of consumption and pollution per person. If we achieve zero population growth such that the number of people living at any given time is constant, but then each person consumes more and pollutes more, our problem is not solved. China had a strict and fairly successful policy of limiting the number of children that a couple could have, and they reduced their population growth greatly. But China also became much more affluent, so even though the number of people was barely increasing, the number of cars is shooting up, the size of houses is increasing, and the amount of food, products, waste, and pollution is rising rapidly. The important point is to adjust the human population growth and human consumption to levels that are sustainable. Earth's resources could be used to support large numbers of people at a miserable standard of living, or a smaller number of people at a more healthy level. Human population growth will be zero at some point in time, either through our own voluntary methods or through scarcity, famine, and disease.

FIGURE 11.3. Harvesting trees by means of clear cutting involves cutting down all trees and shrubs even though only the larger trees are used. All of the plant debris left on the forest floor will quickly decompose as the carbon compounds are converted to carbon dioxide. Often some of the smaller branches and shrubs are chipped and sold as mulch for gardens, but that too causes it to decompose quickly, releasing carbon dioxide to the atmosphere. Soil erosion is a severe problem in many clear cut areas. (© Photodisc.)

involves cutting down all trees in an area of forest, even though only the ones that are large enough to be profitable are hauled out (FIGURE 11.3). All smaller trees are cut down, just to make harvesting easier, and even the branches of the large trees are cut off and left. After clear cutting, the ground is covered with the debris of small trees, branches, and shrubs that were of no interest. This is often allowed to dry out; then it is burned. Even if not set afire, fungi will quickly break down all the debris and convert it to carbon dioxide. In contrast, **selective cutting** takes only specific large trees and leaves the rest unharmed; it is more expensive but it causes less environmental damage and the remaining trees continue to photosynthesize, taking carbon dioxide out of the atmosphere.

Burning forests is not the same as burning grasslands or the leftovers of crop plants after harvest. Grasses and crops are herbaceous material that would decompose naturally in a year or two anyway; burning just converts it to carbon dioxide a few months faster. But forests contain woody material, material that if left alone might last for hundreds of years, keeping the carbon out of the atmosphere.

People also add methane to the atmosphere. A small bit comes from our own metabolism (flatulence) but most anthropogenic methane comes from anaerobic decay of our wastes. Decomposing material in landfills produces large amounts of methane, although many landfills now trap it and burn it to carbon dioxide in order to convert a strong greenhouse gas into a weaker one. Other human waste goes into a sewer system then to a sewage treatment plant or a septic system; both of these produce a sludge that breaks down anaerobically into methane. Cattle, even when healthy, produce a great deal of methane from both ends, and the decomposition of the vast amounts of manure from chickens, pigs, cattle, and sheep account for 37% of all human-related methane (FIGURE 11.4).

Nonhuman Sources of Greenhouse Gases

Molten rock inside Earth itself is a source of carbon dioxide. Magma (molten rock that has not reached Earth's surface) and lava (molten rock that has escaped through volcanoes, fissures, and mid-ocean ridges) are rich in carbon dioxide and release it into the air during eruptions. Even though volcanoes erupt almost daily and the mid-ocean ridges are constantly growing, these release miniscule amounts of carbon dioxide compared to human production. For example, in the months following its eruption, Mount St. Helens released about 1000 tons of carbon dioxide per day, whereas a single coal-fired power plant nearby released 28,000 tons per day.

FIGURE 11.4. Cattle, pigs, chickens, and other livestock are often fed cereal grains or silage (chopped corn plants). If not used for animal feed, the grains would germinate and grow into new plants that take carbon dioxide out of the atmosphere, and the silage would decompose to carbon dioxide. But the digestive system of cattle converts much of the feed to methane; manure from all types of livestock also produces large amounts of methane as it decomposes. (© Thoma/ShutterStock, Inc.)

Much of the methane in the air comes from natural sources and is not produced by people. It is the major component of natural gas, which seeps to the surface in geologically active areas. Numerous bacteria produce methane as a product of their respiration, and such microbial activity in marshes and seafloor sediments produce methane.

One of the most worrisome aspects of methane is that frozen soil in the Arctic and cold seafloor beds contain vast quantities of a substance called **methane clathrate**, which is basically ice with methane trapped in it. As long as clathrates remain cold or under great pressure, they are stable and hold the methane in a harmless form out of the atmosphere. But warming of Arctic regions might cause those deposits of clathrate to break down, releasing their methane into the air. Much more study of seafloor clathrates is needed, but it is suspected that from time to time certain regions suddenly break down, releasing vast amounts of methane that then bubbles up through the seawater and escapes into the air. The amount of methane locked up in clathrates is so extensive that if a large portion of it decomposes, enough methane would be released to cause major global warming.

The Greenhouse Effect Causes Global Climate Change

An increase in greenhouse gases will lead to increased temperatures on Earth, but the term "global warming" understates the many problems that will be caused (**BOX 11.2**). Rather than all parts of Earth becoming slightly warmer, the main effect is that circulation patterns in the oceans and atmosphere will be altered. As was discussed in Chapter 10, these circulations transfer heat from the equatorial tropics toward the polar Arctic and Antarctic regions, preventing Earth from having much more extreme differences in temperature from place to place. But oceanic and atmospheric currents circulate in complex patterns affected by Earth's rotation, the positions of the continents and ocean basins, as well as the input of solar energy at the equator. Increasing concentrations of greenhouse gases in our atmosphere will alter circulation patterns, producing significant changes in many aspects of the biosphere. At present the average temperature of the polar regions is rising faster than other parts of the

BOX 11.2. Natural Climate Change

The focus of this chapter is the global climate change occurring now because of human activity. But this change is not the first. In the billions of years before humans had evolved, Earth's climate changed drastically several times, and our planet alternates between periods when extensive ice sheets are present (called **glacial ages** or **ice ages**) and periods when there is little if any ice anywhere. Just as now, each climate change seems to have been the result of just one or two key factors such as concentration of carbon dioxide in the atmosphere, the drifting of continents to the poles where they accumulate ice (or drift away and lose it), the circulation patterns of atmospheric and oceanic currents, and the presence or absence of plants and algae that sequester carbon as wood or calcium carbonate.

We know of five major glacial ages:

1. The Huronian Glacial Age occurred between 2.4 and 2.1 billion years ago. We know little about it because very few rocks from that time still exist.

2. The Cryogenian Glacial Age lasted from 850 to 630 million years ago. This was the most severe of all known ice ages, and it may be that ice covered not only most of the continents but also that the oceans were frozen over almost to the equator. This is referred to as Snowball Earth. This ice age ended when massive volcanic eruptions released so much carbon dioxide into the atmosphere that levels were more than 400 times higher than today. This ice age occurred long before plants evolved, around 460 million years ago.

3. The Andean-Saharan Ice Age took place between 460 and 430 million years ago. It was relatively minor and coincided with the evolution of land plants.

4. The Karoo Ice Age may have been caused by plants. It occurred 360 to 260 million years ago, and began not long after woody trees evolved and extensive forests existed. Much of Earth's surface consisted of warm, humid marshes with extensive forests. Because of the marshy conditions, many trees that died fell into stagnant water where they did not decay, but instead turned into most of the coal that is found today. The forests and the resulting coal deposits were so vast that the geological age is called the Carboniferous Period. Photosynthesis by the trees removed so much carbon dioxide from the air, and preservation as coal sequestered it for so long, that the levels of carbon dioxide dropped greatly, almost to the levels present today. The atmosphere became much more transparent to infrared radiation; Earth cooled and entered a mild ice age.

5. We are presently in an ice age, the Pliocene-Quaternary Glaciation. It started about 2.8 million years ago, and is ongoing. Like other ice ages, this one has relatively cold periods that last several tens of thousands of years; these are called glacial periods. Between the glacial periods are relatively mild interglacials. Even though we are in an ice age, we are in a warm interglacial called the Holocene Interglacial. It began only 11,000 years ago, when the ice sheets that covered northern North America and Europe began retreating, and about the time humans migrated from Asia to the Americas. The Holocene Interglacial may last longer than normal because of the greenhouse gases we are producing.

globe. Already there is so little sea ice near the North Pole that polar bears have had difficulty hunting seals, and others have died in the open ocean without any ice to climb onto as a place to rest (**FIGURE 11.5**).

Water Can Be Solid, Gas, or Liquid

Water exists in many forms on Earth, all of which are important to living organisms. Think beyond the terms "solid," "gas," and "liquid."

As a solid, water is present as ice, snow, frost, hail, and sleet. Ice occurs in vast ice fields covering the entire large continent of Antarctica, much of Iceland, and Greenland; the tops of high mountains; and as glaciers flowing down valleys and producing icebergs (**FIGURE 11.6**). Ice also completely covers rivers in winter, cutting off oxygen to fish and other animals. Then in spring it breaks up into ice flows that scrape trees off the river banks. Ice can form inside plants if the weather is cold enough; if the plant cannot produce "antifreeze" compounds, ice crystals inside living cells break membranes, killing the cells. And because water expands as it freezes, ice can cause tree trunks to shatter. Freezing rain and snow accumulate on plants as ice, often becoming heavy enough to break large limbs off trees. Solid water, falling as hail, smashes plants and injures animals if the hailstones are large enough. Crops of wheat, citrus, apples, and others can be destroyed by hail storms.

One of the most important aspects of ice, snow, and other solid forms of water is that they have a high **albedo** (reflectivity). Ice and snow appear white because they reflect so much light and absorb so little. Sunshine that falls on ice and snow is mostly reflected back out through Earth's atmosphere as visible light. Little of the light's energy is absorbed, little remains to warm Earth. Ice and snow are virtually always more reflective than whatever they cover: dark rock or vegetation absorbs light and warms the biosphere, but when even a thin layer of snow covers these, the energy is reflected away (FIGURE 11.7).

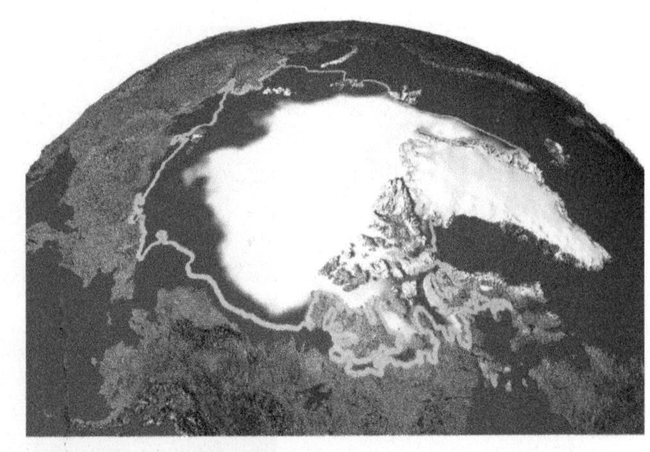

FIGURE 11.5. In this photograph of the Arctic region surrounding the North Pole, the yellow line indicates the average boundary of ice from 1979 until 2004. This photograph was taken in 2005, and it shows that the ice cap is now much smaller. A great deal of open ocean now separates land and ice, affecting many species in addition to polar bears. (Courtesy of Josefino Comiso and NAA/Goddard Space Flight Center Scientific Visualization Studio.)

Water as a gas is water vapor, which is completely invisible to our eyes. Water vapor enters the air when water evaporates from oceans and lakes, is transpired from plants, exhaled from lungs, and sublimes (goes from solid to gas) from ice and snow. Winds then carry the water vapor from place to place until it changes into rain, snow, fog, dew, frost, and so on. We can easily see the vast amounts of water that flow in rivers or that make up ice fields, but it is important to remember that all of that water got there as water vapor that was deposited as precipitation. And huge amounts of rain and snow fall on the oceans themselves. It may seem inconsequential that water evaporates from one part of an ocean and falls on another part of the same ocean. But ocean water is salty and water vapor is not: in regions where water is evaporating,

FIGURE 11.6. It is difficult for us to comprehend the amount of water that occurs in the world's ice sheets. The United States of America has a surface area of 3,794,101 square miles (9,826,721 square kilometers), but Antarctica alone is much larger: 5,300,000 square miles (13,727,000 square kilometers). Imagine flying completely across the United States (about a 6 hour flight) and seeing nothing but ice that is more than a mile thick in most places. Antarctica has more ice than that; Greenland and Iceland also have vast amounts, and there is ice permanently covering the Himalayas, the Andes, and the northern Rocky Mountains. (© Vitaly Korovin/ShutterStock, Inc.)

FIGURE 11.7. The plants and rocks in this area have a very low albedo: most of the energy that strikes them is absorbed and heats them (the same is true of the jacket). The snow and ice on the Alaska Range in the background have a very high albedo and reflect energy away without significant heating. In summer when this photo was taken, the plants and rocks were absorbing large amounts of visible light energy and reradiating it as infrared energy; when covered in snow in winter, this entire area would reflect away most light energy and cause almost no heating.

the remaining water becomes even more salty, something vitally important to fish, algae, and other sea life. And in ocean regions where precipitation is falling, the ocean water is being diluted, becoming less salty, also important to sea life. In addition, water only evaporates if it absorbs heat, and it precipitates only if it loses heat. We can think of the atmosphere's movement of water vapor as a process that transports gigantic amounts of pure water and heat but not minerals.

Water vapor is a greenhouse gas, and at present it contributes more to global warming than does either carbon dioxide or methane. But people have almost no influence on the amount of water vapor in the air.

Water as a liquid has multiple forms. It is salty ocean water, brackish water in marshes and lagoons, and freshwater in lakes and rivers. It is clouds (clouds are droplets of liquid water, they are not water vapor), rain, fog, and dew. Water is also the liquid in protoplasm, in xylem, and in our sweat and urine. Unlike gaseous water and solid water, liquid water has such a strong tendency to dissolve things that it is almost never pure. Even as raindrops are forming in clouds, they begin to absorb nitrogen, oxygen, and carbon dioxide. If the air is polluted with nitrogenous or sulfur compounds, those dissolve in the raindrops also and form **acid rain** (**FIGURE 11.8**). As soon as rain falls onto land, plants, animals, fungi, or anything else that is solid, it begins to dissolve whatever is hydrophilic.

Unlike water vapor, liquid water can transport lack of heat (coldness) as well as heat. Ocean water is warmed by sunlight in the tropics then flows poleward, carrying unimaginable amounts of energy. In contrast, as ice and snow melt, they produce cold water at 0°C (32°F). This warms as it flows downstream through creeks, rivers, and aquifers toward the oceans, but it becomes warmer only by absorbing heat from the riverbeds and surrounding habitats, making them colder. Water transports heat and material from land to oceans.

FIGURE 11.8. Certain types of industrial processes produce waste chemicals that combine with water in the air, resulting in acids that then fall to earth as acid rain, acid snow, or acid fog. The acids damage plants directly and alter soil acidity and nutrient balance. Such pollution is now illegal in many areas, but it has not been eliminated entirely. (© JoLin/ShutterStock, Inc.)

Because water has three states, each with very distinct properties related to energy, the equilibrium between the three is important. If more water is converted to snow and ice, it becomes more reflective and cools Earth, which is likely to cause even more snow and ice to form. This would be a positive feedback loop: the effects of a process cause the process to speed up or last longer or be more intense. As water evaporates, it becomes water vapor, a greenhouse gas that causes Earth to become warmer, which causes more evaporation, another positive feedback loop. One chemical, three forms, two positive feedback loops, one of which cools the planet, the other warms it.

Changes in Water Cycles

One effect of global warming that is easy to predict is that the oceans will become warmer and more water will evaporate. As mentioned, water vapor is a greenhouse gas, and it will cause further warming. But beyond that, we can predict that the more humid air will produce more precipitation: rain in summers, snow and ice in winters. This will have multiple effects, including flooding, mudslides, and erosion. These are obviously detrimental, causing damage to rivers, towns, farms, and riverside plants and animals, many of which drown if flooding is prolonged. Some deserts and other dry areas will receive more rain, which may seem like a good thing, but the plants and animals that inhabit those areas are adapted to specific patterns and amounts of rainfall; increased precipitation will change the habitat, harming them and allowing other species to invade. Extra moisture is just another form of habitat destruction.

Increased snowfall causes not only misery for drivers and electric utilities but also death from accidents, hypothermia, and fires when homes are left without power. Extra snow on mountains adds to the snow pack there, but at present it seems that warm summer temperatures more than balance extra snow in winter: rather than gaining more total alpine snow, there is more melting and runoff throughout the year. For example, Glacier National Park in the United States will soon have no glaciers or

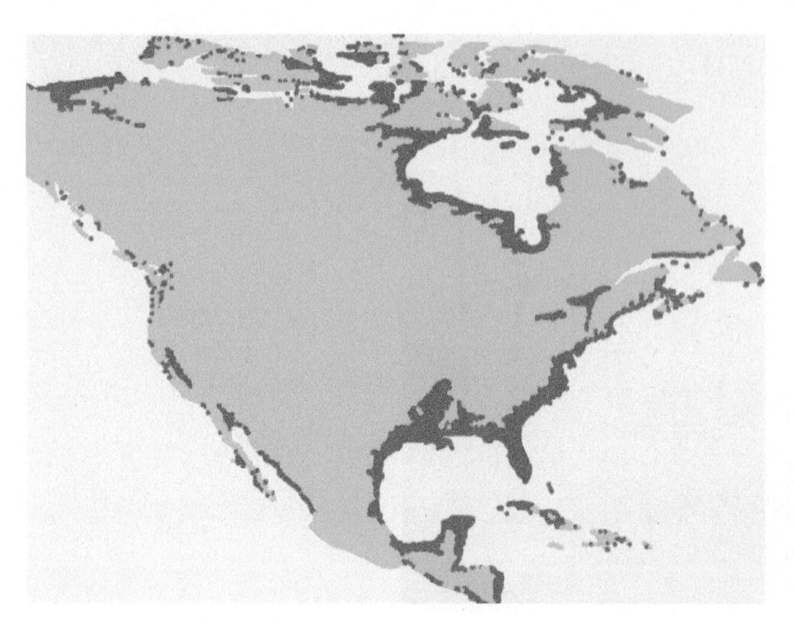

FIGURE 11.9. If all the ice caps on land were to melt due to global warming, sea level would rise about 66 m (about 200 feet), flooding low-lying coastal areas and islands. This map is an estimate of what the coast of the United States would look like if all land-based ice were to melt.

permanent ice fields left, despite extra snowfall in winter.

A portion of precipitation seeps into the ground and enters **aquifers**, systems of cracks and channels and porous rock that run below ground. Aquifers finally emerge at the soil surface, usually as a natural spring or marsh. Having adequate water in aquifers is important for the wildlife that lives in the marshes or the streams that flow from the springs. It is also necessary to the millions of people who depend on wells for their drinking water and irrigation water. In many agricultural areas, so much water has been pumped out of aquifers to irrigate fields that springs, marshes, and wells downstream have gone dry. Increased precipitation would help recharge the aquifers.

The freshwater of most rivers finally flows into an ocean somewhere. Fresh water is much less dense than salty water, so the river water usually floats as a cap on top of ocean water until the churning of waves gradually mixes the two.

Another change in water cycles will be the increase in the volume of the oceans. Water expands when warmed, and because there is so much ocean water, even a slight temperature increase will cause the water to expand and sea levels to rise. Where the shorelines are rocky, vertical cliffs (as in Washington, Oregon, and most of California), this will not matter much. But for low-lying, flat shores (such as all of Florida and Hawaii, and the states along the Gulf of Mexico), a slight rise in sea levels will be enough to flood far inland, and coastlines will be drastically altered (FIGURE 11.9).

Rising sea levels will also cause the lower ends of Antarctic glaciers and ice sheets to float. Glaciers move slowly when flowing across rocky land, but when they finally reach oceans, their ends float and they break up into icebergs. As sea levels rise, more of each glacier's end will float and each will then flow a bit faster. Ice will be transferred more rapidly from Antarctica to waters that carry it northward where it will melt. The warming of the oceans will cause Antarctica to lose ice more rapidly which will cause the oceans to rise even more; this is another positive feedback loop.

Mountains throughout the world already have less of their tops covered in ice and snow than they did just a few years ago. Many mountains are now ice-free in late summer whereas they used to be permanently snowcapped. Iceland, Greenland, and much of northern North America and Europe have smaller ice fields. As soon as an area becomes free of ice, it changes from reflecting sunlight to absorbing it (the area changes from having a high to a low albedo). Fortunately, most of the newly ice-free areas are near the Arctic Circle, a region that receives little solar energy. Mountain tops that now have less snow are in the temperate or tropical zones, which receive more light per square kilometer, but these regions are a relatively small fraction of Earth's surface. The ice on Antarctica is so thick that even with more rapid melting, it will stay covered in ice and be reflective for years.

Changes in Land Temperatures

Earth's land areas will, on average, become warmer. But as ocean and air currents change and alter their heat distribution, some land areas will become cooler even as the overall temperatures rise. Luckily for us, unlike water, none of the land will either melt or turn into a gas; it will simply become slightly warmer or cooler but will remain rock or soil.

But moderate changes in land temperatures are significant to organisms. As winter gives way to spring in temperate regions, an important date is the last day of frost: after this date, any seeds that germinate or buds that expand will be safe from damage by frost. Similarly, the day of first frost in autumn is critical (FIGURE 11.10). These two dates are the effective boundaries of the **growing season** of a region. In areas that become warmer, the growing season becomes longer. This will cause changes in vegetation that are difficult to predict, because some plants are sensitive to frost and others are not. Seeds and buds of many species become active as soon as late winter temperatures become mild, and an occasional frosty night or two do not harm them at all. Such plants should benefit from more frost-free days of growth. But the seedlings or newly active buds of other species are severely damaged by even one night of frost, so the date of the last winter frost (the beginning of the growing season) is crucial to them. Warmer temperatures and a longer growing season should help them greatly. The dormancy and activity of other organisms are governed by day length and these cannot take advantage of the extra days of growing season. They will continue to be dormant while other plants have become active, so they may be outcompeted by their neighbors (see Box 6.2). In addition, in areas where winters have many frost-free nights, plants called winter annuals germinate in autumn, grow through the winter, then flower, fruit, and die in late winter or early spring. Winter annuals benefit from the fact that most other plants are dead or dormant or leafless in winter, allowing plenty of sunlight to reach them. As temperatures rise and the start of the growing season occurs earlier, winter annuals will face more competition and shading just as they need maximum energy for seed and fruit maturation.

FIGURE 11.10. Fruits of kumquats (*Citrus japonica*) ripen in late autumn, but usually well before the first killing freeze occurs. In this particular winter, an early ice storm caught both mature and immature fruits on the tree, killing all the fruits and destroying the crop. The tree itself survived.

An unusual feature of global warming is that Earth's average temperature is increasing mostly because nighttime temperatures are increasing more than are daytime temperatures. Days are slightly warmer but night temperatures are already significantly warmer in many places. Temperature strongly affects the rate of plant respiration at night, so plants are tending to respire away more of the carbohydrates that they are producing during the day.

Many people assume that the increased levels of carbon dioxide will automatically cause increased rates of photosynthesis, so as carbon dioxide builds up in the atmosphere, plant growth will increase proportionately thus solving the problem automatically. But in many areas, carbon dioxide is not the limiting factor of plant growth; instead it is a lack of water or of nutrients such as nitrates and phosphorus. The extra carbon dioxide in the atmosphere may cause increased plant growth only if there is also increased rainfall or if people apply artificial fertilizers. An important aspect of this is that many agricultural crops are grown for their fruits or seeds, organs that need amino acids and minerals: increased photosynthesis might lead only to increased levels of starch, not of essential vitamins and other nutrients.

In addition to these direct effects on the plants themselves, changes in temperatures and the dates of frost-free nights are important to the other organisms in a plant's habitat. Many insects that rely on plants for pollen, nectar, or fruit are just as sensitive to an unexpected frost. If some insects hatch from their eggs too early, they will be killed by a frost, while others are not affected. Birds time their migrations based on day length and other factors. With a longer growing season, some plants might be able to germinate and grow earlier but their pollinators or seed dispersers might not be present when the plant needs them.

Processes that Might Lessen or Reverse Global Warming

Processes that Remove Carbon Dioxide from Air and Store It

In order to combat global warming, it is necessary to both remove greenhouse gases from the atmosphere and to keep them out. This is called **carbon capture and sequestration**. Fortunately, several of the best methods occur naturally. Photosynthesis takes more carbon dioxide out of the atmosphere than any other process. In woody trees, much of that carbon dioxide is converted to wood, which locks it away from the atmosphere as long as the tree is alive. After a tree dies, its sapwood may be attacked by fungi right away and be quickly respired to carbon dioxide within 4 or 5 years, but heartwood is more resistant to rot and can persist for dozens of years before finally being completely broken down (FIGURE 11.11). An attractive feature of this method of carbon capture and sequestration is that it is free. A problem is deforestation; we have cut down so many forests (and we are still doing so), that there are not as many trees left to carry out photosynthesis.

A type of unicellular marine alga called coccolithophorids (see Figure 11.1) also captures and sequesters carbon. These algae are photosynthetic and they build shells of calcium carbonate. As the algae die, their shells sink to the bottom of the ocean where the water is cold and decomposition is slow; the shells last for hundreds or thousands of years. In fact, the White Cliffs of Dover and other immense deposits of coccolithophorids occur around the globe. All blackboard chalk used in classrooms is made of these shells. Mollusks too make their shells of calcium carbonate, locking more carbon away from the atmosphere. Sequestering carbon as calcium carbonate shells is also free.

FIGURE 11.11. Even though this pine tree has died, the carbon in its wood will be sequestered for many years. The bark and sapwood decay most rapidly, perhaps within 10 to 15 years, but the heartwood is durable and solid parts of the center of this log might still be intact 50 years from now. The rate of decay varies with the type of tree and the moistness of the environment. This is in Bastrop State Park near Austin, Texas, where annual rainfall is only about 81 cm (32 inches) per year, so this is rather dry and decomposition is slow.

In business programs called **carbon offsets**, companies are allowed to build factories that release carbon dioxide if they pay for the planting of new trees that will take up an equal amount of carbon dioxide. Theoretically, the factory plus the new trees would be **carbon neutral**. Such a program has great potential, but it can also be abused. The trees might not be planted, or the program might plant short-lived trees that grow quickly, die early, and then decay away rapidly. They do take up carbon but they do not sequester it for long (FIGURES 11.12 and 11.13). In some areas, such carbon plantations are planted but local people cut the trees down for firewood while they are still young. In another type of carbon offset, rather than planting new trees, the company would pay to preserve existing forest.

FIGURE 11.12. These are cottonwoods (*Populus angustifolia*) growing along a stream in New Mexico. Cottonwood trees grow to a large size very quickly but their wood is weak and has few antimicrobial chemicals in it. Once a branch falls off or an entire tree dies, it decays quickly, releasing its carbon to the atmosphere. Planting forests of cottonwood trees would not be a long-term solution to global warming.

FIGURE 11.13. In contrast to cottonwood trees, bald-cypress trees (*Taxodium distichum*) have wood that is extremely hard, heavy, and so saturated with antimicrobial chemicals that even after a tree dies, its wood remains intact for many years. Trees such as this are good for long term sequestration of carbon dioxide.

But not all forests are equal in the amount of carbon they store; one square kilometer of tropical rainforest contains much more wood than does an equal area of pinyon pine forest in the southwestern United States.

It is possible to accelerate carbon capture by trees and algae. Part of the acceleration will be automatic: carbon dioxide is the raw material of photosynthesis, so as atmospheric levels increase, so will photosynthesis. Also, in some cases trees and algae grow faster if fertilized, usually with a nitrogen or phosphate fertilizer. Theoretically, large regions of forest or ocean could be fertilized, photosynthesis would accelerate, and carbon dioxide would be locked away as wood or calcium carbonate. But fertilizers are extremely expensive, and producing nitrogen-based fertilizers requires large amounts of energy.

Artificial methods of carbon capture and sequestration also exist. One is to capture carbon dioxide as it goes up the smokestacks of electricity generating plants, then pump it underground and store it in empty oil wells. Several pilot plants have been built but so far the process is both extremely expensive and energy intensive: the power plants would have to burn more fossil fuel just to capture the carbon dioxide. Estimates are that the cost of electricity would have to double, even after the technology has been improved.

Using Less Carbon-Based Energy

One way of slowing the increase in greenhouse gases is to use less carbon-based fuels. We could rely more on **alternative energy sources**, such as wind turbines, solar panels, and even generators that are powered by ocean waves and tides (FIGURE 11.14). We already use hydroelectric dams and nuclear power plants, but both of these have significant problems. Hydroelectric dams must have voluminous reservoirs, which require the flooding of hundreds of miles of upstream river valleys.

FIGURE 11.14. Wind turbines are being used in many places to convert the power of wind into electricity, thus reducing the need to burn fossil fuels. These wind turbines in Denmark are located between fields where they are out of the way, and their bases need only a small plot of land. Because the turbines are in populated areas, long-distance, high-voltage transmission lines are not needed.

FIGURE 11.15. This is only one of the three nuclear reactors at Fukushima, Japan, that were damaged by a tsunami. For several weeks, thousands of tons of water were sprayed onto the reactors to cool them, and most of that water ran off into the Pacific Ocean (visible in upper right corner), contaminating it heavily with radioactive materials. (© Air Photo Service, file/AP Photos)

The dams disrupt natural water flow and alter wildlife in and along the river; bottomlands are destroyed. Some dams in the western United States are being torn down because they have done such severe damage to populations of salmon, which must swim upriver to spawn. Nuclear power plants expose us to the risk of accidental radiation leaks (as occurred at Three Mile Island, Chernobyl, and Fukushima), nuclear terrorism, and the need to store radioactive wastes for thousands of years. A really big problem with all alternative energy sources is that the direct cost to consumers is always more than that of fossil-fuel based energy. Electricity-generating power plants that burn coal, natural gas, and petroleum are the majority because they supply consumers with the cheapest energy. But the price per kilowatt hour is not necessarily the true cost of the energy. Coal also has the costs of the ecological damage done by strip mining and dumping mining wastes, numerous workers killed in mines, and of course the cost of damage done by global warming due to the release of carbon dioxide. Nuclear plants also appear to provide cheap electricity, but that is mainly because governments rather than consumers pay the cost of storing nuclear wastes and safeguarding the plants. The cost to Japan of dismantling the reactors and cleaning up the environment at Fukushima (FIGURE 11.15) will be tens of billions of dollars, perhaps even approaching 100 billion dollars. That cost, however, will be paid by the Japanese government, not by people who buy electricity, so the cost of nuclear-generated power will appear to be much cheaper than it actually is.

One argument against alternative power generation is that those technologies need large amounts of land. Hundreds of square miles of solar panels or "wind farms" are needed to replace one coal-fired power plant that occupies only a square mile or less. If large amounts of sunny desert areas are covered with solar panels, this would cause more habitat destruction. But solar panels do not need virgin land; they can be placed on roofs of buildings, which would have the additional benefit of cooling the buildings by shading them.

We can also use less energy. Many appliances and processes can be redesigned to be more efficient. On a personal level, each of us can drive less or use a more fuel-efficient car; we can use fewer lights and instead depend more on natural lighting. Shade trees not only reduce our need for air conditioning, they also sequester carbon. We can eat less, and we can focus on foods that are grown locally and which do not need energy-intensive transportation or processing.

There are many ways to construct buildings such that they need less energy; these are **green building techniques**. In colder areas, buildings can be designed such that their windows provide both light and warmth by being placed on the south-facing side of the building, whereas in warmer areas windows only admit light if placed on the north-facing side. Ventilation can be designed such that natural air flow keeps a building cool. Various materials absorb heat and help to warm a building in cold areas, and other materials reflect heat away and keep buildings cooler in hot areas.

Reducing the amount of packaging is a major goal for many conscientious people now. Much of packaging is unnecessary, and if it is paper based, it may end up in a landfill and decompose to methane. If it is plastic packaging, it will never decompose, remaining in the landfill forever. Recycling and reusing materials also save energy costs.

Biofuels

An especially popular concept now is that of using **biofuels**. Biofuels are made from plant products and the concept is that the carbon dioxide produced as the fuel is burned is merely replacing the carbon dioxide that the photosynthetic plant had recently taken from the atmosphere. Plant material can be converted to a form appropriate for fuel by several methods. At present, corn seeds are fermented to ethanol (ethyl alcohol) and this is mixed with gasoline, usually in a ratio of one part ethanol to nine parts gasoline; this is called **gasohol**. Unfortunately, this corn seed is the same as that used as a food for both humans and livestock, so as part of a crop is diverted to biofuel, the price of food increases. Corn is used because it is abundant and cheap, but any other grain such as wheat or rice could be fermented, as can be the sugar from sugarcane, but those too raise the price of important foods. Research is underway to develop methods that use stalks, leaves, or roots of crop plants—parts that are not now used for food—as a way to avoid the need to choose between fuel and food (FIGURE 11.16). The main problem is that grains contain starch that is easily fermented whereas the rest of a plant consists mostly of cellulose, which can be fermented only slowly. Plant cells that are lignified are broken down even more slowly. Alternatively, plants can be simply dried and then burned as fuel in factories or generators that do not require liquid fuel. Such plants typically burn more cleanly than coal and thus produce fewer pollutants, but they do not burn as cleanly as natural gas. Considerable research is investigating the use of algae as biofuel, especially algae that store oils rather than carbohydrates. It is hoped that the oils can be more easily extracted and burned directly without the need for fermentation.

FIGURE 11.16. To obtain sugar from this sugarcane, the entire shoot is cut near soil level, then pressed between heavy rollers to squeeze the sugary juice out. All the rest of the plant is then discarded, sometimes by burning it, sometimes by chopping it and feeding it to livestock. Almost all the residue of these plants consists of cellulose; if a method could be found to economically depolymerize the cellulose to glucose, then virtually the entire plant could be fermented to ethanol and used as a biofuel.

Biofuels appear to be carbon neutral, but in reality they are not. We must consider the energy needed to produce the biofuels. All crop plants require energy to till the soil, sow and tend the plants, then to harvest the plants. If the plants are harvested from nature rather than being cultivated, that too will require energy. Algae would have to be cultured in extensive ponds and tended such that unwanted types of algae do not outcompete the desired species. After harvest, plant or algal material must be converted into a form that can be used, and then it must be distributed. All of these steps require energy and carbon that the plant itself is not producing by photosynthesis. In addition, all of these alternatives require large tracts of land, and we would face the choice of either using farm land to produce fuel rather than food, or we would have to clear natural areas, destroying more habitat and wildlife.

Important Terms

acid rain
albedo
alternative energy source
anthropogenic
aquifer
biofuel
carbon capture and sequestration
carbon neutral
carbon offsets
clear cutting
coccolithophorid
fossil fuel
gasohol
glacial age
green building techniques
greenhouse effect
growing season
ice age
methane clathrate
selective cutting
zero population growth

Concepts

- The temperature of Earth's surface changes until outgoing infrared radiation matches incoming solar energy.
- The amount of infrared radiation trapped by Earth's atmosphere is governed by the presence of greenhouse gases.
- Most greenhouse gases have both natural and human sources; we have greatest control over carbon dioxide and methane.
- Using carbon-based fuels such as coal and oil adds carbon dioxide to the atmosphere, leading to global warming and climate change.
- Carbon capture and sequestration by woody plants and certain algae remove carbon dioxide from the atmosphere.
- Snow and ice have a high albedo (are very reflective and absorb little energy) whereas soil and vegetation have a low albedo.
- As individuals, we can reduce the amounts of greenhouse gases by altering our behavior.

Agriculture and the Biosphere

Prior to the development of agriculture, our ancestors were hunters and gatherers. They obtained food by hunting animals (a good source of protein and lipids) and by gathering fruits, vegetables, seeds, and leaves (sources of carbohydrates, vitamins, and energy). We humans are **omnivores**, eating both plants and animals as food. And we are the only animal that practices agriculture: some animals such as squirrels and birds stockpile extra seeds they have gathered, but no other animal besides us sows and tends crops. Being an omnivorous hunter/gatherer is a successful way of surviving for many animals, just as is **herbivory** (eating only plants) and **carnivory** (eating only animals). Consequently we are not certain why agriculture developed at all. Why did people begin going through the trouble of saving some of their food through winter so that they could plant it in the next springtime? Why did they give themselves the extra burden of digging holes in the ground and dropping in perfectly edible seeds, tubers, bulbs, or rhizomes, then covering them with dirt? As those first crops grew to maturity, people had to defend them, not only against insects, birds, rabbits, and mice, but also against neighboring tribes. Obviously, agriculture could not develop until our ancestors had evolved to the point of having foresight and long-term planning.

Several theories have been proposed as to why people began to practice agriculture. One postulates that the climate became drier, causing suitable habitats to shrink. People were forced to live closer together in the small areas that still had food plants

This wheat field in springtime is an example of how beautiful farmlands can be, yet compare it with the photograph of an oil refinery at the beginning of Chapter 11. The refinery required the clearing of about one square mile of land, but this one wheat field is much larger than that, extending as far as we can see, and all native plants, animals, and other organisms were cleared away to cultivate one species, wheat. The oil refinery gives off chemicals that pollute air and water, but this wheat field is probably fertilized and sprayed with herbicides and insecticides, which also pollute air and water. We must think carefully when we assess the environmental costs of various activities.

(this is the oasis theory). With this concentration, it would have been necessary to cultivate plants intensively instead of merely gathering food from widely scattered wild plants. There is little evidence to support this theory, and a more likely outcome would probably have been warfare.

A second theory is that in areas where cereal grains grew well naturally, humans could settle in one spot and still have enough to eat; they did not need to move from place to place. As they foraged for food, they would have accidentally dropped seeds in areas near their camps, inadvertently sowing an area with extra seed, which in turn would give greater abundance. The people would have started to protect these areas and tend them.

A third theory is that even without climate change, human populations became dense enough that more food was needed to prevent starvation. But recent history shows that population densities tend to be low in nonagricultural societies. For example, at the time Lewis and Clark explored the Missouri River area and the Pacific Northwest, many Native Americans had comfortable lives as hunter/gatherers: the picking of wild berries, bulbs, roots, and tubers supplemented by game and fish provided a balanced, adequate nutrition for many people.

Several things are certain about the origin of agriculture: it developed independently several times in widely scattered areas, beginning about 11,000 years ago and involving diverse plants. The earliest agricultural settlements are found in the area known as the **Fertile Crescent**, an arc-shaped region that stretches from the eastern end of the Mediterranean eastward through the valleys of the Tigris River and Euphrates River (FIGURE 12.1). The modern

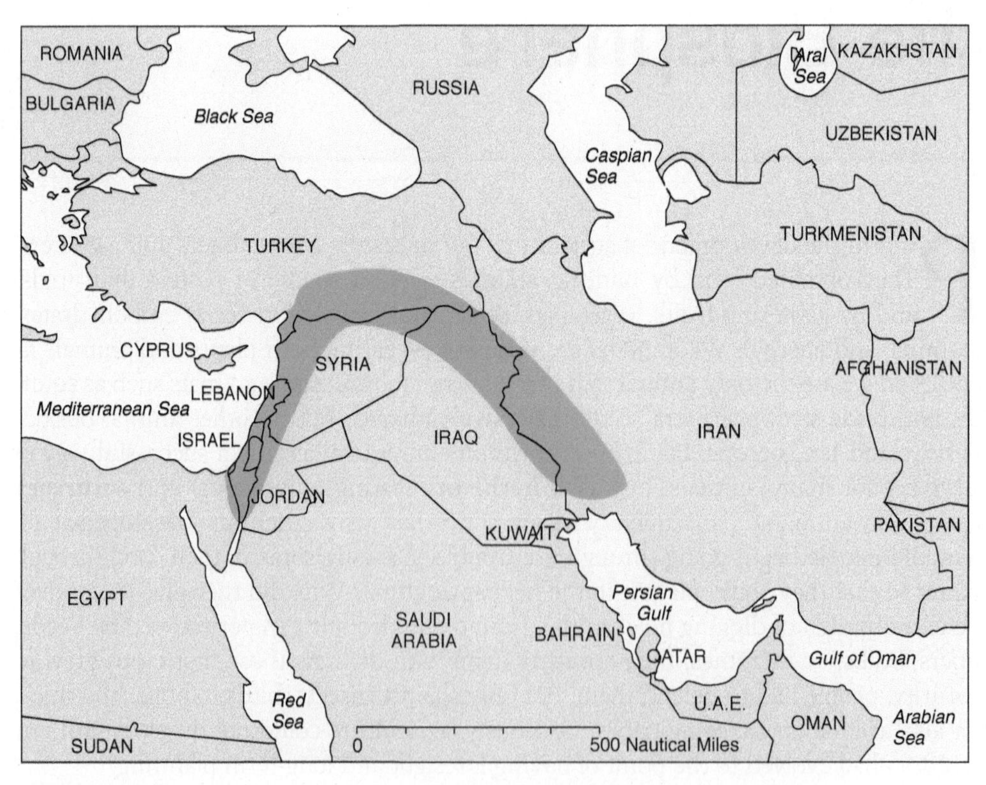

FIGURE 12.1. The earliest cities and agricultural area were built in the Fertile Crescent. At the time, the area was not as arid as it is now.

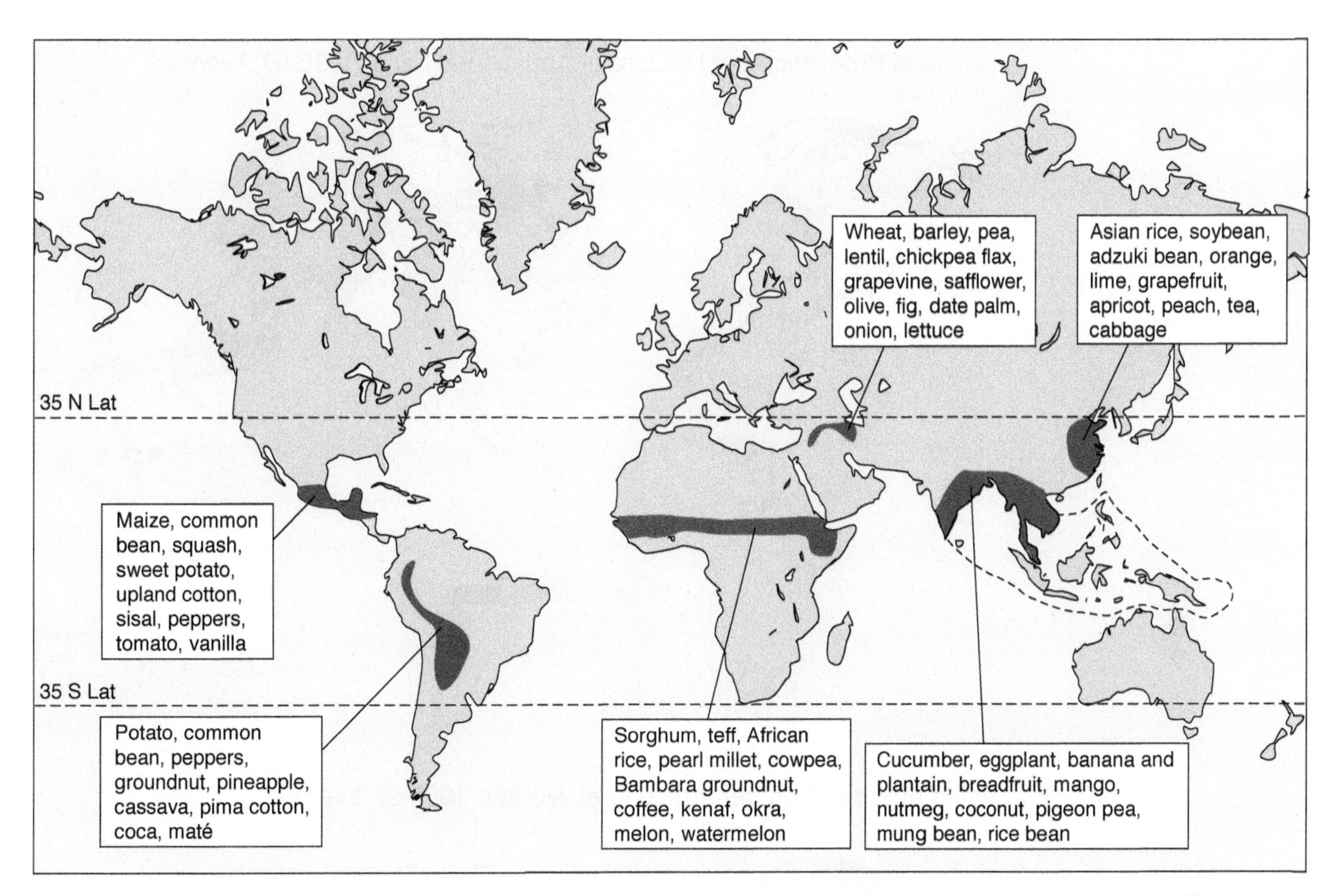

FIGURE 12.2. Crop plants have been domesticated in at least six parts of the world, all of them tropical regions. Different crops were domesticated at different times, but all occurred after the end of the last glacial period. (P. Gepts, 2001. Origins of plant agriculture and major crop plants. In M. K. Tolba, ed., *Our Fragile World: Challenges and Opportunities for Sustainable Development*. Oxford, UK: EOLSS Publishers, pp. 629–637.)

countries of Syria, Iraq, and Iran are in this area. This area has the remains of the earliest towns and cities known to have ever existed: Uruk, Ur, Sumer, and Babylon were large cities surrounded by extensive irrigation systems that maintained crops. The timing of the earliest cities coincides with the retreat of the glaciers in northern Eurasia and North America; agriculture, cities, and civilization began at the same time that our current interglacial period, the Holocene Interglacial, started. This agriculture quickly spread from Sumer southwestward to the Nile River Valley and became the basis for the Egyptian dynasties.

Agriculture also developed independently a bit later in China, based on the cultivation of rice. In the Americas cultivation of corn (maize; *Zea mays*), tomatoes, peppers, and squash began in Mexico about 5000 years ago, and potatoes were domesticated in South America at about the same time (FIGURE 12.2).

Another thing that is known for certain about agriculture is that it has dramatically altered the biosphere (FIGURE 12.3). A few species such as wheat, rice, corn, soybeans, potatoes, and sugarcane now occupy a large percentage of the land surface area. Farming and herding are so extensive that few totally natural areas are left anywhere on land. Grasslands, forests, and wetlands have

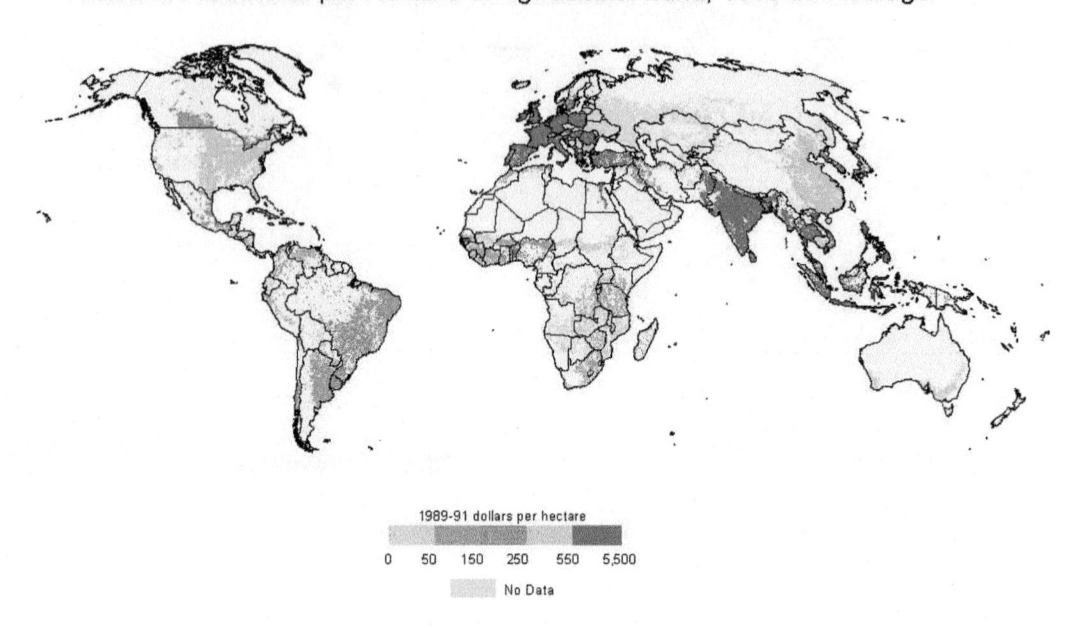

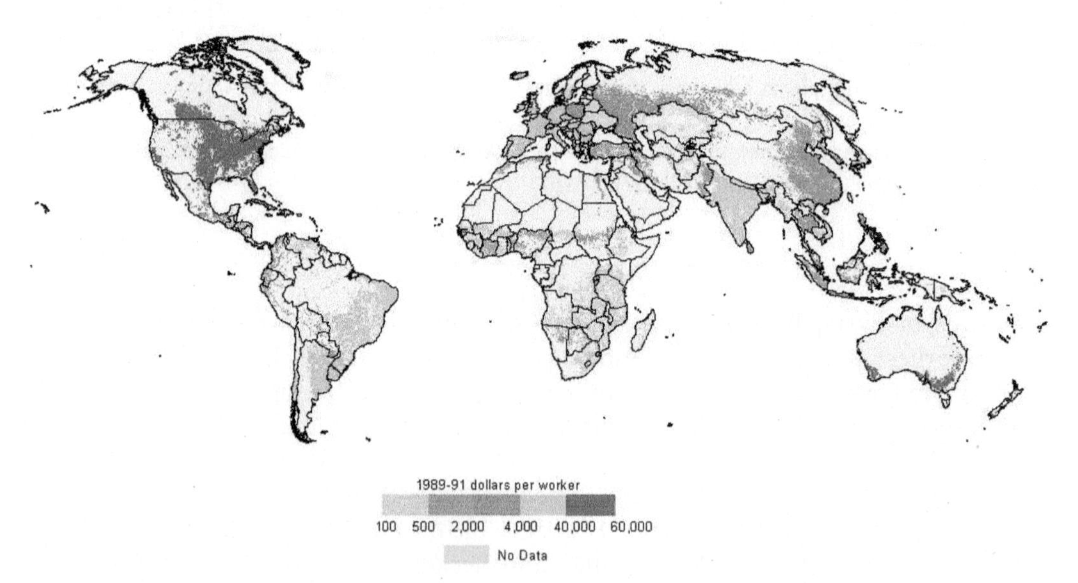

FIGURE 12.3. Agriculture has profoundly affected not only biogeography (the distribution of plants and animals on Earth's surface), but many other aspects of biology as well. Almost no part of the world is untouched by agriculture; even marginal areas with little vegetation are often used as pastures for goats. (Data from Food and Agriculture Organization of the United Nations, 1997 and 1999, "FAOSTAT Statistical Databases," available at http://apps.fao.org; and World Bank, 2000, *World Development Indicators*, Washington, DC: World Bank, cited in S. Wood, K. Sebastian, and S. J. Scherr, 2000. *Pilot Analysis of Global Ecosystems: Agroecosystems*. Washington, DC: International Food Policy Research Institute and World Resources Institute.)

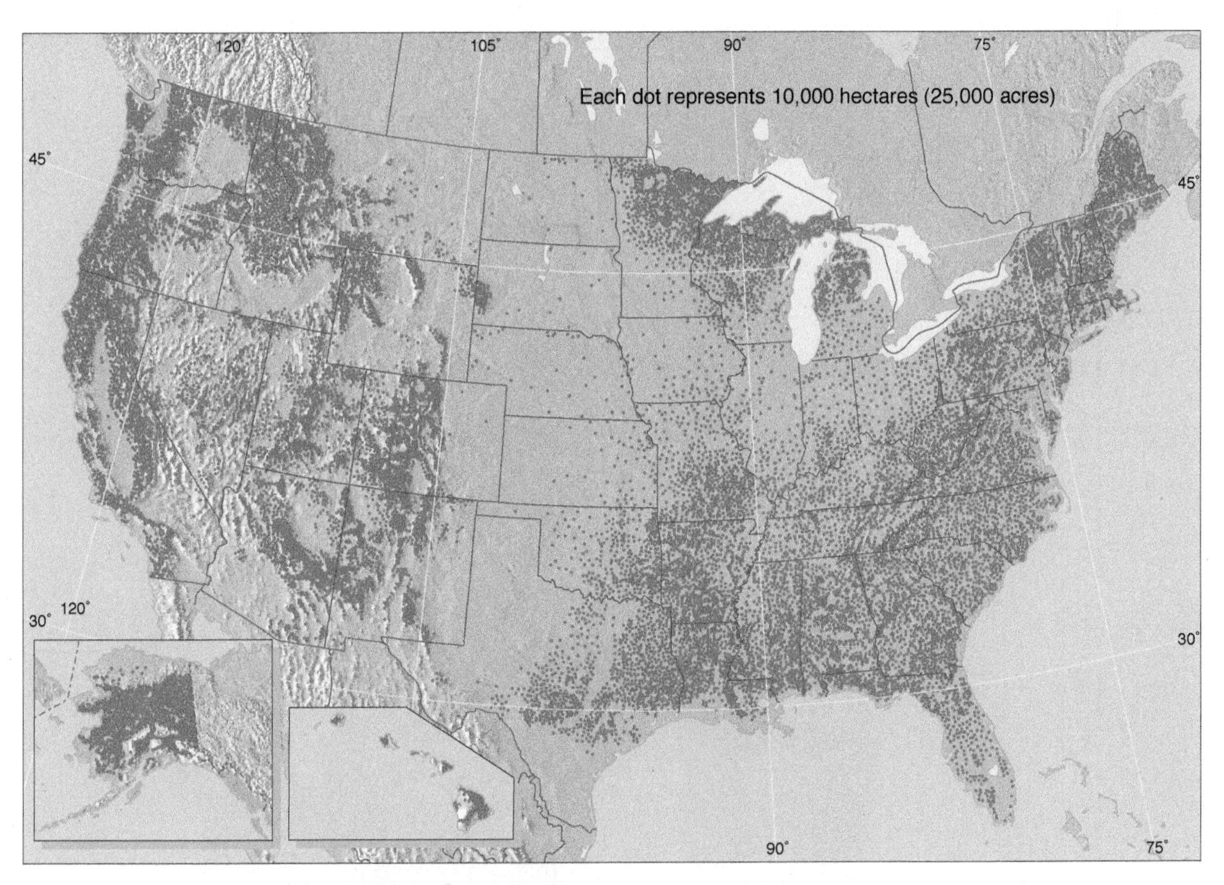

FIGURE 12.4. This map indicates that forest land in the United States is extensive, but forests are not nearly as widespread as they were before European settlers arrived. Also, the nature of the forests has changed, with much of the current forests being relatively young trees that have grown up since the last time the forest was logged. Almost every forest indicated on this map has been altered because it was logged for lumber or paper; others were cleared for farming but the farms have since been abandoned and forests are reestablishing themselves. The diversity of species in many of our present forests is not as great as that in the original forests, and few ancient trees remain.

been cleared of all natural vegetation and converted to agriculture (FIGURE 12.4). Even steep mountain sides have been terraced into fields, and entire rivers have been diverted for irrigation projects. Just as the presence of our crops allows us to live in many parts of the world, it also allows insects and other herbivores to expand their ranges (FIGURE 12.5). Water polluted with fertilizers, herbicides, and pesticides washes off of farms and into rivers; the fertilizers stimulate the growth of some algae and plants, the herbicides harm others. Pesticides affect the animal life of the waterways. And the large amount of food produced by modern agriculture has allowed the human population to grow to 6.8 billion people (and increasing), making us so numerous that we consume huge amounts of resources. Without agriculture, there would be fewer people and much less ecological damage to the biosphere.

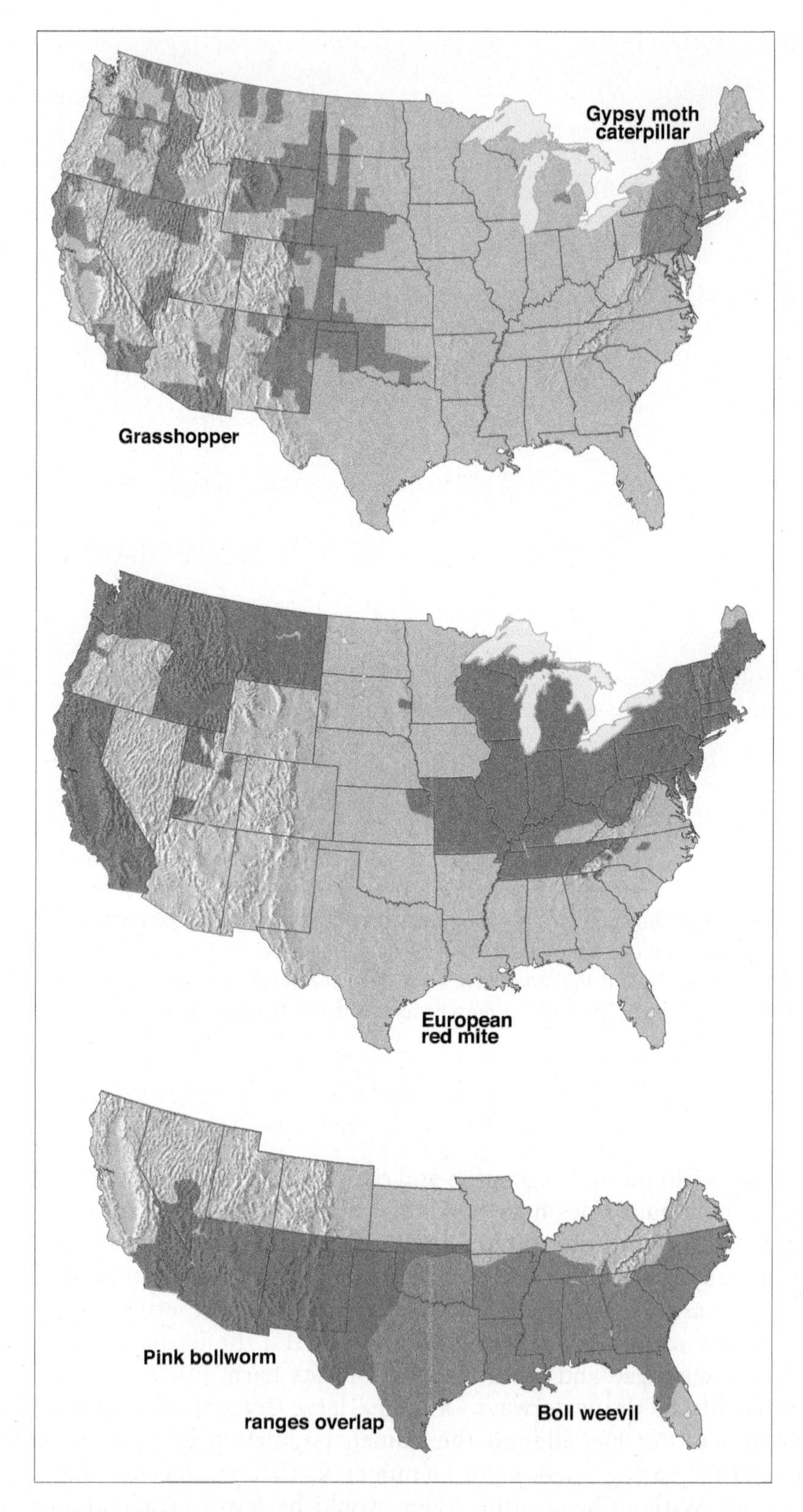

FIGURE 12.5. Our crop plants are preyed upon by a variety of pests. Many weeds grow only in land disturbed by agriculture, and many animals and fungi specialize on crop plants. As we spread agriculture across an area, we alter the habitat such that these pests can expand their ranges and flourish in new areas.

Early Agriculture

Prehistoric Agriculture

Prehistoric agriculture encompasses the time between the beginning of agriculture and the development of writing and historical documentation. Because there are no written records of this period, all evidence is archaeological. Scientists study the garbage dumps that always occur near human settlements. These are examined for bones to see what types of animals and fish the people ate, and inedible plant remnants like seeds, husks, pits, rinds, and shells reveal the vegetable part of the diet. Special attention is given to features that are typical of cultivated plants but not wild ones. For example, some wild cereal grains are difficult to collect because the grains fall off at the slightest touch (they are said to "shatter"), but in cultivated varieties the grains are attached more firmly and so the plants can be cut and carried to a threshing area without losing the valuable grains. An especially clear example is corn (*Zea mays*); it does not occur in nature at all, it grows only in cultivation. Its closest relative is a grass called teosinte, which looks quite different (FIGURE 12.6). Archeological sites that have remnants of corn are definitely those of agricultural societies.

Prehistoric peoples were extremely successful at domesticating plants and animals. Eight fundamentally important crops were in cultivation as early as 9500 BCE: barley and two types of wheat, several protein-rich legumes (peas, lentils, chick peas [garbanzo beans], bitter vetch), and flax. Figs and olives were domesticated shortly afterward.

The Effects of Early Historical Agriculture

As agriculture developed, it had profound impacts on human civilization. It was responsible for the development of cities, writing, surveying, mathematics, exploration,

FIGURE 12.6. Grasses called teosinte (several species in the genus *Zea*) are the closest wild relatives of corn (*Zea mays*). Teosintes differ so greatly from corn that archeologists and anthropologists can easily tell which ancient peoples were using corn and which were using teosinte. (Courtesy of N.R. Fuller/National Science Foundation)

FIGURE 12.7. Food plants were so important to ancient peoples that they were often depicted with a religious significance in art. This particular image occurs frequently in Sumerian and Babylonian art. The person is either a priest or a king and is sometimes accompanied by winged figures. Can you recognize the plant? In earlier depictions it was recognizable as a date palm, but through the years this important food tree became a symbol for life itself and gradually became very highly stylized with mystical symbols surrounding it. Notice that the man is holding a cone-like object in one hand and a pail in the other. Date palms grow as carpellate ("female") plants and staminate ("male") plants. To ensure a good crop, pollen can be collected in a pail and then applied to the carpellate plants with a brush to increase the rate of pollination and fruit production. This scene appeared in Sumerian art before 2000 BCE, indicating that the very earliest farming societies understood this sophisticated aspect of botany.

and warfare, among other things. Because so much food could be grown in such a small area, large numbers of people lived close together, permanently, and the first cities of Ur, Sumer, and Babylon came into being. People had to learn to cooperate not only in farming and herding, but also in construction of irrigation systems and defensive walls to protect stockpiled food. Such construction projects in turn required organization, the development of government, of bureaucracy, of record keeping. When food was shipped or stored in containers, a system was needed to identify the contents of each container, and a system of identifying marks was developed. These marks gradually became more elaborate and evolved into the first writing system, cuneiform. The ancient Sumerians were consummate record keepers, and they wrote by pressing an angle-shape stylus into a soft, wet clay tablet. Once the tablet was dry, the writing was permanent. Archaeologists have collected thousands of these clay tablets, almost all of them being records of harvests and the contents of storerooms (FIGURE 12.7).

Several tablets however are letters, poems, or stories, the most important of which is a poem called *The Epic of Gilgamesh*. It records a legend from the time of the earliest cities in the Fertile Crescent, and tells the adventures of King Gilgamesh and his companion Enkidu. This is significant for us because one of the adventures involved collecting the beautiful wood of a cedar tree so that it could be used to build the doors of a temple: *The Epic of Gilgamesh* is the oldest poem in the world and one of its topics is the concept of collecting exotic plants.

It is easy for us to underappreciate the importance of searching for exotic plants. Until recently, almost everything that people used was made from plant or animal material, plus a bit of metal or mineral. Lumber, cloth, food, drugs, almost everything came from plants, and any new plant might provide some benefit not yet known. After *The Epic of Gilgamesh*, the next major recorded plant collecting expedition was that of the Egyptian Pharaoh Hatshepsut who ruled Egypt from 1479 to 1458 BCE. One of her most significant accomplishments (and which was carved across the entire facade of her mortuary complex) was the successful organization of a large plant collecting expedition of five ships, each with over 200 men, to the Land of Punt (probably Eritrea or Sudan; FIGURE 12.8). The expedition returned with myrrh (tree resin used for incense; FIGURE 12.9) and with living trees of frankincense (two of the gifts mentioned in the Bible). Each tree was potted in a basket, and the trees were successfully transplanted in Egypt. This is the earliest description of an expedition to transplant living trees from one country to another.

The Egyptians cultivated fields along the broad bottomlands of the Nile River. Each spring, floods brought a fresh layer of nutrient-rich silt that kept the field fertile but that also destroyed landmarks. Egyptians invented techniques of surveying to reestablish farm boundaries each year, and along with this, they developed sophisticated mathematics that became the basis for geometry.

The search for new plants and attempts to bring them into cultivation continued to be critical aspects of agriculture. Every major civilization explored for useful plants,

FIGURE 12.8. This is part of the wall carvings depicting the expedition of Queen Hatshepsut to the land of Punt to collect plants. This expedition was so important it was the main carving on her gigantic tomb complex. (© The Art Archive/Alamy Images)

FIGURE 12.9. This is myrrh, the resin of a small desert tree *Commiphora myrrha*. Just as today, early people used myrrh as fragrant incense in religious ceremonies.

and the Romans moved large numbers of plants between the various farming areas of their extensive empire. Plants that could not be cultivated anywhere within the empire had to be imported from Egypt, Persia, India, and even East Asia. All the exotic, expensive spices from the East arrived at Rome after dangerous caravan journeys of thousands of miles.

In the early 1400s, Prince Henry the Navigator of Portugal sent out numerous ships and expeditions, each exploring farther down the coast of Africa, searching for plants (and also searching for a mythical river that was believed to run from one coast of Africa to the other). This series of explorations ultimately resulted in European ships sailing around the southern tip of Africa and proceeding to the Far East. It was quickly realized that this was a faster, less expensive way for Europeans to obtain Asian spices rather than having to rely on overland caravans through the Middle East. It is difficult to imagine now, but Asian spices such as black pepper, cardamom, cinnamon, cloves, ginger, and nutmeg were typically so valuable that, ounce for ounce, they were more costly than gold. Sailing for months through dangerous seas between southern Africa and Antarctica was worth the risk and cost if a ship returned to Europe filled with spices.

When Christopher Columbus sailed west from Spain in 1492, he was not trying to discover America or to prove the world is round. He was convinced he would discover a shorter trading route for obtaining spices from East Asia. Ancient Greek mathematicians and philosophers had realized that Earth is a sphere like the sun and moon, but they underestimated its circumference by thousands of miles. Most people in the Middle Ages had forgotten the Greek discovery, and assumed not only that Earth was flat but also that it was possible to fall off its edge. Columbus trusted both the Greek concept and the underestimated size, and concluded that he could sail to Asia very quickly by sailing directly west; he would not have to take the dangerous route around southern Africa. Just about the time his three ships were expecting to reach China, land came into view: several Caribbean islands and Central America. Columbus discovered America while looking for plants! And amazingly, he was disappointed when he had to go back to Spain with stories of new lands and new peoples, but no spices. Columbus sailed to America four times, always convinced that it really was part of Asia, never realizing that he was still thousands of miles from his goal.

But other sailors and mapmakers realized that the Americas were new continents, and even before Columbus died, the Americas were being explored and exploited. Ships were carrying gold, silver, and plants to Europe, while other plants were brought to the New World for cultivation. Before Columbus's voyages, the Americas had been isolated from Europe and Asia by the thousands of miles of Pacific and Atlantic Oceans, but after Columbus there was heavy traffic of plants, animals, diseases, and culture in both directions. This has been called the "Columbian Exchange" and it has profoundly affected the biogeography of all parts of Earth.

After the initial exploration of the Americas, it was realized that it would be necessary to sail around these two continents to get to the Asian spices. In 1519, Ferdinand Magellan sailed out of Spain, southwest around the southern tip of South America, and on to the Philippine Islands. Although he was killed, his crew continued the trip around Africa and back to Spain, being the first people to circumnavigate Earth. Once again, plants were the goal: Magellan wanted to sail directly from Peru (a Spanish colony on the west side of South America) to the Spice Islands (the Moluccas), thereby proving that the islands and their spices belonged to Spain and not to Portugal.

Many other expeditions were sent out to search for plants. One of the most famous within the United States is that of Meriwether Lewis and William Clark from 1804 to 1806. President Thomas Jefferson sent them to survey the newly acquired lands of the Louisiana Purchase, and in addition to mapping, they also were to collect specimens of plants, animals, and minerals. The newly acquired lands were immediately made available to homesteaders, so the great American migration westward was initiated by farmers and ranchers, and later aided by the discovery of gold in California.

Other expeditions explored other parts of the world, and many teams scoured China, India, and Indonesia for plants that might be of value. Perhaps more than any other country, Great Britain has been responsible for altering biogeography on a global scale: they often moved plants from one area to another so they could be cultivated in areas that were more accessible, had better growing conditions, or cheaper labor. For example, the *HMS Bounty* had picked up living plants of breadfruit trees in Tahiti with the sole purpose of transporting them to the West Indies, where they would be planted to establish breadfruit orchards to feed the slaves there. The expedition was cut short by a mutiny: the sailors were running out of fresh water to drink because it was being saved to keep the plants alive. Other plants that were moved around the world like this are rubber tree plants, sugarcane, coffee, and *Cinchona* (the source of the antimalaria drug quinine), as well as wheat, rice, corn, and many other familiar food crops.

Even today, numerous expeditions are in progress in any given year, searching various parts of the world for new species of plants or for new varieties of known plants. When Americans find new plant species of potential interest, they are shipped to one of several Plant Introduction Stations run by the U.S. Department of Agriculture. The plants are checked for disease, then propagated and studied to evaluate their potential as crop plants for use in providing fruit, wood, chemicals, pharmaceutical, or as ornamentals. Most new specimens are also checked to see if they contain phytochemicals that might be effective against cancer, HIV, and other diseases.

Types of Modern Agriculture

At the beginning of the twenty-first century, agriculture exists in many forms. Traditional farms with fields of herbaceous crops are still widespread, and in many respects these are identical to the farms used by Sumerians 11,000 years ago. They vary in size from small plots less than a hectare (about the size of a city block) to larger ones of hundreds of hectares. Most traditional farms are diversified, that is, each is divided into several fields, each field having a different crop (FIGURE 12.10). For example, a single farm might have a field of wheat, another of alfalfa, and another of potatoes. Crop diversification provides some safety in case one of the crops fails or its price falls to an unprofitable level before harvest. This reduces risk of total failure of a farm, but because many crops require specialized machines, it increases the cost of farming. Many large farms, often run as corporations, specialize in just a single crop, such as soybeans.

The types of crops grown on traditional farms are diverse. Most produce foods such as the cereals wheat, rice, corn, oats, and barley, or legumes like peas, beans, lentils, soybeans, and peanuts, or various vegetables like carrots, lettuce, and onions. Others produce forage for domesticated animals such as alfalfa or grasses for cattle, oats for horses, and corn for chickens and swine (FIGURE 12.11). Other crops include herbs and spices that provide special flavors such as mustard, mint, ginger, cilantro, and so on, or medicinal plants from which drugs are extracted. Many traditional farms produce bulbs, tubers, and rhizomes of ornamental plants like tulips, lilies, and daffodils; others produce flowers for the cut-flower market, used in flower arrangements. Some farms supply seedlings and container-grown plants for nurseries and landscapers. Two of our most important fibers for cloth—cotton and flax—are grown on traditional farms.

FIGURE 12.10. Many modern farms are similar to the very earliest ones. This farm has several crops that are cultivated on cleared land and which rely on natural rainfall rather than irrigation. To protect the topsoil from erosion, small dikes have been built so that each field is more or less level (the dikes follow the contours of the land so that all parts of each dike are at the same level). This farmer now uses tractors and other petroleum-powered machinery, but the first settlers of this farm relied on oxen, mules, or horses to pull plows and other implements. (Courtesy of Tim McCabe/NRCS USDA.)

FIGURE 12.11. This farm specializes in food for animals (forage crops). The hay in this field has been cut and compressed into bales that can be hauled, stored, and then used more easily than loose hay.

Tree farms specialize in trees, of course. In the past, the main product was Christmas trees, but recently many acres have been planted to fast growing trees such as poplars that can be harvested for their wood. The trees are cut just above the roots then the trunks and branches are either chopped into wood chips or are digested to obtain fibers for paper (**FIGURE 12.12**).

Orchards are farms that specialize in fruit or nut trees. Important fruit crops are apples, cherries, peaches, citrus, and olives. Nut crops are pecans, walnuts, and almonds.

FIGURE 12.12. This is a pulpwood farm in which fast-growing trees have been planted very close together, which encourages them to grow straight. Although they are still young and not very large, they are almost ready to be harvested. They will be chipped or shredded, then treated to dissolve their middle lamellas so that the cells separate from each other into a slurry of xylem fibers.

Ranches provide rangeland for grazing cattle and horses, among other animals. Some ranches now specialize in newer, more exotic animals such as buffalo, ostrich, antelope, and zebra. In general, rangeland needs little cultivation; the animals eat grasses and herbs that are either perennial or reseed themselves. Smaller ranches are fenced, often with electric fences for small areas, or with barbed wire for extensive ranches. In very dry grassland areas, vegetation is so sparse and ranches cover such large areas that fencing is not practical; this is open rangeland.

Forestry is a branch of agriculture that manages forests, extensive regions much larger than tree farms. Typically, forestry involves harvesting large trees to be cut into lumber, plywood, and other building material, or processed into wood pulp for making paper. In the western United States and Canada, the most extensive forests are dominated by conifers, and trees are harvested for lumber or paper (see Figure 12.4). Midwestern and eastern forests have more broadleaf trees and provide woods that are prized for their beauty: maple, walnut, and cherry. Most of our extremely valuable artisan woods such as teak, ebony, mahogany, and zebrawood come from tropical forests.

Growing plants **hydroponically** is a newer form of agriculture. It involves cultivating plants with their roots in fertilizer solutions rather than in soil. In some cases, each plant has its roots in its own container of solution, but in others cases hundreds of plants are suspended above one large tank (FIGURE 12.13). It is even possible to have the roots in air, being periodically sprayed with fertilizer solution without ever letting them dry out. Hydroponic culture requires expensive pumps and other machinery, and typically is done in a greenhouse. It is so costly that it is used only for high value crops such as fresh tomatoes.

Modern farming can also be classified as being either organic or conventional farming. In **organic farming**, virtually no synthetic chemicals can be used. For example, synthetic fertilizers, insecticides, fungicides, antibiotics, and artificial hormones cannot be used. This applies not only to a particular crop, but also to the land and machinery: for a crop to be certified as organic, it must have been cultivated on land that had not been treated with any artificial chemicals for at least 3 years (longer in certain countries). Composted organic material can be used as fertilizer with the exception that human waste and sewer sludge cannot be used. Implements that come in contact with the crop also cannot be contaminated with any such artificial chemicals, and if they have ever been used for a conventional crop, the implements must be thoroughly cleaned. The same is true of any storage or transporting containers. For almost all of human history, all agriculture was organic (with the exception that human waste was occasionally used). But in the last century or so, people discovered how to manufacture artificial fertilizers, and then later pesticides and herbicides. Even though many of these are expensive, they improve crop yield so much that they became widely used. Any agriculture that uses any of these artificially produced chemicals is considered **conventional farming**.

FIGURE 12.13. These plants are being grown hydroponically: their roots are suspended in a nutrient solution rather than soil. Fertilizers can be added to the water in the solution to precisely adjust the growth rates of the plants. An acre of these containers is much more expensive than an acre of soil; hydroponics are used only for extremely valuable crops. (© Khoo Si Lin/ ShutterStock, Inc.)

The chemicals used in conventional farming improve crop yield so much that typically conventional foods are less expensive than organic foods. An important question is then "Is there a benefit to eating organic foods that outweighs their cost?" The answer to this question is very easy with regard to insecticides; many of the early synthetic insecticides were poisonous to people as well as insects (see below). Modern ones seem to be less toxic, at least to people, but still many of them target the nerve cells of insects, so if they contaminate our food, those chemicals could find their way to our nerves cells and brain. Beyond this, the use of artificial insecticides means that we are putting all the insects of the environment at risk just so that we can save some money on food. Synthetic fertilizers are less clear cut, it is rare for them to cause any obvious toxicity in people (spinach becomes dangerous to babies if it is highly overfertilized with nitrate fertilizer). But there is always runoff of excess fertilizers from fields, runoff that pollutes streams and groundwater. It may be that there are dangers to conventional farming that will not be discovered for many more years: there have been many instances in which the harmful side effects of certain medicines or processes was discovered only many years after the medicines were first introduced. A particularly convincing argument in favor of organic farming is that foods go into our bodies, and if they contain anything harmful, even in trace amounts, those harmful substances are inside us. Our intestines are extremely effective at transporting things out of our intestines and into our blood, but our kidneys are not as good at excreting exotic substances. Of all the things we can spend our money on, safe food may be the best. On the other hand, if your diet consists mostly of highly processed foods, fried foods, high-fat foods, and carbonated drinks, then organic food will probably not improve your health significantly.

Agricultural Processes

All forms of modern agriculture require four fundamental processes: preparing the land, tending the plants, harvesting the crop, and maintaining land until the next planting season. Many of these impact noncrop plants and animals, and even affect the entire biosphere. A great deal of energy is required to run tractors, irrigation pumps, and harvesters, and to transport crops to food processing factories. Other processes involve chemicals that are carried by water or wind away from the agricultural area, causing impacts miles away.

Preparing the Land

Preparation of the land varies depending on whether the land has already been farmed or if undisturbed natural areas are being "developed." If it is natural land, then existing vegetation and animals are removed with chainsaws, bulldozers, machetes, or fire. Typically, all natural vegetation is removed, even stumps and roots of trees. Animals that depend on those plants either die or flee. Once cleared, the land is plowed: basically the top 10 to 15 cm (4 to 6 inches) of topsoil are turned upside down (FIGURE 12.14). Plowing loosens the soil, buries weeds and other vegetation, and breaks up remaining large pieces of roots or debris from previous crops. It also brings deep-lying seeds to the surface where they geminate and can then be eliminated. Plowing is often done several weeks or months before planting season, giving weed seeds a chance to germinate. The ground may then be disked, which turns over only the uppermost inch or two of soil, killing young weeds with shallow roots (FIGURE 12.15).

Farm crops need water, and farms are either irrigated or they rely on natural precipitation. In rainy areas, such as the eastern United States, natural rainfall is usually sufficient (see Figure 12.10). In the Midwest and parts of the western United States, rainfall is less but is adequate for certain crops such as dryland wheat. In much of the western and midwestern United States, rainfall is not at all sufficient and farms must be irrigated artificially. When natural land is converted to irrigated farmland,

FIGURE 12.14. This land has just been plowed, a process that cuts soil at a depth of several centimeters (several inches) and then turns it upside down. This smothers weeds at the surface and brings buried weed seed to light causing them to germinate so that they can be killed before the crop is planted; it also loosens the soil matrix such that the crop's roots penetrate the soil more easily and extensively. Turning over large amounts of heavy soil requires a great deal of energy, and much diesel or other petroleum-based fuels is needed to plow even a small field.

FIGURE 12.15. Several farm implements (disks, harrows) are used to eliminate weeds by disturbing only the uppermost inch or two of soil. This breaks the roots of young weeds, and it also causes the upper layers of soil to dry out, further harming the weeds. This can be used only before the crop is planted or if the crop is planted in rows separated far enough that the tractor's tires can go between the rows without damaging the crop plants. (Courtesy of Maarten J. Chrispeels.)

an irrigation system must be installed. The land must usually be contoured into very gentle slopes or leveled into terraces. One of the newest irrigation systems is a center pivot overhead sprinkler system (FIGURES 12.16 and 12.17). This is a long irrigation pipe mounted on wheeled, mobile towers and fixed to a water supply at one end; sprinkler heads are placed several meters (several yards) apart along the pipe. The motorized wheels on the towers run such that the outermost tower moves fastest and the innermost slowest, and the system covers a circular field centered at the point where the water supply is attached. A single center pivot system can irrigate a very large field and requires almost no manual assistance.

Rill irrigation uses numerous, closely spaced narrow "ditches" (rills) only a few centimeters (an inch or two) wide to carry water between individual rows of plants (FIGURE 12.18). Rill irrigated fields must slope gently away from the water source so that water will flow down the rills slowly enough to not cause erosion yet rapidly enough to reach the bottom of the field. Water is brought to the top of the field by pipes or a larger ditch.

Terraced fields such as those used on steep slopes or for rice must be perfectly level and surrounded by dikes (see Figure 12.10). The fields are irrigated by flooding the entire terrace all at once to a depth of several centimeters. At times, the terraces are drained, either for the benefit of the rice or to allow the soil to dry out and become

FIGURE 12.16. Center pivot irrigation systems spray water onto a crop from overhead. This requires very little manpower, and often the machine is controlled by sensors and a computer that determine when the crop is dry enough to need irrigation. Fertilizers, insecticides, and other chemicals can be added to the water so that they are applied easily to the crop. (Courtesy of Gene Alexander/NRCS USDA.)

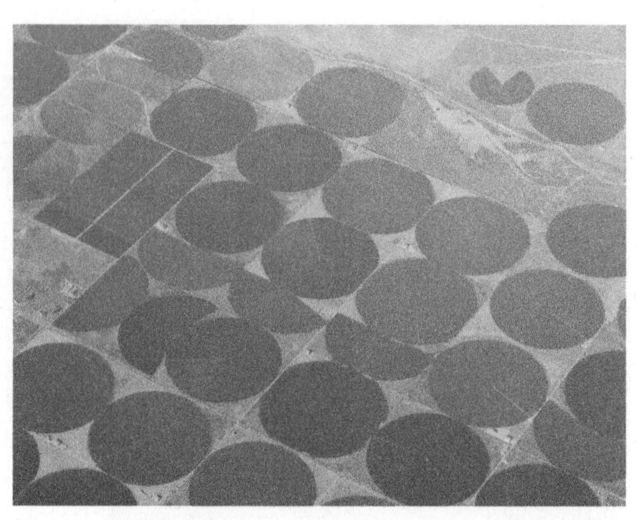

FIGURE 12.17. Center pivot sprinklers as viewed from the air. A circular field might be planted to a single crop (circles of uniform color), or smaller amounts of a crop can be planted in just a wedge-shaped area. The corners of the farm are often not cultivated, but they cannot be considered natural habitat for wildlife because they are so small and are surrounded by non-natural areas. (© Glenn Young/ShutterStock, Inc.)

FIGURE 12.18. If land slopes gently, then water can be distributed from a pipe or ditch at the upper end and allowed to run down the rills between the rows of crop plants. This type of irrigation requires much less energy than that needed to run the high-power sprinklers of a center pivot system, but as water runs down the rills, it causes at least some erosion, moving soil from the upper end of the field to the lower. Also, the water that reaches the bottom end of the field must be collected and drained away somehow, and that water will be contaminated with any fertilizers or agricultural chemicals that have been applied to the crop. (Courtesy of Maarten J. Chrispeels.)

firm enough for tractors and harvesters. Both terraced and rill irrigation need a system of ditches to remove used water from the field; these ditches drain into larger ones that carry the waste water away.

For many crops, plowing, disking, and setting up irrigation is sufficient preparation, but support wires and trellises are needed for vining crops. Grape vineyards are probably familiar to you; a field contains rows of posts 2 or 3 m (6 to 9 feet) high with several wires strung from post to post from one end of the field to the other (FIGURE 12.19). The grapevines attach to the wires with their tendrils and the branches grow horizontally, with the clusters of grapes being low enough to be picked easily. Hops (used in brewing beer) are very long vines that must grow vertically; they require trellises 5 or 6 m tall. Kiwi fruits also grow on vines, typically supported by wires high enough to allow workers to walk under the plants and harvest the fruits from below.

Tending the Crops

After the land is prepared, the crop must be planted and tended. Many annual herbaceous crops such as wheat, beans, and corn are planted as seeds (FIGURE 12.20). A machine called a seed drill makes about ten shallow furrows, drops seeds one by one into each furrow, then covers them as a tractor pulls it along. Potatoes are grown, not from seed but by planting pieces of tuber that have been cut into several pieces, each piece with at least one "eye" (an eye is an axillary

FIGURE 12.19. Grapes vines must be supported on wires or trellises such that the fruits do not lie on the ground where they would be more susceptible to fungi.

bud). Sugar cane and asparagus are started by planting pieces of rhizome, again each piece having an axillary bud or two.

For most crops, the distance between individual plants is important. If they are too close to each other, their roots do not develop fully and the plants do not grow as well as possible, but if they are too far apart, land is wasted. Spacing is especially important for valuable root crops such as carrots, beets, and sugar beets, and for these crops, workers manually thin the crops after the seedlings have grown to an appropriate size.

Weeds must be controlled in any crop either by physical or chemical methods. Physical weed control is done both before the crop is planted and while it is growing. Before planting, the ground is plowed or disked to kill all plants (see Figure 12.15); if the crop is one that is planted in rows (most crops are), the uppermost 2 or 3 cm of soil between rows is disturbed every few weeks with a harrow as the crop grows. This is deep enough to damage the roots of newly sprouted weeds but not deep enough to harm the more well-established roots of the crop plants.

Chemical control is done with **herbicides**, chemicals that kill plants. Almost all herbicides are **nonspecific** (also called **broad-spectrum** herbicides) and kill all plants, so they must be used before the crop seeds have germinated or by careful spraying between rows of plants. Nonspecific herbicides are also used to clear all vegetation away from buildings and roadways. Several herbicides are specific to monocots; for example 2,4-D (dichlorophenoxyacetic acid) kills broadleaf plants (dicots) but does not harm grasses, so it can be used in cereal crops (FIGURE 12.21). Because so many herbicides are nonspecific and can harm almost any plant they encounter, an important concern is to develop herbicides that breakdown quickly: if they are not taken up by a plant, they decompose before they can be washed into the groundwater and enter aquifers. For example, a widely used herbicide called atrazine is long lived if used on alkaline soils, and either rain water or irrigation water washes it off the fields and into streams or ground water. Although diluted, it still affects plants growing along the stream or at any spring where an aquifer comes to the surface.

FIGURE 12.20. These are wheat "seeds" (actually each is a whole fruit that contains a single seed). These are so smooth and uniform in size that they can be planted reliably with a machine. The seeds of some other crops are rough or irregular, but they can be coated with an inert material to make them uniform and smooth enough for machine planting.

FIGURE 12.21. A weed is defined as any plant that grows where it is not wanted. Weeds that grow slowly and die easily are not a problem in agriculture, but weeds that are hardy and grow vigorously can be serious threats to crops. Crop plants typically cannot grow as rapidly or as vigorously as weeds because the crops have been artificially selected to expend more of their energy and resources on fruits, seeds, or fibers rather than for fast growth.

Herbicides are classified as being either contact or systemic herbicides. A **contact herbicide** often kills only the parts of the plant that it is sprayed onto, whereas underground rhizomes and tubers may survive. **Systemic herbicides** are taken up by the plant and distributed throughout the body, some in phloem, some in xylem; systemic herbicides are more likely to kill all parts of the plant.

A new method of chemical weed control uses the nonspecific herbicide glyphosate (marketed as Roundup®) and genetically engineered crop plants. Glyphosate inhibits a plant enzyme needed to produce three amino acids. It is lethal to plants, but because this enzyme is not present in animals, it causes them no harm. Several crop plants have been genetically engineered to use bacterial genes to make these amino acids with the bacterial form of the enzyme, a form that is not inhibited by glyphosate. Consequently, after a crop has grown for several weeks and weeds have started to appear, the field can be sprayed with glyphosate and all plants will be killed except for the crop. Also, glyphosate binds tightly to soil particles, so there is little chance of it being washed out of the field and into groundwater.

In small farms, crops can be weeded manually. People walk between rows of crop plants and either pull weeds out by hand or uproot them with a hoe. Manual weeding is extremely labor intensive.

Insects are a major problem for any crop. Most are controlled by chemicals called **pesticides**, or more specifically **insecticides**, most of which are nonspecific and harm all insects, even ones that are not damaging the crop. Insecticides act in various ways, but in general they interfere with metabolic processes that are unique to animals, such as nerve impulse metabolism. Because we are animals as well, many pesticides are toxic to us, and scientists are continually searching for chemicals that affect only a particular group of insects. Until the 1950s, chemicals called organochlorines (such as DDT, dichlorodiphenyltrichloroethane) were widely used, but unfortunately they were **persistent** (did not break down quickly) so animals that ate the dead or dying insects would themselves be poisoned. DDT affected all animals in the food web, and was responsible for interfering with reproduction of birds that eat fish; their egg shells were so fragile they would break before the chicks could hatch. The book *Silent Spring* written by Rachel Carson in 1962 described this and raised public awareness of the dangers of pesticides, and organochlorines were banned from use in most countries. They were replaced by two other types of pesticide, organophosphates and carbamates, both of which also alter nerve function. Organophosphates are extremely toxic to vertebrates (such as ourselves), and their use is avoided now.

It is almost impossible to coat every bit of surface of a plant by spraying an insecticide onto it. If the plant has a thick layer of hairs, the spray collects on them rather than the regular epidermis cells. Leaf axils typically have narrow crevices that a spray will not enter, and it is difficult to coat the underside of leaves. Insects will find the unprotected parts and enter the plant through them. In contrast, systemic pesticides are absorbed by the plant and taken into its tissues. Most parts of the plant become toxic and the plant is fully protected. At present, safer methods of pest control are being developed. They are discussed below.

Perennial trees and vines must be pruned. Cutting off the ends of long branches stops apical dominance and causes several axillary buds to become active: one old branch is replaced by several young ones. The plant is more compact, bushier, and has more branches on which to produce flowers and fruits.

Harvesting the Crops

Methods of harvesting are extremely varied and depend on the nature of the crop; only a few examples will be given here. Products that are hard and dry when mature, such as wheat and lentils, can be harvested by machines, whereas soft, watery crops like peaches and strawberries must be picked by hand. Cereals like wheat, corn, and rice are annuals that die naturally after the grains are ripe, and most remain standing. A machine called a combine (originally called a combined mower/thresher) moves through the field, cuts the shoots from the roots, then takes the shoots inside itself where they are agitated violently enough to knock the grains off the plants (FIGURE 12.22). The grains fall through a sieve and are collected; the shoots (usually called straw once the grains are gone) are discarded out of the back of the combine. Peas, beans, lentils, and many other annual crops are harvested the same way. Other annuals, such as mustard, do not die quickly enough; they must be cut (mowed) while they are still alive so that the plants become dry enough to go through a combine (FIGURE 12.23).

FIGURE 12.22. This combine is harvesting wheat. In the front is a cutter (similar to the ones used to trim hedges) that cuts the shoots away from the roots. The shoots are then brought inside where they are beaten with rotating bars, knocking the grains free of the shoots. The grains fall through a sieve and the shoots are carried out the back end of the combine. Most crops that have hard seeds attached to dry shoots are harvested by combines; examples are other cereals such as barley and oats, as well as crops like peas and beans. (Courtesy of Chris McCoy.)

Root crops and potatoes are harvested with digging machines. Each machine has one or several large blades that are pulled through the soil below the crop, loosening

FIGURE 12.23. In some crops, such as this mustard, the fruits become dry enough to open while the shoots are still so green and wet that they cannot be run through a combine. To avoid loss of seeds, the shoots must be cut just as the fruits mature. As they lie in the sun, the stems die and dry out, but the fruits do not open. Once dry enough, a combine can be used to separate the seeds from the shoots.

(a)

(b)

FIGURE 12.24. **(a)** Root crops such as potatoes are harvested with mechanical diggers. The front of the machine has blades that run underground deeply enough to pass below the potatoes, then the potatoes and soil are carried up onto broad chains that allow the dirt to fall through but carry the potatoes up into a truck. **(b)** Sugar cane is cut with a machine that has rollers that strip the leaves off the stem and then cuts the stems at ground level. Canes are then cut into short lengths and transferred to the carriage behind the tractor. Canes are then moved from the carriage to a truck that transports them to a refinery as soon as possible to avoid loss of sugar. (Photo a courtesy of Maarten J. Chrispeels; photo b courtesy of Tommy R. Navarre.)

the soil and carrying it and the crop up on to a screen (FIGURE 12.24a). Dirt falls through and the roots or tubers are carried upward by the screen and collected. Sugar cane is harvested by first burning off the leaves then cutting the tall (2 to 3 m) canes into shorter pieces that can be hauled to a processing plant. The underground rhizomes are left undisturbed and will produce a new set of canes in the following year (FIGURE 12.24b).

Many perennial crops are harvested manually. Fruits that grow on trees, such as apples and peaches, are each collected by hand, not machine. The same is true of grapes and other crops that are delicate and might be bruised by a machine. Most of these do not ripen simultaneously, so any particular tree or vine must be examined and harvested several times.

Forage plants that are used to feed animals also have diverse harvesting methods. When alfalfa plants have grown to a height of about half a meter (about a foot and a half), the field is mowed to cut off the tender shoots. Once these have dried for 2 or 3 days, they are collected and compressed into large bales, which are easily shipped and stored (see Figure 12.11). A single alfalfa field can usually be mowed 3 or 4 times per year.

Caring for the Land Between Growing Seasons

After a field has been harvested, the land must be cared for. For annual crops, the field usually contains dead stems, leaves, and roots, plus some weeds. The field is often disked to mix the plant material into the soil so that it will decay more quickly (FIGURE 12.25). This adds organic material to the soil and improves its capacity to hold water. Some fields are burned after harvest, usually to prevent growth of fungi on the dead plant matter and to kill any plant pathogens and harmful insects. In windy areas, a cover crop is planted; this is an inexpensive grass that will hold the soil through the winter and prevent erosion. The cover crop is usually just plowed under in the following spring rather than being allowed to mature for harvest; its main function is to stabilize soil between seasons. A field might be allowed to lie

(a)

(b)

FIGURE 12.25. After a crop is harvested, the field typically has the residual plant parts that are not needed, such as the stems, leaves, and roots of crops grown for their fruits or seeds. Crops like beans and peas produce only a small amount of residual material, but cereal grains produce a lot of straw. **(a)** This wheat straw has been disked to break it up somewhat and to partially bury it so that it will decompose more quickly. After weed seeds have germinated, the field will be disked again to kill the weeds and to bury the straw more completely. By allowing the residual plant matter to decay, all of its minerals are returned to the soil, and the decomposing plant tissues help hold water and nutrients and give the soil better texture. **(b)** In Louisiana and eastern Texas, many rice fields are flooded after the rice has been harvested, then they are "planted" with crawfish eggs. The eggs hatch, and in just a few months the crawfish grow large enough to be harvested and used for food. The small red dots in this flooded field are crawfish traps. Once the crawfish have been harvested, the traps are removed and the field is planted to rice again. Both the water and the crawfish help the rice straw decompose more quickly. Unlike the field in part a, this land is never out of production.

fallow (unplanted) for 1 or 2 years before being replanted. In areas where there is scarcely enough water for agriculture, the fallow years allow the soil to accumulate moisture for the following crop year. Fields for dryland wheat are often planted only in alternate years, with one fallow year between crop years. In areas with severe problems with weeds that are difficult to control, the fallow years are used for weed control: after the weeds germinate and before they reproduce, the field is either tilled or treated with a broad-spectrum herbicide to kill the weeds. After a few weeks, seeds that had been dormant will sprout and begin to grow; then they too are killed. This process is repeated several times until the number of unsprouted seeds in the soil is low. This method is necessary in efforts to control *Striga*, a parasitic plant that attacks crop plants, and which is extremely difficult to eliminate. Many fields have to be abandoned once they are infected with *Striga*.

Plant Breeding and Biotechnology

A critical aspect of agriculture is to discover and domesticate plants that can be used as crops. It is rare that a wild plant can be immediately cultivated. Usually certain features must be altered so that the plants grow more rapidly or produce more fruits or fiber, or are easier to harvest and process. Altering plants to be better crops is the process of **domestication**, and it involves changing the plants genetically, either by searching for and using those with the best traits or by crossing plants to combine various traits into a single individual. Biotechnology is the latest set of tools for modifying plants to be better crops (**BOX 12.1**). Because domestication changes the gene pool of the crop species and because it is done by people, it is known as artificial selection instead of as natural selection.

BOX 12.1. Genetically Modified Organisms (GMOs)

A genetically modified organism (GMO) is any organism (plant, animal, fungus, or microbe) whose DNA has been altered by using DNA technology, often called **recombinant DNA technology** or genetic engineering. Typically this involves obtaining one or several genes from one organism and inserting them into a second organism of a different species, producing a **transgenic organism.** Occasionally, an organism's own genes are modified with DNA technology, and this is a **cisgenic organism.** Using classical techniques of cross-pollinating plants or interbreeding animals and so on, it is possible to move genes from one organism to another, but this works only for organisms that are so closely related they can reproduce sexually with each other. With DNA technology, genes can be transferred between any organisms whatsoever: bacterial genes, jellyfish genes, or human genes can all be moved into plants.

Numerous types of transgenic plants have been genetically modified to have desirable traits, such as resistance to herbicides like glyphosate, or to have high levels of the important amino acid lysine. Transgenic BT plants produce a bacterial toxin that protects them against insects. GMO rice has been altered to prevent vitamin A deficiency in people whose diet is high in rice.

Many people have concerns and objections about GMOs. One objection is that this is an unnatural interference in the biology of organisms, a lack of respect for the organism. This may be true, but the same could be said about the artificial selection that produced the wide range of dog and cat varieties that exist, or the livestock that have been modified by classical crossing techniques to produce exceptional amounts of meat, milk, eggs, or wool.

Another objection is that it is difficult to be certain of the long-term effects of new technologies. This is definitely true. For example, the early days of nuclear power were full of optimism for cheap and clean energy, but now we realize that radiation has been more harmful than realized initially, and many more people have died from it than had earlier been suspected. And many new medicines seem wonderful at first but after long-term use, some have unforeseen side effects. With GMOs, this concern is especially real because we are dealing with genetic material that can mutate and evolve. DNA technology in a lab produces only a few genetically modified organisms, but these are then used as the breeding stock to generate millions of progeny. GMO corn already is planted on 200,000 km^2 (50 million) acres in the United States alone. Each acre contains thousands of plants, each with hundreds of flowers undergoing meiosis and crossing-over with the potential to generate mutations in the artificial gene. Of course, all genes have the potential to mutate, but in the situation of transgenic organisms, we need to think carefully about how mutated bacterial or animal genes might react now that they are in plants. Plants have different DNA regulating molecules and safety mechanisms than do bacteria and animals, and they may not be able to control an anomalous, mutated transgene. The concern becomes a bit more serious when the GMO is something we eat: milk, rice, meat, flour, and so on. However, GMOs have been cultivated for more than 20 years now, and there seem to be no credible reports of significant problems.

An additional concern is that GMOs might reproduce sexually with wild relatives and thus transfer the exotic gene to the species at large. Many of our food crops are members of the grass family (Poaceae), legume family (Fabaceae), and mustard family (Brassicaceae), and there are dozens of wild species of these that grow in or near fields and cities. The likelihood of pollen being blown or carried by pollinators from the GMO crop to a related wildflower near the field may be low but it is not nonexistent. We could envision the wild population suddenly gaining immunity to insects and thus being able to survive better than other species in its habitat. The plant could become a noxious weed as it outcompetes its neighboring plants.

Cross-contamination of food crops is another concern. Many people take special efforts to have a healthful diet of organic foods, ones that have not been treated with herbicides, pesticides, and other artificial chemicals. Many of these people prefer to have their foods also be non-GMO foods. It has become difficult to guarantee that: a harvester might be used on a GMO corn field one day and then on a non-GMO corn field without being carefully cleaned between the two. A corn processing factory might handle a GMO crop one day and a non-GMO crop the next. Cross-contamination is almost inevitable, and minute traces of GMO foods have been detected many times in non-GMO foods.

We do know that GMO crops have many benefits in being more nutritious or requiring less herbicide or pesticide. We do not yet know if they have unforeseen dangers. Some people would argue that rather than using an uncertain technology to produce increased crops, we should give more attention to reducing the growth of the human population (see Box 11.1 Zero Population Growth).

Plant Breeding

People have modified plants and animals since the earliest days of agriculture, cultivating the best plants, and interbreeding plants with most favorable traits. While our ancestors were still gatherers, they would collect the seeds or fruits that were easiest to pick or carry; for firewood or lumber, they would use the trees that were easiest to cut. This type of harvesting took the best plants and left fruits, seeds, or whole plants that were less desirable; the gatherers were unknowingly selecting against the plants with desirable traits. While human populations were small, this probably had little effect. Once people realized they could sow seeds or transplant rhizomes, bulbs, and tubers, they would have used the ones they liked the best; just as today, people share cuttings of their favorite flowers, not their least favorite.

Until recently, all harvesting and planting was done manually, so people could see the individual seeds and fruits they were dealing with, and careful selection could occur. Even though early farmers had no theory of genes or heredity, they still knew that seeds from good plants grew into good crops. But once the concept of Mendelian inheritance of genes was understood (Chapter 8), breeding and selecting plants and animals became carefully planned and scientific. Accurate records are kept of the traits of specific plants, which are grown in special plots of ground away from other plants of the same species to avoid accidental cross-pollination. Pollen is collected manually, often with a small paint brush, from plants with a particular set of traits and then applied to the stigmas of others; when seeds are produced by this method, the genotypes of both the pollen parent and the ovule parent are known. Records must be kept of every cross, and of every parent of particular plants and seeds, so that new combinations can be tried. Artificial mutagenesis is used to create more alleles of genes to work with, and also people search natural populations for plants with unusual traits that might now be useful. These techniques are still the ones used most commonly today. Seed companies and research stations have huge repositories of seeds of many varieties of particular species, and extensive sets of greenhouses and field in which the genotype of each plant is known (FIGURE 12.26).

FIGURE 12.26. Seeds of many types are maintained in long-term storage facilities. By keeping the seeds cold, they will remain alive longer than if maintained at room temperature, but periodically some must be removed, planted, and then harvested for a fresh batch of seeds for storage. The seeds here are not only many varieties that have been produced by artificial hybridization, but also seeds of wild relatives of crop plants, which may have genes that might be useful for breeding new crops in the future. (Courtesy of the the U.S. Department of Agriculture.)

Finding and Moving Genes with DNA Technology

The term "DNA technology" is broad and covers several topics, and has several important techniques. An important agricultural use is in the identification and manipulation of the alleles of specific genes. As discussed in Chapter 8, when we look at a plant, we see its phenotype, but we can discover its genotype only by careful study. In a single species, some plants might have red flowers, other plants have white flowers; the two plants differ both phenotypically and genotypically. The same is true of tall and dwarf plants of a particular species, or of plants that differ in lignin content. How can we identify which genes are responsible for these traits, and how can we manipulate the alleles of these genes?

One of the first steps is to make a map of the genome. This may be done by crossing plants that differ in two characters of their phenotypes (a dihybrid cross) and noting the percentage of recombinant F_1 progeny that are produced, as described in Chapter 8. As a further refinement, portions of DNA can be mapped by using a class of enzymes called **restriction endonucleases**. Each type of endonuclease recognizes a short sequence of nucleotides and then cuts the DNA in the middle of that sequence. If one endonuclease is used on a batch of DNA, it will cut it into several fragments corresponding to the number of places where that sequence occurs in the DNA. For example, the enzyme EcoR1 recognizes and cuts the sequence GAATTC, whereas the enzyme BamH1 recognizes GGATCC, and so on with many more enzymes. If a molecule of DNA contains, for example, six EcoR1 sites and eleven BamH1 sites, treating the DNA with EcoR1 will produce seven fragments, and treating an identical molecule of DNA with BamH1 will produce twelve. By treating the DNA of a species with many different restriction endonucleases, we build a **restriction map** of the species.

Restriction maps are extremely useful. If the restriction maps of two closely related species are compared, they should be quite similar. On the other hand, the restriction maps of species that are not closely related will differ greatly. At present, this is used in reverse: restriction maps of many species are prepared, and then those that are the most similar are assumed to be more closely related than those that are very dissimilar. This has helped identify some of the wild relatives of domesticated crops. Restriction maps also help us locate genes that control particular phenotypes. If two plants of a species differ in just a single trait, such as red and white flowers, then their restriction maps may differ at just a single point, and that is probably the area in which the gene is located. Once this area is identified, other techniques can be used to actually determine the sequence of each nucleotide in that region (**DNA sequencing**). DNA sequencing is becoming faster, easier, less costly, and more reliable every day, but still it is more practical to sequence just the DNA in a particular restriction map as opposed to sequencing all the DNA in every chromosome of the plant.

An alternative to searching for a gene with restriction maps is to search for related genes in data bases of genes that have already been studied. Such databases are called **gene banks**, and they are accessible online. Each time someone sequences, studies, and then reports on a gene, they are required to deposit the gene sequence into a gene bank. At present, the entire genomes of several plants have been sequenced and are available online. Consequently, if we are interested in the genes and alleles that control red and white pigmentation of flowers of roses, for example, we can search a gene bank to see if someone has already found a similar gene. If so, we can then synthesize a short piece of artificial single-stranded DNA (called a **probe**, and usually made to be radioactive or fluorescent so that we can detect it easily) that has the same sequence as part of the known gene. We apply this single-stranded DNA to DNA from roses, and if roses have a similar gene, the probe will bind to it, indicating the location of the gene on one of the chromosomes of the roses. Because the probe is just a short piece of DNA, the gene is longer than the probe, but by cutting the rose DNA with several restriction endonucleases, we will find one that cuts in just the correct sites to isolate all the gene and not too much surrounding DNA.

Once genes for certain traits have been identified, we can then attempt to use them to precisely modify a species. An artificial gene is made by adding certain necessary components, like recognition sites, termination sites, and control sites, then the artificial gene is inserted into the genome of a new plant. This is not such an easy part. If the plant can be regenerated from cells grown in culture (see below), then we just treat all the cells with a solution of the DNA, transform them, and trigger some of the cells to grow into new plants. This will result in many plants: some that have taken up the gene, many that have not. To distinguish between these, we usually do the experiment with a piece of DNA that contains not only the gene we are studying, but also one that makes a plant resistant to a particular herbicide. Once the cultured cells have started to grow into plants, they are treated with the herbicide, and those that have taken up the DNA will be resistant and survive, those that have not taken up the DNA will be killed.

Some plants cannot be grown from cultured cells. For those species we put the DNA into a bacterium, usually *Agrobacterium*, and let it infect the plant. If enough plants are treated, the apical meristems of at least a few will become infected and express the gene we are studying. All new shoot tissue produced above this point will have the genetically engineered gene. In this method also, herbicide is used to obtain only the plants that have taken up the genes and are expressing them properly.

These techniques are often called **genetic engineering**, and it is necessary to use them only to obtain a few plants with experimentally altered genes. Once we have produced those plants, they can be grown to maturity and then used in classical genetic crosses by transferring the pollen of some to the stigmas of others (FIGURE 12.27).

FIGURE 12.27. Many of the seeds planted by farmers are F_1 hybrids (see Chapter 8), and seed companies must produce each year's seed crop by crossing particular parents to obtain just the right combination of characters in the F_1 hybrids. Here, one of the sunflower parents has smaller shoots and many inflorescences; the other parent has larger shoots and just a single inflorescence. The larger plants are the female parent: they do not produce viable pollen but instead rely on bees to carry pollen from the smaller plants to them. After pollination is complete, the smaller plants are destroyed so that the larger plants grow better, then they are harvested for the seeds. (Courtesy of K. J. Bradford.)

Producing New Plants with Cell and Tissue Culture Techniques

Genetic engineering techniques typically only introduce the altered genes into one or two cells of a plant, and not even into all the meristem cells. It is necessary to induce the altered cells to grow into a new plant. Young leaves and stem tissues of the treated plants are cut into small pieces and placed into a culture medium of salts, sugars, some vitamins, and hormones. The medium may be liquid or have a gel added to make it more or less solid. With the correct balance of hormones, the cells proliferate, usually into a **callus**, a mass of cells that grow irregularly, some dividing, some quiescent (FIGURE 12.28). If the altered cell is dividing, then it is creating copies of itself. The callus is then broken apart and pieces are spread onto a new culture with a different mix of

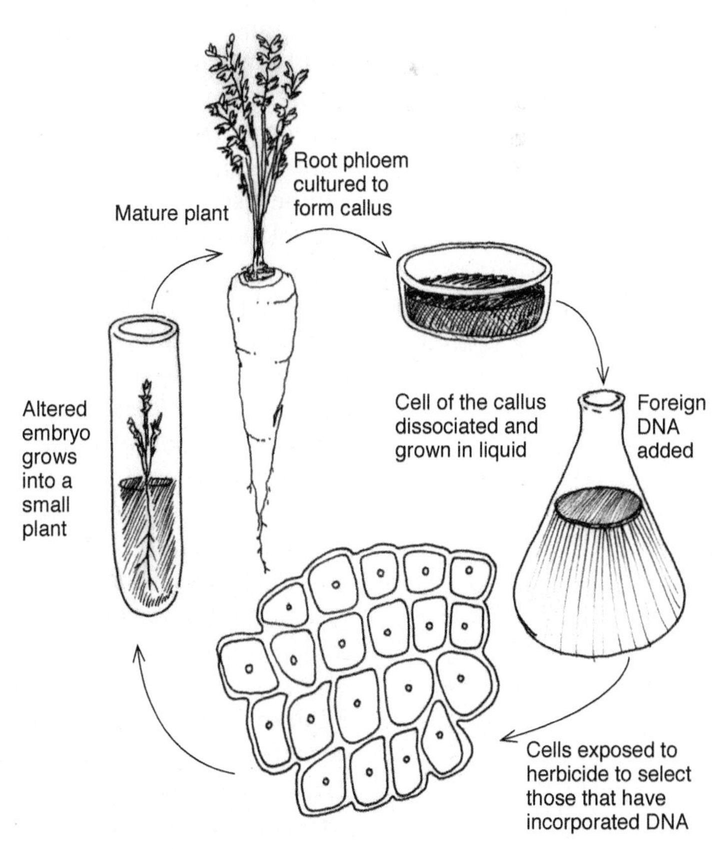

FIGURE 12.28. Diagram depicting the production of plantlets by tissue culture. The method is described in the text.

FIGURE 12.29. The small irregular clumps are callus produced in tissue culture. These have been treated to produce plantlets, which will be allowed to grow in culture for a short while and then be transplanted to soil to continue their growth. Production of plantlets like this is used to obtain genetically engineered plants. (Courtesy of James E. Irvine and T. Erik Mirkov.)

hormones, and in certain species, this will cause some of the cells to either form embryos or form shoot apical meristems (FIGURE 12.29). These are nurtured and allowed to grow until the culture has many plantlets. These can be tested to identify any that has the altered gene. Tissue culture works well and easily for some plants, especially dicots, but it is less effective for others. For some species we have not yet found any combination of hormones that will trigger organized growth into a plant. For these, genetic engineering is much more difficult.

Goals of Plant Improvement Programs

Many traits of crop plants affect the overall quality of the crop. Here are just a few examples.

1. *Plants that need less water and that can be grown with less irrigation or in areas that have less rainfall.* This trait can be affected by selecting for plants that have larger root systems, or root systems able to pull water from drier soil. Also, if plants close their stomata quickly when the air is dry, they lose less water and probably will not wilt, whereas similar plants that keep their stomata open will lose more water and wilt, and may die in just moderate droughts. Unfortunately, varieties of plants that are drought-adapted due to closing their stomata quickly also cut off their access to carbon dioxide, so they grow more slowly and are less productive than ordinary varieties.

2. *Plants that are more resistant to fungi.* For many crops, especially cereal grasses, fungi are a constant concern. There are hundreds of species of fungus that can attack these crops and cause them to sicken and die, or at least produce fewer seeds. Fungi called rusts and smuts are especially damaging. These fungi can be a serious problem because each produces hundreds of thousands of spores that blow away and possibly infect new plants. When the plants are being grown close together in a field, the chances are very high that fungal spores from just one infected plant can be blown onto many of the other plants in the field. On the other hand, when plants grow in natural populations, they might be so widely scattered that some will—just by luck—not have any fungal spores fall on them. They will survive, even though they are just as susceptible as other plants in the population. The dense fields of most types of agriculture greatly increase the problems of fungal infection. To combat fungi, research facilities grow as many varieties of the crop as possible, exposing all to the fungus and looking for any that have a natural resistance. The resistant plants are used as parents to breed new seeds that also are resistant. But fungi mutate, just as crop plants do, so not long after a new, resistant crop has been established, some mutant fungus somewhere will be able to attack it and the scientists will need to once again search for a resistant variety of the crop species.

3. *Plants that are more resistant to insects.* Similar to fungi, but more often farmers use insecticides to kill insects or chemicals to make the plant unpalatable. Treating whole fields with insecticide has been more effective than breeding in resistance. However, DNA technology has allowed genetic engineers to insert a gene from a bacterium (*Bacillus thuringiensis*) that codes for a protein that poisons only insects. The genetically modified crop plants are called **BT plants**, and they are poisonous only to insects that eat the plant; neither mammals nor plants are affected by the bacterial toxin (FIGURE 12.30). It had been feared that pollen of BT crop plants would also be poisonous and might affect butterflies, but there is little evidence that the pollen harms any nontarget animals.

4. ***Plants with seeds, fruits, tubers, and so on that have "improved qualities."*** People disagree about what "improved qualities" are. For those of us who eat foods, improved qualities are better flavor, texture, color, and aroma, but for many growers and processing companies, fruits and seeds must be tough enough to resist being bruised or smashed during harvest, processing, and transport. In some cases, the result is that peaches and pineapples at the store are hard and flavorless, but not bruised. You probably think of pineapple as being sour and hard, but if you are in an area where they are cultivated and if you can eat some that have ripened in the field, they are amazingly sweet and juicy—but they are so soft it is impossible to ship them very far or store them for more than a few days.

FIGURE 12.30. This field contains two types of cotton; that on the left has been genetically engineered to produce the anti-insect toxin of *Bacillus thuringiensis* (so it is known as *Bt* cotton), that on the right lacks the *Bt* gene. Caterpillars have damaged the flower buds of the natural cotton so badly that almost no cotton bolls have been produced, whereas the *Bt* cotton is still healthy. A cotton boll is the mass of seeds with long hairs used to make cotton thread; the white spots in the photo are bolls. (Courtesy of Monsanto Company.)

5. ***Plants that are better for our health.*** A less controversial aspect of "improved qualities" is increasing the vitamin content of the food, or altering its proteins such that they have a more nutritious balance of amino acids, or more beneficial lipids. For example, new types of corn, called high-lysine corn, have been bred to have higher levels of the amino acids lysine and tryptophan than are present in old-fashioned corn. The older versions of corn had such low levels of these amino acids that people whose diets relied heavily on corn suffered a deficiency of these amino acids. The new high-lysine corn varieties are much better for human nutrition.

6. ***Grains that are easier to harvest.*** Wild relatives of many cereal grains have such fragile fruit stalks that the seeds "shatter": they fall off the plant before harvest starts, or before the threshing machine can get them inside itself where the seeds can be captured. Genetic modifications adjusted the stalks so that fruits and seeds stay on the plant until they are fully ripe and can be harvested. With wheat, rice, and several other cereals, another problem is that the natural plants have tall, spindly stalks that are easily blown over (called "lodging"). If this happens early, the seeds may not ripen; if it happens late, the seed heads are lying so close to the ground that the harvester cannot pick them up without the risk of taking in dirt and rocks that damage the machinery. These cereals have been bred to have short, strong stalks that resist lodging.

7. ***Rice that is resistant to flooding.*** Rice fields are flooded intentionally when the plants are young: at this stage, they can survive by anaerobic respiration but weeds drown. But if older plants are flooded accidentally, most rice then falls over or the base of the stalks suffocate. Plant geneticists have now come up with a strain of rice that can survive being flooded even when it is older.

8. ***Reducing lignin in plants that will be used to make paper.*** Most paper is made from the tracheids of softwood trees such as pines. These cells have high

amounts of lignin in their cell walls, a chemical that makes the walls strong and allows the trees to grow tall. But it also interferes with the processes that convert wood into paper, and the lignin must be extracted. This tends to be one of the most ecologically damaging aspects of making paper. If trees could be bred that have a lower lignin content, this might reduce the ecological damage. Of course, it might also make the trees too weak to stand up, but typically these trees are grown as crops on plantations, and are harvested before they grow very tall anyway (see Figure 12.12).

Lignin also interferes with processes that convert plant stems, leaves, and roots to biofuels. At present, the starch in corn seeds or the juice of sugarcane is fermented to alcohol. Much of the plant body that is left over after seeds or fruits are harvested (wheat straw, sugar cane stalks) consists of vascular bundles containing lignified fiber cells. Their thick cell walls consist mostly of cellulose that can be converted to sugar and fermented, but the lignin that impregnates the walls makes conversion and fermentation slow and incomplete. If the lignin content could be reduced, these plant parts could be used to produce alcohol.

Concepts for Reducing Environmental Impact of Modern Agriculture

Reducing Damage to Other Organisms

One of the first things to consider is the loss of organisms as natural land is cleared to turn it into fields, pastures, stockyards, or processing plants. All existing plants, animals, and fungi may be killed outright as the land is burned or graded and then planted into a **monoculture** (an area with just a single species). Throughout the North American Midwest, thousands of acres of grasslands with a high diversity of plant and animal species were plowed and turned into wheat fields. In other areas, rich forests with thousands of species were chopped down, the stumps pulled out, and then they were turned into fields. Even now in much of the American West, old growth forests are being cut down for wood and replanted with just one species of tree; a causal observer would think that a new forest is there, but there is no species diversity, most of the original wildflowers and understory plants are gone, and stream sides are damaged. Many squirrels and birds nest in holes that occur only in old trees, and birds such as woodpeckers eat insects that live in old or weakened trees (FIGURE 12.31). Many types of fungi need old trees as well, being unable to grow on young, vigorous trees. Even dead trees, whether standing or fallen, are necessary for all sorts of insects and fungi. Certain mosses and liverworts live only on dead tree trunks and branches after they have lost their bark (FIGURE 12.32).

One of the goals of crop research is to breed or engineer plants that need less water. If such plants were to be cultivated on existing farmland and thus would lead to lower use of irrigation, this would be beneficial. But often it means that natural areas in dry regions can now support a crop: developing crops that need less water may lead directly to the loss of natural habitats in our arid and semi-arid areas. Rather than being a good thing, it could cause more habitat destruction.

Another way that agriculture harms organisms other than the crop itself is through the use of herbicides and insecticides to control pests. By using herbicides, the fields are kept clean of weeds, and that means that plant biodiversity is kept at almost zero—no diversity at all. Larger weeds can interfere with crop production by competing for water and mineral nutrients, and by clogging up harvesting machinery with wet leaves and stems. But many small plants could grow among

(a) (b)

FIGURE 12.31. Many animals such as this spotted owl **(a)** need old-growth forests, forests that have not been disturbed for hundreds of years and that have many ancient trees **(b)**. Old trees and old-growth forests differ greatly from the trees in tree farms or in areas that have been logged (especially if logged by clearcutting): there is a different mix of species, and old trees differ from young ones in the type of bark they have, the amount of seeds, the presence of dead branches that contain insects and wood-rotting fungi, and so on. (Photos courtesy of John and Karen Hollingsworth/U.S. Fish & Wildlife Services.)

FIGURE 12.32. Old-growth forests have many fallen and partially decayed trees, which add to the complexity of the ecosystem. Certain types of fungi, insects, and bryophytes are extremely specialized in the microhabitats they require: some live only on the bark of fallen trees, others only on the wood after the bark has decayed away, others only on wood that is partially decomposed by fungi, and so on. None of these can live in forests that consist only of young trees. The green material on this decaying tree is a small liverwort *Nowellia curvifolia*.

the crop plants without harming the harvest, and broad-spectrum herbicides often damage them as well.

Many insects definitely harm crops. But there may be many other insects in fields causing no harm at all; perhaps just flying by, or living on the weeds, or just using the plants as a shelter without eating them. And many crop plants need insect pollinators, such as honeybees. A broad-spectrum pesticide might kill many types. And with insect populations held low by pesticides, that means insectivores like birds, bats, and small mammals have less food to eat. If insectivores eat insects that have been poisoned but which have not yet died, then they take in the poison as well. Scientists are searching for narrow-spectrum pesticides that control only the pest insect and no others. Pesticides with a short residence time break down quickly, so animals that come by a few days or weeks later are unharmed.

One means of reducing the need for herbicides, insecticides, and other chemicals is **intercropping** (also called heterocropping), in which two compatible crops are cultivated in the same field (FIGURE 12.33). They usually are planted in alternate bands with 5 or 10 rows of one crop alternating with several of the other. The benefits are that insect pests adapted to one crop may not be able to attack the other, so the food resources for the pests are cut in half. Insects may waste some of their time and energy searching the wrong crop. Similarly, if a fungus infects some plants of one crop, many of its spores will land on the other crop; those spores will not be able to establish an infection and will not be able to produce new spores to attack the susceptible crop.

An alternative to the use of pesticides is the use of **integrated pest management**. Techniques include encouraging insectivorous animals that will eat pest insects without the need to poison them. For example, the small red or orange beetles we call ladybugs are beneficial because they eat aphids and scale insects that harm plants. By cultivating ladybugs along with the crop plants, at least some bothersome insects can be controlled without the use of chemicals (FIGURE 12.34). Pests can

FIGURE 12.33. Rather than merely tilling the ground between the almond trees, the farmer here has planted clover. This not only protects the soil from erosion and control weeds, but it also adds nitrogen to the soil (clover has root nodules with nitrogen-fixing bacteria), so the orchard does not need to have as much artificial nitrogen fertilizer added to it. This is one technique of sustainable agriculture. (Courtesy of R. Bugg, University of California, Davis.)

FIGURE 12.34. Ladybugs (a type of beetle) attack and eat a variety of insects that damage crop plants. By introducing ladybugs into the crop, certain insect pests such as cottony-cushion scale can be controlled without artificial insecticides. (Courtesy of J. G. Morse, University of California, Riverside.)

(a)

(b)

FIGURE 12.35. **(a)** This field is being drained into a ditch that carries the waste water to a nearby stream or river, then to an ocean. The water in the ditch is so rich in fertilizers that water weeds are flourishing: so many plants are growing in the run off that the water surface is completely covered, making it impossible for fish, turtles, and frogs to get to the surface. As these plants die, they sink, and the decomposition of their bodies uses up oxygen in the water, depriving animals of this important element. Because all fields in this area are drained like this, this is called nonpoint source pollution. **(b)** This crop of carrots has been mulched with a mixture of straw and compost. The straw blocks sunlight so it keeps the soil cool, and it smothers any weed seeds that might germinate. Both the straw and the compost break down slowly, gradually releasing mineral nutrients to the crop.

also be controlled by traps rather than poisons; physically removing insects by hand or with vacuums; and intercropping with plants that have aromas or phytochemicals that repel the pests.

Fertilizers also may harm noncrop organisms. The crop plants do not take up all of the fertilizer chemicals, so some are washed into the soil water, into streams, lakes, and rivers. This causes damage away from the cultivated area: the fertilizer may cause riverside plants to grow more than they normally would (FIGURE 12.35a). But the main thing is eutrophication of rivers and lakes: algae proliferate, forming a scum on the surface of slow-moving water, which makes it difficult for water animals to come to the surface. The algae die and sink to the bottom where they are broken down by bacteria that grow so rapidly they use up much of the oxygen in the water, causing fish to suffocate. The fertilizer continues down rivers, and in the United States, much of it ends up in the Mississippi (simply because it is our largest river and drains the most area). Once all the fertilizer-rich water empties into the Gulf of Mexico, it encourages rampant growth of certain algae (an **algal bloom**), resulting in a dead zone as well as red tides (see Box 5.2). Pollution caused by runoff from many farms and ranches across a broad area is called **nonpoint source pollution** in contrast to a **point source of pollution**, a source of pollution such as a single factory, feedlot, or processing plant. Fertilizer runoff can be minimized by using slow-release fertilizer, pellets that dissolve over weeks, releasing the fertilizer at about the same rate that plants use it. Organic fertilizers such as **compost** (partially decayed plant material) release nutrients slowly (FIGURE 12.35b).

On rangeland, cattle, sheep, goats, or other livestock harm plants in several ways. Overgrazing is one of the most typical: if too many animals are put onto a patch of land, they eat so much that the plants cannot recover. But if a small enough number of animals are placed in a pasture or range, their grazing is **sustainable**: the plants either grow back as quickly as the few animals can eat them, or the animals are moved to another area for several months so that the pasture can recover. In fact, as animals eat old leaves and twigs, they digest them and then leave them as manure, so the mineral elements that had been locked up in old leaves are now available to the plant's roots. The presence of livestock increases the speed of mineral recycling.

As the natural pasture plants are harmed by overgrazing, a typical result is that other plants increase in abundance, in particular, spiny or poisonous plants that are resistant to being eaten, such as sagebrush, cacti, and spurges. The ratio of plant types in the area changes due to overgrazing, and because the plants change, the associated pollinators, seed dispersers, and pests will also change. In some areas, exotic forage grasses have been planted on purpose—grasses that are more nutritious or drought resistant—and these crowd out the native grasses. The exotics allow more animals to use the land, but these grasses are not natural to the area. In some instances, the exotic grasses have escaped and invaded otherwise natural areas.

Livestock often damage streams and creeks. They come to the water's edge to drink, and their hooves cause the banks to cave in or crumble, harming plants along the bank and making the water muddy. Fish are harmed, as are small animals that nest along stream banks.

Draining wetlands and converting them to agriculture causes enormous damage to many species. Wetlands, either freshwater wetlands next to rivers and lakes, or marine wetlands next to oceans, are some of the richest habitats in the world, with very high species diversity. They support not only many types of plants but also numerous fish, birds, crustaceans, and other invertebrates. On any given day, a wetland might appear to have few birds in it, but during spring and autumn migrations, thousands of waterfowl depend on wetlands for feeding and resting. When wetlands are drained, not only are the immediate species affected, but so are the many birds that no longer can rely on the wetlands.

Salinization is the accumulation of salts in flat areas (see Figure B4.2 in Box 4.2). Many areas are drained by rivers that flow into oceans. When rainwater or irrigation water picks up minerals and flows off the land, it ultimately ends up in the ocean, as do the minerals. But in many areas—for example in the Basin and Range Province of Nevada, in the Central Valley of California, and the Great Salt Lake in Utah—rivers flow into the valley but none flows out: water comes in and accumulates temporarily, but it all evaporates before a lake can form that is deep enough to flow over the edge of the valley and continue on to an ocean. Consequently, all minerals that flow into the valley remain there as the water evaporates, and year after year the minerals become more abundant. At some point, they form a visible white crust on the soil, and may even grow into thick deposits, as in Death Valley. Even before the salts are plentiful enough to be visible, they are making the soil "osmotically dry"—that is, the salts hold water so tightly that most plant roots cannot absorb the water, even if liquid water is present. It is the same reason we cannot use ocean water to irrigate crops: it is too salty. In the Central Valley in California, this was not a problem years ago, because river flow into the valley was low, but now irrigation canals bring in huge amounts of water, and large amounts of fertilizer are added, so salinization has caused once rich farmland to become useless.

Reducing Damage to Land

Both farming and ranching reduce the fertility of land. With farming, crops are harvested and shipped away from the fields, and the crops—usually mineral-rich fruits and vegetables—are consumed in towns or cities. If used as food, the crop is turned into peels, cores, scraps, and digested waste, all of which may end up in a sewer or landfill. Thus there is a flow of mineral elements from the farm to landfill or to rivers where they contribute to eutrophication, and the field becomes depleted. The same happens with pastures because the livestock are shipped off the ranch, taking the minerals they have consumed with them. Wood is not very rich in minerals, but when entire logs are removed from forests, that too reduces soil fertility. The fields may simply be abandoned, as in slash-and-burn agriculture, or fertilizers, either organic or inorganic, must be added. Organic fertilizers are often made from the waste material of various crops, so in a way their use is a type of recycling, of bringing minerals back to the land (FIGURE 12.36).

FIGURE 12.36. In regions with large amounts of livestock, the manure can be used as an organic fertilizer. This reduces or eliminates the need for artificial fertilizer, and it also solves the problem of what to do with large amounts of manure. A few very large feedlots convert manure to methane to be used as fuel, but often manure is merely piled up and can then contaminate streams and groundwater. (© andrew tiley/Alamy.)

Erosion is an extensive and common form of damage caused by agriculture. Erosion is always occurring naturally as glaciers and rivers wear down mountains. Even rain drops splash soil particles downhill, and wind blows loose soil away. But many human activities cause accelerated erosion, or accelerated soil loss. When fields are plowed, there are no plants whatsoever to hold the soil in place, and either wind or rain can carry soil particles away, sometimes in huge quantities. Rain-caused erosion is most severe if the land is hilly such that rain runs off quickly and has a lot of carrying power, but even the water coming off of very flat areas will carry fine silt and clay particles, the ones that are richest in minerals. Wind can carry soil away from either flat or hilly areas.

Several techniques reduce erosion. **Shelterbelts** are rows of trees along the windward side of a field; they slow the wind and reduce its ability to pick up soil particles. For mildly hilly areas, erosion can be reduced by **contour plowing** and planting; this creates small ridges that follow the contour of the land, so water tends to run sideways and more slowly, not carrying so much soil (FIGURE 12.37). If plowing and planting ran straight down a hill, water would run rapidly and its erosive power would be greater. In extremely hilly areas, and where land is at a premium, terracing is practiced.

A technique called **minimum tillage** also reduces erosion. With intensive tillage, a farmer might disk or plow or otherwise disturb the soil many times after a crop is harvested and before the next crop is planted. This kills weeds and prevents them from making even more weed seeds in the few weeks or months between harvest and planting. Multiple disking is used for crops like corn that have thick shoots that break down slowly; the disks cut and fracture the shoots into smaller pieces that decompose more quickly, thus making a smoother field that is easier to plant. Each time the soil is processed though, its surface dries out and the top particles are vulnerable to wind erosion. Also, it continuously disturbs animals, fungi, and nonharmful plants, and it uses great amounts of energy. Minimum tillage emphasizes allowing the soil to rest and permit soil microbes to break down the chaff from the previous crop. Also, in sandy soils, the chaff is an excellent matrix for holding water and minerals...if it breaks down too quickly because

FIGURE 12.37. When these fields were plowed, the tractor followed the contour of the land so that the plow creates ridges that prevent rainwater from flowing rapidly downhill. This reduces erosion and it helps rainwater seep into the soil where it is needed rather than just running off and contaminating streams with mud. (U.S. Department of Agriculture.)

of intensive tilling, the soil is not as nutritious. Herbicides such as Roundup have made it possible to control weeds without disturbing the soil so frequently and without much ecological damage.

Conversion of agricultural land to other uses is a serious problem. It is common to see farmland converted to housing subdivisions, shopping malls, parking lots, and so on. This of course makes it useless for agriculture, and increases the need to convert natural land into new farmland, with all the ecological damage that causes. Strip mining, especially for coal, also destroys agricultural land.

There are several techniques for saving agricultural land, and land in general. One is **land reclamation**; that is, reclaiming or restoring land to a previous condition. In many areas now, strip mine operators are required to map the contours of land before any mining begins, then the topsoil is removed and saved; after mining is completed, the hole is filled back in to recreate the original contours, spread the topsoil over it again and replant the area with native species (FIGURE 12.38).

Some of the land that can be reclaimed consists of vacant lots and underutilized areas inside existing cities. Cities contain lots that are either empty or have abandoned buildings on them. These areas could be used for new homes, stores, or factories; this would reduce the need to convert natural habitat or agricultural land to urban land. The concept is to build **compact cities**, cities in which land is used more effectively and less of it is wasted (FIGURE 12.39). If more people would live in smaller houses with smaller yards, or would live in high-rise housing, urban sprawl could be greatly reduced. Also, the amount of land used for parking lots could be minimized if businesses would share parking areas and if businesses were clustered so that some (such as office buildings) would use the parking areas during the day, others (such as restaurants and movie theaters) would use them at night.

FIGURE 12.38. Reclaiming land is a much more complex process than merely leaving it alone and not using it anymore. This is an abandoned pasture that is being overgrown with fast-growing weed trees, mostly mesquite (*Prosopis*) and cacti. The forest that had been here before the land was cleared for grazing, however, was a much more complex mix of species such as magnolias, oaks, and hollies that are now extremely rare. If this land is to be truly reclaimed, it must be carefully replanted with seedlings of as many of the original species as possible and populated with the appropriate animal species when the plants are able to support them.

FIGURE 12.39. This church parking lot covers acres of land that had previously been used for grazing. Now it is used for only 2 or 3 hours on 1 day a week. Other than the few trees that were spared, it has no plants or animals, it prevents rain from soaking into the ground, and the asphalt leaches harmful chemicals into nearby creeks. There is no other business nearby that could use this parking lot on weekdays or at nights, which would at least provide additional benefit to offset the ecological damage it causes.

Important Terms

algal bloom
broad-spectrum (herbicide or pesticide)
BT plant
callus
carnivory
cisgenic organism
compact city
compost
contact herbicide
contour plowing
conventional farming
DNA probe
DNA sequencing
domestication
Fertile Crescent
gene bank
genetic engineering
herbicide
herbivory
hydroponics
insecticide
integrated pest management
intercropping
land reclamation
minimum tillage
monoculture
nonpoint source pollution
nonspecific (herbicide or pesticide)
omnivory
organic farming
persistent (herbicide or pesticide)
pesticide
point source pollution
recombinant DNA technology
restriction endonuclease
restriction map
shelterbelt
sustainable agriculture
systemic herbicide (or pesticide)
transgenic organism

Concepts

- Agriculture developed independently in several places, with the earliest being about 11,000 years ago in the Fertile Crescent.
- Agriculture is so widespread and extensive that it impacts the biosphere significantly, mostly in deleterious ways.
- Almost all important crop and livestock species cultivated today were domesticated by prehistoric cultures (before written histories).
- Expeditions to search for new plants have been important national undertakings throughout history.
- Agriculture requires preparation of land, tending crops, harvesting crops, and maintaining land.
- As plants are domesticated, they are changed genetically by artificial selection. Both traditional crossing techniques and biotechnology are used to alter the traits of crop plants.
- Agriculture damages land and noncrop organisms, but its harmful effects can be reduced or eliminated by various techniques.

Economic Botany

Economic botany sounds a bit cold and business-like, but think of what it really means. We participate in economic botany when we buy a cup of coffee and a pastry in the morning, and when we shop for food or new clothes. Buying a bottle of wine to celebrate an event is part of economic botany, as is buying flowers for special occasions, or for decorating sacred spaces.

13 Food Plants: Plants that Make Our Lives Possible

All of us eat plants: most of our calories, vitamins, minerals, and other essential nutritional factors are provided by the fruits and vegetables in our diets. Food plants also provide us with wonderful flavors, aromas, and textures. Many plants, such as this artichoke, are best when cooked only lightly and are eaten in an almost natural state. Artichokes are a good example of plant domestication: we have selected them to have larger bracts (the edible parts) around their flowers, but we have not changed them so much that we cannot easily recognize that they are thistles.

What is food? A brief answer is that food is the material we eat and drink and that supplies us with the nutrients our bodies use. But a longer answer is better. The concept of food allows us to reexamine a fundamental difference between plants and animals. A green plant needs only sunlight, carbon dioxide, water, and the essential elements it obtains from the soil. Chloroplasts produce the simple sugar glucose, and plants have such a complete set of constructive reactions they are able to build all the chemicals they need using just sugar and essential elements. They synthesize all 20 amino acids, all nucleotides, all lipids, all vitamins—everything. We humans also have a complex set of constructive reactions, but we cannot make everything we need. Just like plants, we can put an amino group onto an acid and make an amino acid. In fact, we have no trouble making 12 of them, but all organisms need 20 amino acids. We must obtain isoleucine, leucine, lysine, methionine, phenylalanine, threonine, tryptophan, and valine in our food; these are **essential amino acids**. Similarly, whereas plants synthesize all the lipids they need, we synthesize almost all of ours, but we cannot make the omega-3 and omega-6 fatty acids. Those are **essential fatty acids** and must be present in our food; good plant sources of these are soy oil, canola oil, pumpkin and sunflower seeds, walnuts, and leafy green vegetables. Our synthetic metabolism is especially poor at making the vitamins that are absolutely critical to our life. Thirteen vitamins are essential and we must obtain them in our diet (**TABLE 13.1**). Like plants, we synthesize all our nucleic acids; even if our diet lacked them completely, we could make all we need. Thus one fundamental aspect of food is that it must provide us with all the essential organic compounds listed above.

TABLE 13.1. The Essential Vitamins

Vitamin	Source
A	milk and eggs
B1 (thiamine)	whole grain cereals
B2	dark green vegetables and whole grains
B6	bananas, potatoes, green leafy vegetables, nuts, whole wheat products, and green beans.
B12	eggs and dairy products
C	citrus fruits and red sweet peppers
D	milk, soy-based drinks (our bodies do synthesize adequate amounts of vitamin D if we expose enough of our skin to sufficient sunshine, but that can be difficult in dark, cold winter months; having a good dietary source then is important)
E	nuts, leafy green vegetables, and whole grains
K	dark green leafy vegetables, green beans, broccoli, and asparagus
biotin	tomatoes, carrots, almonds, onion, and cabbage
folic acid	fruits, dark green leafy vegetables, and dried beans
niacin	nuts, leafy vegetables, asparagus, and broccoli
pantothenic acid	broccoli, mushrooms, and milk

Food illustrates another difference between plants and animals. Plants obtain their energy from sunlight by means of photosynthesis, but we animals obtain our energy by respiring (oxidizing) part of the food we eat. When studying the complexity of metabolism, nutrition, and a healthful diet, it can be a little perplexing to realize that almost all of what we eat is just respired away to carbon dioxide and water, salvaging the energy as ATP (see Chapter 5). We digest most carbohydrates to glucose and then run it through glycolysis, the tricarboxylic acid cycle, and the mitochondrial electron transport chain. Much of the lipids we eat are converted to acetyl CoA and shunted into the tricarboxylic acid cycle also. Even much of the protein in our food has the amino group split off (and then usually just thrown away in our urine) and the remaining acid is respired in the tricarboxylic acid cycle. We do use some of our food to build our body, but that happens more in growing children than in sedentary adults. If we exercise and are active (running, biking, swimming, etc.), that will increase the amount of tissue building we do, as well as the amount of material we must respire to generate ATP. One way to guarantee that more of our food is used for construction rather than respiration is to overeat: when we take in more food than is needed for generating ATP and for building muscles, bones, nerves, and other tissues, the excess is converted to fat and is stored—we get fat.

Our foods also contain materials that we do not need for our metabolism. The part of our food called fiber in vegetables, fruits, and whole grains is essential to us even though it is not digested and it never enters our bodies. It provides moist bulk that keeps everything moving through our intestines. Our foods, especially plant-based foods, also contain chemicals we do not need but that do enter our bodies. Plants synthesize defensive compounds that are toxic, bitter, astringent, or have some other quality that prevents animals from eating them. Even our food plants contain small quantities of some of these, and they pass through our intestine wall and into our blood stream. It is then up to our liver and kidneys to break them down and get rid of them.

Foods also contain chemicals that stimulate our taste buds and our sense of smell, chemicals that act directly on our nervous system. These are the flavor and aroma molecules, the ones that make peaches, apples, cherries, strawberries, grapes, and blueberries so appealing. Notice that these are all fruits with seeds that need an animal to disperse them. Coevolution has resulted in flavors and aromas that appeal to us, and our eating the fruits benefits both the plant and us; this is a mutualism (**BOX 13.1**). On the other hand, chocolate, vanilla, coffee, and roasted nuts have flavors and aromas that appeal to us but our consuming these products does not benefit the plant; the seeds are killed as we prepare them to be food. It is just that we have learned that these flavors indicate plants that contain some beneficial nutrient. In contrast, our foods do not contain chemicals like cadaverine or putrescine; these are the stenches from putrefying meat, and we have learned to avoid anything with such an aroma.

BOX 13.1. Classification of Fruit Types

There are several ways of grouping or classifying fruits. In one method, emphasis is placed on whether the fruit is **dry** or **fleshy**. A dry fruit is one that is not typically eaten by the natural seed-distributing animals; fleshy fruits are those that are eaten during the natural seed distribution process.

A further classification of dry fruits emphasizes fruit opening: **dehiscent fruits** break open and release the seeds, whereas **indehiscent fruits** do not (**TABLE B13.1**). For the most part, fleshy fruits are indehiscent. Animals chew or digest the fruit, which opens it; if uneaten, the fruit rots and liberates the seeds. Although many dry fruits are dehiscent, others must rot or the seed must break the fruit open itself.

Perhaps the simplest fruits are those of grasses. Each fruit develops from one carpel containing a single ovule. During maturation, the seed fills the fruit and fuses with the fruit wall, which remains thin: "seeds" of corn and wheat are actually both the fruit and its single seed, fused into a single structure. It has little protection and no attraction for animals. The fruit/seed falls and germinates close to the parent sporophyte. The fruits are indehiscent, and during germination the embryo absorbs water, swells, and bursts the weak fruit. These fruits are caryopses.

Beans and peas form from a single carpel that contains several seeds. The fruits, known as legumes, are dry and inedible (unless eaten while still young and green, such as snowpeas and green beans). At maturity, the two halves of the fruit twist and break open (dehisce) along the two specialized lines of weakness. After liberation from the legume, the seeds are protected from small animals and fungi only by the seed coat.

Fruits or seeds carried by wind must be light; they often have wings (maples) or parachutes (dandelions) that keep them aloft as long as possible. Such fruits are dry and weigh much less than 1 g (a dollar bill weighs 1 gram). Fruits and seeds transported by water (ocean currents, streams, floods) can be larger and heavier, but they must be buoyant and resist mildew and rot. Excellent examples are coconuts that float from one island to another.

Animals carry fruit in a variety of ways. Dry fruits with hooks or stickers catch onto animal fur or feathers; sticky, tacky fruits glue themselves onto animals; and fleshy, sweet, colorful fruits are eaten. Edible fruits require particular specialization because the fruit must be consumed but the seeds must not be damaged; they must not be crushed by teeth or gizzard or be digested. Immature fruits deter frugivores (fruit-eating animals) by being hard, bitter, or sour, such as unripe apples or peaches. At maturity, the fruit is soft, sweet, flavorful, and typically strikingly different in color. The enclosed seeds often have hard seed coats or endocarps (peach, cherry) that allow them to pass through an animal unharmed.

Grapes and tomatoes are examples of simple fleshy fruits, and are known as berries because all parts of the fruit are soft and edible. The seed coat provides some protection against being crushed during chewing, but perhaps more important are the size of the seeds and their slippery seed coat, which allows the seeds to slip out from between our molars. Pomes (apples and pears) differ from berries in two ways. First, they develop from inferior ovaries, so the inner tissue is the true fruit and the outer tissue is accessory fruit. Second, the seeds are protected because the innermost fruit tissues, the cores, are bitter and tough. The most elaborate fleshy fruits are drupes (peaches, cherries), in which the innermost tissue, the endocarp, is extremely hard and totally inedible, full of sclerenchyma. The mesocarp becomes thick, fleshy, juicy, and sweet, and the exocarp forms the peel, whose color informs the frugivore of the fruit's ripeness. The endocarp of a drupe is known as the pit and is often mistaken for the seed, but the true seed and its seed coat are located inside the pit. Drupes provide maximum attraction to animals with minimum danger to the seed.

Because frugivores are a major danger to seeds, it is necessary to think about how edible fruits evolved: they must offer some advantage that outweighs the danger. Animals have predictable habits and migration patterns. Ants, birds, and mammals carefully select the sites where they feed, nest, and sleep; therefore, seeds are not distributed at random but are moved through the environment to specific sites. One example is mistletoe: its seeds are sticky and adhere to a bird's beak as it eats the fruit. The bird probably feeds and cleans itself in the same species of tree, so the seeds are rubbed off the beak directly onto a proper host tree. Few seeds are ever lost by dropping to the ground or being deposited in the wrong tree, as would happen with wind or water dispersal. Just as efficient is dispersal of fruits and seeds of marsh plants. The birds and animals that live in marshes typically spend little or no time in other habitats, so seeds carried by these animals are almost certainly distributed to favorable sites.

A special benefit of seed distribution by animals is that the "deposited" seed may find itself in a small (or large) mound of "organic fertilizer." Seeds adapted to pass through an animal's digestive tract are resistant to digestive enzymes and can tolerate the high level of ammonium in fecal matter. Because many seeds of plants cannot tolerate these two factors, the adapted seeds find themselves in a microenvironment that is not only nutrient-rich but also excludes some competitors.

TABLE B13.1. Classification of Fruit Types

1. Fleshy fruits

a. Berry: a fleshy fruit in which all three layers—endocarp, mesocarp, exocarp—are soft (grape, tomato).

b. Pome: similar to a berry except that the endocarp is papery or leathery (apple).

c. Drupe: similar to a berry except that the endocarp is hard, sclerenchymatous (stone fruits: peach, cherry, plum, apricot).

d. Pepo: a fleshy fruit in which the exocarp is a tough, hard rind; the inner soft tissues may not be differentiated into two distinct layers (pumpkin, squash, cantaloupe).

e. Hesperidium: a fleshy fruit in which the exocarp is leathery (*Citrus*).

2. Dry fruits

A. Indehiscent fruits

Developing from a single carpel

a. Caryopsis: simple and small, containing only one seed; the testa (seed coat) becomes fused to the fruit wall during maturation (grasses: wheat, corn, oats).

b. Achene: like a caryopsis, but the seed and fruit remain distinct. Fruit wall is thin and papery (sunflowers).

c. Samara: a one-seeded fruit with winglike outgrowths of the ovary wall (maples, alder, ash).

Developing from a compound gynoecium (a compound pistil)

a. Nut: although the gynoecium originally consists of several carpels and ovules, all but one ovule degenerate during development. Pericarp is hard at maturity (walnut).

B. Dehiscent fruits

Developing from a single carpel

a. Legume: fruit breaks open along both sides (beans, peas).

b. Follicle: fruit breaks open on only one side (columbine, milkweeds).

Developing from a compound gynoecium

a. Capsule: opens many ways:

Splitting along lines of fusion (*Hypericum*).

Splitting between lines of fusion (*Iris*).

Splitting into a top and bottom half (primrose).

Opening by small pores (poppy).

b. Schizocarp: compound ovary breaks into individual carpels called mericarps.

3. Compound fruits

a. Aggregate fruits: carpels of flower not fused, but grow together during fruit maturation (raspberry).

b. Multiple fruit: all the fruits of an inflorescence grow together during fruit maturation (pineapple).

Carbohydrates Provide Most of Our Metabolic Energy

For almost all people, carbohydrates such as bread, pasta, rice, and potatoes make up more than half their food's bulk and calories. Most carbohydrates are simple sugars like glucose and fructose, disaccharides like sucrose, or glucose-based polymers such as starch (amylose and amylopectin) and glycogen. Most carbohydrates are digested to their component monomers such as glucose and fructose, which are then used for the same processes of respiration, construction, or storage. However, the different carbohydrates affect our health in important ways. The simple sugars glucose and fructose do not need to be digested; they can be absorbed by the small intestine and transferred to the blood immediately. Sucrose requires only a simple, very quick cleavage, and the acid in our stomach immediately breaks it into glucose and fructose. In contrast, both starch and glycogen are large molecules with just a few sites where enzymes can act, so digestion is slow compared to the rapid uptake of simple sugars. Also, each may be located in cells that physically impede their digestion until we have chewed our food well enough to break down cell structure. This is especially true when we eat foods that are close to their natural state: baked or boiled potatoes, boiled rice, steel-cut oats, coarse ground flours (FIGURE 13.1). Finely ground flours or potatoes processed to be instant mashed potatoes have had the cell walls physically broken and the starch can be digested very rapidly.

The important concept here is the rate at which carbohydrates are digested and enter our blood stream after we eat them. Our blood sugar level (the concentration of glucose in our blood) affects many aspects of our metabolism; when the level is high, we feel energetic, when low we feel lethargic. Many other aspects of cell- and organ-level metabolism are influenced by blood sugar level. Our pancreas monitors the concentration of glucose in our blood, and as it begins to rise after a meal or a drink, the pancreas secretes the hormone insulin, which causes the liver and muscles to absorb glucose and polymerize it into glycogen for temporary storage. Insulin also enables several of our tissues and organs to use glucose properly. Once our meal is digested and the small intestine is no longer adding glucose to the blood, blood sugar levels drop and the pancreas secretes glucagon, which triggers the liver and muscles to depolymerize their glycogen and release it into the blood.

FIGURE 13.1. The sugar on top of this oatmeal is sucrose, which our bodies digest quickly. The oatmeal contains starch, which is digested and absorbed slowly because our digestive enzymes must first diffuse through the cell walls of the oats before they can start to work. Once liberated from the starch, the glucose molecules must diffuse back out through the cell walls. These oats have been rolled to break open the tough outer layers of the grains and then they were cooked to hydrate the cell walls and the starch grains so that both are more permeable to the enzymes and glucose. If they had not been rolled or cooked, the oat grains would pass through our bodies almost intact.

As long as all this happens slowly, the two hormones and the various organs keep our blood sugar level more or less constant and we have a sense of well-being. But certain foods such as sweetened soda drinks, candy, and children's cereal are rich in simple sugars, and they need almost no digestion, so glucose floods into the blood stream very rapidly, and blood sugar levels rise to unhealthy levels. The pancreas secretes large amounts of insulin, but this takes time, and by then all the glucose has been absorbed and the small intestine is empty. Once the insulin finally causes the liver to absorb glucose, the blood sugar level is normal but is now driven very low by the liver. The pancreas must make an emergency release of glucagon, and our body's metabolism swings from one extreme state to another, which is not healthy. You can experience this yourself if you drink a sugar-rich soft drink (and are not diabetic): wait until you are very hungry and then drink just a can of soda with sugar (not with artificial sweetener) and do not eat anything with it. Within minutes you should feel extremely energetic, and less than an hour later you will be exhausted. It is much more healthful to eat foods that have complex carbohydrates and only low amounts of free sugar.

On food labels, words to watch for are "glucose," "sucrose," "fructose," "corn syrup," and "high fructose corn syrup"; all release simple sugars quickly. Corn syrup is made from corn seeds by depolymerizing their starch into glucose, then converting part of the glucose to fructose. Depending on the ratio of glucose to fructose, the product is either corn syrup or high fructose corn syrup. The benefits of using either type of corn syrup is that it is cheaper than sucrose (table sugar, the typical sweetener), and it does not have "sugar" in its name so some people are fooled into thinking foods with corn syrup are low in sugar when in fact they are not. Similarly, honey is natural but it too is mostly water with glucose and fructose. Many natural fruit juices are also mixtures of glucose, fructose, and sucrose; when eaten as part of a raw fruit, their concentrations are moderate and we must chew the fruit long enough to break down the cells, and this slows the release of the sugars into our blood. But fruit juice drinks do not have to be chewed, they can be gulped.

Plant-based foods also contain carbohydrates that are either difficult or impossible for us to digest. Cell walls contain hemicelluloses composed of a variety of simple sugars such as xylose, mannose, galactose, and others. These are digested and absorbed only slowly if at all; those that are absorbed are usually excreted by our kidneys so they do not supply us with any nutrition. Cellulose, the main component of both primary and secondary cell walls in plants, is completely indigestible for humans and most other animals that lack four stomachs. Even though cellulose is made entirely of glucose monomers, just like starch is, we have no enzyme that attacks the beta-1,4 bond of cellulose. Even ruminants such as cattle cannot digest cellulose themselves; they eat cellulose-rich hay (alfalfa or grass), swallow it, and digest out the proteins and starches, then regurgitate it into their mouths and rechew it as a cud. The cud is then reswallowed into a different stomach where microbes break the cellulose down and live on the glucose. After the population of bacteria has increased greatly, the cattle transfer the bacteria to another stomach and digest them. It is a circuitous process, but cattle ultimately derive much of their energy and glucose from cellulose. Dinosaurs are believed to have lived on cellulose by having such long, large digestive systems that plant material remained in the animals for days, during which microbes fermented the cellulose.

Plant-based foods like lettuce and spinach are rich in vitamins and cellulose but low in sugars and starch. If we tried to live on just leafy vegetables, almost all the carbohydrates would pass right through us and we would starve. Similarly, we never eat ordinary wood because the lignin in it is just as indigestible as is cellulose. Furthermore, ordinary wood is low in starch and sugar; we could chew for hours and obtain virtually no glucose. We do, however, eat certain specialized woods: carrots, beets, turnips, and other root crops consist mostly of an extremely parenchymatous wood filled with starch.

When we need a plant-based food as a source of carbohydrates, we choose plant parts rich in sugars and starches and low in cellulose and lignin. We choose cereal grains, potatoes, and other tubers.

The Grass Family, Poaceae

The grass family sustains most of human life on Earth. If all grasses were to instantly disappear, almost everyone would starve to death within a few weeks. All our cereal grains are grasses: wheat, rice, corn, oats, rye, barley, sorghum, millet

FIGURE 13.2. Three cereal grains are present here: rye on the left, wheat at the bottom, and barley on the upper right. Wheat is one of the most important sources of energy in our diets, and rye, barley, and other cereals provide both good nutrition as well as a diversity of flavors.

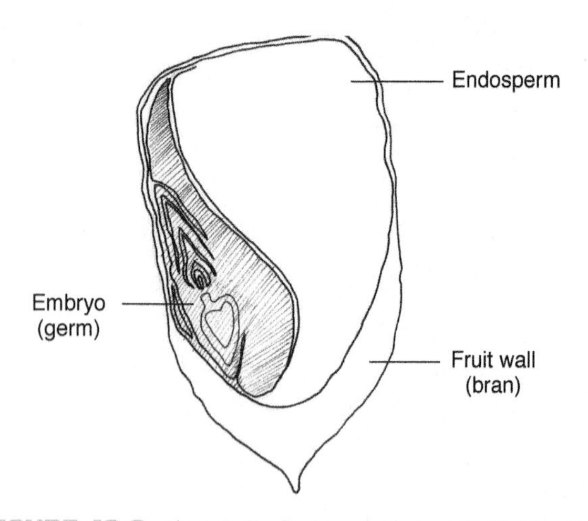

FIGURE 13.3. Almost all of a kernel of grass fruits ("seeds") is endosperm; the embryo (referred to as the "germ") is tiny and located at one end.

FIGURE 13.4. A wheat plant with mature fruits.

(FIGURE 13.2). These are all rich in easily digestible amylose and have only a small amount of indigestible cellulose in their bran (the seed coat and fruit wall), and most have high levels of protein as well, especially the whole grains. Notice that we eat only the fruit and seed of grass plants; all stems, leaves, and roots are not eaten because they consist mostly of lignified cellulose with few nutrient molecules.

The grains of cereals deserve a bit of explanation. Each grain or kernel is often called a seed but in reality each is a dry fruit that contains a single seed so large that the seed and fruit fit tightly together. These fruits are technically known as a caryopsis (plural caryopses). The outermost layer of a grain is the thin, dry fruit wall, which developed from the ovary wall of the grass flower (see Chapter 7). The inner part of the grain is the seed itself, consisting predominantly of endosperm, the nutrient reserve tissue found only in seeds. The seed coat, which developed from the integuments of the ovule, fuse tightly to the fruit wall, and the seed coat plus the fruit are called the bran. Each grain has a very small embryo near its base; the embryo is often referred to as the "germ" (FIGURE 13.3). Even though it is rich in protein, the embryo is often removed when grains are converted to flour or for use as grains in cooking, because enzymes from the embryo attack the endosperm and cause the grains or the flour to become rancid if stored for more than a few months. Whole-wheat flour, which has the germ (embryo) ground up in the flour, can only be stored for a few months whereas white flour (with no embryo) will keep well for years. Similarly, whole grain rice contains the embryo and thus has more protein than does polished rice, which has had the embryo removed.

Wheat is the grain most consumed by westerners (FIGURE 13.4), but we almost always encounter it after it has been ground into flour and baked into bread, pasta, pastries, cakes, cookies, and so on. In the most common white flour, the endosperm is ground until most cell walls break: the flour particles we can rub between our fingers are clusters of starch grains (amyloplasts) held together by a bit of dry protoplasm and cell wall. If ground even finer, as for wedding cakes, the flour particles are mostly just individual starch grains.

Wheat is a popular food because it has an unusual protein called **gluten** that allows it to rise. As wheat flour is moistened, the gluten particles hydrate; as the dough is kneaded, the gluten is stretched into stringy strands that form a network throughout the dough, making it springy and elastic. If yeast has been added (or is naturally present as in sour dough bread), its respiration produces bubbles of carbon dioxide that are trapped by the gluten network, causing the dough to rise (FIGURE 13.5). Allowing wheat dough to rise before cooking it makes it soft, easy to chew, and more readily digestible. If the dough is baked without kneading or rising, it is called unleavened bread. Other grains such as corn and rice have no gluten and will not rise or become airy. Breads made from these flours tend to be very dense, but they can be made lighter by mixing wheat flour in with the other flour, allowing it to rise a bit. Unfortunately, some people are allergic to gluten and cannot eat wheat products.

Many types of wheat have been cultivated in the 9000 years since wheat was domesticated in the Fertile Crescent. Einkorn wheat (*Triticum monococcum*) may have been one of the very first types. It is diploid (each nucleus has two sets of chromosomes), has small plants, and grows well even on stony, dry soil. Wild plants can still be found in Turkey growing on land too rocky and marginal to be farmed. Einkorn wheat is a relict crop, which means that it is rarely cultivated at present. Emmer wheat (*Triticum dicoccum*) was also domesticated very early; it is a tetraploid wheat (four sets of chromosomes in each nucleus) resulting from the interbreeding of two genetically distinct parental species. Emmer wheat is a small plant with a low yield (few grains per plant), but this was the dominant wheat for early people; in fact, this is the wheat cultivated by ancient Egyptians and traded throughout the world. It is still cultivated on a small scale in Italy and Turkey, and it is often referred to as "farro." Bread wheat or common wheat (*Triticum aestivum*) is the most widely cultivated wheat today. Bread wheat is hexaploid (six sets of chromosomes in each nucleus), so it is the result of hybridization between two separate species at first and then with a third species later. We are not yet entirely certain which wild species are the parents. Unlike einkorn and emmer wheat, bread wheat is a large plant that is high yielding, producing many seeds per plant. It is not as hardy as the other two, so farmers must give it more attention; but, bread wheat is so high in protein that it is the world's largest source of plant protein.

Rice (*Oryza sativa*) has been called the world's most important crop because it is the main food source of more than half the people on Earth (FIGURE 13.6). It was the foundation of agriculture in Asia beginning at least 7000 years ago. Just as with wheat, rice is often "polished," removing the bran and germ and thus producing a form that can be stored a very long time and that is easy to eat. Polishing also removes

FIGURE 13.5. Gluten in wheat is a stringy, elastic protein that forms a network as bread dough is kneaded. The network traps carbon dioxide bubbles, causing the dough to rise. If there were no gluten, the carbon dioxide bubbles would merely float to the surface and burst, and the dough would remain flat.

FIGURE 13.6. A rice plant with mature fruits. (© Revensis/Dreamstime.com.)

vitamin B1; lack of this leads to a deficiency disease called beriberi. In many parts of the world where rice is the principle food, vitamin A deficiency is common. Genetically engineered "Golden Rice" was developed to have very high levels of beta-carotene, the precursor needed for vitamin A synthesis. However, concern over the safety aspects of genetically modified organisms (GMOs) have caused people not to accept Golden Rice, and it is not being used for human consumption.

Domestication and artificial selection have produced numerous varieties of rice. *Oryza sativa* subsp. *indica* is called long-grain rice in the United States, and after it has been cooked, the grains tend to have a smooth, dry surface and the rice is fluffy. *Oryza sativa* subsp. *sativa* (often called japonica) becomes more moist and sticky when cooked, and it is more easily eaten with chopsticks. Most rice is still grown on terraces (or paddies) with standing water, but upland rice does not need to be flooded and it can be grown in ordinary fields. Wild rice differs from rice, being in a completely different genus, *Zizania aquatica*.

Most of the wheat and rice crops are eaten by people, which makes us their primary consumers. Of the grains described next, most of the crops are used as animal feeds and then we eat the animals or their milk; we are secondary consumers of these cereals.

Oats (*Avena sativa*; FIGURE 13.7) are familiar as breakfast food and cookies. They may have been one of the last major grains to be domesticated, about 3000 years ago. Oats are high in protein and are a good source of lipids, and recent research shows they help keep our cholesterol at healthful levels. Although oats were mainly used as animal feed in the past, they are becoming much more popular now, especially as oatmeal for breakfast. Steel-cut oats are entire grains that have been cut in half and thus have had little processing and retain most of their nutrients and fiber (FIGURE 13.8). Rolled oats

FIGURE 13.7. An oat plant with mature fruits.

FIGURE 13.8. The oat kernels on the top are "steel-cut" oats: they have been cut but not cooked or otherwise processed, so they retain all their natural nutrition. The oats on the bottom have been soaked or partially cooked to soften them, then they have been "rolled" to break open the surface so that they are more digestible.

FIGURE 13.9. Barley plants with mature fruits. (© LiquidLibrary.)

have been partially cooked to soften them and then they are flattened between rollers to crack the fruit wall and seed coat to make them softer and easier to eat. Oats that can be cooked in just 5 minutes or less, or that need only to have boiling water added to them, have been completely precooked and then dried, steps that reduce their nutritional value as well as their flavor.

Barley (*Hordeum vulgare*; FIGURE 13.9) is believed to be one of the first crops domesticated in the Fertile Crescent, apparently about 9000 years ago. Probably most of us would have difficulty remembering when we last ate barley; it is rarely used in breads or breakfast cereals, although it is used in hearty soups and stews. But most of us drink barley, and we can thank the ancient Sumerians for that: they discovered how to brew barley into beer. At present, about half the barley crop in the United States is used to feed livestock, and much of the rest is used to make beer and whiskey.

We eat only small amounts of rye (*Secale cereale*; FIGURE 13.10), mostly as rye bread. Rye has less gluten than does wheat; it is difficult to separate the bran from the endosperm of rye, so rye flour tends to be coarser than wheat flour. Pure rye bread is often flat and hard, and it can be dried until it is basically a cracker. This form can be stored well, and in the past, it was one of the main foods on board sailing ships. We still do eat rye crackers, but when rye is made into bread, we almost always add equal parts of wheat flour for its gluten so that the rye bread will be softer and easier to eat. Through much of our history, rye had been an important cereal because it grows well in areas that are too cold for wheat. Rye seeds germinate even if temperatures are just 1°C above freezing (34°F), and in cool summers, a temperature of just 12°C (55°F) is warm enough for rye to mature and be harvestable. Consequently, rye

FIGURE 13.10. Rye plants with mature fruits. (© yuris/ShutterStock, Inc.)

has allowed people to live father north than if only wheat was available. Rye seed tends to be so inexpensive that it is often used as a cover crop for land in winter. Parts of the chaparral in California burn each summer (see Figure 10.19), and then rains later cause mudslides; rye seeds sprout so quickly and their roots grow so extensively that rye seed is often spread by airplanes over burned hillsides as an effort to stabilize the soil before the rainy season starts. The rye germinates and grows more rapidly than native plants, but being an annual, the rye dies at about the time that native vegetation has recovered and can stabilize the soil on its own. The rye does not become an invasive weed. In housing developments, rye is used to quickly give the new homes an appearance of having a lawn, and the homebuyer can easily replace the rye with lawn grass or garden plants.

North Americans and Mexicans are almost unique in the world for eating corn (*Zea mays*). Corn was domesticated in Mexico perhaps as much as 9000 years ago, and has been a significant part of the Mexican diet as corn tortillas, masa, and many other forms ever since. The United States now rivals Mexico in consumption of corn tortillas, as well as corn bread, corn muffins, corn fritters, whole kernel corn, corn-on-the cob, and in many other ways. Some of us drink it as bourbon. Most of the world just uses corn as animal feed, both the grain itself, and the entire plant cut while green and chopped into small pieces. Corn kernels are low in protein, and especially low in the amino acids tryptophan and lysine; lack of tryptophan causes the deficiency disease pellagra. Genetic engineering has created varieties of corn that are higher in these two amino acids and which are thus more nutritious, but there is hesitation to eating GMO corn. Much of the corn crop is used to make corn syrup, as described above, and very recently we have begun fermenting it to make ethanol as a biofuel (see Chapter 11).

An ear of corn is a familiar example of grass reproduction (see Figures 7.25 and 12.6). Corn plants are monoecious; they have staminate flowers at the top of the plant, in an inflorescence (a group of flowers) called a tassel. The ears of corn are axillary buds that bear an inflorescence of carpellate flowers. The husks that we peel off and throw away are the large scale leaves that protect the inflorescence. The "silks" are the styles, each one being connected to one kernel of corn, and each one had a pollen tube grow through it from the outer end to the kernel, carrying two sperm cells. By the time we harvest corn to eat, the carpellate flowers have begun to develop into caryopses. The cob is the stem (the nodes and internodes) of the inflorescence. No other grass species has an inflorescence like an ear of corn, it is unique in the grass family.

FIGURE 13.11. Potatoes have been subjected to intense artificial selection and now are available in numerous sizes, shapes, colors, textures, and flavors.

Other Plants Rich in Starches

After cereal grains, the most important starch-rich foods are various tubers (white potatoes and yams) and roots (sweet potatoes and manioc).

All North Americans are familiar with potatoes (*Solanum tuberosum*) as food (FIGURE 13.11). We eat them baked, boiled, fried, and roasted; we eat them whole or mashed or cut into French fries or sliced into potato chips. Few North Americans go more than a day or two without eating potatoes in one form or another.

The potatoes we eat are a special type of tuber. The plants have ordinary upright, bushy shoots with green leaves and small flowers that resemble tomato flowers (potatoes and tomatoes are closely related). Several of the axillary buds of the lowermost internodes of the shoot become active and grow out horizontally as very slender

(2 to 3 mm diameter [less than a quarter inch]) underground shoots. After growing only several centimeters (an inch or two), the tip of the shoot begins to swell and the shoot stops growing; it is a determinate shoot. The last several nodes and internodes develop to have a very large diameter, the diameter of a potato. These are as much as 10 cm (4 inches) wide in the large baking potatoes. This large diameter is worth thinking about; it is all primary body: epidermis, a thick cortex, a thin ring of vascular bundles, and a very broad pith. It is extremely unusual for a shoot apical meristem to produce such a broad stem; cacti, cycads, and a few succulents are the only other plants that do this. Most other broad stems result from a vascular cambium that produces wood and secondary phloem, but that does not occur in potato tubers.

Being a shoot, the potato tuber has a shoot apical meristem, leaves and axillary buds. You have noticed the axillary buds, but probably called them "eyes" (see Figure B2.1b in Box 2.1). The leaf that makes the axil is either extremely tiny and difficult to see or is represented by a slight wrinkle near the eye. At the proximal end (near the plant), you might be able to see a bit of the narrow, horizontal stem that connected the tuber to the plant. Despite being so narrow, the phloem of its vascular bundles was able to transport sugar rapidly enough for the tuber to grow within a few months and fill itself with starch. At the opposite end of the tuber are several small eyes in a cluster; the center one is the shoot apical meristem. If a potato has a slightly thickened, tan or brown skin or peel, that indicates that the epidermis has converted itself to cork cambium and produced cork.

Potato tubers are perennating organs; that is, they are a means by which a potato plant survives a stressful season, usually winter. In late summer, the tubers are mature, turgid, and their cells are filled with starch grains (amyloplasts; see Figure 3.9). Potatoes also store protein and many vitamins, especially vitamin C. In late summer or autumn, the above-ground leafy shoot dies, but its many subterranean tubers remain alive, surviving freezing temperatures and prolonged winter. In springtime, several axillary buds become active and produce shoots that grow upward, guided either by negative gravitropism or positive phototropism. While still growing upward through the soil, they are etiolated, having slender, white stems, and tiny, unexpanded leaves (see Figure 6.9). After the shoots break through the soil surface and perceive light, all further growth is normal, not etiolated: nodes and internodes are green, broad, and strong; leaf primordia develop as typical dicot leaves with a broad lamina and so on.

Until the shoots are photosynthesizing well enough to be self-sufficient, the cells of the tuber digest the starch in their amyloplasts to glucose, convert it to sucrose, and transport it through phloem to the growing shoots. The tuber has enough carbohydrate reserve to ensure that several shoots reach the soil surface, even if the tuber is buried deeply and the shoots must grow without photosynthesis for several centimeters. As the tuber uses its starch, it becomes soft and less nutritious.

Almost all potatoes are propagated by tuber cuttings rather than seed. In the spring, tubers are cut into several pieces, each with at least one axillary bud, dusted with fungicide to protect the cut surfaces, and then they are sown. Potato seed is almost never used except by geneticists when experimenting to produce new varieties. Most potato **cultivars** are **male sterile**; they fail to produce viable pollen.

Potatoes are so ubiquitous in the diet of the United States, it is easy to think of potatoes as quintessentially American. They are, but it is South America where they were domesticated, perhaps as long ago as 11,000 BCE in the Andes of Chile. If that is correct, it would not only mean that potatoes were the first domesticated plant, but also that agriculture arose in the Americas before it did in the Fertile Crescent. Although Chile may have been the site of domestication, highland Peru was the center of preindustrial cultivation of potatoes. Marketplaces in Andean villages of Peru still have numerous varieties of potato, mostly small and of various colors. It was from these that the typical large white potato was developed and then distributed throughout Europe and North America (see Figure 13.11).

Potatoes gave us one of our most memorable lessons in the problem of relying on a single crop, as well as the fact that plants can strongly influence human affairs. In the United States, potatoes are cultivated in warm areas with sandy soil that has good drainage; tubers tend to rot in moist soil. But early cultivars from the Andes were adapted to cool, moist soil and they thrived in Ireland where wheat does not grow well. Potatoes were one of the main crops in Ireland and produced enough food per acre to support a large population. Then in 1843 and 1844, a disease called potato blight, caused by the alga *Phytophthora infestans*, swept through the country, ruining almost the entire crop (see Box 18.1). Potatoes were being cultivated so densely that some spores from any infected plant had a good chance of landing on leaves of another potato plant and infecting it. Because the Irish farms were not diversified, the failure of this one crop meant the almost complete absence of food for the entire country. It is estimated that as many as one million people died of starvation, and many millions more emigrated to the United States and other countries. The influx of Irish into the United States was great enough to significantly affect our culture and was one of the largest mass migrations in history. Even today, many countries are dependent on a single crop, especially rice, and if another devastating disease were to arise, millions of people would perish, even in the twenty-first century.

After white potatoes, yams (several species of *Dioscorea*) are the second most important tuber food worldwide (**FIGURE 13.12**). Yams were domesticated in Africa and are still widely cultivated there; they are not particularly abundant in the United States. Various species have been domesticated in Africa, Asia, and South America and are the most important food crops in those areas. Yam tubers are much larger than those of white potatoes, sometimes being up to 2.5 m (8 ft) long and weighing as much as 70 kg (150 lbs). In addition to starch, yams are rich in vitamin C, vitamin B6, and several minerals.

FIGURE 13.12. This is a species of *Dioscorea*, one of the true yams. In the United States, we use the words "yam" and "sweet potato" interchangeably for true sweet potatoes (*Ipomoea batatas*). It is very rare to encounter true yams in grocery stores in the United States. (© QUAN ZHENG/ShutterStock, Inc.)

Sweet potatoes (*Ipomoea batatas*) are true roots, not tubers, and are not closely related to white potatoes (**FIGURE 13.13**). They are members of the morning glory family (Convolvulaceae), and their shoots are vines. Sweet potatoes were domesticated at least 5000 years ago in tropical South America, and are still cultivated mostly in tropical areas; the plants do not tolerate frost. Sweet potatoes are considered one of the most nutritious foods known. Their dark orange color indicates they are rich in beta-carotene, which our bodies need to synthesize vitamin A. Sweet potatoes also are high in complex carbohydrates, the ones that are digested slowly and thus release glucose more evenly. Sweet potatoes have abundant fiber, vitamin C, and vitamin B6.

Cassava (*Manihot esculenta*) is the most important root crop worldwide, but North Americans do not eat much of it. We know it as tapioca, but many other people call it manioc. It is especially important in tropical regions as it grows well in both dry areas as well as wet ones, and it tolerates poor soil. It is the primary source of carbohydrates for over half a billion people, but unfortunately, it is almost pure starch and is extremely low in protein (only about 1%) and vitamins. Cassava is one of several crops whose domestication is difficult to understand because it requires complex preparation. The roots are extremely poisonous when harvested, containing toxins called cyanogenic glycosides, compounds that release cyanide when a sugar molecule is removed by an enzyme. "Sweet" cassava is less toxic than "bitter" cassava, but farmers prefer the latter because it has less trouble with animals, pests, and thieves. When processed at home, the roots are ground into a flour and mixed with water; the paste is then spread as a thin layer and allowed to stand for several hours. An enzyme in the cells themselves cuts the sugar from the glycoside, releasing the cyanide, which diffuses away. For large-scale production, the roots are ground into flour, mixed with water, and then squeezed dry. This washing and squeezing is repeated several times, with fresh water each time. Finally, the flour is dried and toasted. The wash water has enough cyanide to cause environmental damage where large amounts of cassava are processed.

Cassava can be eaten many ways, but often the flour is added to other ingredients as a thickener. If used by itself to make a pudding, it is tapioca. A new novelty use is in bubble or pearl drinks, in which balls of tapioca pudding are added to tea and drunk with a straw wide enough to suck up one "bubble" or "pearl" of tapioca at a time.

In most of the world, cereal grains, especially wheat and rice, supply the largest amount of carbohydrate in most diets. But in tropical Africa, the tuber and root crops are much more important.

Bananas (*Musa acuminata* [until recently *Musa paradisiaca*]) are one of the few fruits that are starchy rather than sugary or oily (**FIGURE 13.14**). Some types of bananas produce seeds, but the ones we eat are sterile hybrids whose ovules stop developing while still rudimentary. In a banana flower, three carpels are fused together and there are three rows of ovules, as is typical of monocots. After the ovules stop developing, the tissues where they attach to the carpel wall (the placentas) swell and develop into the part of the banana we eat. The carpel wall develops into the true fruit, but we know it as the peel and just throw it away. Bananas are climacteric fruits; they ripen slowly for months, then very quickly become mature in just a few days as the hormone ethylene builds up within them. For shipping, bananas are picked while still green and hard,

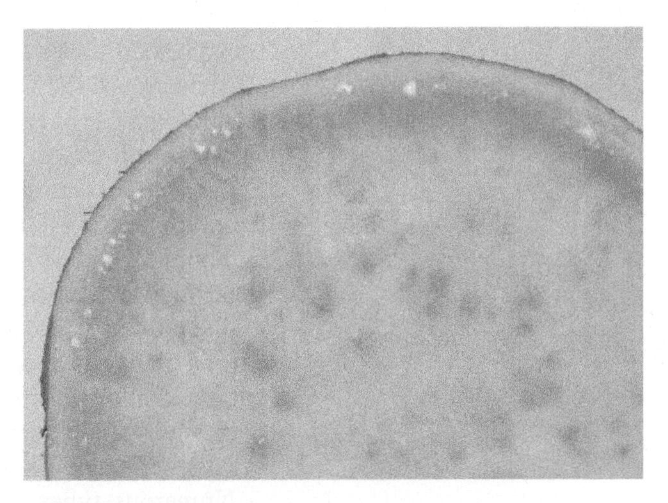

FIGURE 13.13. Sweet potatoes (*Ipomoea*) typically have orange flesh whereas yams (*Dioscorea*) are usually white. It is necessary to say "usually" for sweet potato color because recently new varieties with various colors have been introduced. Uncooked sweet potatoes also exude a small amount of milky sap when they are cut.

FIGURE 13.14. Banana plants are large herbs with giant leaves and a "trunk" that consists of concentric sheathing leaf bases. As is true of all monocots, bananas do not produce wood, and unlike bamboo and palms, they produce only weak, flexible fibers: most parts of a banana plant are too soft to be used for any kind of construction. (©Thirteen/ShutterStock, Inc.)

transported in cool temperatures (10°C, 56°F) with plenty of ventilation to remove ethylene. Once they are at market, they are brought back to room temperature and are given a dose of ethylene gas, which induces them to complete their maturation as they sit on the store shelf.

Because they produce no seeds, bananas must be propagated vegetatively. Each banana plant is a large herb; as is true of most monocots, they have no vascular cambium, wood, or secondary phloem. Most of the plant's trunk consists of many sheathing leaf bases (similar to the construction of an onion), and many of the axillary buds at the base of the plant develop as short branches called suckers. After a plant has flowered, fruited, and been harvested, the main trunk dies and is cut back, allowing the suckers on one side to develop. Each year, a row of banana plants moves a bit sideways across the field. Some plants will produce fruit for as many as 25 years, then must be replanted from fresh suckers.

Numerous types of bananas occur, but most are not shipped long distances. Larger bananas called plantains have a firm texture and are cooked rather than being eaten raw.

Plants Rich in Sugars

All plants produce sugars and transport them through their phloem, but only a few species have concentrations high enough that the sugar can be extracted for food.

Our main source of sugar is the phloem sap of sugarcane (*Saccharum officinarum*). This is a giant perennial grass that grows to 2 to 6 m (6 to 19 ft) tall with thick stems (see Figure 11.16). It is cultivated in warm, rainy areas such as the southern Gulf states and Hawaii in the United States, in Brazil, India, and other tropical areas. Sugarcane is propagated vegetatively by planting pieces of stem that contain one or more axillary buds. Tall aerial stalks grow quickly, and rhizomes spread below ground. A field is harvested by cutting off the aerial stalks and allowing the rhizomes to send up new shoots the following year. A field can be harvested several times, but sugar content gradually decreases and the field then must be replanted. Photosynthesis in sugarcane is one of the most efficient known, capturing as much as 2% of the energy in the light that strikes the plants.

Sugarcane must be processed rapidly after it has been cut during harvest or its sucrose will break down into glucose and fructose. The cane is shredded and mixed with water, then pressed to obtain the juice, which is 10% to 15% sucrose (the phloem sap itself can be 22% sucrose). Lime is added to prevent the breakdown of sucrose and then the juice is concentrated to a syrup of about 60% sucrose. With further concentration, sugar crystals form, and these are separated from the liquid by centrifugation. The sugar is called raw sugar and is yellow or brown, due to various residues of the plant body. The liquid is molasses, and if more sugar is removed from it, it becomes blackstrap.

Raw sugar is refined further. It is dissolved into a heavy syrup that is then treated with several chemicals to trap and remove impurities. The syrup is concentrated and crystallized again, and is again centrifuged. To make the sugar pure white, it is filtered through activated carbon (made from charred bones). This is now the sugar we see on our tables and that we use for baking and cooking. Refining raw sugar into white granulated table sugar produces more molasses as leftover liquid.

Sucrose and molasses are only two of many products from sugarcane. Cane juice is fermented to cachaça, the most popular distilled alcoholic beverage in Brazil. Molasses is fermented to rum. Recently, a great deal of sugarcane juice is fermented for ethanol as a biofuel, with Brazil leading the world in this technology. After juice is extracted, the shredded stems are known as bagasse, which has many uses. It is so fibrous it can be used to make paper, cardboard, and other fiber products. After drying, it burns well and is often burned as fuel to run the sugar refining factory. The bagasse typically produces more electricity than the factory needs, and the excess power can be sold.

Sugarcane was domesticated in Southeast Asia at least 5000 years ago, and its cultivation spread worldwide. Before the discovery of extracting sugar from sugarcane, the only significant sweetener anywhere was honey (which is plant nectar that has been processed by bees; see below). Sugarcane was brought to the Americas by Christopher Columbus on his second voyage, and sugarcane cultivation was established in the Caribbean. Unfortunately, sugarcane plantations needed large amounts of manual labor, and the slave trade between Africa and the Americas was established in large part to cultivate sugarcane.

FIGURE 13.15. Sugar beets grow to a large size (about the size of a football) in just one year. Although the swollen root contains a high concentration of sugar, its flesh is firm and solid, and is not particularly sweet if eaten raw. Sugar beets are white, not red like table beets. (© Marek Pawluczuk/ShutterStock, Inc.)

Sugar beets (*Beta vulgaris*) also provide sucrose, but their cultivation is much smaller than that of sugarcane (FIGURE 13.15). Sugar beets are a biennial crop cultivated as an annual; seeds are sown in spring and by autumn the plants produce a large, swollen root—much larger than that of red beets in grocery stores—that can be as much as 20% sucrose. Sugar beets are processed just like sugarcane; the roots are shredded, the juice extracted and then concentrated, crystallized, centrifuged, and refined as described above.

Sugar beets are an interesting example of an early success story of plant genetics. As early as the 1700s, long before Gregor Mendel had worked out the basics of genetics, people were selecting beets that had higher sugar contents. In 1804, during the Napoleonic Wars, the British imposed a blockade of France, cutting off all sugar imports from French colonies. Sugarcane will not grow in France, but beets will. Napoleon encouraged the breeding of higher-yielding sugar beets, and within a few years, beets with much higher sugar content had been bred and were being cultivated in France.

Syrup and sugar can be obtained by tapping certain trees in springtime. Maple syrup is most well-known, but birch and other trees are also used to a lesser extent.

Honey

Honey is a concentrated sugar solution manufactured by honeybees (*Apis mellifera*) starting with nectar they collect. Flowers that rely on animals to pollinate them must provide some type of reward; certain animals eat pollen, others gather lipids from the flowers, but most animals seek out nectar. The composition of nectar varies from species to species but is always predominantly a dilute solution of sucrose, glucose, and fructose. Most pollinators drink the nectar and digest it immediately, but honeybees instead transfer it to a special part of their digestive system, the honey sac. As the bees fly back to the hive, honey sac enzymes break down much of the sucrose into glucose and fructose. Once back at the hive, the bees that have gathered the nectar transfer it to other bees that concentrate its sugar by repeatedly drinking then regurgitating the nectar. The regurgitated nectar forms a small droplet at the bee's proboscis, allowing water to evaporate before the droplet is ingested again. Each droplet is cycled about 15 or 20 times, until enough water has been lost that the sugar concentration reaches 50% or 60%. At that point, the nectar is placed on a layer of wax, and other bees fan their wings, creating a strong current of air moving across the nectar. This evaporates even more water from the nectar, and, after about three weeks, its sugar concentration has increased to 80% and it is finally honey. Bees then transfer the honey to cells of the honeycomb and seal it in with a layer of wax. The honey is stored for future use by the colony.

It might seem that a sugar solution would quickly be spoiled by bacteria and fungi, but because honey is so concentrated, it acts as if it is very dry. Honey has such a strong tendency to absorb water that most cells of fungi or bacteria that land on honey lose water to it and die from dehydration (the same thing happens to us if we drink seawater). Also, bees add enzymes that convert some glucose to gluconic acid, which makes the honey acidic enough to inhibit microorganisms.

When we harvest honey, no further processing is necessary other than to extract it from the honeycomb. The comb's cells are sliced open and the honey is forced out by centrifugation. Afterward, it may be heated to 155°F (68°C) to kill yeasts that might have contaminated it. Often honey is filtered to remove pollen, a few random body parts of bees, and tiny air bubbles that would make the honey look cloudy. At any point after being collected, the honey is ready for us to eat it.

Although honey is a natural food and, therefore, is often praised as being particularly beneficial, it is just a sugar solution. It has virtually no vitamins or any nutrients other than sugars. There may be a small amount of protein in some honeys (the more protein, the darker the honey's color), but it is always negligible nutritionally.

Various types of honey differ from each other because of their flavors. The nectar of each type of flower contains a few molecules of chemicals other than sugars, and as bees concentrate the nectar into honey, they also concentrate these molecules into flavors and aromas. Many bees collect nectar from any type of flower that has enough nectar to be worth their trouble, so the honey of any particular hive may have come from dozens of different species of flower. Bee keepers often place hives in fields or orchards so that the bees will collect nectar predominantly from just a single type of plant and, thus, will produce a "clover honey" or a "tupelo honey" and so on, each with a characteristic flavor.

The relationships between bees, plants, and people are complex. Bees and plants have a mutualistic relationship in which both benefit: the bees obtain food and the plants obtain a means of cross-pollination. If people collect honey from naturally occurring hives, they obtain food, but the bees do not benefit. This is a predation; the people are preying on the bees. Many honeybees, however, are cared for by beekeepers who build artificial hives and transport both hives and bees to areas where plants are blooming. Many of our fruit trees and other crops must be cross-pollinated by honeybees or they will not produce fruit. In such cases, bees, plants, and people form a three-part mutualism in which all members benefit.

Plant Foods that Provide Protein

The Legume Family

Many of us obtain a large fraction of our dietary protein by eating animal products such as meat, milk, yogurt, eggs, and seafood. But for most people in the world, protein-rich plant foods are a critical part of their diet. Vegetarians do not eat any animal products at all, and they rely completely on plants, algae, and fungi (see Chapter 18) as their only source of protein (**BOX 13.2**).

BOX 13.2. Vegetarianism: Alternatives to Eating Meat

Our ancestors evolved to be omnivores. As a result, we modern human beings can eat seeds, leaves, and fleshy roots because our flat molars allow us to grind and mash them before we swallow them, and our small intestine is so lengthy it ensures that plant matter remains inside us long enough to be digested. Similarly, we can eat meat because our incisor and canine teeth allow us to rip meat from bones (think of barbequed ribs, fried chicken, and pork chops) and our digestive system produces plenty of protein-digesting enzymes.

But just because we have the ability to eat meat does not mean we must eat it. A large number of people have adopted some form of vegetarianism to reduce the amount of meat in their diet or eliminate it altogether. There are several types of vegetarian diet.

1. *Ovo-lacto vegetarians* (often just called vegetarians) eat a diet based almost exclusively on plant products such as fruits, vegetables, greens, and seeds (in the form of foods like bread, pasta, and oatmeal) with some eggs and dairy products (milk, cheese, ice cream, and yogurt) (**FIGURE B13.2**). The concept here is that animals do not have to be killed to obtain eggs and milk. For some vegetarians, the only acceptable animal-based food is eggs, for others only milk-based foods, and for others even fish and seafood are acceptable.

2. *Vegans* avoid all foods obtained from animals, including meat, eggs, and all dairy products. Strict vegans also do not wear clothing made from animals, such as leather belts, shoes, and coats or anything made with fur or turtle shell.

3. A *fruitarian* diet consists of only fruits, nuts, and seeds, items that can be gathered without harming a plant. Fruitarians do not eat roots such as carrots, beets, and cassava because those can be obtained only by killing the plants.

4. *Flexitarians* have a flexible approach to meat and other animal products in their diet; they eat mostly plant foods and merely reduce the amount of meat they eat. For example, flexitarians might set the goal of having meat-free lunches or designating several days of the week as being meat-free. The advantage of a flexitarian approach is that a person can succumb to the occasional hamburger or steak without feeling that they have failed and might as well give up.

There are several reasons for reducing or eliminating animal products in our diet and our daily lives. Many people do so for ethical reasons: they do not want to be responsible for animal suffering or death. There is a range of empathy here. Red meat is muscle tissue that can only be obtained by killing cattle. Cattle are mammals like we are, and they definitely feel pain and suffering. Obtaining milk from cows or goats or other mammals (only mammals produce milk) does not involve killing them, so there is little or no suffering in many dairy operations. Gathering eggs from chickens or other birds also does not involve harming them, although some poultry farms keep the animals crowded together in poor conditions and clip their beaks so they won't peck each other. Eating fish and seafood involves killing the animals, but many people feel little emotional connection to fish, shrimp, clams, and so on.

When choosing your diet for ethical reasons, there are various issues to consider. Raising cattle in the United States and elsewhere is often cruel. Most cattle are raised on factory farms, crowded into large, dirty feedlots where the animals must stand in their own feces; they are given feed that includes antibiotics, growth supplements, and even ground debris from cattle that were slaughtered earlier. When it is time to kill them, the cattle are poked with prods that give them an electrical shock, forcing them into a line into the bright, noisy slaughter house (cattle are terrified by bright lights, noise, and having people moving around them, especially above them). Laws in many countries prohibit the slaughter of any animal

FIGURE B13.2. This is an ovo-lacto vegetarian meal because it contains eggs and cheese in addition to plant-based foods. A dessert of pie and ice cream would fit into such a diet.

(continued)

that is not healthy enough to walk into the slaughterhouse, but that law is sometimes violated. PETA (People for the Ethical Treatment of Animals) has developed a set of guidelines for treating cattle more humanely, including giving them uncrowded pastures for grazing; safe food, and when they are to be killed, the slaughterhouse should be quiet, dimly lighted and with as few people as possible. The animals are still killed, of course, but at least their lives and deaths would not be miserable. For many people, this humane treatment reduces their ethical objections and they eat meat occasionally. To be certain that the meat you eat has been treated humanely, shop at reputable health-food stores or organic food stores; don't be afraid to ask about the treatment of the animals.

Veal is especially objectionable because it is the flesh of calves who are killed while still so young they have never eaten anything other than their mother's milk. Often, they are kept in cages so small that the calves cannot move because movement would cause their muscles to develop, making the meat less tender. The calves also may be kept in the dark to keep them quiet. Treating animals like this just to have a particular kind of dinner is difficult to justify.

Eggs and milk do not require the slaughter of animals, so many people do not feel unethical consuming these products. Here, too, the chickens and cattle should be treated humanely and allowed access to open pastures or pens with fresh air and water. People with the strictest ethical standards try to avoid all animal products because they feel it is unfair to the animals for us to breed and maintain them just so that we can use them. Eliminating the use of all animal products is difficult, though: sugar is whitened by filtering it through the ash of animal bones; most cheese is made with rennet from calves' stomachs; clear juices (apple and grape) and alcoholic beverages are clarified with gelatin from animal hides and bones, and so on.

Other people avoid animal products for economic and ecological reasons. When we feed animals corn, soybeans, oats, and other feeds that we ourselves could eat, we are eating at a high trophic level. The plants are primary producers, the livestock animals are primary consumers, and we are secondary consumers. Typically it takes 10 pounds of food at one trophic level to make 1 pound of food at the next level. Consequently we need to feed cattle or chickens 10 pounds of oats just to get 1 pound of eggs or milk or meat. The other 9 pounds are lost as carbon dioxide and methane (both greenhouse gasses), urine, feces, and waste. If we would eat the plants ourselves we would need to farm only one-tenth the land as we use now, and there would be much less sewage produced. As much as 70% of all the wheat, corn, and other grains we cultivate is used to feed livestock. Also, huge areas of tropical rainforest have been and are being cut down and converted into pasture for cattle just to produce hamburger. Reducing the amount of animal products we eat is an important aspect of a philosophy of living more simply, of trying to minimize the impact each of us has on our environment.

An important consideration for anyone considering a vegetarian diet is whether it is healthful. The answer is easiest for a flexitarian diet: virtually all studies show that diets that are low in meat and high in fruit and vegetables are much more healthful than ones with more meat. And it is now well established that even a strict vegan diet provides all the nutrients we need for a healthy life. We need to be especially careful to obtain adequate amounts of vitamin B_{12}, which is not found in any plant but which can be obtained as a supplement or by consuming yeast products. Many people believe that athletes and bodybuilders need a high protein intake and must have meat, but that has been shown repeatedly to not be true. Most of us Americans consume much more protein than we need, and the plants of a vegetarian or vegan diet provide adequate amounts for anyone. We must be careful to obtain adequate amounts of all 8 essential amino acids. All plants, animals, fungi and other organisms construct their proteins from the same 20 amino acids, and we humans can synthesize 12 of them within our own bodies, but our metabolism cannot make 8 of them. Dairy and egg foods supply adequate amounts, and so do the proteins of soy, amaranth, quinoa, and several other seed plants. But the proteins of many food plants such as wheat, rice, beans, chickpeas, and many others have adequate amounts of 18 or 19 of the amino acids but only sparse amounts of one or two others. We can avoid having a deficiency by eating what are known as complementary foods, which are foods in which one member has plenty of an amino acid that the other member lacks, and vice versa. For example a combination of brown rice and beans, beans and corn, tofu (soy) and rice, or one of hummus and wheat pita bread provides adequate amounts of all 20 amino acids. Think about a typical Mexican meal that includes beans and corn tortillas; these two plant foods together provide all essential amino acids. The same is true of a typical American lunch: a peanut butter sandwich (peanuts and wheat) provides complementary proteins.

The take-home message of this box is that we should eat less meat and animal products. Both our bodies and our environment will be healthier.

Just as the grass family provides us with most of our carbohydrate-rich foods, the legume family (Fabaceae, an old name is Leguminosae) provides almost all our protein-rich plant foods. This is the family of beans, peas, and peanuts; it is a very large family with more than 16,000 species. Many members are adapted to arid and semi-arid regions, places where people tend to be poor and have few food options. Fortunately, it is often possible for them to cultivate one species or another of legumes as a source of protein, either to be eaten directly or by using them as feed for livestock.

Many legumes form symbiotic associations with nitrogen-fixing bacteria (see Chapter 4) and this provides them with a greater amount of organic, reduced nitrogen than is available to other plants. Other plants tend to produce seeds rich in carbohydrates or oils, neither of which contain nitrogen; because legumes have a mutualism with nitrogen-fixing bacteria, they can store extra protein in their seeds. When the seeds germinate, the protein is depolymerized to its amino acids, which are then transported through phloem to the embryo's shoot and root apical meristems, allowing them to synthesize the enzymes they need. This is exactly what makes legumes such an important food for us: we digest their proteins to amino acids in our small intestine, then absorb the amino acids and use them to build our own proteins (all organisms use exactly the same 20 amino acids to build their proteins). Carbohydrates and lipids cannot be used for this because they lack the nitrogen.

The processes of searching for and domesticating new food plants continue today, and the legume family holds great promise. Many legumes are adapted to poor soil so they need little water or fertilizer, and they have their own source of nitrogen. Of the thousands of legume species that exist, many are already cultivated but many more are being studied to determine their potential as crop plants.

Technically the word "legume" refers to the plant in general or to the fruit specifically, which is almost always a pod that is dry when mature. The less familiar word "pulse" refers to the dry seeds of legumes. Legumes provide several unusual cases in which we sometimes eat the fruit, and sometimes the seeds of a plant. For example, we eat young pea pods (the fruits) as snow peas or snap peas (FIGURE 13.16), or we wait until the seeds are ripe then eat them as fresh peas, or wait until they are mature and dry and then harvest them for things like split pea soup. The same is true for beans.

FIGURE 13.16. When we eat peas as "snow peas," we are eating the true fruit, the "pod." When we eat peas as "sweet peas" or in split pea soup and so on, we are eating just the seeds and the fruits are thrown away. Pea pods become fibrous and inedible after the "snow pea" stage.

Peas (*Pisum*) have been found in archeological sites up to 9500 years old, but because those peas are the same size as wild peas, we cannot be certain if they were cultivated or merely collected. Lentils (*Lens culinaris*) too are found in ancient sites, but these lentils were larger than the wild ones that still grow nearby, and that probably indicates that the lentils were cultivated. They are considered one of the eight basic crops that were domesticated early and formed the basis of the diets in ancient Mesopotamia.

There are two basic types of beans. Broad beans (*Vicia faba*) were also domesticated in the Middle East and are known to have been used by early Egyptians. Common beans (*Phaseolus vulgaris*) were domesticated in two parts of the Americas: Mexico and Central America for some, the South-Central Andes of South America for others. Common beans are very familiar to us as the varieties kidney, navy, pinto, black, green, string, wax, and snap beans. The domestication of beans and corn in Mexico is significant because the two types of plants have **complementary proteins**. Corn proteins are poor in certain amino acids that are abundant in bean proteins and vice versa for bean proteins. A diet that contains both corn and beans provides a good balance of amino acids. Rice and beans (especially soybeans) are two other foods that also have complementary proteins. In addition, beans are vining, climbing plants, and they were grown in the same fields as corn; the beans climbed the corn plants and enriched the soil with nitrogen. Early Americans discovered both intercropping and balanced diets.

Soybeans (*Glycine max*) were domesticated approximately 3000 years ago and are one of the most important plant-based sources of protein worldwide. Their protein is **complete**, meaning that it contains all the amino acids essential in the human diet. Furthermore, these bushy annual plants are amazingly productive: they produce twice as much protein per acre as any other major vegetable or grain crop. Compared to a cattle pasture used for meat production, an acre of soybeans produces 15 times more protein than an acre of cattle. If everyone ate soy protein instead of beef, most pasture lands would be unneeded and could be restored to a more natural state. Each soybean seed is 40% protein and 20% oil. In the United States, we mostly extract the oil for various purposes and then use the leftover protein-rich bean meal as animal feed. However, a growing number of Americans now drink soy milk (especially if they are lactose-intolerant and cannot digest the lactose [milk sugar] in cow's milk) and eat tofu products as an alternative to eating animal-based protein.

Soy is also fermented to make soy sauce (shoyu), miso, and tempeh (see Chapter 18). And soy protein can be spun into **texturized vegetable protein (TVP)** used to make plant-based food that has a texture that resembles meat. Such imitation meat is often sold in prepared foods (frozen or canned), and the person buying the food might not realize they are getting artificial meat unless they check the list of ingredients carefully.

Peanuts (*Arachis hypogaea*; FIGURE 13.17) are native to South America but are now cultivated in many areas, especially Africa. The fruits have an unusual development; they develop from yellow pea-like flowers located on shoots, as would be expected, but after self-pollination, the pedicel (fruit stalk) elongates, droops, and puts its tip on the ground, even burying it slightly. At that point the ovary begins to develop into a peanut, which at maturity is a subterranean fruit. Usually two to four ovules develop into seeds inside each fruit, each seed being separated from the others by an obvious constriction in the fruit wall.

FIGURE 13.17. The peanuts of the peanut plant develop underground even though they are true fruits that develop from flowers that had been above ground. (© LIN, CHUN-TSO/ShutterStock, Inc.)

Peanut seeds are eaten in many ways. The most common method in the United States is to grind them into peanut butter and eat that on bread, in cookies, in candy, and many other ways that are familiar to you already. We also eat peanuts roasted, and a few of us, especially in the southeastern states, eat them boiled. Large amounts of peanuts have the oil extracted (see below) and then the remainder of the seed is ground into peanut flour, which is added to many foods as a source of flavor and protein.

Peanuts have two health concerns. Slightly less than 1% of Americans are allergic to peanuts, with symptoms ranging from mild to life-threatening. The second concern is that peanuts can be attacked by the fungus *Aspergillus flavus*, which produces a toxic compound called **aflatoxin**. Every truckload of raw peanuts is tested, and if a threshold level is detected, the entire load is destroyed. All peanuts, other than those sold raw, are heat-treated to destroy any residual fungi.

Chickpeas (*Cicer arientinum*; also called garbanzo beans) are protein-rich pulses that are especially popular in vegetarian regions of India, Pakistan, and Bangladesh, where they are often the main source of protein (FIGURE 13.18). As the cuisine of these countries becomes more popular and well known in the United States, chickpeas themselves are more familiar. We also eat ground chickpeas as the main ingredients in falafel and hummus, two foods that are already popular here.

FIGURE 13.18. The main ingredient of these cookies is chickpea flour, so they are high in protein, and they are gluten-free because they contain no wheat flour. Americans still eat very few chickpeas, but as we incorporate more international foods into our diet, the diversity of plants we consume increases.

Food for Domesticated Animals

In the same way that many cereals are cultivated primarily to feed livestock, many legumes are too. It might seem as though horses and cattle can eat just about any grass, but just like people, their food must have an adequate amount of protein. The shoots and leaves of some grasses are adequate, but plants in the legume family provide the most nutritious **forage** (animal feed derived from shoots rather than fruits and seeds).

Alfalfa (*Medicago sativa*) produces the most nutritious forage of any crop, and is cultivated worldwide (see Figure 12.11). It is high in protein (it has root nodules with symbiotic nitrogen-fixing bacteria), and the fibers in its stems are soft, easily chewed by cattle, and are at least partially digestible. Its high protein makes it especially well suited for dairy cattle, each of which might produce several gallons of protein-rich milk every day that they are lactating (cows only produce milk for several months after they have given birth to a calf). Alfalfa is also fed to horses, beef cattle, and any other livestock whose value warrants the rather high price of alfalfa compared to other types of hay.

A field of alfalfa is harvested three or four times per summer. Once the shoots are a little more than a foot tall (less than half a meter) and just starting to bloom, their carbohydrate levels are high and the shoots have not become tough. The field is mowed and the cut shoots are allowed to dry for several days, then they are compressed into hay bales for ease of shipping and storage. Some alfalfa is chopped and stored as silage (in a large pile, where it ferments slightly, much to the delight of the animals), and some alfalfa is chopped and highly compressed into dense pellets. A field of alfalfa can be harvested for years, and when it is finally plowed under and planted to a different crop, the soil is high in nitrogenous compounds.

Several species in several genera are called clover (*Trifolium*, *Melilotus*, and *Lotus*) and are also used as forages. Fields of clover are often used as pasture, where the animals are allowed to graze freely and eat the shoots as they are growing. Clover can also be mowed for storage as hay.

Oils and Fats Are Essential to Human Health

Plant oils, in their natural state, tend to be extremely nutritious for people. Plants synthesize virtually no cholesterol and no *trans*-fats (see Box 4.1). Plants supply us with several fatty acids that our bodies need but cannot synthesize. Other than avocadoes and coconut meat, plants tend to be so low in oils that it is almost impossible to become overweight on a diet centered on plants. We obtain a small amount of lipids in any plant food we eat because all cells have membranes composed of lipid, but because plant cells have such large central vacuoles and so little cytoplasm, leafy vegetables, tubers, and roots tend to have very low lipid content. For larger amounts of lipid, we eat nuts such as walnuts, pecans, almonds, Brazil nuts, and so on; these are rich in oils. Other good sources are avocadoes; their pulp can be up to 30% oil, and they are the most energy dense fruit known, having up to 2800 calories per kilogram. Sunflower seeds, peanuts, and coconut are other rich sources of oil that we eat directly.

Much of the plant lipids we eat are extracted oils that we add to other foods when we bake, fry, or add salad dressing. Numerous plants provide extractable oils, and familiar examples are olive oil, peanut oil, corn oil, safflower oil, and canola oil. Among these, safflower oil has the highest concentration of linoleic acid, an essential fatty acid for us. Many of these oils are familiar because of their distinctive flavors, the oil of olives, peanuts, coconut, and sesame being especially valued. Most of the flavors of all plant oils come from non oil phytochemicals accidentally extracted with these oils. Because of their appealing flavor, these oils are not refined to the point of removing the flavor. However, oils from soybean, cotton seed, rapeseed (canola oil), and several others are purified to the point where they have almost no flavor and thus are more or less interchangeable for food manufacturers.

Plant oils are extracted either by pressing or with solvents. Pressing involves applying pressure to rupture the cell walls and force the oils out. In hot-pressing, heat is applied along with pressure; heat makes the oils more fluid and more easily extracted, but it also causes many other components to come out as well, most of which must later be removed. In cold-pressing, no heat is applied, the oil comes out with more difficulty but it tends to be purer. For olive oil, the word "virgin" indicates that the oil was obtained by pressing only and has undergone no chemical treatment at all. Mashed plant material can be mixed with solvents such as hexane, which dissolves the oils; the solvent/oil mix is separated from the mash, then the solvent is removed by evaporation. After either type of extraction, the residue is used as animal feed.

After oil is extracted, several other steps may be necessary. Some oil, such as sunflower oil, must have waxes removed; otherwise the oil turns cloudy when refrigerated. Other oils must be bleached and deodorized by removing various chemicals that are extracted along with the oil. Not all plant oils are edible. For example, castor oil is contaminated by ricin, which is extremely poisonous. Careful purification removes enough ricin to make castor oil edible, but it is such a strong laxative that it is rarely consumed. Tung oil itself is also toxic.

Many plant oils, especially those that are polyunsaturated, are used as chemicals rather than as foods; they are discussed in Chapter 16.

Plants that Provide Vitamins, Minerals, and Antioxidants

In addition to carbohydrates, proteins, and lipids, our diets must also include vitamins, minerals, antioxidants, and numerous phytochemicals. All the foods discussed above provide at least small amounts of vitamins, and the oil-rich foods are especially important for containing four vitamins that are lipid-soluble (A, D, E, and K). Many of the seed-based foods have high amounts of minerals in the cotyledons and endosperm.

Antioxidants are necessary for all organisms. When any organism uses oxygen in its metabolism, a dangerous molecule called a free radical is occasionally produced. Free radicals are extremely reactive and they cause extensive damage to almost any kind of molecule in a cell. One widely accepted theory postulates that many aspects of aging in people are due to the accumulation of damage caused by free radicals. **Antioxidants** are chemicals that inactivate free radicals and prevent any further damage. Examples of antioxidants are vitamin C (ascorbic acid) and vitamin E; carotenoids (red and yellow pigments of fruits and the orange of carrots); phenolic compounds in tea and coffee as well as many savory leaf-based spices; and flavonoids present in many berries. It is important that our diets have good sources of antioxidants, and the easy rule to remember is to eat red or yellow fruits and vegetables and dark green leafy vegetables. But a more complete list of plants rich in antioxidants includes beans (red, pinto, and kidney); blueberries and cranberries; blackberries, raspberries, and strawberries; and apples. People have known for hundreds of years that a diet rich in fruits and vegetables and having less meat was healthful.

"Phytochemical" is a vaguely defined term that encompasses any compound found in a plant and suspected of being beneficial to our health. Most tests and surveys of specific compounds have been inconclusive so far, but this does not mean that none is valuable in our diet. The principle of skepticism in the scientific method requires us to keep an open mind and be willing to accept new evidence if some is presented. We can say at this point that several of the compounds proposed as being beneficial phytochemicals must not be required in large amounts, that a diet with very low amounts is adequate, just as is true of some vitamins and essential mineral elements. Several compounds that have been proposed as being beneficial phytochemicals are beta-carotene (yellow vegetables), diindolylmethane (broccoli, cabbage), lycopene (tomatoes), lutein (leafy green vegetables), and zeaxanthin (red and yellow fruit and vegetables).

Numerous families provide us with foods rich in vitamins, minerals, antioxidants, and potentially beneficial phytochemicals.

The citrus family (Rutaceae) is famous for its fruits, high in vitamin C and an important part of our diet (FIGURE 13.19). The genus of greatest economic importance is *Citrus*, having oranges (*C. sinensis*), lemons (*C.* x *limon*; the "x" indicates this is a hybrid), grapefruit (*C. paradisi*), and lime (*C. aurantifolia*). All of these are small trees and are cultivated in orchards in warm areas where the danger of frost is low. The main citrus cultivating areas in the United States are Florida, south Texas, and California. Several times in recent years, Florida has experienced such cold nights in winter that many citrus orchards were devastated and the trees were so badly damaged that the orchards had to be replanted. If these cold nights were a part of global climate change, then we might lose this region as a center of citrus cultivation, and the cost in money and in the hardship of farm families will be immense.

We eat citrus as fresh fruit, juice, and cooked into many foods. The fruit is basically a berry (but technically known as a hesperidium), and the skin or peel is a combination of both the exocarp and the mesocarp. Oil cavities filled with volatile, aromatic oils occur in the outermost exocarp, which can be used as "zest" in cooking (FIGURE 13.20). The fruit's endocarp develops in an unusual way: it is covered with hairs that enlarge greatly into "juice sacs" and project into and fill the fruit lumen, surrounding the seeds. When we eat an orange or grapefruit, we eat only the endocarp's juicy trichomes.

The rose family (Rosaceae) is very large and diverse (95 genera with 2830 species). It is familiar to you as rose bushes and cut roses, but this is also the family of raspberries, blackberries, boysenberries, as well as the stone fruits (fleshy fruits with a hard endocarp, the "pit" or "stone"), such as peaches, cherries, apricots, plums (prunes are dried plums), and almonds (almonds are the stones with the fleshy

FIGURE 13.19. Kumquat fruits are small, about the size of a thumb, and are usually eaten whole, rind and all. Although many species of citrus are sensitive to cold weather and are damaged by frost, kumquats survive freezing weather. Even immature fruits can be covered in a solid case of ice and will then continue to develop after the weather warms above freezing. Kumquat trees are a beautiful, healthful addition to gardens in areas that receive mild freezes in winter.

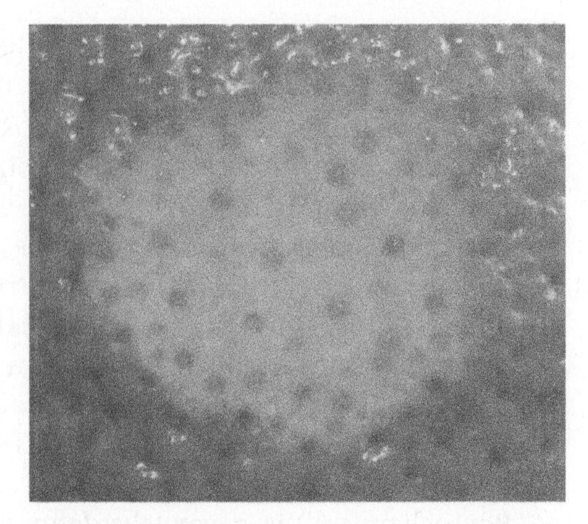

FIGURE 13.20. The bumps on the rind of citrus fruits are oil cavities. The outermost layer of the rind has been cut away on this orange, revealing the cavities. The oil has a strong citrus flavor and is used in cooking by scrapping the outermost, oil-rich rind (called the zest) off an orange, lemon, or other citrus fruit.

mesocarp removed). The rose family also supplies us with apples, pears, quinces, and loquats. Next time you are in a grocery store, notice that about half of the fresh fruit section is filled by members of the rose family (FIGURE 13.21). Notice also that these are mostly all sweet and most can be eaten fresh or cooked.

Whereas citrus plants are trees, the rose fruits listed above are produced by diverse types of plants. Many are trees, but blackberries and raspberries grow on slender, spiny shoots that sprawl and clamber and form thickets. Strawberries are small herbs with stolons. Each of these crops requires a distinct type of cultivation and farmland. Apples and the stone fruits must have cold winters; if cultivated in warm areas, they will grow but not produce fruit. However, the date of the last frost of spring is important; peach and apple flowers are badly damaged by a last frost if it occurs after the flowers have opened. Orchards of these fruits could be strongly affected by climate change. In contrast, strawberries are very susceptible to frost damage at all stages, and must have warm climates.

The potato family (Solanaceae) provides us with tomatoes, chili peppers (not related to black peppers), eggplant, sweet peppers, and tomatillos, as well as potatoes. Tomato sauce is almost synonymous with Italian cuisine, but tomatoes were domesticated in the Americas and were not introduced to Italy or the rest of Europe until after Columbus. Even then, tomatoes were not eaten for many years because this family is famous for having many highly poisonous members, such as deadly nightshade. Even potato tubers will become poisonous if exposed to sunlight long enough to turn greenish-purple. One of the newest foods to be discovered in this family is the goji berry (also called wolfberry), which is high in both nutrients and antioxidants.

The cucurbit family (Cucurbitaceae) provides many nutritious fruits, although unlike those listed above, many must be cooked to be edible. Examples are squashes of

FIGURE 13.21. So many of our most popular fruits come from the rose family that almost any produce section in a grocery store has an extensive selection of them. Everything in this view is from the rose family: pears, apples, peaches, apricots, and more.

many kinds (Hubbard, buttercup, butternut, acorn, summer and more), pumpkins, gourds, cucumbers (often eaten as pickles), and zucchinis. These may seem like vegetables to you rather than fruits; they are not sweet and are too hard and dry when mature to be eaten raw. However, notice that they all contain seeds, and that means they developed from a flower ovary. They are indeed true fruits, just as are the dry pods of beans and peas. Cantaloupe, watermelon, and honeydew melon are also cucurbits, but are eaten fresh and are more easily recognized as fruits. Seedless watermelons are a variety in which ovules abort early, just as in bananas and other seedless fruit.

Grapes (*Vitis vinifera*) in the grape family (Vitaceae) are nutritious fruits, with red grapes being especially high in antioxidants. We eat grapes in many ways: fresh, dried into raisins, as juice, and as juice that has been fermented into wine (see Chapter 15).

The next two families provide us with nutrient-rich foods that are not fruits. The family Brassicaceae provides the "cruciferous vegetables" (a common name for the family is "crucifer" because the flowers have four petals in the shape of a cross). The term "dark green leafy vegetables" encompasses many of the cruciferous vegetables, examples being kale, broccoli, cabbage, collard greens, brussels sprouts, cauliflower, kohlrabi, arugula (rocket), watercress, radish, daikon, wasabi, and bok choy (FIGURE 13.22). You might have noticed that none of these was your favorite food as a child; many are not popular with adults either. They have strong, pungent flavors due to diindolylmethane, sulforaphane, and other chemicals related to a family of compounds called glucosinolates. Many are believed to be beneficial phytochemicals, so breeding the plants to have a milder flavor might lessen their nutritional value.

FIGURE 13.22. Kale is one of the "green, leafy vegetables" that is rich in vitamins but low in calories, oils, and proteins. It would be difficult to have a well-balanced diet without eating leafy vegetables like this.

This family shows the power of artificial selection, and by extension, the power of natural selection. All of the first seven vegetables listed above (kale to kohlrabi) are varieties of exactly the same species: all are *Brassica oleracea*. From early times, various members of this species were chosen for their leaves (kale), others for their compact shoots (cabbage), others for their axillary buds (brussels sprouts), others for their tightly packed inflorescences of unopened flowers (cauliflower, broccoli), and the result is that the single ancestral species, through artificial selection, quickly diversified into these very different vegetables.

The family Apiaceae gives us carrots (rich in carotene) and celery. It also provides us with many spices that are discussed in Chapter 14.

Important Terms

aflatoxin
antioxidant
complementary proteins
complete protein
dehiscent fruit
dry fruit
essential amino acid
essential fatty acid
fat
fleshy fruit
forage
gluten
indehiscent fruit
male sterile cultivar
oil
texturized vegetable protein (TVP)

Concepts

- Food is the material we eat and drink and that supplies us with the nutrients our bodies use.
- The amino acids, fatty acids, and vitamins that we cannot synthesize ourselves are called essential.
- Simple sugars enter our blood stream quickly; complex carbohydrates release glucose more slowly.
- Familiar food plants have been domesticated in the Middle East, Asia, Africa, and both Americas.
- Cereal grains supply most of our carbohydrates; potatoes are a second significant source.
- Depending primarily on a single crop can result in starvation if that crop fails.
- Legumes, with their symbiotic association with nitrogen-fixing bacteria, produce protein-rich seeds.
- A large percentage of the plants we cultivate is used to feed animals. If we relied more on plant-based foods, we would not need to use so much land for agriculture.
- We obtain most of our vitamins and antioxidants from fruits, seeds, and green, leafy vegetables.

Spices and Herbs: Plants that Make Eating Fun

Introduction

Chocolate. What else is there to say?

The development of agriculture, especially the domestication of plants, had many consequences, two of which are important here: it made eating more reliable, and it made food boring. Having crops of wheat, rice, barley, potatoes, and corn allowed a more certain food supply, populations increased, and cities grew. Certainly fresh whole wheat bread is delicious, as are corn tortillas and rice all by itself. But if that is your only food, day after day, eating becomes monotonous. Such a simple diet may seem far-fetched, but even today in the twenty-first century, poor people throughout the world, and especially those in refugee camps, often face a diet that has no variety and just routine flavor. The same would have been true for poor people crowded into the first cities.

But herbs and spices change the flavor of foods, and by their very definition, they make foods more enjoyable, more interesting. Most herbs and spices have such strong flavors and aromas they are used in only small quantities, and do not really add either calories or any significant nutrients. Hot red chili peppers and jalapeños are high in vitamin C, but no one would try to eat enough of them to stave off scurvy. For a less dramatic example, consider going to the spice rack and eating a single bud of cloves or a leaf of oregano; they seem mild in foods because we use just a "pinch" so they are greatly diluted. Despite the lack of nutritional value, spices are so well liked they are our most expensive food items if measured by weight. A single "bean" of vanilla weighs only 4 g (0.12 oz) and costs $5.00 (that is, $660 per pound), but who would give up vanilla?

Having a pungent flavor and aroma such as those of mints, rosemary, dill, and mustard protects the plants from herbivores and perhaps even microbes. The spicy chemical of mustard does not even exist until the plant is damaged, for example by being chewed. Clove oil, which is produced even in healthy undamaged flowers, strongly inhibits microbial growth. An herbivorous mammal, with its grinding teeth, would probably damage the flat seeds of chili peppers, but the burning flavor protects them from all mammals except us. Birds, however, are insensitive to the flavor and eat the fruits with no problem, then disperse the seeds undamaged.

We know from archeological evidence, such as wall paintings and food remains at camp sites, that people used spices thousands of years ago. And spices were a main commodity and driving force for long-distance trade (**BOX 14.1**). Dried spices are not as heavy as gold or silver, and many were far more valuable on an ounce for ounce basis. When loading a mule, camel, or ship, spices would be more profitable than precious metals. Cities at the crossroads of trading routes grew rich by controlling international movement of spices, and many wars were fought to control these plants. Innumerable exploratory expeditions were sent out to find new spices or new sources of already popular spices. The plants in our spice racks have been the driving force in exploration, international trade, and geopolitics.

BOX 14.1. Spices, Global Exploration, and World Politics

Think of the pepper you put on your food, or the cinnamon you undoubtedly enjoy. Where do they come from? Neither is very expensive, and in modern times they are easy to obtain and certainly cause no significant international problems. They are just two spices in the kitchen. But in the past, these two spices and several others were the reason that nations went to war, that the world was explored, and that many aspects of our modern world are the way they are today.

These spices are native to India and Southeast Asia, formerly referred to as the Far East. Even in ancient times, some of these spices reached Egypt and Europe; we know that black pepper was imported into Egypt as early as 1213 BCE because it was used in that year in the preparation of the mummy of Ramses II. In Europe, as in Egypt, these spices from the Far East were so rare and exotic they were used mostly in religious ceremonies or as medicines rather than to simply augment the flavor of food. Ancient European—mainly Greek—explorers and traders knew a little about the lands to the east, which were controlled by the Persian Empire, but the first real expedition that established maps and solid knowledge of the lands all the way to northwestern India was the campaign of conquest by Alexander the Great. He reached India in 326 BCE. While he was young, one of Alexander's teachers was Theophrastus, known as the father of botany. During Alexander's extensive travels, he sent plant specimens back to Theophrastus, introducing many plants to Greek and thus western study.

The result of Alexander's campaign was a more active trade between Europe and the Far East. During the time of the Roman Empire, both black pepper and cinnamon were expensive but abundant enough to supply the wealthy classes. After the fall of Rome, Europeans paid little attention to exotic spices, and trade languished. But during the middle ages and then in the Renaissance (starting in the early 1300s), European wealth and sophistication increased, as did a desire for spices. At this time, all Far Eastern spices still had to be imported, but now the monopoly was controlled by Arabia, not the Persian Empire. The Arabs had established trading relations with the spice-producing regions in Indonesia, and had complete control over which spices were sold in Europe and at what price. The city-state of Venice formed an alliance with the Arabs, and as a result, both became extremely wealthy and the amount of spices was carefully controlled.

At about this time, in 1295, Marco Polo returned to Venice after a 15,000 mile journey of exploration to China. He brought back many valuable objects, but the most crucial were maps of the Far East and a knowledge of where spices were cultivated. Until that time, Arabian traders had made up fantastic stories of nonexistent lands, so that potential rivals would be sent off searching in the wrong direction.

Everything began to change in the early 1400s. Prince Henry of Portugal ("Prince Henry the Navigator") had a desire to explore, and began to send ships down the west coast of Africa. At the time, Europeans well knew that the world becomes colder as you go to the north, and finally is frozen over in the northern parts of Scandinavia. Likewise, temperatures rise as you go south, and it was assumed that the southern parts of Africa were probably too hot for life. But Prince Henry's explorers pressed southward, never very far on any one voyage, and by 1462 they reached Sierra Leon, the southern edge of the western bulge of Africa. At this point, they could hope that they had reached the end of Africa and only had to sail east to reach India, China, and all the spices. They would not have to continue south (or so they thought), and would not have to risk entering a land of fire.

The role of myths, legends, and outright fraud is often important for our advances in knowledge, as well as for holding us back. One of the reasons Prince Henry was so confident and urgent is that he was also searching for "Prester John," a mythological king of an empire that did not exist. At this time, all of Europe believed Prester John was a superhero ruler of a lost Christian empire somewhere in Africa, probably in Ethiopia. Not only would Prester John be able to supply gold and spices, he would lead a Christian army from the south while Europe mounted an attack from the north to liberate Jerusalem from Muslim control. Also, a fabulous river was believed to flow from Prester John's kingdom to the west coast of Africa, so Prince Henry's ships would have easy sailing as soon as they found the mouth of the river. Needless to say, this was all just legend, but from time to time, one scam artist or another would show up at one of the courts in Europe, claiming to be an emissary from Prester John. The emissary would be wined and dined lavishly, as the hosts listened to every tall tale. This fraud kept hope alive, and Portuguese ships continued farther down the African coast until Bartolomeu Diaz really did reach its southernmost point, the Cape of Good Hope, in 1490. In 1498, Vasco da Gama sailed all the way from Portugal to India, and Prince Henry's dream of circumventing the Arabian/Venetian monopoly was finally achieved: Portugal could import spices directly.

In 1492, impatient with the slow progress of the Portuguese exploration, Christopher Columbus convinced the Spanish court that he could reach the Far East by sailing west. Early Greek mathematicians and geographers had concluded that Earth is a sphere, so you can get to the east by sailing west. Here, there was no myth or fraud, but there was a mathematical error, a big one. The true circumference of Earth is 24,000 miles but the Greeks had miscalculated it as being only 2000 miles. Columbus decided that such a journey would take only a few weeks and the small ships of his day could carry enough supplies to manage that. Off he went, and right on schedule, he discovered land. He was convinced he had reached Asia and was perplexed that these Caribbean Islands had none of the spices he was seeking. Columbus was a die-hard, and continued to believe that he had reached the Far East, even after he made three more voyages to America. But other explorers realized he had discovered a new world rich in gold and silver, if not pepper and cinnamon. They realized that if they could just sail around one end or the other of the Americas, they could still reach the land of spices. In 1519, Ferdinand Magellan set sail from Spain with the plan to reach Peru on the west side of South America, then sail from there to the spice islands. At the time, Peru was a colony of Spain, so a short distance between Peru and the spice islands would provide Spain with a basis to challenge Portugal's trade. Magellan's ship did reach the islands and loaded up with spices, then continued on around Africa and back to Spain in 1522; Magellan himself died in the Philippines. This quest to establish a right to trade in spices led to the first complete circumnavigation of Earth. All of this effort by Portugal and Spain just to obtain pepper, cinnamon, cloves, and others spices led to our modern understanding of the shape of Earth and all its land masses.

The Dutch, British, and French quickly realized that sailing routes were feasible and that there was money to be made. By 1599, Dutch ships were trading in the Far East, and quickly battles were fought and blood was shed

(continued)

to control the trade. A lot of blood. Each European country established trading relations with local kings or villages, and would then aid in attacking any other king trading with a different country. Dutch, Portuguese, British, and Spanish soldiers were involved, but most of the victims were local people that cultivated and harvested the spices. Eventually, the Dutch East India Company gained control of most of the spice-producing lands and established a monopoly. They immediately raised the price of spices very high again in Europe, occasionally burning whole warehouses of cinnamon if the price was too low, rather than selling it cheaply.

Much of the spices are native to Indonesia, especially to a group of islands now known as the Maluku Islands. They were long known as the Moluccas or Spice Islands. They are mountainous, have a hot climate, and high rainfall. Because some of the spices were native to only some islands, the Dutch were able to prevent smuggling. For example, cloves grew only on a few small islands, so the Dutch chose two that could be easily patrolled, then they destroyed all clove trees on all other islands. Every tree was chopped down and burned, and they made it a crime to possess cloves anywhere. This had been the livelihood of the people on these other islands, and resistance began. As many as 60,000 indigenous people were killed. These islands were almost completely depopulated, and then new inhabitants were brought in from other areas, changing the ethnicity and culture of the islands. Other spices were more widespread and difficult to control. The Dutch would sterilize all nutmeg seeds by soaking them in lime or citric acid; thus they could be exported with no worries that someone would be able to germinate a seed and start a new plantation outside of Dutch control. From time to time, seeds or plants of various spices were smuggled out and grown elsewhere, and so gradually the spice monopolies collapsed one by one and a more ordinary trade was instituted.

Although this may sound like a case of the "bad old days," the effects are still felt now; the colonial aspects did not end until World War II. The British had taken control of India during the spice wars and did not grant it independence until 1947. Indonesia did not become independent of the Netherlands until 1949. France fought to maintain control of its colony in Viet Nam until 1960, at which point the United States stepped in and continued the war and bloodshed for many more years.

Chemistry of Flavors

The chemicals most commonly responsible for flavors and aromas are terpenes and phenolics. A terpene is any chemical built up from a particular 5-carbon compound called an isoprene unit. There are thousands of terpenes, and an important character they share is volatility: they evaporate easily so they are often the first molecules we smell in a plant or a food. The needles and resins of conifers, as well as citrus fruits, are rich in terpenes, and these will give you an idea of the scent of terpenes. They are described as pine-like, citrusy, leaf-like, or "fresh." Because they are so volatile, herbs and spices that depend on terpenes must often be added to food only near the end of cooking, or as a garnish, otherwise the flavor is lost.

Phenolics are another very large class of compounds, this group being based on the chemical phenol. Whereas terpenes have a more generic quality to their aromas and flavors, each phenolic is more distinctive. Thus thyme, oregano, cloves, cinnamon, and vanilla have completely different flavors despite being based on the same chemical family.

Other chemicals such as alkaloids (the flavor of black pepper) and lipids (cloves) also contribute to flavor and aroma. Somewhat surprisingly, our taste buds register only five sensations: sweet, sour, salt, bitter, and savory (sometimes called umami). Consequently, it is not the actual chemistry in the mouth that is important, but rather that of the aromas in our nose. The scent receptors in our nasal passages distinguish between thousands of chemicals, and any component of a food that is volatile and gives off molecules into the air will contribute to its flavor. For example, in studies of vanilla extract, 171 compounds have been identified so far. All of this does us no good if we have a cold with a stuffed up nose; without our sense of smell, all foods, herbs, and spices are tasteless.

The Predominant Families of Herbs and Spices

Any plant that has an interesting flavor and which is not too toxic can be used as a spice. But two families have biochemical pathways that permit them to produce numerous flavors that appeal to us, and a large number of our spices come from these families.

In the mint family (Lamiaceae) the leaves carry most of the flavor. They have oil glands that synthesize and store fragrant, flavorful oils (FIGURE 14.1). Most plants of Lamiaceae grow in sunny, hot, dry regions around the Mediterranean Sea, and they are at their most flavorful if cultivated in sun and heat. Plants that have milder growing conditions are less aromatic. Examples are basil, marjoram, oregano, rosemary, sage, and thyme. Spearmint and peppermint are exceptional in preferring some shade.

We use whole fruits of the carrot family (Apiaceae) as spices (FIGURE 14.2). Their fruit walls have pockets that fill with oils that contain the favors we enjoy. Examples are dill, anise, caraway, celery, cumin, coriander, and fennel. We also use leaves of dill (often called dill weed) in cooking, and of course we eat celery petioles and the roots of carrots. Leaves of coriander are also eaten, but then they are called cilantro. In addition, the family contains several species, such as poison hemlock (*Conium maculatum*) that are extremely deadly.

Pungent, tangy compounds are produced by a variety of plants in several other families as well. The mustard family (Brassicaceae) gives us two pungent spices: mustard and horseradish (FIGURE 14.3). Hot peppers (chili peppers) are in the nightshade family (Solanaceae), the same family that provides potatoes, tomatoes, and the medicinal plant *Datura*, whereas black pepper is in the pepper family (Piperaceae); the two types of pepper are not closely related, nor is their heat based on the same compound. The ginger family (Zingiberaceae) gives us ginger, whose rhizome is the part we use for gingersnaps, gingerbread, ginger ale, and crystallized ginger.

(a)

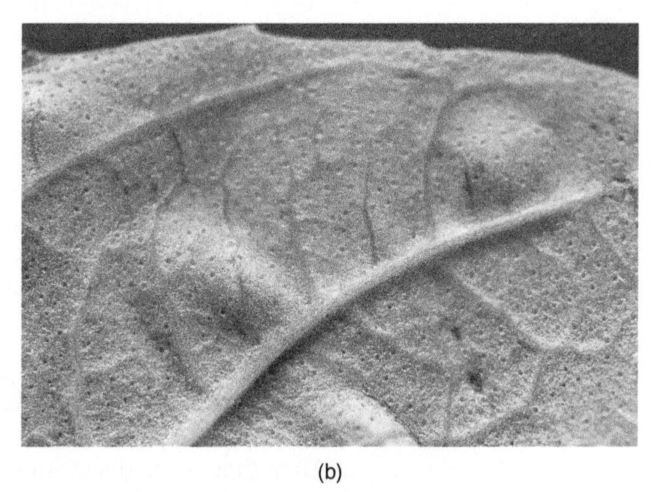

(b)

FIGURE 14.1. **(a)** Rosemary is a member of the mint family, Lamiaceae, and its leaves produce abundant aromatic compounds. Just lightly touching the leaves will give your fingers a wonderful fragrance. However, the chemicals are so potent that you—and most other animals—would not be able to eat more than a small bit. **(b)** This is a close-up view of the underside of a basil leaf. The numerous small dots are glands that contain the characteristic flavor of basil; the lines are leaf veins. You can see the glands yourself if you look at the underside of a basil leaf with a loupe or strong magnifying glass (the glands are easier to see if the basil is wilted).

(a)

(b)

FIGURE 14.2. Plants of the umbel family Apiaceae are easy to recognize when in flower **(a)** or fruit **(b)**: they have a characteristic organization called an umbel, as shown here. Many flower or fruit stalks emerge from one point, then usually branch again, and all flowers and fruits are borne in a more or less flat disk. The fruits contain canals filled with essential oils that give these fruits their strong flavors.

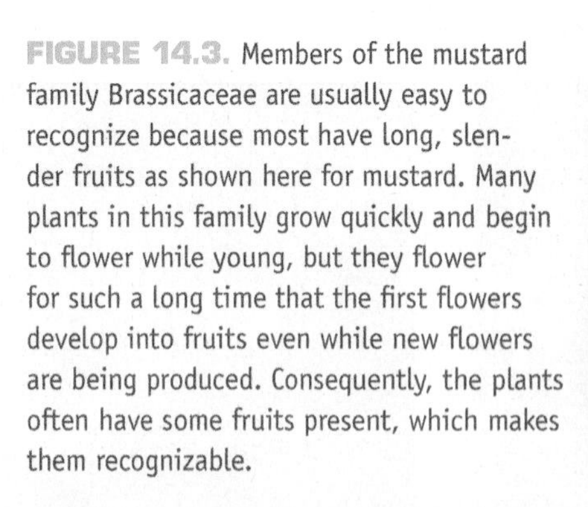

FIGURE 14.3. Members of the mustard family Brassicaceae are usually easy to recognize because most have long, slender fruits as shown here for mustard. Many plants in this family grow quickly and begin to flower while young, but they flower for such a long time that the first flowers develop into fruits even while new flowers are being produced. Consequently, the plants often have some fruits present, which makes them recognizable.

Spices from the Americas

After Columbus discovered the Americas in 1492, many plants and plant products were taken to Spain and the rest of Europe. Although many of our most valuable modern crops are American, such as corn, potatoes, tomatoes, and squash, only three spices originated in the Americas: vanilla, hot peppers, and allspice. It is difficult to imagine life without vanilla.

Vanilla

Vanilla flavoring comes from an orchid, *Vanilla planifolia*, native to Central America and southern Mexico (FIGURE 14.4). Although the orchid family Orchidaceae is very large, with more than 20,000 species, many of which are valuable as ornamental plants, *V. planifolia* is the only one that provides a food product. Like many orchids, *V. planifolia* is a vine with thick, succulent, fibrous stems and leaves; it climbs trees by means of adventitious roots, with one root occurring opposite each leaf. Once old enough, the plant produces groups of white flowers, and after pollination the fruits develop into the vanilla "bean," a pod that strongly resembles a long (15 cm [6 in]), slender green bean when mature.

Vanilla is now cultivated in many parts of the world that are warm, tropical, and rainy. Although native to the Americas, more than 95% of vanilla is produced in Madagascar. A plantation usually consists of small, sparsely branched trees used to support the orchid vine. All vanilla orchids are propagated vegetatively by cutting a vine into pieces, each with at least six nodes. The lowermost leaves are cut off, then the lower end of the vine is placed in the soil at the base of the tree, covered in soil and mulch, and the vine is tied to the tree. After a few weeks, one of the uppermost apical buds becomes active and grows out as a new vine that climbs upward, attaching itself by adventitious roots and leaning on the tree's branches. After the top of the orchid is too tall to be reached from the ground, it is bent back downward so that it is easy to work with. It starts growing upward again, and orchid vines often become 30 m (100 ft) long. The bending induces the orchid to flower. Once about 3 years old, the plant flowers every year with many clusters of about 20 flowers each, and each plant may have a thousand flowers each year.

Flowers are pollinated by hand, and usually only the first five or six flowers in each cluster are pollinated; the rest are allowed to wither so that the plant does not stress itself from trying to support too many fruits. The natural pollinator is a small bee native to Mexico, but it is very ineffective, and even in Mexico, all commercial orchids are hand pollinated. After vanilla plantations were established in other parts of the world, attempts were made to introduce the bees into the plantations, but for some reason the bees never survive.

Vanilla pods require about 9 months to mature. They resemble long, slender French beans, and when they begin to turn yellow at one end, they are harvested. Because timing is critical, each is picked by hand, so every vanilla orchid must be checked many times per season. The pods are placed in boiling water for a few minutes to kill the fruit and prevent any further growth. This also causes enzymes to begin modifying the fruit such that aroma and flavor develop. The beans are laid in the sun to warm for several hours, then are tightly wrapped in wool blankets, placed in wooden boxes overnight to "sweat," and the next day they are placed in the sun again. This is repeated for 10 days, and then they are stored in airtight wooden boxes for a month or more to finish the development of the flavor (FIGURE 14.5).

The aroma and flavor of vanilla are primarily due to a compound called **vanillin**, and as much as 2.5% of a vanilla bean's weight is vanillin. A second compound piperonal is also important; altogether, vanilla pods have 171 chemicals that contribute to their distinctive flavor and aroma. Vanillin can be synthesized artificially, starting with, of all things, residue from the paper and pulp industry. It is possible to synthesize pure vanillin, and this is what is sold as artificial vanilla as a substitute for vanilla extract or for entire vanilla beans. The synthetic vanillin has a strong flavor, but is missing all the complexities that occur in real vanilla due to the many other flavor and aroma compounds in the natural product.

FIGURE 14.4. Vanilla flavoring comes from the orchid *Vanilla planifolia*, which, like many orchids, is a vine that grows epiphytically on trees. (© Lantapix/Dreamstime.com.)

FIGURE 14.5. After processing and when ready for use, *Vanilla* fruits are long and slender, dark brown or black, and very fragrant. They are often called vanilla beans, but the word "bean" typically refers to a fruit of the legume family Fabaceae. (© arteretum/ShutterStock, Inc.)

"Mexican vanilla" is often sold as being more authentic because it is produced in the area where *V. planifolia* is native. But in some cases, it is a mix of real vanilla extract and extract of the tonka bean, which does smell like vanilla but that contains compounds called coumarins in concentrations high enough to cause liver damage. "French vanilla" is not a particular type of vanilla but instead refers either to a particular method for preparing the custard base for vanilla ice cream, or it is a mix of flavors that includes hazelnut, caramel, or butterscotch in addition to vanilla.

Hot Peppers, *Capsicum*

The peppers in this section are the whole fruits of large herbs or small bushes in the genus *Capsicum*, including bell peppers, jalapeño peppers, and chili peppers (**FIGURES 14.6** and **14.7**). The black pepper we use every day in the kitchen and at the table is ground peppercorns from a completely different plant (*Piper*) and will be discussed later.

Hot peppers are such an integral part of so many different cuisines around the world. They occur in so many sizes, shapes, and degrees of "hotness" or pungency that it is important to be clear that they are all *Capsicum* pepper from the tropical Americas. None existed in the Old World before Columbus, even though they are now cultivated worldwide. The pungent dishes of China, Thailand, India (even curry), and the paprika of Hungary did not exist in those countries 500 years ago. All rely on *Capsicum*, and almost all use varieties of just *C. annuum*, even if we include Mexican and Tex-Mex foods with their amazing diversity of peppers. Although there are many species of *Capsicum*, variations of *C. annuum* are the ones used in almost all cooking and have been bred to have a breathtaking range of shapes, colors, and pungency.

FIGURE 14.6. The four large peppers here are bell peppers and have no capsaicin, the chemical that gives hot peppers their flavor and pungency. The small peppers, however, have a high concentration of capsaicin and must be eaten carefully. After handling hot peppers, you should be careful to not touch your eyes until after you have washed your hands thoroughly.

FIGURE 14.7. Jalapeño peppers are one of the most familiar hot peppers, at least among people who enjoy Mexican food. Their flavor when fresh is much stronger and more pungent than after they have been cooked.

The burning sensation of *Capsicum* peppers is due to **capsaicin**, a lipid-soluble chemical whose synthesis is controlled by a single gene. In bell peppers (also called sweet peppers), both copies of the gene are inactive and no capsaicin is present at all. In the other varieties, at least one copy of the gene is active and various amounts of capsaicin are produced in the "membranes," the white tissues (placentae) that bear the seeds and that project into the fruit (FIGURE 14.8). The seeds themselves do not produce capsaicin. The amount of capsaicin and the amount of heat produced by *Capsicum* peppers is measured on the Scoville scale. A tiny amount of an alcohol extract of the pepper is added to water, and then several people are asked if they can taste it. If not, more is added just to the point were it is detectable; the degree of dilution is then called the Scoville heat units (SHU). Bell peppers have a value of 0 Scoville heat units, and the hottest pepper known, a variety called Bhut Jolokia, has 1 million Scoville heat units; that means that if 1 million milliliters (1000 liters) of water contains more than 1 mL of the extract, people will detect the capsaicin (in American terms, that would be 20 drops in 264 gallons of water). Anaheim peppers have 500–2500 SHUs, jalapeño (2500–8000), serrano (10,000–23,000), and the others fall somewhere in between. Pepper spray used by police has enough capsaicin to have 5 million SHUs, and pure capsaicin itself has 15 million. The most pungent substance known is resiniferatoxin, an alkaloid present in some *Euphorbia* (spurge) plants, rated at 16 billion (not million) SHUs; it is 250,000 times hotter than the hottest jalapeño.

FIGURE 14.8. Fruits of hot peppers are hollow and have several strips (placentae) where the seeds are attached. The placentae have the highest concentration of capsaicin and are the most pungent part.

It seems odd that bright red or yellow fruits would be spicy rather than sweet like apples or strawberries: what good does it do a plant to produce fruits that, when ripe, have a color that attracts animals but a flavor that repels them? The key is that birds are unaffected by capsaicin and eat the fruits with no trouble. The juiciness of the fruit combined with the capsaicin allows the plant to target one type of seed distributor: birds rather than mammals.

Processing of *Capsicum* peppers is much simpler than that of vanilla. Capsaicin develops automatically and there is no need to ferment the fruits or do anything special to enhance the flavor or pungency. We eat bell peppers completely fresh with no processing at all. The green variety are merely ones picked while immature; if left on the plant, the green peppers would continue developing and as their chloroplasts change to chromoplasts, their color would change to red, yellow, or orange. We, at least some of us, also eat jalapeño peppers fresh, the less brave of us removing the seeds, membranes, and innermost white core which are the most pungent parts; even so, the remaining green fruit wall (pericarp) still has enough capsaicin for most people. Always remember that capsaicin is lipid soluble, so it is easiest to cool your mouth with queso (cheese) sauce or milk; a mild alcohol solution (such as beer or a margarita) will also help. A teaspoon of sugar is reported to help cool off the mouth. Both bell peppers and jalapeños are used in cooking as well as being eaten fresh. Dried, smoked jalapeños are known as chipotle. All are extremely high in vitamin C. Anaheim chilis are often stuffed and cooked, and most other *Capsicum* peppers are dried, then chopped or ground and used for cooking.

The total number of species of *Capsicum* that are cultivated and used as chili peppers is still unknown because of their long history of use and their variability, but at least five have been identified for certain. Other than *C. annuum*, the only other widely used species is *C. frutescens*, and you might have encountered in two forms: it is used to make Tabasco sauce, and it is also dried, chopped, and mixed with *C. annuum* to make the red pepper flakes found in pizza restaurants and Mexican food restaurants. The other three species are *C. baccatum* var. *pendulum*, *C. pubescens*, and *C. chinense* (also from northern South America and the West Indies, not from China despite its name). These last three are, for the most part, used only regionally in South America and the Caribbean.

Several terms sound like they refer to specific peppers but are instead actually mixes. "Cayenne pepper" indicates dried, ground pungent *Capsicum* peppers, regardless of species. "Paprika" consists of dried, ground *C. annuum*, and is made from peppers that vary from yellow to red and mild to hot. "Red pepper" on the label of a bottle of spice in a grocery store can be any mix of *Capsicum* peppers and have any degree of pungency. "Chili powder" varies greatly depending on the manufacturer's recipe, but generally contains oregano, cumin, and garlic powder, as well as dried peppers.

Allspice

The third spice from the Americas, allspice (*Pimenta dioica*) got its name because its flavor resembles that of a combination of cinnamon, cloves, and nutmeg. It is in the family Myrtaceae, the same family as cloves. The plant is a tall tree that grows widely in the tropical Americas, especially Jamaica. There have been attempts to cultivate it in other areas such as the Maluku Islands and Sri Lanka (formerly known as Ceylon), but that has not been successful, and Jamaica still produces most of the world's supply of allspice. The pea-sized berries hold two seeds each, and must be picked while green because they lose their flavor when completely mature. After harvest, the berries are dried and ready for use.

In cooking, allspice is rarely used by itself, but instead is combined with other spices in pies, desserts, pickles, preserves, and sauces like ketchup. In addition, allspice oil is extracted and used to flavor liqueurs such as Chartreuse and Benedictine. It is also used in cosmetics.

Spices from Asia

Black Pepper, *Piper nigrum*

A pair of salt and pepper shakers. What home is without them? Black pepper, *Piper nigrum*, is our most commonly used spice, and it is a rare meal that does not have some dish garnished with black pepper (FIGURE 14.9). It is popular worldwide, and although relatively inexpensive, it is traded in such large amounts that it is the most important commercially.

Black pepper is an ancient spice, native to India and used there since at least 2000 BCE. It was part of the Egyptian embalming process, and black peppercorns were found in the mummy of Ramses II, buried in 1213 BCE. Pepper was known to ancient Greeks and was used extensively in Rome; it has been part of the Western diet ever since. All black pepper used in Europe had to come from India by caravans through Arabia and was extremely costly. A desire to find an alternate route to the Far East and its pepper was one of the driving forces that propelled exploration around Africa and South America in the fifteenth and sixteenth centuries.

Black pepper is still cultivated in India and Indonesia, but Brazil is now a major producer as well. Plants of *P. nigrum* are perennial, woody vines that need either a tree or a trellis of poles to climb on (FIGURE 14.10). As with many crops in which it is important to maintain a particular set of genetic traits, pepper is propagated clonally by cuttings that are rooted then planted at the base of something they can climb. Fruits can be harvested starting when the vine is only 4 years old; it will continue to bear for another 20 or 25 years. The flowers hang in long spikes, and each vine has about 20 or 30 clusters of fruit.

The fruits are harvested at various stages of maturity, resulting in different types of pepper. "Green pepper" results from harvesting the fruits (drupes) while they are still immature and have a green color. They are then freeze-dried or pickled in brine or vinegar to maintain the color.

FIGURE 14.9. Black peppercorns (*Piper nigrum*) are the starting point of our black pepper. Each peppercorn is a fruit that has been allowed to ferment briefly, after which they are dried. Fermenting develops the flavor and softens the fruit wall, which then wrinkles as it dries. There is a single seed inside each peppercorn.

FIGURE 14.10. Plants of black pepper (*Piper nigrum*) can grow to be small trees, but in cultivation they are pruned to maintain a smaller, bushier shape that produces more fruit that is easier to harvest. (© Davinci/Dreamstime.com.)

For "black pepper," the fruits are allowed to ripen slightly more until they start to turn yellow, then they are picked and cooked briefly in hot water to rupture the cell walls in the fruit. They are piled together and allowed to ferment slightly, which changes their color from green to black and helps the flavor become more intense. The flavor of black pepper is mainly due to an alkaloid called piperine, and the aroma is caused by an essential oil. After fermenting, the peppers are dried and then known as peppercorns, which consist of the seed and entire fruit. In stores, we buy either whole peppercorns or ground pepper. As with most spices, grinding exposes the tissue to oxygen, leading to loss of flavor. Buying whole peppercorns and grinding the pepper at home as needed gives the best flavor.

"White pepper" is just the seed and white endocarp (the innermost layer of the fruit wall) with none of the mesocarp or exocarp. It is prepared from fully mature fruits by retting: the fruits are placed in water for a week or two, during which the fleshy fruit tissues are softened and degraded by fungi. After drying, the fruits are agitated, causing the fragile fruit tissues to break apart and fall off, leaving the seed and endocarp.

When fully mature, fruits of *P. nigrum* are red. If harvested and treated like green peppers, they retain their red color and are known as "rose pepper" (sometimes "red pepper" but this is not the same as *Capsicum* red pepper). You may see "pink pepper"—it is the fruit of an unrelated plant, *Schinus molle* from South America.

Cinnamon and Cassia

It is very likely that cinnamon is one of your favorite spices, and it is just as likely that you have never or only rarely tasted true cinnamon. The genus *Cinnamomum* in the laurel family (Lauraceae) has several species: *C. zeylanicum* is considered the "true cinnamon" whereas *C. cassia*, *C. burmannii*, and *C. loureirii* are known as cassia. All produce bark that has a cinnamon flavor, but that of *C. zeylanicum*, native to Sri Lanka, is more delicate and subtle whereas the bark of the cassias is more robust. The U.S. Department of Agriculture (USDA) permits the bark of all four species to be labeled and sold as cinnamon so even if you see "cinnamon" on the label, the flavoring may be that of cassia rather than true cinnamon.

Trees of *Cinnamomum* are evergreens that grow to be rather large, up to 15 m (about 49 ft) tall, but on plantations they are kept short to make harvesting easier. When plants are about 3 years old, they are trimmed back almost to ground level leaving only 6 to 10 strong shoots; all the rest are cut away. During the next 2 years, the remaining shoots grow as much as 2 m (6 ft) and will be ready for their first harvest. With true cinnamon, the outer bark is scraped away, and then the inner bark is tapped evenly with a hammer. Afterward, it is peeled off and allowed to dry immediately so that microbes do not grow on it. As it dries, it rolls into quills (FIGURE 14.11). With the cassias, the whole bark is used by first cutting through the bark then peeling it from the shoot. This gives a much coarser, harder bark, and some may be sold as thick quills, others as smaller pieces. After its bark has been harvested, a shoot must be allowed to recover for several years before it can be stripped again.

FIGURE 14.11. Cinnamon quills are strips of bark that roll inward as they dry. The thickness of these quills indicates that these are cassia.

Cloves

Cloves are the unopened flower buds of trees native to Indonesia (FIGURE 14.12). Their wonderful aroma and flavor are primarily due to the essential oil **eugenol**, which is based on an old name for the species, *Eugenia aromaticum*; the modern name is *Syzygium* *aromaticum*. Although eugenol makes up 80% of the oil extracted from cloves, many other aromatic compounds also are present, including vanillin, the flavor of vanilla. The flower buds must be picked by hand before they open, just as they have changed color from green to light pink but not red. If harvested too late, the flowers fall apart during processing and have very low value. Because timing is so important, each tree must be examined and harvested at least four times each season. The buds are dried immediately to prevent fermentation. There is little worry about contamination by microbes during drying because clove oil is effective at inhibiting microbial growth. A single tree produces about 4 to 5 kg (8.8 to 11 lbs) of dried cloves per year.

We use cloves in many ways. Whole cloves are pressed into hams as they bake, and onions can be baked with cloves pressed into them to add flavor to surrounding foods. Ground cloves are an essential ingredient in many sweets and desserts. Clove oil is used both as a flavoring and for its antimicrobial properties in mouthwashes and dental products. It has the ability to temporarily numb our nerves and is often used in dentistry; you may have noticed its aroma as you sat in the dentist's chair. Clove oil is also used to dissolve certain stains used in histology; almost any plant anatomy laboratory has the aroma of clove oil.

FIGURE 14.12. Cloves are unopened flower buds that have been allowed to dry. The flower stalk and all the petals have a sponge-like texture because they are filled with cavities that produce clove oil. The word "clove" comes from the Latin "clavus" for nail, although these buds resemble nails only slightly.

Other Spices from Tropical Asia

Rhizome-flavored cookies. Sound delicious? Rhizome bread? How about rhizome ale to drink? It is all in the name. These are wonderful if the rhizome is ginger (*Zingiber officinale*). Ginger plants for spice grow just like the ornamental gingers or bamboos; they have thick, fleshy, fibrous rhizomes that grow just below the soil surface, with aerial shoots that bear leaves and flowers (see Figure B2.1d in Box 2.1). Fresh ginger is available in most stores (it may be labeled "ginger root" but it is really the rhizome), so it is easy to look at the spice in its natural form. Ginger is propagated vegetatively by transplanting pieces of rhizome, and you could start your own ginger crop by using the rhizomes in a grocery store. Ginger flowers rarely produce viable seeds.

Ginger requires the least processing of the spices considered so far. The rhizomes are dug up, cleaned, and are ready for market. They are grated as they are needed, so the flavor is completely fresh. Ground ginger is also available, but it loses its flavor after several months, as do most ground spices. In recipes that call for dried, ground ginger, freshly grated ginger can be substituted at a ratio of 6 to 1 (6 teaspoons of fresh ginger for each 1 teaspoon of dry ginger).

The aroma and flavor of ginger are due to a variety of chemicals. These can be extracted as a light, yellow oil used to make ginger ale, ginger beer, and other foods. Because of its aroma, it is also widely used in cosmetics and shaving lotions.

Turmeric (*Curcuma domestica*) is also a rhizome in the ginger family. After harvest, it is dried and ground, and is an important component of curries, giving them their distinctive yellow color.

Cardamom (*Elettaria cardamomum*) is also in the ginger family; it is made from the dried seeds rather than the rhizome. You are most likely to encounter it in Indian food.

Nutmeg and mace both come from the same species, *Myristica fragrans*. Nutmeg is the seed and mace is a bright red aril that surrounds the seed; an aril is an outgrowth present in seeds of some species, absent in most (FIGURE 14.13). Nutmeg is a dioecious species: some plants have staminate ("male") flowers; others have carpellate ("female") flowers. A single male tree provides enough pollen to pollinate many female trees, and only the female trees bear fruit and seeds. There is no way to distinguish male trees from female trees before they flower. In the past, orchards were planted with seedlings, and then once the plants started to flower—when they are about 7 years old

FIGURE 14.13. This fruit provides two spices: the red network is mace, the seed inside is nutmeg. (© Elena Schweitzer/Dreamstime.com.)

or so—most of the male plants were pulled out and replaced with new seedlings. After 7 years, most of the male trees in the new set were also taken out and replaced. This caused much of the orchard to be unproductive, but now most orchards are planted with cuttings, in the ratio of one male tree for every 10 females.

When fruits are mature, the aril is removed and sold as mace. The remaining seed, which is mostly endosperm, is sold as nutmeg. This gives us a good lesson in the value of knowing our botany. Reportedly, an administrator in Amsterdam long ago noticed that mace had a higher price than nutmeg, so he ordered the plantation managers to cut down all their nutmeg trees and replace them with mace trees. Hopefully, the managers had better knowledge (and sense) than he did.

Nutmeg can be purchased as a powder, but whole seeds are available and people often grind nutmeg as they need it. Its flavor is so ephemeral, it should be ground directly onto the food. In large doses, nutmeg is toxic; in intermediate doses it is hallucinogenic, but not recommended.

Herbs and Spices Native to Europe and the Mediterranean Area

Even without herbs and spices imported from the Americas or tropical Asia, Europe had many flavorful culinary plants. Interestingly, most of these are savory (basil, tarragon, and others) or slightly sweet/spicy (mints), and consequently, most are used to flavor meats, breads, and sauces, but few other than mints are used for desserts. This may just be a consequence of the region being dominated by the mint family (Lamiaceae) and dill family (Apiaceae), so most of these herbs share two basic types of chemistry.

Mints are the most popular of the Mediterranean herbs (FIGURE 14.14). Spearmint (*Mentha spicata*) and peppermint (*Mentha piperita*) are both important in cooking (especially with lamb) as well as for the oils they produce, which are used to flavor all sorts of products from candy to herbal tea to toothpaste. Mints also add an essential flavor to mint juleps, mojitos, and the liqueur crème de menthe. The flavor of mints is predominantly **menthol**, with spearmint also having *R*-carvone.

Mints are low, herbaceous perennials that prefer cool, moist, partly shaded areas. They spread by runners, and grow so vigorously it is best to cultivate them in pots or with some sort of barrier to prevent them from invading surrounding flower beds. A small pot of spearmint or peppermint will be enough for most households, but mints can be attractive as groundcovers and they provide a wonderful, unexpected fragrance to a garden.

Two other plants should be mentioned with the mints. Wintergreens (several species of *Gaultheria*) are not related to mints; they are in the family Ericaceae (blueberries and cranberries), but they produce oil of wintergreen (methyl salicylate) with an aroma that evokes memories of mint. Oil of wintergreen is often used with the menthol from mints as a flavoring. The other plant to mention here is *Nepeta cataria*, catnip. It is astonishing to watch a cat with catnip; the plant has only a mild aroma for us, but cats become euphoric with it. They will rub against a catnip plant and roll in it until it is crushed or uprooted. It is thought provoking to see how a plant produces a chemical that has so little effect on us but is so mesmerizing to cats. It does not take long to think of aromas and flavors that have similar effects on us (start with chocolate).

Other Mediterranean members of the mint family have more savory flavors. Basil (*Ocimum basilicum*; see Figure 14.1b) is used in pesto, its fresh leaves being ground with garlic, olive oil, and pine nuts. Basil is also a frequent component in Italian foods, especially those with tomatoes.

Sage (*Salvia officinalis*) is almost synonymous with Thanksgiving for Americans; it is an indispensable ingredient in stuffing for turkey. The aromatic oils of sage mix well with the fats of meat, so sage is often used in sausage and other processed meats. Culinary sage, *Salvia*, is not related to the strongly fragrant sagebrush widespread across the western plains of the United States.

Rosemary (*Rosmarindus officinalis*) is a small dense shrub with stubby, dark green leaves (see Figure 14.1). The plants are so attractive they are now popular in xeriscape landscapes and gardens. When grown in a home garden, twigs can be used for flavoring, and go very well with roast chicken and other meats.

We know oregano (*Origanum vulgaris*) almost exclusively as an herb for Italian food, especially pizza. It is a robust perennial herb, growing to about 1m (3 ft) tall.

The dill family gives us numerous savory seeds (really whole fruits) often used in baking: anise, caraway, celery, coriander, cumin, dill, and fennel (FIGURE 14.15). Anise (*Pimpinella anisum*) has a licorice flavor but is not the plant used to make licorice. Anise is used in a variety of ethnic dishes, such as Greek dolmas and German Pfeffernusse. An essential oil, anethole, gives anise its flavor, but it is also present in the unrelated star anise (*Illicium verum*; FIGURE 14.16). Because star anise is

FIGURE 14.14. Mints have opposite leaves that bear glands (not visible at this magnification) that produce the essential oils that give the plants their aroma and flavor.

FIGURE 14.15. These are fruits of anise but they are usually called seeds. Each fruit is very thin and has between 20 and 30 canals filled with essential oil, which in this case has the flavor of licorice. The embryo itself does not have any canals and relies on the fruit and its chemicals for protection. Related spices such as fennel and cumin have fewer, larger canals.

FIGURE 14.16. Fruits of star anise do not resemble those of true anise (see Figure 14.15), but they have an almost identical flavor and are much less expensive.

less expensive, it has replaced anise as a flavoring in many foods. Licorice the candy is made with roots of *Glycyrrhiza glabra* (in the legume family Fabaceae), which also has anethole. In addition, *G. glabra* contains a compound called **glycyrrhizin** that is 50 times sweeter than table sugar (sucrose). Glycyrrhizin is often used as a sweetener to avoid sucrose; it has a similar flavor but its sweetness lasts longer.

Mustard (*Sinapis hirta* for yellow mustard, *Brassica juncea* for brown mustard) and horseradish (*Armoracia rusticana*) provide us with a nice pungency without resorting to the heat of *Capsicum* peppers. Mustard the spice is prepared from seeds of mustard plants (see Figure 14.3). Mustard fruits contain the enzyme myrosinase that can act on its substrate sinigrin to produce allyl isothiocyanate, also known as mustard oil. In whole, intact fruits, the enzyme is located away from sinigrin so the two do not mix and mustard oil does not occur in an undamaged plant. But when the cells are damaged, such as when an animal chews on them or they are being ground to make mustard, the enzyme and the substrate come in contact and mustard oil is produced. This is an excellent defensive strategy for the plants because they only make the mustard oil when it is needed, as they are being attacked. For us, it means that preparation of mustard at home is a bit complex; the fruits must be ground with a liquid, preferably a cool liquid because heat inactivates the enzyme. Water is often used, and ground mustard should be allowed to stand in the water about 30 minutes to develop pungency.

Vinegar or wine vinegar can be added to stop the action of the enzyme and hold the pungency at a certain level; this is the method used commercially so that we can buy prepared mustard in bottles. For a stronger mustard, the ground seeds should be allowed to set in water for a longer time before the vinegar is added. Although our familiar "yellow mustard" is indeed yellow, that is due to the addition of turmeric, not the natural color of mustard fruits.

Horseradish is a root that, when fresh, has almost no aroma or pungency. It too contains myrosinase and sinigrin, and the point of our grating horseradish is to damage the cells so that enzyme and substrate come together and make mustard oil. The same reaction occurs when wasabi (*Wasabia japonica*, also in the mustard family) root is grated. Anyone who has eaten wasabi with sushi is familiar with just how potent mustard oil can be (**FIGURE 14.17**). As its name indicates, wasabi is from Japan, so it is an Asian spice that is not tropical.

FIGURE 14.17. Sushi is always served with thinly sliced ginger, soy sauce, and a bit of bright green wasabi. (© Fanfo/ShutterStock, Inc.)

Important Terms

capsaicin
eugenol
glycyrrhizin
menthol
vanillin

Concepts

- Herbs and spices are used in such small quantities they contribute few calories and nutrients to food.
- Strong or pungent flavors and aromas deter herbivores.
- Many flavors and aromas are based on terpenes and phenols; alkaloids and lipids are also important.
- Several families with spices and herbs are the mint (Lamiaceae), dill (Apiaceae), mustard (Brassicaceae), nightshade (Solanaceae), pepper (Piperaceae), ginger (Zingiberaceae), and orchid (Orchidaceae) families.
- Many flavors and aromas occur naturally in the plant, but others develop after harvest, during processing.
- The search for spices was the driving force for the exploration of much of the world, including Columbus's expedition that discovered the Americas and Magellan's voyage that was the first to sail completely around the world.

Plants as Sources of Medicines, Drugs, and Psychoactive Compounds

These peyote cacti have no spines, but they are not defenseless. Their bitter, toxic, defensive alkaloids make most animals avoid eating them, but some people seek them out because they cause hallucinations. Psychotropic chemicals (those that act on an animal's nerve cells) are some of the most effective defensive compounds that can evolve in a group of plants; unfortunately, many of those chemicals are the basis of recreational drugs that cause great harm to many people.

Introduction

Datura stramonium is a small herb related to potatoes and tomatoes, and although people have been familiar with it for a long time and it is easy to cultivate, it has never been domesticated (FIGURE 15.1). In our search for foods, spices and herbs, *D. stramonium* has been passed over. Why? Its common name will give you a clue: devil's trumpet. All parts of the plant contain a poison called atropine: leaves, stems, fruits, and seeds will all kill you after giving you tachycardia (a rapid heartbeat), hallucinations, and seizures. The list of plants that are too toxic to eat is a long one: we don't eat even tiny amounts of poison ivy (*Toxicodendron radicans*), strychnine tree (*Strychnos nux-vomica*), wolfbane (*Aconitum* species), deathcamus (*Zigadenus*), poison hemlock (*Conium maculatum*), or suicide tree (*Cerbera odollam*). The total list would contain thousands of plant species.

Features that make plants inedible are essential to their survival; otherwise a plant would be an immobile source of food for any animal that encountered it. Mutations that cause defensive phenotypes in plants are very advantageous. Some plants protect themselves with physical barriers: cacti, roses, and many others use thorns; leaves of oaks, magnolias, and palms are sclerified and tough; many grasses make their leaves inedible by having silica cells filled with glass. And large numbers of plants have evolved to have defensive chemicals, compounds that are poisonous, caustic, nauseating, astringent (they make your mouth pucker), or bitter.

A diverse variety of defensive chemicals has evolved. In most cases, only a small amount of the plant compound is needed to cause a significant reaction in animals because the defensive compound interferes with fundamental aspects of animal metabolism. For example, the atropine in *Datura stramonium* binds to a particular type of receptor in certain nerve cells, interfering with nerve transmission. Strychnine alters the movement of chloride in the nerve cells of our spinal cord and brain, and aconitine interferes with sodium movement in the heart. Colchicine from autumn crocus (*Colchicum autumnale*) prevents mitosis and meiosis by blocking the assembly of microtubules into a spindle. Ricin, the poison in castor bean (*Ricinus communis*), prevents ribosomes from reading mRNA and building proteins.

FIGURE 15.1. *Datura stramonium* grows easily, has large soft leaves, and produces abundant fruit and seeds. It would make an excellent crop except that it is too poisonous to eat. Common names are devil's trumpet and jimson weed.

As it turns out, many of these chemicals are potentially valuable as medicines because they are a means by which physicians control particular aspects of our metabolism. Taking atropine as an example again, your ophthalmologist may use it as eye drops to dilate your pupils; your anesthesiologist might treat you with it before anesthesia to minimize your production of saliva while you are unconscious and to counteract the tendency of anesthetics to slow the heart beat. With modern synthetic chemistry, we can alter the natural plant defensive compounds and create new medicines. By modifying atropine in various ways, we now have several related drugs to treat psychosis, asthma, morphine addiction, pain, and diarrhea. A recent survey found that modern medicine uses 88 specific chemical compounds from 72 medical plants, and the search for new drugs in new plants continues.

At present, plants are used as drugs in many ways. The plant may be chopped or ground with a liquid and applied to an injury on the outside of the body; this is a poultice. Or plants can be steeped in water or alcohol to make a tea that is drunk to treat internal problems. Alternatively, specific compounds are extracted and then used as medicines after being purified or altered. In several cases, plants produce only trace amounts of important medicinal compounds, but we now know how to either synthesize the compound completely artificially, or we can extract a similar but abundant compound and then convert it to the medicinally useful one. Most recently, many people have postulated that plants contain numerous undiscovered phytochemicals that promote good health and that rather than trying to isolate one or two specific compounds, it is best to use the whole plant or unrefined extracts that contain many compounds. These are called functional foods or medical foods.

Stimulating Beverages

We begin with tea and coffee because these are familiar plants that require little processing of the plant material and most of us do the final preparation ourselves in our own homes. Tea and coffee may not seem like medicinal plants, but the chemicals they contain do exert strong effects on us.

FIGURE 15.2. Fields of *Camellia sinensis*, cultivated for tea, cover entire mountainsides in tropical regions. The plants are kept low enough that they can be harvested easily, because every leaf must still be picked by hand. (Courtesy of Allegro Coffee Company.)

Tea

Tea is the most widely consumed beverage in the world, other than just water. More cups of tea are drunk per day than all other beverages combined. Tea is prepared from the leaves of a large bush, *Camellia sinensis*, native to Southeast Asia (FIGURES 15.2 and 15.3). The bushes are pruned to keep them short, because all leaves are gathered by hand. Pruning also overcomes apical dominance and encourages the growth of many axillary buds so that each plant makes many new leaves each year. After picking, the leaves are processed, then dried, and we prepare the drink by making an **infusion**: we steep the leaves in hot water for several minutes, then discard the leaves and drink the water.

FIGURE 15.3. These are the new leaves of *Camellia sinensis*. Only the newest, freshest leaves can be used for tea; older leaves carry out the photosynthesis necessary to maintain the plant. (Courtesy of Allegro Coffee Company.)

During the steeping, numerous water-soluble chemicals move from the leaves into the water. The brown color and slight astringency of tea are due to tannins that are extracted by the hot water. The stimulating effect of tea is caused by two alkaloids, **caffeine** and theophylline, both of which are classified as psychotropic stimulant drugs, that is, drugs that act on the brain. Caffeine is able to stimulate our central nervous system, quicken our reaction time, and relieve both fatigue and sleepiness. In high concentrations, caffeine prevents sleep and causes feelings of nervousness and restlessness. Tea leaves have a higher concentration of caffeine than do coffee beans, but because tea is usually prepared as a weaker drink, a cup of tea has much less caffeine than a cup of coffee.

The flavor of tea depends on numerous chemicals, which are in turn affected by the processing of the leaves. When freshly picked, tea leaves are extremely bitter and astringent. After being picked, they are steamed for about 1 minute to cause them to wilt, then they are rolled or pressed to disrupt the cells and allow enzymes in the leaves to oxidize various compounds and create the proper flavors. Leaves that are to be used for **green tea** are processed the least to maintain as much of the natural flavor as possible; the leaves are allowed to oxidize only briefly, then they are dried, chopped, and then ready for use. The oxidation step is called "fermentation" in the tea industry, but it is not really fermentation because oxygen is present. Leaves for **black tea** are also rolled or pressed, but the enzymes are allowed to ferment for much longer than green tea, up to 3 or 4 hours before the leaves are dried. This longer time allows many more changes than occur with green tea. First, the plant's enzymes cut sugar molecules off of various types of complex molecules; with the sugar removed, the remainder of each molecule becomes volatile and thus contributes to the aroma of the tea. Second, small bitter, astringent molecules called phenols react with oxygen and bind together into very large molecules called polyphenols. The polyphenols are neither bitter nor astringent, and they have a deep red/brown color. This long "fermentation" gives black tea its dark color, rich aroma, and more robust flavor. Leaves for **oolong tea** are also rolled or pressed but are fermented for an intermediate amount of time (FIGURE 15.4).

FIGURE 15.4. These are the three types of tea that result from different fermentations. The lightest is green tea, fermented for only a few minutes; the darkest is black tea, fermented for several hours. The medium tea is oolong, fermented longer than green tea but less than black tea. All three are the same species, *Camellia sinensis*, and the differences in their colors, aromas, and flavors result entirely from the fermentation.

These three teas are pure tea; many other teas have other plant material added to give a variety of flavors. For example, Earl Grey tea is black tea with oil from the rind of bergamot oranges added. Lapsang Souchong tea gets its smoky flavor because the leaves are dried over smoldering pine needles. Chai tea (also called masala chai) is flavored with numerous spices including ginger, cardamom, cinnamon, and several others; there are diverse recipes.

The word "tea" is often used for a hot water infusion of almost any plant material. Herbal teas are made from various plants other than *C. sinensis*. One of the newest types of tea is rooibos tea made from the South African plant *Aspalathus linearis*; it is becoming popular due to its flavor, lack of caffeine, and high levels of antioxidants.

The main tea cultivating countries are China, India, Kenya, and Sri Lanka. Although each plant can be harvested several times per year, the demand for tea is so great that large amounts of land are used for its cultivation. *C. sinensis* is grown primarily in areas that receive at least 50 inches of rain per year and where temperatures are cool, never hot.

Despite being such a ubiquitous crop plant, tea does not have the long history of most of our food plants. It was first consumed in China, perhaps as long ago as 1000 BCE, but despite China's long recorded history, the first certain written references to tea do not appear until 59 BCE.

Coffee

Coffee is prepared by passing hot water through the ground, roasted seeds of two species of a small tree (FIGURE 15.5). *Coffea arabica* is the species used most often and which provides the richest flavor, but seeds of *C. canephora* var. *robusta* have about 50% more caffeine and better body. As with brewing tea, the hot water extracts various chemicals, many of which contribute to coffee's flavor. It is the caffeine that makes coffee a stimulant.

FIGURE 15.5. Although coffee trees can grow to be small trees, in some plantations they are pruned to keep them low enough so that individual berries can be picked by hand. These trees are being cultivated in the shade of naturally occurring forest trees. (Courtesy of Allegro Coffee Company.)

FIGURE 15.6. Unroasted coffee beans are dry and pale green; roasting breaks down their starches into simple sugars and then caramelizes them, resulting in the dark color. Roasting also develops the oils and brings them to the surface. Beans as dark as this are "dark roast" coffee beans.

After the beans are separated from the berry pulp, they are carefully dried, cleaned, and sorted. The result is green coffee beans that are stable enough to be shipped and stored for several months. Roasting is a critical step in the preparation of coffee. Beans are roasted at temperatures ranging from 188°C to 282°C (370°F to 540°F) and for as little as 3 minutes to as long as 30 minutes. With short roasting times, more of the natural flavor of the coffee is preserved, and the differences between coffee beans from different growing areas are apparent. With longer roasting, many of the chemicals that make each region's coffee distinctive are lost, but the extra roasting depolymerizes starch in the bean, converting it to sugar, which then caramelizes, giving the beans their darker, richer flavor and aroma (FIGURE 15.6). Longer roasting also affects the oils, bringing them to the surface, converting some into new forms, one of which is caffeol, the oil that gives coffee its distinctive flavor and aroma. Some of the caffeine is also broken down during roasting: dark, rich coffee has less caffeine than light, mild coffee. Once roasted, the flavors of coffee are more labile and can deteriorate rapidly, especially if exposed to air at room temperature. Many stores and coffee shops roast and grind their coffee every day.

Like all forms of agriculture, cultivation of coffee trees involves some environmental damage and habitat destruction. In the last few years, methods of cultivating coffee trees have changed. In the past, all coffee trees were grown in the shade of larger rainforest trees; many of the smaller trees of rainforests were cleared out, then *Coffea* was planted in the open ground below the remaining tall trees. This disturbed the habitat greatly, but at least many of the native trees survived. At present, many coffee plantations are on ground where all natural vegetation has been removed and the *Coffea* is grown in full sun, at high density, and using fertilizers and pesticides. The trees grow more rapidly and produce greater yields, but the ecological damage is much greater. If you drink coffee, you can request "shade-grown" or organic coffee.

Harvest methods have also changed. In the past, all coffee fruits, called coffee berries or cherries, were picked by hand, and only when ripe. Each tree was harvested several times so that each berry was picked at its optimal time; some plantations still do this. A newer technique is to wait until most berries are ripe, and then pick them all; afterward, they are sorted and the immature berries are discarded.

After picking, coffee berries must be stripped of their pulp so that just the seed, the "coffee bean" is left. The berries may be laid in the sun for several weeks to allow the fruit pulp to decompose. Alternatively, the berries may be washed in water, pressed to break them open, then forced through a screen to separate most of the pulp from the bean. A residual layer of mucilaginous pulp is removed either by machine washing and scrubbing, or by letting the beans ferment for about a day. Fermenting adds to the flavor of the coffee. In the past, removing pulp always required large amounts of freshwater, and produced large amounts of polluted waste water. Newer methods have reduced the amount of water needed.

Chocolate

Chocolate comes from the roasted, processed seeds of a small tree, *Theobroma cacao*, native to a region of tropical America stretching from Mexico to the Amazon River Basin (FIGURE 15.7). Trees of *T. cacao* prefer shaded conditions, and they can begin flowering and fruiting while still young, as little as 4 years old. They are unusual in that they bear small flowers on their trunk (this is called cauliflory) rather than on twigs as most trees do. Although each tree may have thousands of flowers, only about 20 develop into large pod-like fruits, each fruit having about 30 to 60 seeds. Trees are grown on large plantations as well as on small family farms in most countries with tropical areas; currently West Africa is the largest producer of cacao.

FIGURE 15.7. The seeds that will eventually be processed into chocolate are borne in large pods that develop from flowers that occur on the trunks of *Theobroma cacao*, not on the twigs as is typical in many plants. Each pod contains many large seeds and only a small amount of pulp, but it is the fermentation of the pulp that is crucial to the development of the chocolate flavor. (Courtesy of Allegro Coffee Company.)

It is important to distinguish between several stages in the production of chocolate because some of the names can be confusing. The raw seeds are *T. cacao*; after they are dried, cleaned, and roasted, they become **cacao nibs**. After grinding, the spelling of the name changes to cocoa, and the word "chocolate" can be used: ground cacao becomes **chocolate liquor** which is immediately separated into **cocoa solids** (which contain the chocolate flavor) and **cocoa butter**. The production of all types of chocolate then involves recombining cocoa solids and cocoa butter in various proportions and with other ingredients, especially sugar and milk. Cacao is an old food; sweet chocolate and milk chocolate are new foods. Archeological evidence shows that cacao was being consumed by Mesoamericans at least as early as 1100 BCE and was well known to Olmecs, Mayans, and Aztecs. Sugar was unknown in the Americas, and cacao was drunk as a bitter beverage known as xocoatl, which contained chili peppers, achiote, and vanilla as well as cacao. Europeans first encountered cacao in the courts of Aztec royalty, and it quickly became fashionable in Europe, still as a bitter drink. The types of chocolate that we love today were not invented until the 1800s.

The conversion of seeds into chocolate is a complex process and begins immediately after harvest. First, the fruit skin is removed and the seeds with their surrounding pulp are dumped into a pile and allowed to ferment; this is a crucial step. The pulp but not the beans are acted on first by yeasts, which change sugars to alcohol; after a few days, lactic acid bacteria flourish, displacing the yeasts and causing a different set of enzymatic reactions. After several days, the pile is turned to allow oxygen into it, and this inhibits the lactic acid bacteria while encouraging acetic acid bacteria. These three distinct microorganisms acting one after the other alter the flavor and nature of the pulp and beans, and allow various enzymes to create new flavors and aromas while eliminating some undesirable ones. After fermentation, the beans are dried until they are stable enough to be shipped and stored.

Whereas fermentation usually occurs on the plantation, the next steps are all performed in factories with expert supervision. The beans are roasted for up to 60 minutes at a precise temperature. Just as with coffee, proper roasting is essential if flavor is to develop properly. Once roasted, the beans are then ground, and it is at this point that modern chocolate becomes distinct from Aztec chocolate. Grinding produces the coarse, oily chocolate liquor mentioned above. Cacao beans are 55% fat, so grinding cacao beans is similar to grinding peanuts: solid seeds go in, an oily, gritty paste comes out, but cacao is much more oily than peanuts. Aztecs and early Europeans would mix this chocolate liquor with water, but it produced a bitter, oily drink that was extremely filling. In 1828, it was discovered that pressing the chocolate liquor would force the cocoa butter out as a liquid, leaving behind the dry cocoa solids that could then be used as cocoa powder for making drinks as well as for cooking and baking.

At first, the cocoa butter was just a waste product that was discarded, and only the cocoa solids were used. But later it was discovered that if cocoa butter was added back to the cocoa solids in just the right amount and with sugar and powdered milk, a wide range of chocolates could be produced. It might seem like an unnecessary step to separate chocolate liquor into cocoa butter and cocoa solids and then recombine them again, but the important point is that the separation allows us to recombine them in numerous controlled ratios and with other ingredients.

Dark chocolate is made by mixing cocoa butter, cocoa solids, and sugar without any powdered milk. If large amounts of sugar are added, we get sweet chocolate; medium amounts of sugar produce bittersweet chocolate and if no sugar is added, the result is bitter chocolate used in baking. Probably all of us had the childhood experience of taking a bite of baking chocolate and discovering that it was not the same as a candy bar; however, that is the true flavor of just chocolate without other ingredients. *Milk chocolate* is made by mixing powdered milk solids with the cocoa butter, cocoa solids, and sugar. Milk chocolate is the most popular kind for candy, and generally it has more milk and sugar than actual cocoa butter and cocoa solids. Mixing these four ingredients together is not easy; if the proportions are not correct, the ingredients separate or become pasty or dry. Also, they must be mixed by grinding them for hours, often more than a day, just so that the remnants of the cells in the cocoa solids are ground so fine that our tongues cannot feel them as coarseness or grittiness. *White chocolate* is a mix of cocoa butter, sugar, and milk solids, but with no cocoa solids at all, so it has no actual chocolate flavor.

The physiological effects of chocolate are still not known in detail, especially the basis for it being almost addictive to many people. The most prominent alkaloid is theobromine, which is chemically similar to caffeine and is a mild stimulant. Chocolate also contains serotonin, a chemical that some of our own nerve cells produce as they transmit nerve impulses from one cell to another. It is suspected that the serotonin in chocolate may thus affect our nervous system and be part of the reason chocolate gives us a sense of well-being. Dogs, cats, horses, and many other animals cannot metabolize theobromine; it just stays in their bloodstream. If fed chocolate, these animals may suffer seizures, heart attacks, internal bleeding, and finally death.

Just as coffee cultivation and processing have several negative aspects, so does the cultivation of cacao. As much as half the work is done by child labor, usually slave labor.

The practice is most widespread in the Ivory Coast, and although many calls have been made to abolish this practice, it is still common, and it is still used by certain large international chocolate companies.

Colas and Caffeine-Based Energy Drinks

Modern soft drinks with the word "cola" in their name were developed using ingredients based on the kola nut (with a "k"; FIGURE 15.8) that comes from one of several species of the genus *Cola* (with a "c"), a small tree closely related to *Theobroma cacao.* Kola nuts are extremely bitter and have a high level of caffeine. In parts of West Africa, the nuts are chewed for their stimulating quality, and their ability to ease hunger. The original cola soft drink (Coca-Cola) was invented in 1886 as an alternative to alcoholic beverages; its two main ingredients were cocaine (from the plant *Erythroxylum coca*) and caffeine (from *Cola*). After several years, all cocaine was removed from the formula; more recently, artificial ingredients have replaced the kola nuts also. Other cola drinks have been invented and are still extremely popular, but none actually use kola nuts any longer. The caffeine in most of these beverages is the caffeine that is obtained when coffee is decaffeinated.

In the last few years various stimulating drinks based on caffeine, such as Red Bull and Jolt, have entered the market. These are completely artificial and can thus set the level of caffeine very high. They are popular with students studying late at night, long-haul truck drivers, and athletes needing extra energy and alertness. These drinks are effective and do keep people awake, alert, and at least feeling energetic.

FIGURE 15.8. Trees in the genus *Cola* have large, thin-walled fruits that contain the kola nuts. Formerly, these were an important ingredient in the original cola drinks, but now most such drinks are made with artificial ingredients. (© WILDLIFE GmbH/Alamy Images.)

Beverages that Depress Our Central Nervous System

Fermentation of plant material produces **ethyl alcohol (ethanol)**, and this is classified as a depressant. This seems counterintuitive, because light drinking causes people to be more lively and less inhibited, but even moderate drinking reduces people's ability to focus their attention, and it slows their reaction speed.

Drinking ethanol has both beneficial and harmful consequences. Moderate consumption, about two drinks per day (two bottles of beer or two glasses of wine or two ounces of liquor) lower the risk of heart disease. Red wine seems to be the most beneficial because it has resveratrol (see below). In contrast, deaths related to ethanol—including driving while under the influence of alcohol (DUI, drunk driving)—is the fifth greatest cause of death among North Americans. And of course, drunk drivers don't just harm themselves; they kill and maim innocent bystanders. Ethanol is not at all addictive to many people, but it is to others and results in alcoholism. Ethanol easily passes from mother to fetus through the placenta and can cause fetal alcohol syndrome if the mother consumes too much ethanol while pregnant. The effect of ethanol on us depends on the amount we drink in proportion to our body size, which in turn determines how much ethanol passes into our blood. When the concentration of ethanol in our blood (our blood alcohol content or BAC) reaches 0.02%, it starts to interfere with both our coordination and our reaction time, and it may cause us to have impulsive behavior. At a BAC of 0.15%, people are drunk and have trouble walking or doing other simple things. A BAC of 0.4% is usually fatal. In the United States, it is illegal to drive if your BAC is 0.08% or higher. As a general rule, a person can metabolize the ethanol in one standard size drink in about 60 to 90 minutes; if you drink faster than that, chances are good that you will get drunk.

Ethanol is only one of many types of alcohol. Other types we encounter commonly are isopropyl alcohol (isopropanol) used as rubbing alcohol, and methyl alcohol (methanol) as wood alcohol. All alcohols are poisonous, but our bodies are able to detoxify ethanol if we drink only small amounts. With larger amounts, we develop a hangover, and even larger amounts cause death. One to several college students die every year from drinking so much ethanol (usually high-strength liquor) that it disrupts the lipids in their cell membranes (remember that biologists kill and preserve specimens of plants and animals in a mixture of ethyl alcohol and formaldehyde). Methyl alcohol is very toxic by itself, and our livers convert it into the even more poisonous substances formaldehyde and formic acid. Even at doses too low to kill us, methyl alcohol causes blindness. Isopropyl alcohol is converted to acetone in our bodies; you may be familiar with acetone as fingernail polish remover.

The ethanol of beverages is produced by the fermentation (anaerobic respiration) of glucose by yeasts. Sugars present in fruits can be fermented immediately but starches in seeds and tubers must be depolymerized to glucose before they can be fermented. Ethanol produced by fermentation finally kills the yeast if it builds up to a concentration of about 18%. To obtain a stronger concentration of alcohol in the beverage, the fermented mixture must be distilled, in which it is heated to concentrate the alcohol while removing some of the water. Alternatively, extra alcohol can be added to a beverage rather than distilling it. These factors combine to give us three basic types of alcoholic beverages: **beers** and **sake** result from partially fermenting starchy seeds; **wines** are produced by a more complete fermentation of sugary fruits; and **spirits** have their ethanol concentration increased by distillation or by adding alcohol.

Beer

Beer is made by fermenting starchy cereal grains, especially barley, wheat, corn, or rice. Barley is by far the most common ingredient in beer, and if a different grain is used, it is typically specified. For example, beers made from wheat are becoming more popular, and are called "wheat beer." Before a cereal grain can be fermented, at least part of its starch must be converted to glucose. This is done by moistening the grains and allowing them to germinate. The moistened embryo secretes enzymes that break down the starch located in endosperm (cereal grains are mostly endosperm; see Figure 13.3). Barley has the greatest number of enzymes and converts starch to sugar more quickly than do other grains. Once the seedling root is visible, sprouting has proceeded long enough, and the grains are dried to prevent any further enzymatic reactions. Sprouted, dried barley grains are called **malt** (this is also the malt in a malted milkshake). To brew the beer, malt is mixed with water in a large vat; the mixture is called **mash**. If malt is the main ingredient, the beer will have a strong flavor, and the quality of the sprouting and roasting of the barley will be very important. But often, at least in the United States, unsprouted grains or even corn syrup or potatoes are added. These are very cheap sources of extra starch that can be acted upon by the enzymes in the malt. By adding these adjuncts, the beer is less expensive, and typically has less flavor.

Hops are the dried carpellate inflorescences of a tall, vining plant, *Humulus lupulus* (FIGURE 15.9). *Humulus* is closely related to *Cannabis* (marijuana). Hops have a bitter taste and aroma; the amount of hops added to the mash influences the taste of the beer: the greater the amount of hops, the more bitter the beer. Hops also have enzymes that clarify the beer and keep it from being cloudy when finished.

(a)

(b)

FIGURE 15.9. **(a)** Hops are produced by vines that must be grown on trellises. Whereas grapes grow well on low, horizontal wires, hops are cultivated on very tall vertical wires. **(b)** Hops are a dioecious species, meaning that some plants produce staminate flowers ("male" plants) whereas others produce carpellate flowers ("female" plants). The hops that are added to beer to reduce its sweetness and to clarify it are clusters of carpellate flowers. Most fields of hops are planted with cuttings so that the entire field will be just carpellate plants; the staminate plants are not needed in brewing beer. Also, by not having any staminate plants in the field, the carpellate plants do not produce fruits or seeds, both of which would reduce their value. (Photos courtesy of Linda Wong Bruno.)

Once all the ingredients are in the mash, it is warmed to 68°C to 73°C (154°F to 163°F) and allowed to stand for several hours while the various enzymes are active. The chaff and solid materials are then strained out, and the sugar-rich liquid, called **wort**, is boiled to inactivate the enzymes and kill any microbes. Once it has cooled, yeast is added. For beers that will be ales, bitters, or stouts, *Saccharomyces cerevisiae* is used, but for lagers and Pilsners, *S. uvarum* is added. Most beers in the United States are lagers. The yeasts are allowed to ferment the wort from 1 week to 12 days. The beer is filtered and pasteurized to stop all further fermentation, then it is aged for 2 or 3 weeks. The carbon dioxide produced by fermentation is allowed to escape during brewing, so to make beer frothy when it is poured, new carbon dioxide is added to the beer artificially just before it is bottled.

Various types of beer result from controlling the malting process, the adjuncts, hops, and the type of yeast used. *Saccharomyces cerevisiae*, the yeast used for ales, floats to the top of the fermenting vats (called top fermentation), and it works best at a warm temperature. This species of yeast produces compounds called esters that add light, fruit-like flavors to the beer. Common types of ale are brown ale, pale ale, and porter (or stout, made with extra hops and slightly more caramelized barley). *Saccharomyces uvarum*, used for lager beers, sinks to the bottom of the vats (bottom fermentation) and functions at a cool temperature. Cool, bottom fermentation does not produce esters, so lager beers have a crisper taste. Most of the beer consumed in the United Stated is pale lager, with an alcohol content of only 3% to 6%. The concentration of alcohol depends on the amount of starch converted to fermentable sugar and whether fermentation is artificially stopped at a particular point. With a greater conversion of starch to sugar and a longer fermentation, the concentration of alcohol increases. Many states strictly limit the amount of alcohol a beverage can have and still be called a beer.

Light beers have fewer calories than ordinary beer. This can be achieved in either of two ways. One is to begin brewing with a mash that contains fewer carbohydrates. When brewing is halted, regular beer still has many starches and unfermented sugars present, but light beer has fewer of them, giving the beer fewer calories. Many people believe that most of the calories in beer comes from the alcohol; while this may be true for light beers, it is not true for regular beers and certainly not for stouts. These have relatively high amounts of starches and sugars, giving the beer its body and fullness. The second way to make light beer is to convert more of its starches to sugars; this will ferment to a beer with much higher alcohol content but lower in starch. Water is added at the end to lower the alcohol content to the normal 3% to 6% level for beer, but it also dilutes the flavor.

Sake

Sake is a fermented cereal grain made from rice. It is usually called rice wine, but because it is a cereal, technically it should be classified as a beer. However, sake is made by a unique process that differs from brewing beer. The rice is polished to remove all fruit and seed coats, and then is steamed to cook and soften its starch grains, as if it were to be eaten. Instead, it is inoculated with the fungus *Aspergillus oryzae* and allowed to set for several days. Enzymes from the *Aspergillus* depolymerize the starch to fermentable sugars. After the rice gets a distinctly moldy aroma, it is mixed with warm water and then *Saccharomyces cerevisiae* is added. No hops or adjuncts are added as they would be for beer, and instead of fermenting for only a week, sake is fermented for almost a month. During this time, the *Aspergillus* enzymes continue to convert more starch to sugar, and the yeast ferments that to ethanol. At the end, the sake is separated from the solids by filtration through cloth, and extra ethanol is added to bring the final concentration up to about 20% to 22%; this makes it a fortified beer. Sake is thus much stronger than beer, and it is served hot.

Wine

Wines are fermented fruit juices that are rich in sugars. When the word "wine" is used by itself, it typically refers to fermented grapes of the species *Vitis vinifera*; if other fruits are used, then they are named, such as elderberry wine or peach wine. As anyone who has shopped for or selected wine knows, there seems to be an endless variety of wines, such as Chardonnay, Pinot Noir, Cabernet Sauvignon, Merlot, and many others. It may seem that these represent numerous species of grape, but actually all are varieties of just one, *V. vinifera*. Wine was first produced about 6000 BCE, and through those many years since, farmers have selected numerous varieties (there are now 15,000 varieties), each with its own special flavors, aromas, and colors, but the farmers always cultivated *V. vinifera*. This is a little surprising because there are 60 species of grape in the genus *Vitis*. These others are typically used for jams, jellies, juices, raisins, or for eating fresh (table grapes are *Vitis labrusca*), but they are not used for wine.

Compared to the brewing of beer and sake, producing wine is rather uncomplicated. Grapes are harvested when they reach proper maturity and their sugar content is suitable. Typically 20% to 30% of the grape's weight is sugar (mainly glucose and fructose) when harvested. They are washed then crushed to obtain the juice. For white wines, the skins (exocarps) are removed, but for red or rosé wines, the skins are mixed in with the juice. The skins not only provide the color of red wine, they produce various flavors and aromas. The chemical resveratrol is produced when yeasts attack the skins; there are claims that resveratrol lowers blood pressure, prolongs life, and has anticancer activity, but none of these has been proven definitively in humans. *Saccharomyces cerevisiae* is added to the juice, and the mixture is cooled slightly; white wines are fermented at 10°C to 15°C (50°F to 59°F) and red wines slightly warmer at

FIGURE 15.10. Wine production. Grape juice is being fermented into wine in these vats. Each vat has valves that permit the outward flow of carbon dioxide while preventing the inward flow of oxygen. Although beer can be fermented in tanks with open tops, wine must be protected from oxygen at all times otherwise it will be converted to vinegar. (© David H. Seymour/ShutterStock, Inc.)

25°C to 30°C (77°F to 86°F). Fermentation vats have valves that allow carbon dioxide to escape without allowing oxygen to get in (FIGURE 15.10). If oxygen were to enter the tank, certain bacteria would convert the ethanol to acetic acid, which would turn the wine into vinegar. After about 8 to 10 days, the liquid is removed from the skins and allowed to continue fermenting for up to a month. During fermentation, dead yeast cells and other particles settle to the bottom of the fermentation tank, and crystals of tartaric acid (cream of tartar) also sink to the bottom. The wine must be drawn out of the tank carefully so as not to disturb the sediment because wine with sediment is less appealing. The tartaric acid crystals are a hallmark of grape fermentation, and archeologists search from them in old pots to ascertain whether particular peoples were producing wine at a certain time or place.

Fermentation continues as long as sugar is present and the concentration of ethanol has not reached lethal levels (18% to 20%). If all the sugar is fermented, the wine is a dry wine, but if some sugar remains, it is a sweet wine. Most wines produced in the United States have an alcohol content of between 12% and 14%, and have some sugar left, so their fermentation must be stopped artificially. This is usually done by microfiltering the wine to remove all yeast; the alternative is to heat the wine enough to kill the yeast (to pasteurize the wine), but that damages the flavor. Sulfites may be added to the wine at this point as a preservative, to prevent the growth of microbes as the wine sits in the bottles. Some people are allergic to sulfites and have mild reactions after drinking wine. Unfortunately, wine can never be completely free of sulfites because a small amount occurs naturally on the grapes, and sulfites are used to sterilize oak aging barrels between batches of wine.

At this point, the wine begins an aging process. White wines are aged only briefly, between 12 and 18 months, but red wines may be aged for up to 5 years. Typically wine is aged in large oak barrels, where chemicals in the wine interact with compounds in the heartwood. These reactions alter the flavor and aroma of the wine, and even create new types of longer-chain alcohols that give the wine a more viscous, substantial feel in the mouth. When the wine master has decided the wine is ready, it is transferred to bottles and either shipped to market or set aside for more aging in the bottles. Certain types of wine are aged in stainless steel vats so as to avoid the changes in flavor and aroma caused by oak barrels. Although this gives the vintner greater control over the quality of the final product, the resulting wine has few complexities and less individuality. Stainless steel tanks are more appropriate for mass produced, inexpensive wines.

Champagnes and sparkling wines are made by adding sugar to a bottle of wine that still has a few live yeast cells in it. The wine then continues to ferment, but all the carbon dioxide is trapped inside and builds up pressure. When the bottle is uncorked, the carbon dioxide forms bubbles and makes the wine effervesce, or if not done properly, the wine shoots across the room. To be called "champagne," the wine must be produced in the Champagne region of France. All other effervescent wines are "sparkling wines," although those of Italy are usually referred to as prosecco and those of Spain are known as cava. Inexpensive "sparkling wines" are made by simply carbonating white wine in the same way that soft drinks are carbonated.

If extra ethanol is added to a wine, it becomes a fortified wine. Examples are sherry, port, and Madeira. Such wines are then aged, either in bottles with no exposure to oxygen, or in oak barrels that permit a slight oxidation and evaporation, giving the wine a richer feel in the mouth. For example, tawny port is made by stopping fermentation while half the grape sugars are still present. Distilled spirits are added to bring its alcohol level to 20%, then the wine is aged for ten years in oak barrels that allow a very small amount of oxygen to act on the mix, converting it from wine to port. During this time, the bright red pigment of the grape skins precipitate and change color to a tawny brown.

Many aspects of the flavor of wine are affected by the conditions of its cultivation. The chemistry of the soil, the climate, the amount and timing of rainfall and of warm and cool periods, and the number of sunny days as opposed to cloudy ones are all critically important. These are referred to as the terroir of the wine, but we mostly refer to the wine's region and its year of harvest (its vintage). In all important wine producing countries, each region is given its own name (appellation), for example, Bordeaux, Burgundy, Champagne, and Savoy in France, and Napa Valley, Sonoma, Columbia Valley, and many others in the United States. Many new vineyards have been established in very sunny areas with artificial irrigation, so those factors are more or less constant from year to year, and vintage is not too important. But where grapes are cultivated without irrigation and where some summers are cool and cloudy but others are hot and dry, the quality of the grapes varies greatly from year to year, so vintage matters. Few people have the time or interest to remember which year was good for a particular wine and instead rely on the knowledge of a reputable merchant.

The effects of global climate change could be devastating for vineyards and wine production. Even with irrigation and a sunny climate, weather still plays a crucial role in the quality of the grapes. Vineyards are typically some of the most costly land used for agriculture, and the production facilities are often located close to the vineyards; juice is rarely shipped more than a few miles. If an area changes to being slightly warmer or wetter, if the date of the last frost of spring changes, if rains fall at a different time, any or all of these could make wine grapes useless. Most existing vineyards have already been carefully chosen to be suitable for a particular variety of grape and for the production of a particular type of wine, so any change in weather or climate will almost certainly hurt the crop rather than improve it.

Chapter 12 discussed the problem of introduced, nonnative species, and viticulture provides an excellent example. Sometime during the 1850s, grape plants were brought to Europe from America, and their roots were infested with a root louse called phylloxera (*Viteus vitifoliae*). American plants are resistant to phylloxera and it causes little trouble here, but the European plants were susceptible to it, and entire vineyards began dying in the 1860s. Although wine is an important food and it does supply significant amounts of calories and nutrients, loss of the vines did not cause mass starvation as occurred during the potato blight in Ireland. But it was an economic disaster, and a remedy was needed. France sent teams to the United States seeking species that were both resistant to phylloxera and that would grow in the French soils and climate. Many candidate species suffered from chlorosis in French soils and failed to thrive. Finally, several Texas species, *Vitis cinerea* var. *helleri* and *V. vulpina*, which grow in limestone soils here that resemble the limestone in France, were found to provide excellent rootstocks to which the French grape vines could be grafted. At present virtually all European grapes are grafted onto American roots, most being from Texas. Although phylloxera has been an introduced, nonnative pest for over 150 years in Europe, it has never been possible to eradicate it; it is still widespread there.

Spirits

Spirits are alcoholic beverages with an ethanol content above 20%. Once fermentation has produced enough alcohol to raise its concentration to 20%, the solution is toxic to the yeast and they die. To obtain higher alcohol content, either strong alcohol (always ethanol) must be added, or the solution must be distilled. During distillation, the solution is heated, which causes most of the alcohol to evaporate, along with some water and some of the flavoring compounds. Distillation is possible because different liquids boil at different temperatures, and ethanol boils at 78°C (173°F), which is much cooler than the boiling point of water at 100°C (212°F). By keeping the mixture above 78°C and below 100°C, ethanol and certain flavors evaporate whereas most water remains behind. The vapors are carried away to a pipe that is chilled so that the vapors condense into a new liquid, one that has a higher concentration of alcohol, less water, and an altered set of flavorings. This second liquid can be redistilled to raise the alcohol even higher and to achieve a different flavor. Distillation can raise the alcohol content only to 95.6%. This would be 191 proof liquor; the "proof" number is twice the alcohol content.

The type of spirit that results from distillation depends upon the initial fermented material. Distillation of grain-based fermentations produces whiskies/whiskeys, vodka, and gin. For example, Scotch and Irish whiskies (spelled with "ie") use primarily or only malted barley, but American whiskeys (spelled with "ey") use up to 80% corn as well, and rye and wheat may also be added. Gin also begins with the fermentation of malted barley, corn, and rye, but during distillation, juniper berries are added for their flavor and aroma. Vodka is basically just pure ethanol and water: the objective of distillation is to raise the concentration of ethanol and to eliminate all flavors. Consequently, the initial fermentation for vodka is made with whatever carbohydrate is cheapest: wheat, corn, potatoes, even sugar beets. Brandy results from distillation of grape wine (if fruit wines are distilled, they produce fruit brandies). Distillation of fermented molasses creates rum. Tequila and mescal are distilled from fermented juices of agave plants: tequila from *Agave tequilana* and mescal from *A. americana*. Both of these agaves are unusual in that they store their carbohydrates not as starch (a polymer of glucose) as most plants do, but as inulin, a polymer of fructose. The agaves must be cooked to break the inulin down to simple sugars before fermentation can start. Tequila agaves are steamed, while those for mescal are roasted slowly (despite its name, mescal does not contain mescaline; **FIGURE 15.11**).

FIGURE 15.11. Mescal is made from *Agave americana*, which has long, spiny leaves (see Figure 2.22) but which stores its carbohydrate in its broad, short stem. The leaves must be cut off, leaving just the stem (the "heart") that you see here. These hearts are being transferred from the truck to a stone-lined pit; burning charcoal will be added and the entire pit will be covered to allow the hearts to roast slowly, converting their carbohydrates to simple sugars that can be fermented.

Psychoactive Plants

The toxins in many poisonous plants affect an animal's nerves. The plants in this section are several in which the toxins are dilute enough that some people consume the plants specifically for their effects on our brain and its perception, either its perception of the person's own body (relaxation, stimulation, reduced anxiety) or its perception of the world (hallucinations).

Tobacco

The tobacco used for smoking and snuff is made from the leaves of several species of *Nicotiana*, mostly *N. tabacum* and occasionally *N. rustica* (FIGURE 15.12). These are both in the family Solanaceae, along with potatoes, tomatoes, and deadly nightshade. Cultivated tobacco is an annual herb that is planted as seed in spring and grows quickly into a plant about 3 m (almost 9 ft) tall with very large, simple leaves. At the proper size, leaves are harvested, dried, and cured. Curing is often over a low fire, and it removes some of the moisture from the leaves, converts starches to sugars, and breaks down carotenoids. Cured leaves are then aged for as long as a year to allow their flavor to continue to develop. For final processing, veins and petioles are removed, and the leaf blade is cut into strips for cigarettes, cigars, or snuff. Additional flavors, such as menthol, rum, or licorice, are often added.

Leaves of *N. tabacum*, even while they are alive on the plant, have numerous toxic alkaloids that deter insects from attacking the plant. The principle alkaloid is nicotine, which is so toxic it is often extracted and sold as an insecticide to be used on other crops. Nicotine acts on an animal's nervous system in many ways. In us, it binds to several receptors in our central nervous system (especially in the brain), as well as our sympathetic nervous system (particularly the adrenal glands). In the brain, it alters the levels of dopamine (one of our own natural neurochemicals), and in the adrenal gland it causes release of extra epinephrine, a stimulating hormone.

When a smoker inhales, nicotine-rich blood arrives at the brain within 7 seconds. Blood vessels in the brain differ from those in the rest of the body; their walls are much less permeable and typically, only natural chemicals that our own body synthesizes can move across the vessel walls and into the brain cells. This is called the blood–brain barrier, and in addition to protecting the brain from exotic chemicals, it often prevents medicines and beneficial drugs from entering brain cells. We can administer many drugs to most parts of the body by giving someone an oral medication that moves from stomach to blood to the rest of the body, or we give an injection into the bloodstream, or we use an intravenous drip. But for many drugs, these circulate to and through the brain without ever leaving the blood, and the blood–brain barrier makes it difficult to medicate brain tissues. Nicotine has no such problem: it readily passes from bloodstream to our brain cells. Within seconds of smoking, a person feels relaxed, alert, and calm.

FIGURE 15.12. Tobacco grows quickly into tall plants with very large leaves. Each leaf is selected and collected by hand, so each plant must be checked repeatedly. (Courtesy of Ken Hammond/USDA ARS.)

If these were tobacco's only effects, it would probably be considered a beneficial, healthful plant. Unfortunately, use of tobacco greatly increases risk of cancer of the lungs, throat, larynx (voice box), and mouth, which all come into direct contact with tobacco during smoking (chewing snuff focuses the damage to mouth, tongue, lips, and teeth). Pancreatic cancer is also increased. Tobacco use is a major cause of heart attacks, strokes, pulmonary disease, and emphysema. The World Health Organization estimates that tobacco causes more than 5 million deaths each year, and that more than 100 million people died of tobacco use in the twentieth century. The United States Center for Disease Control describes tobacco use as "the single most important preventable risk to human health in developed countries..." This means that if a smoker wants to improve his or her health and length of life, the most effective thing to do would be to stop smoking. Unfortunately, nicotine is extremely addictive, and giving up smoking is not easy. However, it is easy to not start: if you don't smoke, don't start. ***The World Health Organization reports that tobacco kills one half of all its users***. For more information, go to the Quit Smoking website of the Centers for Disease Control at http://www.cdc.gov/tobacco/quit_smoking/index.htm.

All crops cause some environmental problems, but tobacco seems particularly bad. Just as with *Theobroma*, the source of chocolate, child labor is especially common in tobacco farming. Tobacco leaves are usually harvested one at a time, only as they are mature, so harvesters must walk through the fields daily, checking each plant. Nicotine in tobacco leaves actually passes from the leaves into bare skin, and harvesters often get "green tobacco sickness," a form of nicotine poisoning by this direct contact. If the harvesters are children, then they are being exposed to toxic levels of nicotine while their bodies and nervous systems are still developing. All crops take up minerals from the soil, but tobacco removes phosphorus, nitrogen, and potassium faster than any other major crop, making use of fertilizers necessary. Finally, the drying and curing of tobacco typically relies on wood smoke; this contributes to deforestation and increased carbon dioxide in the atmosphere. Each year, Brazil alone cuts more than 60 million trees for processing tobacco.

FIGURE 15.13. Marijuana leaves are distinctive, being long and narrow and having toothed edges. Several common houseplants, especially false aralia, however, have similar leaves and are often mistakenly assumed to be marijuana. (© Mitchell Brothers 21st Century Film Group/ShutterStock, Inc.)

Marijuana

Marijuana is made from the large, leafy herb *Cannabis sativa* (FIGURE 15.13), which contains the psychoactive drug **tetrahydrocannabinol (THC)**. *Cannabis sativa* is dioecious; that is, some plants produce inflorescences with carpellate flowers ("female" plants), other plants have staminate flowers ("male" plants). Carpellate inflorescences and the nearby bracts are covered in a resin that has much higher levels of THC than occur in the resin on the male plants or other parts of the female plants. Chemically, THC is an aromatic terpenoid that binds to specific receptors, called the cannabinoid receptors, in our central nervous system and cells in our immune system. The binding is so specific that researchers hypothesized that our bodies must naturally make a compound that fits into and activates these receptors. This hypothesis was correct, and soon our own natural endocannabinoids were found.

The effect of THC on our bodies results in relaxation, euphoria, and introspection. Beyond these quick and temporary psychoactive effects, use of *Cannabis* has many beneficial health effects, the most well-known being its ability to reduce the nausea caused by chemotherapy during cancer treatment. It is also effective in treating pain, certain symptoms of multiple sclerosis, and glaucoma. *Cannabis* and THC have few proven negative effects. It is not addictive like nicotine, and there has never been a documented fatality from overdosing on either THC or *Cannabis*.

Cultivation of *Cannabis sativa* has serious negative impacts on ecology, but mostly because it is illegal to grow it in regular farming areas. Fields are often hidden in remote areas, and unfortunately, many fields have been planted in some of our most beautiful national parks. Areas that have been preserved and protected for more than a century have been destroyed just to grow pot. In other countries, clandestine fields are cultivated in forests, and when discovered they are usually sprayed by airplanes with herbicides and poisons, the first to kill the plants and the second so that any material harvested and sold will be toxic when smoked.

The trichomes that produce the THC-containing resin can be separated from the leaves and flowers by rubbing them and then sifting. The trichomes and resin particles are called hashish or just hash. Afghanistan is the world's largest producer of hashish.

Cannabis sativa is also the source of hemp. The tall stalks are rich in long fibers that are harvested and made into ropes and rough cloth. At present, strains of *C. sativa* that are grown for hemp have been bred to produce almost no THC.

FIGURE 15.14. After opium poppies are pollinated, the carpels develop into a large capsule filled with latex ducts. By making shallow cuts in the capsule, the ducts are opened and the opium-rich latex can ooze out and be collected. The capsule recovers and produces more latex, so each capsule can be harvested several times. (© Mafoto/Dreamstime.com.)

Opium Poppies

The poppy family (Papaveraceae) is large, with many genera and species, but only one concerns us here: *Papaver somniferum*, the opium poppy. The immature fruits of *P. somniferum* contain latex ducts filled with a white sticky liquid called **opium** (FIGURE 15.14). These are the large red poppies often cultivated in gardens, and indeed garden poppies do actually have some opium in them.

Use of opium occurred as early as 5000 years ago: Sumerian tablets from 3000 BCE describe the use of small balls of poppy latex to relieve pain. Opium is harvested by slashing the fruits deeply enough to make the latex ducts bleed, but not so deeply as to harm the fruit. Opium oozes out and dries into a sticky mass that can be collected the next day. A single fruit can be tapped at least three times, occasionally 10 times. Poppies are cultivated as a crop for drug production in Australia, Turkey, and India, as well as Afghanistan and numerous other countries. The United States buys 80% of its legal, medical opium from India and Turkey; other countries supply illegal opium for recreational drug use.

Opium contains numerous alkaloids, but the important ones for us are **morphine**, codeine, and thebaine. Morphine is the principle drug, being able to reduce severe pain (it is an **analgesic**), cause drowsiness, and give feelings of pleasure. It binds to specific receptors in our brain, and it helped lead to the discovery of our own natural painkiller/pleasure signals, the endorphins, that we produce during trauma, excitement, orgasm, and strenuous exercise. Morphine is an extremely valuable medical drug, used after surgery to block the pain of deep incisions, broken (or sawed) bones, and other trauma. Some people can become addicted to morphine, so it is used carefully in hospitals; various techniques have reduced that risk, and it is used more frequently now. The codeine in opium is used as a cough suppressant and to treat diarrhea; it is not addictive.

Morphine can be altered chemically to convert it to **heroin**. This is an even stronger analgesic than morphine, but it is extremely addictive. Injecting heroin every day for as little as 2 or 3 weeks is enough to cause total dependence on the drug. People quickly develop a tolerance to heroin and need to use stronger and more frequent doses to achieve the same level of high. Typically heroin is dissolved (either in acidified or heated water), then injected intravenously. The drug itself is extremely damaging to our central nervous system and it also leads to collapse of blood vessels and has a high risk of infection with HIV and hepatitis. It can also be taken by snorting the powder into the nose.

From an agricultural standpoint, opium poppies are an extremely valuable crop. Where they are cultivated for the illegal drug trade, their value is far greater than that of any other crop that could be grown on the same ground. Because they are grown for illegal sale and use, their cultivation is often organized by drug cartels or terrorist organizations.

Some actual food is harvested from opium poppies: poppy seeds and poppyseed oil. The poppy seeds we eat in baked goods do have very low levels of opium-related chemicals and can trigger a positive result in a drug test.

Coca Plants

Cocaine is an alkaloid extracted from the leaves of small trees of *Erythroxylum coca* and *E. novogranatense*, both called **coca**. These grow at low to moderate altitudes in the Andes Mountains. Starting at least 3000 years ago, people of the region chewed coca leaves as a mild stimulant and to relieve hunger, thirst, pain, and fatigue. Mummies from the area often had small bags of coca leaves buried with them. When the Spanish conquered the Inca Empire, they allowed the enslaved laborers to continue chewing coca, as it permitted them to work longer and harder with less food. Coca leaves are still widely available in regional markets and supermarkets, and the tea is served in restaurants to help prevent altitude sickness in high Andean regions. Few health problems are found among people who regularly chew coca leaves or drink the tea, because they absorb only miniscule amounts of cocaine, as little as 0.002 g per day, whereas addicts may put as much as 2 to 3 g (1 g is about 3 one-hundreds of an ounce) into their bodies every day.

Cocaine is easily extracted from the leaves of *E. coca*. It is used either as the hydrochloride salt, known as coke, or is mixed with baking soda and boiling water to form crack. "Freebase" is the base form (not the salt form) of cocaine. Cocaine is used by either snorting the powder up the nose, by smoking it, or by injecting it into the bloodstream. Other than the toxicity of the cocaine itself, each method presents dangers; cocaine causes blood vessels to constrict, so snorting cuts off blood flow and causes damage to nasal passageways. Injecting always carries the risk of infection.

Cocaine acts in various ways. When a nerve impulse must be transferred from one nerve cell to another, the upstream nerve cell releases chemicals that diffuse across a slender gap between the two cells, causing the downstream nerve cell to fire. Unused chemical is taken back into the upstream cells so that it is not wasted. Cocaine prevents this uptake in cells that use dopamine, and because dopamine is not reabsorbed, it stays in the gap between the two cells and causes the downstream cells to fire repeatedly. An ordinary nerve impulse is turned into a strong, repeated one. Dopamine is used by nerves that carry sensations of well-being, so cocaine enhances this feeling. Unfortunately, our bodies quickly adjust to this abnormal biology, so a cocaine user must increase the dosage and frequency of usage, just to maintain the same level of high.

Peyote

Most cacti produce an abundance of alkaloids, and in two species, they are hallucinogenic. **Peyote** (pronounced pay OH tea; *Lophophora williamsii*) is the more well known. It is a small blue-green cactus that looks like a soft, spineless cushion, barely protruding above the soil (FIGURE 15.15). Below ground is a long tapering shoot and a thick taproot. The top part (a "peyote button") can be cut off slightly below ground, and an axillary bud on the remaining shoot will grow into a new top that can be harvested a few years later. This method allows a sustainable harvest of peyote. An alternative method is to dig up the entire plant, but of course nothing grows back afterward. Overcollecting like this has made peyote an endangered species. It is not cultivated on farms, all peyote buttons are collected from wild populations. *Lophophora williamsii* has always grown in just a small area along the Rio Grande River between Texas and Mexico, but overcollecting has exterminated almost all populations in the United States.

FIGURE 15.15. Plants of the peyote cactus, *Lophophora williamsii*, are always short and remain at ground level, even if allowed to become very old: as new tissues are produced at the shoot apex, old tissues contract. The low stature keeps the plants out of sight, but still collectors have found almost every population in Texas and have dug them out. Very few plants still occur in nature.

The principle hallucinogenic alkaloid in peyote is **mescaline**. Peyote buttons are chewed, and usually they cause nausea, so they do not stay in the stomach for long. But a few hours later, the person has visions of vivid colors, altered time, and sensual hallucinations. It is illegal to eat or even to own a plant of peyote, with the exception that members of the Native American Church may use peyote in their ceremonies. Before the Spanish conquest of the Americas, peyote was used in Aztec religious ceremonies; later, members of the Plains Indians adopted peyote as well.

A slender, small columnar cactus called San Pedro cactus (*Echinopsis pachanoi*, also called *Trichocereus pachanoi*) is native to western South America, and it too has enough mescaline to cause hallucinations. The shoots, which have five to seven ribs, are cut into short cross sections for consumption. This produces flat disks that resemble stars with five to seven arms, which is a common decorative motif on pottery and textiles dating back at least 2000 years, to the Moche culture. In the United States, it is legal to own San Pedro cactus as an ornamental plant, but not to eat it.

Plants Used as Sources of Pure Medicinal Drugs

As the section on stimulating beverages and psychoactive plants show, many plants contain compounds that strongly affect our metabolism. Only a few plants are simulating enough to have been turned into recreational drug plants; the majority instead provide medical benefits without noticeable psychoactive side effects. These medicinal plants have been and still are being used in a variety of ways. In some cases, the medicinal drug is extracted and used as is, or it may be purified first. In many cases, it is no longer necessary to use the plant extract at all; chemists have identified the active compound and invented methods to synthesize it artificially. Some medical compounds are too complex to be synthesized starting from simple chemicals, but we can extract complex ineffective compounds and then alter them into the medically active form we need. In some cases, the compounds found in plants are only moderately beneficial in their natural form, but chemists synthesize variations that are much more effective or have fewer negative side effects.

The following section briefly mentions some well-established drugs that were initially discovered through the health benefits of the plants that contain them. The next section highlights a few of the newest discoveries of plant-derived drugs that are being tested as medicines. Although people have always scrutinized plants for beneficial medicines, at no time in the past has the search ever been as intense as it is today. Thousands of species are being examined with dozens of techniques. Botanical teams are exploring all accessible parts of the world for new, useful plants more carefully than ever before.

Well-Established Plant-Derived Medicines

Aloe vera may be familiar to many readers and is an example of a medicinal plant that needs little processing (FIGURE 15.16). Its thick, juicy succulent leaves can be sliced lengthwise, and the cut, moist side applied directly to skin that has been sunburned or received a mild (first-degree) burn. The inner tissue of the leaf is sterile, moist, and it promotes healing. *Aloe vera* juice is promoted as being beneficial for a variety of ailments, but those claims have not been proven.

Aspirin is a medicine most of us have used for relief from headaches and other pain. The original source for this is bark of willow trees (various species of *Salix*). Chewing willow bark for pain relief is an ancient practice, but now aspirin is synthesized artificially.

Cephaelis ipecacuanha is not a familiar plant, but ipecac syrup is often present in first aid kits as an emetic: it makes you vomit. That doesn't sound pleasant, but when children have eaten poison berries, vomiting gets the poison of the stomach quickly and can save lives.

Vinblastine and vincristine are two powerful anticancer drugs obtained from a familiar garden plant, the common periwinkle (*Catharanthus roseus*; FIGURE 15.17). Another plant, mayapple (*Podophyllum peltatum*) provides alkaloids used to treat testicular cancer as well as breast and lung cancer.

Quinine, present in the bark of cinchona trees (*Cinchona officinalis*) was the first, and for a long time the only, remedy for malaria. Malaria is still a worldwide scourge, but in the past, it made large parts of the tropics completely uninhabitable. Malaria is caused by a parasitic protozoan, *Plasmodium falciparum*, that lives in our blood after we are bitten by an infected mosquito. Quinine kills the protozoan without killing us. But quinine is extremely bitter and many soldiers and workers stationed in tropical areas refused to take it. It was discovered that quinine could be dissolved in water (the result

FIGURE 15.16. The long fleshy leaves of *Aloe vera* contain phytochemicals that relieve the pain of minor burns and help the skin to heal. (© Mirenska Olga/ShutterStock, Inc.)

FIGURE 15.17. Two of our most important anticancer drugs are extracted from this periwinkle (*Catharanthus roseus*) that is cultivated as an ornamental in many home gardens. (© Monkeystock/Dreamstime.com.)

FIGURE 15.18. Foxglove (*Digitalis purpurea*) is a beautiful garden plant, cultivated by many people. It is also the source of drugs used to treat the heart. (Courtesy Tommy R. Navarre.)

is tonic water), and then mixed with gin: the popular summertime beverage gin and tonic is a mild version of the early antimalaria medicine.

Purple foxglove, *Digitalis purpurea*, cultivated in many gardens (FIGURE 15.18), is the source of digitoxin and digitalin, two important drugs for treating the heart. They stabilize the heartbeat when it becomes irregular.

Atropine, from *Atropa belladonna*, was mentioned in the Introduction to this chapter.

Newly Discovered Plant-Derived Drugs

Years of testing are required before any new drug is approved for use in human beings. If it is a plant-derived drug, the process begins with the screening of extracts of thousands of species for some indication of beneficial biological activity. Compounds will be extracted in various ways: using alcohol or oils to extract hydrophobic compounds; water to extract hydrophilic ones. Some extracting solvents should be acidic, others neutral, others alkaline. A beneficial compound might be extracted with a toxic one that masks the benefits of the first, so it is typically necessary to separate crude extracts into their various components and test each individually. The initial tests are often performed on cultured cells so that live animals do not have to be harmed needlessly. The process is long, complex, and costly, and a recent review lists only five new drugs that have been approved since 2001. Galantamine (from *Galanthus*) is an alkaloid prescribed for Alzheimer's disease. Nitisinone (from *Callistemon*, the bottlebrush tree) is used against a disease that causes liver and kidney damage in children. Apomorphine is a modified version of morphine that is used against Parkinson's disease. Tiotropium bromide treats pulmonary disease and emphysema, and valenicline (from *Cytisus*) helps people stop smoking. Numerous other drugs are in clinical trials and some may be approved soon.

Plants as Sources of Drug Precursors

Some plant extracts themselves are not active, but they are the starting point for synthesis of drugs. Bark of yew trees, *Taxus brevifolia*, provides the drug paclitaxel (Taxol), which is extremely effective against ovarian cancer. But bark contains only very low amounts, too little to save many women. However, a chemical called 10-deacetylbaccatin III is present in large amounts in the needles of other yew species and it can be converted easily into paclitaxel.

Diosgenin, obtained from the tubers of various species of *Dioscorea* (the same genus as yams) can be converted to progesterone, which is used in oral contraceptives. Progesterone can also be converted to cortisone, an anti-inflammatory drug. Diosgenin is an important starting material for production of various steroid hormones.

Tamiflu (oseltamivir phosphate) is one of our most effective drugs against influenza viruses A and B. It is synthesized from shikimic acid, a chemical extracted from star anise (*Illicium verum*). Two anticancer drugs, podophyllotoxin and camptothecin, are too toxic or not water soluble enough in their natural form for use as medicines, but analogs have been developed that are effective. Guanidine from *Galega officinalis* is too toxic for clinical use, but one derivative, metformin, is suitable for treating type II diabetes.

Important Terms

analgesic
beer
black tea
cacao nibs
caffeine
chocolate liquor
coca
cocaine
cocoa butter
cocoa solids
ethanol
ethyl alcohol
green tea
heroin
hops
infusion
malt
mash
mescaline
morphine
oolong tea
opium
peyote
sake
spirit
tetrahydrocannabinol (THC)
wine
wort

Concepts

- Many defensive chemicals are effective in small amounts.
- Most stimulating beverages contain caffeine as well as other alkaloids.
- Cultivation of many plants for stimulating or psychoactive chemicals involve problems such as child labor, drug cartels, or terrorist organizations.
- All ethanol-based beverages rely on fermentation of plant sugars by yeasts. Plant starches must be converted to sugars before they can be fermented.
- The flavor of wine grapes is strongly affected by growing conditions, and global climate change could damage the world's important wine producing regions.
- Most plants with psychoactive drugs cannot be cultivated legally, so plants are overcollected from wild populations or are cultivated in hidden fields that damage otherwise natural areas.
- The search for new plants and plant-based drugs is more intense now than ever before.
- Even if a plant does not contain a medically effective drug, it might contain compounds that can be converted to a useful form.

16 Fibers, Wood, and Chemicals: Plants that Clothe and House Us

The bark of many trees contains long, soft fibers that can be extracted and then made into either paper or a soft cloth. We obtain fibers from many different plants and from many plant parts, such as leaves, roots, wood, fruits, and seeds. Shown above is the bark of juniper (*Juniperus*), which is common in semi-desert areas and which has been an important source of soft fibers for Native Americans.

Introduction

Plants tend to be fibrous. Animals have hides and shells and feathers but plants excel at making fibers. Leaves, stems, and roots are usually reinforced by strands of fiber cells that run along their veins or just below their epidermis. Tracheids in xylem are long, fiber-like cells, and many plant hairs resemble fibers.

It did not take long for early humans to notice that some stems and roots are "ropey" and could be used to tie bundles together, woven into sheets, or crafted into bowls, baskets, and sandals. At some point, our ancestors learned to extract the fiber bundles and twist them into string finer than could be made from whole stems or roots, string that could hold animal skins together as clothing. Later, in various parts of the world, people learned to spin fibers into fine thread and weave that into cloth. Although we now have the technology to produce innumerable types of synthetic fibers and sheets of material, many people prefer natural fibers for clothing, furniture, and fabrics that we use personally. Much of agriculture is dedicated to cultivating plants that provide us with fibers.

BOX 16.1. Wood as Fuel

Unfortunately, more than half the world's population cook their food and keep themselves warm by burning wood. This is the main cause of deforestation. In poorer regions, especially semidesert areas, all shrubs and trees surrounding villages have been cut and burned. Many of these areas are being "reforested" by planting thousands of seedlings or saplings to replace the trees that have been cut and burned. Unfortunately, in many cases the trees being planted are not the native, natural species but instead are foreign species of *Pinus* (pine) and *Eucalyptus*. These two genera are chosen because their trees grow quickly and their wood is good firewood. But both pine forests and *Eucalyptus* forests tend to be monocultures because their leaves produce chemicals that prevent other plants from growing (they are allelopathic). These planted forests have no botanical diversity and consequently they have little diversity of animals, fungi, or even microbes. They produce firewood but offer few benefits of natural forests.

In the United States and Europe, especially in forested areas, it is becoming popular to have wood-burning stoves or fireplaces to heat one's home. But home fireplaces tend to burn wood very incompletely, resulting in more ash and smoke than would be produced if the same amount of wood were burned in a commercial heating plant. The wood smoke hanging in the air of our forests may be beautiful and fragrant, but it is not healthful, and the inefficient use of wood means that each home is putting more carbon dioxide into the atmosphere than if it were being heated by conventional means.

Early people also learned to use wood. Perhaps the first use was as a weapon, as clubs, or for fire (**BOX 16.1**). But people also learned to build with wood, even though cutting and processing wood was difficult while our ancestors had only stone tools. But with the discovery of copper and then of bronze, people could make knives and saws that allowed them to shape wood to fit their needs better. They began to build wooden benches, buckets, buildings, and so on. One of the earliest uses was to express their spirituality through art: wood was carved, painted, or polished into beautiful representations of spirits and gods. Just as with fibers, wood too is still highly prized today, not only as a utilitarian construction material but also as a medium for visual and musical arts. Imagine what our world would sound like if there were no wooden musical instruments such as pianos, violins, guitars, clarinets, and so on. Imagine what our homes would look like with no carved or finely worked woods, woods we use to add beauty to our lives through furniture, cabinetry, serving dishes, religious symbols, and art objects.

As with foods, spices, and medicines, people in the twenty-first century still use plants—still need plants—to clothe us, house us, and to help us express our concepts of spirituality and beauty.

Fibers

Fibers derived from plants are remarkably diverse and even more remarkably useful. How do you use plant fibers in your life? Cotton cloth is woven from thread spun from fibers that grow out of the epidermis of cotton seeds. What else? The paper in the page you are reading is made of fibers, probably from pine trees. String and rope are made of plant fibers, and to avoid Styrofoam® peanuts, many things are packed in plant fibers for shipping. The "welcome" mat at your front door, your brooms, mops, and brushes are made of plant fibers, especially if you are trying to avoid plastics in your life. And as this is being written, bundles of plant fibers are being used to absorb some of the oil from the 2010 British Petroleum Deepwater Horizon oil spill in the Gulf of Mexico. Many new composite materials are being manufactured by mixing hard but brittle resins with tough but flexible plant fibers.

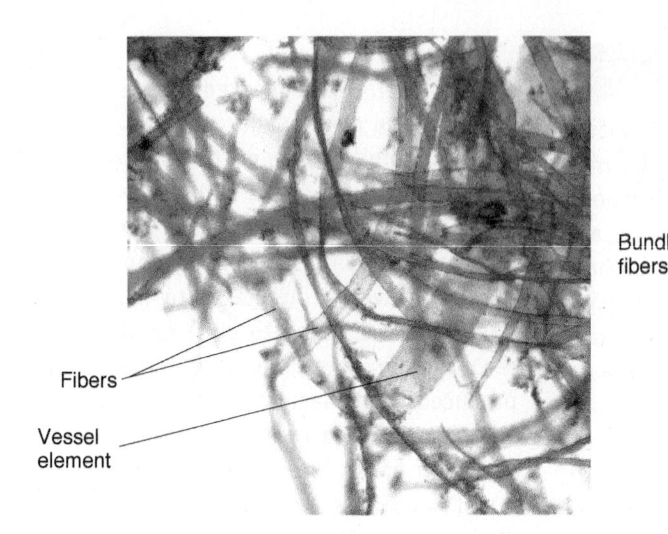

FIGURE 16.1. These are fibers from a Post-it Note®. The paper was moistened in a drop of water on a microscope slide, and then the fibers were teased out of it. The long slender fibers are single cells; the one broad "fiber" is a vessel element, so botanically speaking, it is not a true fiber (and because conifers do not have vessels, this paper must have been made from angiosperm cells). The many small particles that resemble dirt are particles of clay used to "size" the paper, to fill in gaps between the fibers so that the paper is smooth enough to write on (×50).

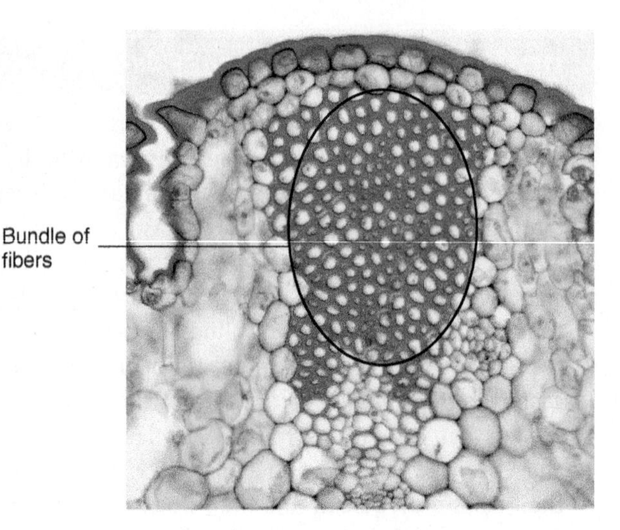

FIGURE 16.2. This *Yucca* leaf has a bundle of many fibers, each of which appears circular because the bundle was cut in transverse section (see Figure 3.16). The entire bundle can be pulled from the leaf and used as a single very long, thick "fiber," so in the textile industry the term "fiber" sometimes refers to single cells, sometimes to bundles of cells (×200).

We must be careful of the word "fiber" in this chapter. Its botanical meaning always refers to a single cell (see Chapter 3), but in textiles, "fiber" often refers to entire sets of cells, usually an entire column of fiber cells (**FIGURES 16.1** and **16.2**). A single fiber cell provides elastic strength: after being bent or stretched, the fiber returns to its original size and shape. Within a plant's body, one single fiber cell, surrounded by parenchyma cells, would be of little use, and instead plants always make masses of fiber cells, just as our bodies make masses of bone cells and tooth cells. In the primary plant body (the herbaceous body), the masses of fiber cells occur as a cap exterior to the vascular bundles, in a position where they protect phloem from sucking insects (see Figure 3.13). In monocots, a sheath of fiber cells usually surrounds each vascular bundle completely (**FIGURE 16.3**). In all types of plants, it is not unusual to have either bundles of fiber cells or an entire layer of fiber cells located just interior to the epidermis. These types of fiber cells grow as the entire stem or leaf is elongating, so these are some of the longest plant cells known. Because these fiber cells occur in masses surrounded by parenchyma, the entire mass usually comes out of the plant as one very long "fiber" when the plants are crushed to harvest the fibers. Such fibers are entire vascular bundles or entire fiber bundles.

In the secondary plant body (the wood and bark), fiber cells are produced by vascular cambium, and consequently are rather short. Those produced in the wood are **xylary fibers**, those in the phloem are **phloem fibers** (**FIGURE 16.4**) and in both, fiber cells are clustered in masses, they never occur as isolated cells. In species that have tough masses of fibers surrounded by soft phloem parenchyma, entire masses of fibers can be extracted and used. Because wood is so much harder, masses of fibers

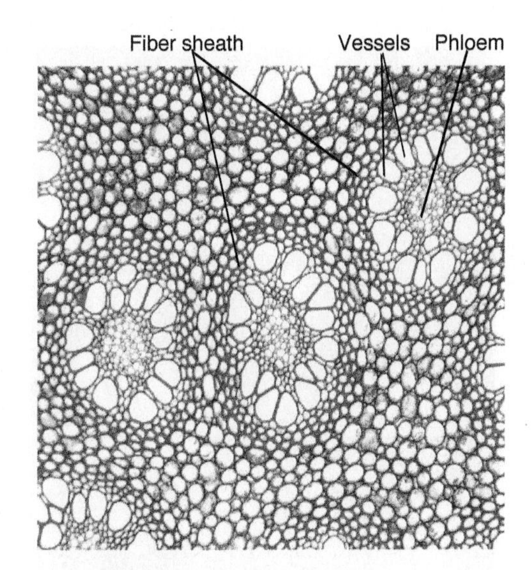

FIGURE 16.3. This transverse section of a rush (*Juncus*) shows several fibrovascular bundles. Each bundle has a small mass of phloem at its center (sieve tube members and companion cells are visible), surrounded by a ring of broad, red-stained vessels, and the vascular tissues are ensheathed by a layer or two of fibers. When fibers are extracted from rushes and similar monocots, very often the "fiber" is an entire fibrovascular bundle (×100).

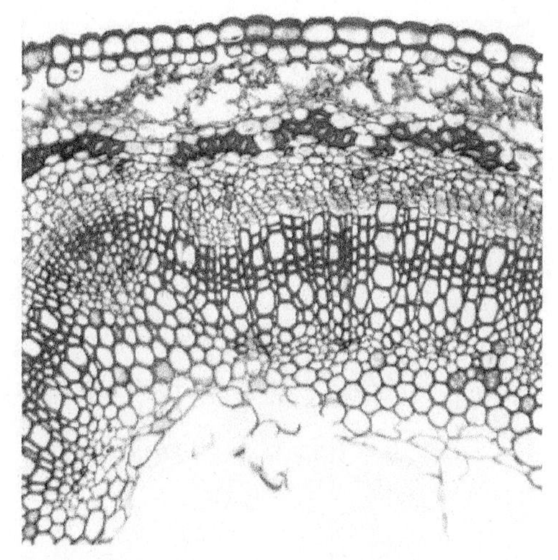

FIGURE 16.4. The uppermost layer of cells in this transverse section of a flax stem (*Linum*) is the epidermis, interior to it is a thin layer of cortex and then a layer of green-stained fibers located in the outermost part of the phloem. The broad band of red-stained cells is xylem, and pith is at the bottom of the image. Because the green-stained fibers are in the phloem, they are phloem fibers, also called bast fibers. When flax fibers are used for textiles, the material is called "linen" (×100).

cannot be pulled out, and the wood must be treated chemically to break down the middle lamellas, causing the fibers to come out as individual cells.

Plant fibers are usually classified according to where they occur in the plant. **Bast fibers** are those in the phloem, cortex, or bark; these are usually called **soft fibers**. Fibers in leaves are **hard fibers**, and those in wood are **wood fibers**.

Discovering which plants are fibrous is not difficult. The fibers are visible, as in cotton or cottonwood seeds (FIGURES 16.5 and 16.6), or the plants are stringy and tough when we try to eat them. We all know from experience that asparagus, bamboo shoots, and rhubarb are fibrous plants. And plants rich in fibers are difficult to cut, especially with stone tools, so early humans would have been well aware of which plants are fibrous and which are not. The earliest definitive evidence of human use of plant fibers is from 36,000 years ago: flax fibers have been found with stone tools in a cave in the Republic of Georgia.

Extracting and using plant fibers would have been different and more challenging for early humans than merely using whole plant parts that are fibrous. Tough, fibrous leaves like those of palms can be used as roofing or shelter simply by cutting them off the plant and propping them overhead or leaning them on a framework of interwoven branches. Slender fibrous stems or roots can be twisted together to make crude rope, or woven together (FIGURE 16.7). Bundles of fibrous plant material were crafted into bowls and baskets early in our history, and such items have never lost their artistic appeal (FIGURE 16.8). Probably the fibers that were easiest to extract and use are those of agave leaves in the Americas. The fiber bundles in their succulent leaves converge on a stout spine at the tip of the leaf, and if the leaf is beaten to break down the parenchyma, the spine tip can be pulled off, taking some fiber bundles with it: the leaf provides both needle and thread. Plants such as jute also provide easily extractible fiber bundles that are up to 4 m (12 ft) long and that can be used as string.

FIGURE 16.5. When cotton seeds are mature, the fruit dries and opens, revealing the very hairy seed coats.

FIGURE 16.6. Cottonwood seeds are extremely hairy and fluffy, and these can be collected and used to stuff things such as pillows and toys, but their fibers are too short to be spun into thread. This species is named cottonwood because of these seeds, but this species (*Populus deltoides* in the willow family Salicaceae) is not at all closely related to cotton (*Gossypium hirsutum* in the mallow family Malvaceae); the hairiness of the seeds is the result of convergent evolution.

FIGURE 16.7. Weeping willows have branches that, at least while young, have wood that is so flexible the branches can easily be woven into baskets, chair seats, and other things. It is not necessary to extract the fibers or fiber bundles, whole branches can be used as "fibers." Photographed at the garden of Claude Monet in Giverny, France. (Image provided by Chris McCoy.)

FIGURE 16.8. This basket is woven from a material you might not think of as being fibrous: pine needles. Bundles of very long pine needles are held together by wrapping them with the light-colored material, which is made from grass leaves. (Courtesy of Tommy R. Navarre.)

Bast Fibers

Burlap is a rough cloth you might be familiar with (FIGURE 16.9). It is woven from long, soft shiny bast fibers, called **jute**, extracted from stems of plants in the genus *Corchorus* (related to linden or basswood trees). The plants are tall (up to 5 m [16 ft]) herbaceous annuals that are cultivated in wet tropical areas. The stems are cut and placed in shallow, slowly flowing or stagnant water to allow stem parenchyma to rot and release the fiber bundles. This process of rotting in water (or rain and dew) is called **retting**. During retting, some of the fiber cells come partially loose from the fiber bundles, giving the bundles a coarse, rough texture. Jute is widely used, being second only to cotton in the amount of fibers harvested each year.

Jute cloth (burlap) is tough and flexible, resists being torn or cut, and can be made into bags (often called gunny sacks) that have a tight enough weave for holding cereal grains and other seeds, but is "breathable" so that moisture does not accumulate in the bag and cause the seeds to mildew. Burlap bags are used on farms and ranches to hold seed, animal feed, and other bulky supplies. Burlap is also used as a strong, wear-resistant backing for carpets, area rugs, chair coverings, and linoleum, as well as wrapping large bales of cotton. The fibers are generally too rough to be used for clothing, although very fine fibers can be separated out and made into a soft imitation silk.

Because jute is a natural fiber, burlap can be used where a biodegradable cloth or rope is needed, such as wrapping the roots of young trees when they are being moved from a nursery to a planting area. The burlap wrap can be left in place as the tree is planted, and by the time the roots need more space, the burlap will have decomposed (FIGURE 16.10). Burlap is also used for **geotextiles**, rough, strong cover cloths that

FIGURE 16.9. Much of the cloth we use is not fine, smooth material suitable for clothing; instead it is tough and resists tearing or being cut. Burlap bags such as this are useful on farms or anywhere that small bulky items must be carried or stored in bags that will not rip, and which will also allow oxygen, carbon dioxide, and moisture to pass through.

FIGURE 16.10. This shrub of holly has been "balled and burlapped." That is, it has been dug out of the ground then the remaining ball of soil is held in place around its roots by a layer of burlap material. This keeps the soil from falling away and leaving the roots exposed and dry. When the shrub is planted, the burlap will be left in place and covered with soil. The plant's roots can penetrate the loose weave of the burlap, and within a year, the burlap will have decayed away, so the roots will be able to grow freely.

FIGURE 16.11. These are flax plants, *Linum usitatissimum*. Their long, straight stems provide bast fibers that are used to make linen cloth. (© Sylvie Lebchek/ShutterStock, Inc.)

FIGURE 16.12. Fibers of ramie become even stronger if they are wet, so they are especially useful for weaving the fabric used in fire hoses. (© Jupiterimages/Photos.com/Thinkstock.)

are laid down over large areas of ground to stabilize loose soil and prevent erosion after a fire or during land restoration. As seeds or rhizomes sprout and grow, the burlap decomposes. Plastic nets and polyester-based cloth are unsuitable for such uses.

Jute fibers, as well as those from many other plant species, are now being used in composite materials. The fibers can be mixed with liquid resin, placed into a mold to give a specific shape, and then the resin is allowed to harden. Such composites are being used increasingly in the automobile industry.

Flax fibers are extracted from flax plants (*Linum usitatissimum*; FIGURES 16.4 and 16.11) and woven into cloth called **linen**. Flax is the earliest known fiber to be used by humans; Egyptian mummies were wrapped in linen cloth 5000 years ago. As with jute, flax fibers are obtained by retting tall annual stems. The resulting fibers are smooth (not rough like jute), straight, and up to 150 cm (59 in) long and about 16 μm in diameter. They are two or three times stronger than cotton fibers. Flax fibers are lustrous and beautiful, and linen cloth is highly appreciated but expensive due to a large amount of hand labor needed to extract and clean the fibers. Even now, linen fabrics such as tablecloths, napkins, and clothing are a sign of elegance; linen for clothing was only supplanted by cheaper cotton in the 1800s. At present, most flax is cultivated on conventional farms in Europe.

Ramie (*Boehmeria nivea*) is another fiber used since prehistoric times, at least 6000 years ago (FIGURE 16.12). Its harvest is more complicated because surrounding cells contain gums that must be removed chemically. However, ramie is one of the strongest plant fibers known, and it becomes even stronger when wet. It is used for fabrics that must be tough, such as canvas, upholstery, industrial sewing thread, fishing nets, and filter cloths. Because of its wet strength, it is used in fire hoses, and it had been the primary material for fishing nets. When polyester and nylon fibers were developed, they were used to make many fishing nets because the synthetics do not mildew or rot and do not have to be dried before storage. But now we are concerned about the large amount of nylon and plastic netting that has been lost or abandoned by fishing boats; thousands of nets float freely in the ocean, trapping and killing fish, marine mammals, and other marine life. Because these never decompose, they will continue harming animals for centuries. Many people now believe that fishing nets should be made of natural, biodegradable fibers such as ramie again.

Hemp fibers are obtained from varieties of *Cannabis sativa* (marijuana), varieties that have almost no psychoactive tetrahydrocannabinol. In the past, the fibers were extracted without much cleaning, resulting in rough, dark-colored fibers that were used mostly for sailcloth, canvas, and rope. But if cleaned carefully, hemp fibers are soft, long, creamy white and have a silky sheen; they are now blended with flax, cotton, or silk to be used for clothing or home furnishings. In contrast, the rough fibers produced with minimum cleaning (the entire bundles of fibers) are strong and very long, up to 4.6 m (15 ft); these can be used to strengthen concrete and adobe for construction. Hemp can also be used for paper, but it is still much more expensive than the standard paper fibers derived from wood. Interest in hemp as

FIGURE 16.13. Fields of kenaf (*Hibiscus cannabinus*) are very productive: the plants grow quickly and produce an abundance of fibers, and are easier to harvest than are pine trees. By using kenaf fibers to make paper, less forest needs to be cut down. (© AGStockUSA/Alamy Images.)

an environmentally friendly crop has been increasing. It grows rapidly and produces biomass quickly, up to 25 metric tons of dry matter per hectare per year (1 metric ton is 1000 kg, or 2205 lbs; 1 hectare is 10,000 m^2).

One of the newest bast fiber plants is **kenaf**, from *Hibiscus cannabinus* (FIGURE 16.13). It shows great promise as being a raw material for making paper because its fibers are naturally whiter than those of pine so they need less bleaching, and they have less lignin. Removing lignin from pine fibers is one of the most polluting steps of making paper. Kenaf also produces fibers more rapidly than does pine: a given plot of land will produce about twice as much fiber if planted with kenaf as compared to the same plot planted with pine. Kenaf fibers can also be used for rope, twine, and coarse cloth, as well as composites when mixed with resins.

Fibers from Leaves

Sisal is a good example of the changing fortunes a crop plant can experience. Sisal fibers are harvested from long, thick succulent leaves of a desert plant, *Agave sisalana*, which is native to Central America (FIGURE 16.14) and is related to the agaves that produce tequila and mescal. Sisal fibers were plentiful in the past, and although they are rather coarse, they were used by early Mesoamericans and later by people of the region who had few other resources. As other fibers became available by international shipping, and then as synthetic fibers were developed, use of sisal declined greatly. But recently, many people have developed an appreciation for its more natural look and feel as compared to artificial fibers, and it has a more rustic character than do finer plant fibers. Sisal has become popular as carpeting, wall coverings, and other household fabrics. Its other uses show us the range of ways plant fibers are employed: it is a component in buffing cloths, filters, macramé, craft and art papers, and dartboards as well as twine for binding bales of hay or straw. Most sisal is now cultivated in Africa, especially Kenya, rather than in Central America.

FIGURE 16.14. Although *Agave sisalana*, the source of sisal fiber, is native to dry regions of Central America, plants are now cultivated on large farms in Africa, mostly in Kenya. The edges and tips of the leaves are extremely sharp, but all leaves (the source of the fibers) are harvested by hand. (© mediacolor's/Alamy Images.)

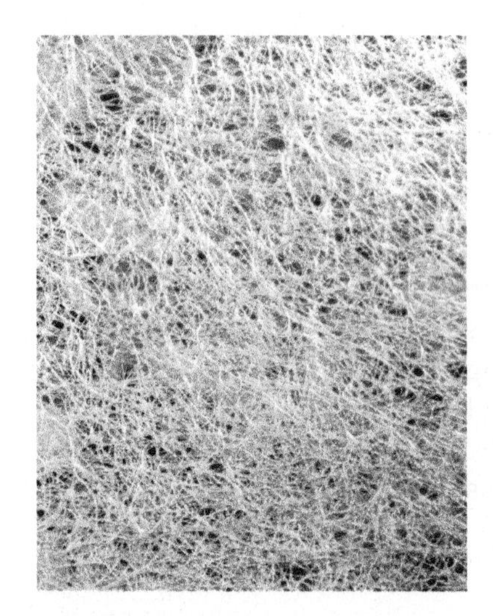

FIGURE 16.15. This paper is so thin and has such an open weave it would be difficult to write on it, but this is the paper of a tea bag. This paper must allow water and flavor molecules to pass through quickly, and it must withstand being stored dry for months, then not fall apart when placed in hot water and stirred. It is made of abacá fibers, extracted from a relative of bananas. Notice how clean these fibers are compared to those used for writing paper (see Figure 16.1): clay and other sizing chemicals are not added to paper for tea bags.

Agave sisalana is cultivated on marginal ground in desert areas where few other crops would thrive. This is good because it provides income for people with few other resources, but it is unfortunate that this crop encourages the clearing of otherwise natural desert areas. The plants grow slowly, so each acre of farm land produces only a small amount of fiber per year, whereas jute and kenaf produce large amounts and therefore need much less land.

Abacá is a fiber you have probably never heard of but have undoubtedly used. It is the fiber used for tea bags (FIGURE 16.15), manila envelopes, bank notes including dollar bills, and German and Italian sausages that have a cloth-like casing. Abacá fibers are extracted from the leaf bases of *Musa textilis*, a relative of bananas.

Fibers from Fruits and Seeds

Cotton is by far the most commercially important of all plant fibers. It is cultivated more than any other fiber plant, and more pounds of cotton fiber are harvested than any other plant fiber. Cotton fibers have the greatest number of uses, and of course cotton cloth for clothing is extremely important. Cotton fibers have numerous features that make them especially suitable for clothing. They are soft; they can be spun into a smooth, fine thread and woven into cloth that is either thin, airy, and comfortable in hot weather, or thick and warm for winter weather. Cotton fibers are naturally white or light tan and can be dyed a wide range of colors using many different dyes. Cotton fibers, and more importantly the thread and cloth, can be washed repeatedly without breaking down; they can be ironed to flatten or pleat the cloth, and since 1970, they can be Sanforized, which minimizes the shrinking that natural cotton experiences when it is washed and dried. Cotton cloth can also be treated to chemically link cellulose molecules, resulting in permanent press fabric.

Cotton fibers differ from all others described in this section. They are actually trichomes (plant hairs) on the epidermis of cotton seed coats (see Figure 16.5).

There are several species and hybrids of *Gossypium* that produce cotton; about 95% of world production is *G. hirsutum*, with *G. barbadense* (pima or Egyptian cotton) making up most of the rest. All are perennials, but they are grown and harvested as annuals, thus keeping the plants at a uniform size, which makes harvesting by machine easier. When the cotton bolls (fruits) are mature, the plants are sprayed with defoliants to remove the leaves (**FIGURE 16.16**). The seeds are then ginned to separate the fibers from the seeds: slender teeth on a rotating drum pull the fibers through a grating with slits so narrow that the seeds cannot pass through. Afterward, the fibers are repeatedly pulled through combs to stretch the fibers and make them lie parallel to each other (this is carding). The fibers are then spun into thread, which may be thick or thin, hard or soft, depending on the rate at which cotton is fed into the spinner and the tension applied to the thread.

Cotton thread undergoes several more processing steps. First it is cleaned by being boiled in sodium hydroxide (caustic soda), and bleached with hydrogen peroxide. These treatments remove waxes and pectins so that the cotton will accept dyes more readily. Cotton is next stretched in cold sodium hydroxide, which causes the cell walls to swell and any fibers that had flattened become round again, with a circular cross section. Finally, the thread is **sized**: starch or gel is added to strengthen the fibers and fill in any surface irregularities, so that the thread is now very smooth. Most of these steps are unique to cotton simply because it is used to make cloth that is comfortable enough to be worn next to our skin.

Cotton must be cultivated in areas with long growing seasons and high rainfall or with irrigation. China and India are the largest producers, but cotton is cultivated worldwide. International trade in cotton exceeded $12 billion in 2009. Major problems with cotton are that it often needs large amounts of fertilizers and pesticides. Some areas now grow organic cotton, which is safer for cloth that will be used for babies. Although most cotton cultivation is mechanized, some regions, especially Uzbekistan, are reported to use forced child labor. Of course, cotton plantations in the southern United States relied heavily on enslaved Blacks, which was a major factor in our Civil War.

Coir is the fiber that makes up the bulk of a coconut fruit (**FIGURE 16.17**). The round, hard, dark almost rock-like coconut we see in stores is only the endocarp and seed; the mesocarp, which consists of short, thick fibers, and the exocarp have been removed. The coconut "meat" (**copra**) is the most valuable part of a coconut, so coir is usually not harvested until the coconuts are completely mature. By that time, the fibers are tough, heavily lignified, and dark brown. They cannot be spun into thread but coir can be used for doormats, brushes, and mattress stuffing. In the last few years, coir has become popular

FIGURE 16.16. Although cotton is a perennial plant, it is treated as an annual to make harvesting easier. Under natural conditions, the plants remain alive even as the fruits and seeds are mature; fields are treated with defoliants (chemicals that kill the leaves) so that the plants do not have green, wet leaves that would clog the harvesting machinery and make the cotton wet (which would cause it to rot while being stored). (© Alaettin YILDIRIM/ShutterStock, Inc.)

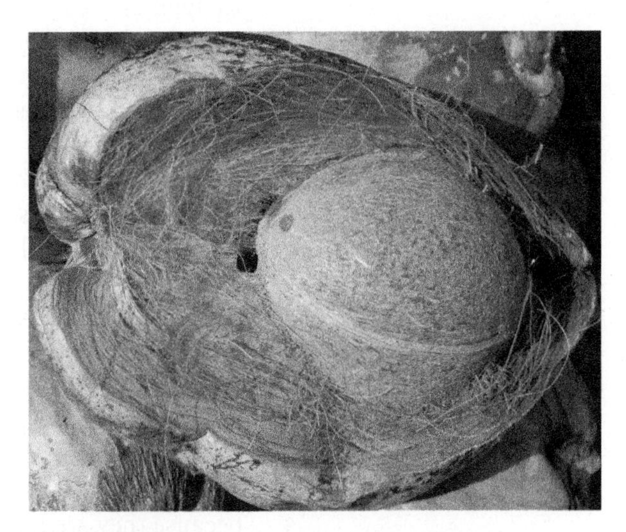

FIGURE 16.17. Most people who buy coconuts in grocery stores have no need for the leathery epicarp or the fibrous mesocarp, so those are stripped off before the endocarp with its tasty coconut meat is shipped to the store. This photo shows the mesocarp fibers, known as coir; these are extremely rough and coarse, unsuitable for clothing, but excellent for door mats and brushes with stiff bristles.

as a component of potting soil for both hobbyists as well as commercial nurseries. It can be used as a replacement for sphagnum moss in potting mixes, and thus by using coir, many sphagnum bogs can be saved from exploitation.

Coir is a byproduct of cultivating coconuts for their milk and copra. After these are harvested, the rest of the fruit could be thrown away or used for coir. The demand for coir is so low, as is its price, it would not be feasible to maintain a coconut plantation only to harvest coir.

Kapok is obtained from seed pods of a large (up to 70 m [230 ft] tall) tropical tree, *Ceiba pentandra*. Kapok fiber is light and fluffy, and is used as filling for stuffed toys such as teddy bears as well as mattresses, pillows, and upholstery.

Wood Fibers and Paper

To make paper, individual fibers are suspended in water as a slurry, and then poured onto a fine-mesh screen that traps the fibers and allows the water to pass through. The wet paper is allowed to dry and then is peeled from the screen. The texture, thickness, and smoothness of the paper depend on the nature of the fibers used, the thickness of the slurry on the screen, and the amount of pressure applied to the paper to squeeze out the water and make the paper thinner. Also, chemicals can be added to size the paper, that is, to fill in the gaps between the fibers, to smooth the surface or make it glossy, and to make the paper accept dyes and inks more readily.

This method of making paper was invented in China in about 100 CE, and rags and old cloth were the source of fibers. Being made of cotton or linen, old cloth could be easily separated into fibers by soaking it in water and beating it gently. The fibers were long and made a smooth, strong paper that could be written on easily. As paper became ever more popular, a new source of fibers was needed, there simply were not enough old rags.

Wood fibers were an obvious choice, but cells in wood are bound so tightly together that obtaining individual fiber cells is not easy. Also, wood fibers have three compounds that must be dealt with: cellulose, lignin, and hemicelluloses (see Chapter 3). Cellulose molecules are long, unbranched, and very strong, and they bind to each other as fibrils in plant cell walls. Hemicelluloses cross-link various fibrils so that they cannot slide past each other; they keep the strong cellulose molecules locked into the proper position to hold a cell wall's shape and size. Plants add lignin to a finished secondary wall of sclerenchyma cells, making the wall elastic, waterproof, and almost inert chemically. When making paper, the only component we actually want is the cellulose in the form of fibrils. They are strong, white, and long-lasting. Lignin and hemicelluloses interfere with the paper's ability to be bleached white and accept dyes, inks, and sizing agents. Much of the expense of making paper is the need to remove lignin and hemicelluloses, and one of the appeals of alternative fibers such as kenaf and hemp is that they have lower lignin content.

Keep in mind that the term "fiber" here is used in the broad sense as it is for textiles. Hardwoods from dicot trees produce fibers that are single cells, but softwoods from conifers produce tracheids rather than fibers in the botanical sense. Actually, the tracheids of conifers are better "fibers" for paper than are the real fibers of dicots because they are longer (about 2.0 to 4.0 mm [0.08 to 0.16 in]) than the true fibers of hardwoods (only about 0.5 to 1.5 mm [0.02 to 0.06 in]).

The first method developed for obtaining fibers from wood involved mostly just grinding wood with hot water. The fibers are just torn from each other; this not only damages the fibers, it also leaves the lignin and hemicelluloses bound to the fiber fragments. This paper yellows rapidly and quickly breaks down, but the method is inexpensive and causes little chemical pollution. It is still used to produce paper that does not need to last long, such as newsprint, the paper of newspapers. World demand for newsprint is about 37 million metric tons (4.08 million tons) per year, but demand is decreasing in the United States as more people use the internet rather than printed newspapers and magazines.

Two chemical methods were developed in the 1800s for separating wood fibers, the **sulfite method** (also called the acid method), and the **sulfate method** (the **kraft method**). The kraft method is the most widely used and today accounts for more than 90% of all wood pulp (fibers mixed with water). The kraft process does less chemical damage than does the sulfite method, so the resulting fibers are stronger, and numerous methods have been developed so that kraft pulping mills recycle most of their chemicals; they have made great strides in minimizing pollution. Especially important is that the kraft method produces **acid-free paper**, which has a very long life if kept dry. The acid papers produced by the sulfite method turn yellow and crumble even if kept dry and in the dark. Whole libraries full of books and magazines printed on acid paper are in danger of disintegrating unless costly remedies are undertaken, such as treating each item one by one with diethyl zinc to neutralize the acid. The Library of Congress and other major libraries have programs to rescue every one of their items that were printed on acid paper.

At present, the sulfite process is used to make specialty papers such as tissues, glassine, and fine papers, and it produces cellulose fibers that can be made into rayon, cellulose acetate, and other products that require very pure cellulose as a starting material.

In a kraft pulp mill, logs are debarked and the wood is chipped to small pieces that will dissolve quickly. A liquid called "black liquor," which is actually the chemicals salvaged from a previous batch of pulping, is mixed with the chips, along with "white liquor," which is fresh chemicals, in particular sodium hydroxide and sodium sulfide. Delignification requires several hours at high temperatures, about 180°C (356°F). The lignin and hemicelluloses break down into fragments that dissolve in the strongly basic solution. The pulp is brown and called "brown stock." Much of the liquid is drawn off, concentrated, and dried until it can be burned as fuel for the mill. The fibers and a small amount of liquid, under great pressure, are "blown" by shooting them into a container at ordinary pressure: the water in the fibers flashes into steam, blowing the fibers away from each other explosively. The volatile terpenes from resins ("pitch") that had been present in the wood are collected and used to make turpentine or used as the raw material for other chemical processes.

The wood fibers are screened to separate out any undigested chips, especially knots, and then the fibers are washed. The number of washings depends on the nature of the finished paper. For brown paper shopping bags and cardboard boxes, fibers receive minimal washing and bleaching, so using this sort of paper whenever possible minimizes the amount of chemicals needed and the pollution produced. Fibers for finer paper must be washed more thoroughly and then bleached. They are then applied to a belt-like screen that moves as the pulp is poured onto it. The paper is made as a sheet hundreds of feet long (FIGURE 16.18). As the paper moves through rollers to dry and compress it, chemicals such as resins, glues, and starch might be added to size the paper, reduce its abrasiveness, reduce its tendency to become fuzzy, and improve its ability to accept inks and dyes. Rollers can also give the paper various surface textures and adjust its hardness.

The kraft process is capable of handling almost any kind of plant fiber. It was designed to work with wood, but it can process new fiber plants such as kenaf, hemp, and bamboo to provide "tree-free" paper.

FIGURE 16.18. In factories, paper is made as a continuous sheet thousands of feet long. A slurry of water and fibers is drizzled onto a continuously moving screen that passes through rollers that squeeze the water out, compressing the paper to the proper thickness and density. The sheet continues through other rollers that give it the desired texture, fillers, and dyes, then the paper is dried and collected into giant rolls. If used for newspapers, magazines, or books, the paper goes through the printing presses as a single, long sheet and is cut into individual pages only after all printing is finished. (© Moreno Soppelsa/ShutterStock, Inc.)

Great advances have been made in recycling paper. High quality paper still must use mostly fresh fibers because fibers in recycled paper are weaker and many break into short fragments, which give a weaker paper. The main problem with recycling paper has been the need to remove dyes and inks, many of which are oil-based and become gummy, sticky blobs as the paper is broken down into a pulp again. But now virtually all inks and dyes can be removed, so just about any paper can be recycled. Newsprint can be recycled up to five times before the fibers become too damaged and too shortened to be useful for paper. Keep in mind the need for soft, low-strength, low-quality papers used in egg cartons, packing material, building insulation, and so on: recycled paper is completely suitable for that.

Non-Food Chemicals Obtained from Plants

Oils

In addition to the oils we use for food, plants provide numerous oils used as chemicals for many purposes. The most familiar of these are the **drying oils**, oils that appear to dry out and form a hard film when exposed to oxygen. Tung oil, linseed oil, and walnut oil are used to seal wood and give it a satiny, waterproof finish. Drying oils do not actually dry out; they do not have any water to lose, but they are highly unsaturated (they are lipids that have several or many double bonds), and oxygen causes the double-bonded carbons of one fatty acid to react with and attach to the double bonds of other fatty acids. The drying oils are applied as liquid oils, then they cure into one single, highly interconnected sheet that protects the surface they cover (FIGURE 16.19). Although drying oils can be used by themselves, often they have resins dissolved in them, which give a harder film. For example, **varnishes** are made by adding resins such as amber, dammar, copal, or rosin to a drying oil along with a thinner such as turpentine. Drying

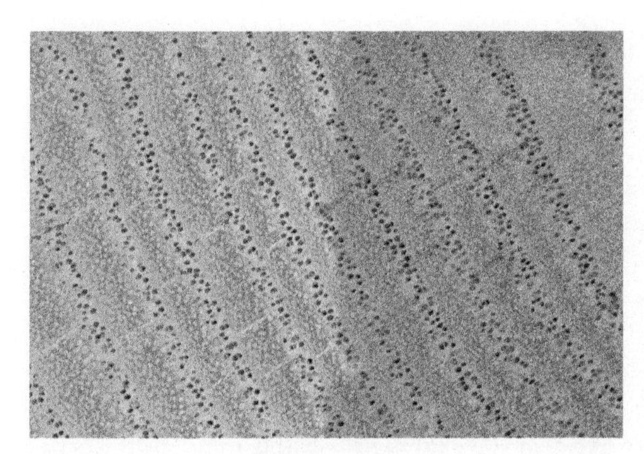

FIGURE 16.19. Half this wood is dark because it has been coated with linseed oil, a drying oil, to protect it. Drying oils are liquids that can be brushed or wiped onto a wood surface, then the double bonds in their molecules cross-link to each other, converting the liquid coating into a single gigantic molecule that protects the wood. Linseed oil is so fluid that it penetrated the vessels (the dark dots in the earlywood of each growth ring) then "dried" to such a thin film that when this sample is viewed with a microscope, details of the cell wall are still visible. Cells in the left` side (the uncoated side) will gradually deteriorate as humidity in the air attacks the cells.

oils are also used to carry pigments: they are the oil of oil-based paints. The use of drying oils has declined recently with the invention of synthetic alternatives such as polyacrylamides and polyurethane that cure more quickly (a few hours instead of several weeks), and provide an even more durable finish. Hardwood floors, unpainted wood cabinets, and furniture are now typically finished with these synthetics whereas not long ago they would have been protected with drying oils from plants. Fine wood utensils such as salad bowls and rolling pins are still treated with plant oils instead of synthetics, but a word of caution: tung oil is too poisonous to be used on anything that comes in contact with food.

Plant oils are also used in cosmetics such as skincare products as well as perfumes and candles. They, along with animal fats, are the basis for soap; converting fats to soap with an alkali (a base) is called **saponification**. Vegetable oils used for making soaps are called saponifiable oils, and include olive, coconut, palm, hemp, and shea butter. The oil is treated either with sodium hydroxide to make a hard soap, or with potassium hydroxide to make a softer soap. The nature of the soap is also affected by the oil used. Coconut oil soaps provide plentiful lather, whereas palm oil soaps are hard, and olive oil produces a mild soap. Soaps made from just olive oil without any other oils are called Castile soap or Marseille soap.

A recent use of vegetable oils is as an environmentally friendly insulating fluid in electrical equipment. Synthetic insulating fluids, especially polychlorinated biphenyls (PCBs) are extremely dangerous and a spill causes significant pollution. Vegetable oils are not toxic, but they are more difficult to use because they break down more quickly than synthetics, and they cannot be exposed to oxygen. Intense research is being focused on plant oils as biofuels (see Chapter 11).

Waxes

Waxes are similar to lipids, being complex mixtures of long-chain fatty acids, but in addition, waxes contain fatty alcohols, various acids, and other hydrocarbons. All plants produce some waxes in their epidermis cells, either mixed in with the cutin or as a layer of pure wax on the surface of the cuticle. Depending on the chemical composition, plant waxes are oily liquids, soft like candle wax and beeswax, or extremely hard like a tough plastic.

Only a few plant waxes are used in large amounts. **Carnauba wax** comes from the leaves of a Brazilian palm, *Copernicia prunifera*; FIGURE 16.20. The wax is present as hard yellow flakes, obtained by collecting the leaves and beating them to make the wax fall off. Carnauba wax produces a hard, glossy surface and it is the wax used to polish automobiles, shoes, floors, and furniture. Carnauba wax is also used in products you

FIGURE 16.20. Carnauba wax is obtained from the young leaves of these *Copernicia prunifera* palms. The wax is secreted to the top of the leaf epidermis in a thick layer, and can be knocked loose and collected as large flakes. (© GM Photo Images/Alamy Images.)

FIGURE 16.21. Although plants of *Euphorbia antisyphilitica* do not look especially waxy, their gray color is due to a substantial layer of candelilla wax. If it is scraped off, the plants appear bright green. Candelilla plants were heavily overharvested and numerous populations were damaged during World War II, but many of the remote, desert areas of west Texas (especially in the protection of Big Bend National Park) and adjacent Mexico are thriving. The plants produce small pink flowers and are attractive in xeriscape gardens.

might not expect: dental floss slides between your teeth because of its carnauba wax coating, and the hard shiny shells on many candies is also carnauba wax. Because it is hypoallergenic (it causes almost no allergic reaction in people), it is used to thicken cosmetics such as lipstick, eyeliner, deodorant, and skincare products. Carnauba wax is a hard wax, but if mixed with beeswax, it can be used on flexible material such as leather coats and belts to make them waterproof.

Candelilla wax is obtained from candelilla plants (pronounced can da LEE ya; *Euphorbia antisyphilitica*; FIGURE 16.21). These are small shrubs with slender, leafless stems that grow in desert areas of west Texas and northern Mexico. At present, candelilla wax is no longer used for many things, being a minor component in cosmetics, lubricants, coatings and sizings for paper, chewing gum, and floor polish. Large amounts of candelilla wax were used in the manufacture of phonograph records, and during World War II, harvests were greatly increased because the army used it to waterproof tents, canvas, and wood products. At one point, demand was so great that the species was in danger of being overexploited to extinction. Candelilla plants live only in hot, rocky desert regions and have never been domesticated; all wax was obtained by pulling up entire plants, roots and all, boiling the shoots in water with sulfuric acid, then skimming off the wax as it floated to the surface. Because all material was obtained by collecting wild plants in remote areas, entire regions had every plant removed. Limits were placed on the harvest, both in Mexico and the United States, but enforcement was lax; the only thing that saved the species from extinction was the development of alternatives and the increasing price of the wax as plants became more difficult to find. At one time, *Euphorbia antisyphilitica* was used as a medicinal plant to treat syphilis, but despite its name it is not effective.

Jojoba wax (pronounced ho HO ba) is harvested from another desert shrub, *Simmondsia chinensis*, native to the desert southwest of the United States and adjacent Mexico. Jojoba wax is liquid, not solid, and it was developed as a replacement for sperm whale oil for use as a lubricant for fine machinery. Plantations of jojoba have been established in dry areas around the world, and now it is plentiful enough to be used in cosmetics and as a carrier for specialty fragrances. It has been mentioned for use as biofuel, but the plants produce too little wax per acre to make that practical.

Bayberry wax, from wax-myrtle (*Myrica faya*), is used for the scent it gives to candles.

Resins

Resins are thick, viscous sticky liquids produced by specific trees. The most familiar is probably the "pitch" that oozes from pine, juniper, and other coniferous trees (FIGURE 16.22). The liquid components of resins are chemicals called terpenes, chemicals you know as turpentine. Turpentine is a mixture of two terpenes, alpha-pinene and beta-pinene, both from the resin of pine trees. Depending on the species, various other compounds are dissolved in the terpenes while in the tree; thus resins vary in their color, aroma, stickiness, and other characteristics. As trees are processed into pulp to obtain their fibers, the resin is released, floats to the surface, and is skimmed off and collected. The terpenes are volatile and easily extracted; some are saved to make turpentine and some are used for the synthesis of other chemicals, especially fragrances that have a chemical structure related to that of terpenes. There is usually a large amount of terpenes left over, and these are used as fuel to run the pulp-making mill. Turpentine is used as a thinner for oil-based paints, and as a solvent for varnishes. It is mixed with beeswax and carnauba wax to make polish for fine furniture.

FIGURE 16.22. Many but not all conifers have ducts that produce the resins we call "pitch." This is the base of a juniper tree (*Juniperus*) after its top was cut off: resin has oozed out of the cut ducts and formed a sticky mass on the cut surface. Although only a small amount of resin flows from cut ducts here, if resinous trees are ground up and digested to obtain fibers for paper, large amounts of resin are obtained.

If all terpenes are removed from resin, the result is a sticky solid called rosin. This is used on the bowstring of stringed musical instruments to increase friction as the bow moves across the strings. Dancers also use rosin to stabilize their shoes on a slippery floor.

Various trees have chemicals that give their resins distinctive qualities. Fossilized resins of conifers, especially the members of the southern hemisphere family Araucariaceae (the monkey puzzle trees), are known as amber, a semiprecious yellow, glassy material used for jewelry. Many resins are hard and transparent rather than fluid and sticky, and an example is copal. This comes from a Central American tree in the genus *Copaifera*, and it resembles amber, but hasn't been fossilized. Copal is used both as a substitute for amber in jewelry, but it is also burned for incense and aromatherapy. Dammar is also a hard resin, obtained by tapping large East Asian trees in the family Dipterocarpaceae. Dammar is also used as an incense, but it can be dissolved in turpentine as a constituent of varnish. Some resins are soft and gum-like; for example frankincense (*Boswellia*) and myrrh (*Commiphora*) and are used for their aromas (FIGURE 16.23). These latter two are familiar from the bible; myrrh was used as an embalming ointment to treat bodies before burial, but it is no longer used for that. Frankincense has always been used for its aroma; it is one of the main types of incense used in temples and religious rites, starting at least 5000 years ago. It is still used for its fragrance both in religious settings and for personal use.

FIGURE 16.23. This is frankincense, the resin of *Boswellia*. It can be obtained at many stores that specialize in herbs, scents, and oils; by burning it, you will smell an aroma that has been important in religious ceremonies for thousands of years.

Rubber

Natural **rubber** is made from the latex of the large tropical tree *Hevea brasiliensis* (FIGURE 16.24); by the way, this is not the tree popular in homes and gardens called the "rubber tree" (FIGURE 16.25). The trees are slashed diagonally to cut open long, tubular latex canals. The white, sticky latex flows out of the canals, down

FIGURE 16.24. The latex used to make natural rubber is harvested from large trees of *Hevea brasiliensis*. Slashing the trees must be done carefully so that the incision is deep enough to open the latex tubes but not so deep as to damage the vascular cambium. (© phoenixman/ShutterStock, Inc.)

FIGURE 16.25. Plants of *Ficus elastica* are often called "rubber trees" but this is just a confusing common name, they are not the source of latex for rubber. Rubber comes from *Hevea brasiliensis* in the spurge family (Euphorbiaceae) whereas *Ficus elastica* is in the fig family (Moraceae).

the slashes, and into cups that are attached to the bottom of the tree. The latex must be refined to make rubber, and now it is possible to synthesize rubber artificially. However, natural rubber is more elastic and more resilient than artificial rubber, and natural rubber is preferred for many uses. Large plantations of *Hevea brasiliensis* still are cultivated, mostly in Southeast Asia where they are free from a leaf blight fungus that attacks the trees in their native Brazil.

Cellulose-Based Products

Pure cellulose can be chemically modified into a liquid form and then treated to have various properties. If it is extruded through pin holes into a bath of sulfuric acid, the liquid cellulose solidifies into fibers called **rayon**. Rayon fibers can be chemically and physically altered in many ways to give them particular physical characteristics. Many rayon fibers are smooth and silky enough to be woven, either pure or mixed with other fibers, into cloth for clothing. Other types of rayon are so strong they are woven into the cloth used as the base material that holds the rubber of tires together.

Cellulose can also be modified such that it can be extruded in sheets of cellophane, the familiar wrapping material, or as cellulose acetate, the film used in photography and movies. Methylcellulose is another derivative of cellulose; it is used as a solvent for many dyes in histology labs, and as a raw material in the synthesis of other chemicals.

Wood as Construction Material

Review of Wood Structure

Wood is a composite material, composed of long cells oriented parallel to the trunk, branch, or root, mixed with bands of shorter ray cells that run radially from the vascular cambium inward toward the center of the wood (see Chapter 3). The long cells make up the axial system of the wood; all the rays constitute the ray system (see Figures 3.33–3.37). Among softwoods (that is, conifers), the long cells consist only of tracheids and fiber-tracheids, but in hardwoods (dicots, broadleaf plants), the long cells are typically more diverse, with an abundance of fibers (technically, xylary fibers), tracheids, and vessel elements. All these cells have both primary walls and thick, lignified secondary walls. If a wood has an abundance of thick-walled fibers or fiber-tracheids, the wood is dense, tough, heavy, and usually resistant to rot, for example oak and bald cypress. If there are only a few fibers or fiber-tracheids, or if their secondary walls are relatively thin or only lightly lignified, then the wood will be light and weak, and will probably decay quickly if in a moist area; examples are balsa and cottonwood. Vessels are often too wide to provide much of the wood's strength, although in many desert shrubs and trees, vessels can be so narrow they resemble fibers and then they do provide strength.

Mixed among the long cells of the axial system are shorter cells, almost always parenchyma cells. These store water and starch in living wood, but when wood is being used as lumber, they make the wood lighter and weaker than it would be if the parenchyma had been fibers instead. Conifer wood has few or no axial parenchyma cells. The ray cells too are almost always parenchyma and contribute to wood's weakness, but they add to its texture and beauty by introducing cells that differ from the fibers in size, shape, and color.

Wood is produced by a vascular cambium that usually is active for only a few months in spring and summer. Consequently, the cross-section of wood reveals concentric growth rings; these are sometimes called annual rings, and many trees of temperate areas do reliably produce one growth ring each year. But in very dry or cold areas, the cambium may not produce any growth ring, or in very good years it might produce two, so they are not always annual.

Typically, the first wood produced in a growing season differs from that produced at the end. The first wood, the earlywood, has more conductive cells (more and wider vessels in hardwoods, wider tracheids in softwoods), and the last wood, the latewood, has more fibers or fiber-tracheids. Earlywood and latewood often differ strongly, especially in color; if so, growth rings are easily visible and contribute to the look of the wood (elms, ashes, oaks). When wood is sawn lengthwise, especially in tangential sections, the latewood and earlywood form attractive stripes, as is often visible in baseball bats and the handles of shovels, hoes, hammers, and so on (**FIGURE 16.26**). In tropical areas where trees have only a brief dormant period, or in deserts where the cambium may produce only a few cells even in a good year, growth rings might be difficult to detect even with a microscope. Even some trees of temperate climates, such as maple, dogwood, and holly, have indistinct growth rings.

FIGURE 16.26. The dark streaks in baseball bats and other very strong woods used for tool handles are the latewood portions of growth rings, whereas the light streaks are earlywood. Woods that are strong enough to withstand hitting a baseball or pounding a nail obtain their strength from abundant latewood fibers that have extremely thick walls that are heavily lignified. (© Monkey Business Images/Dreamstime.com)

After wood has been formed, its conducting cells function for several years, and its parenchyma cells remain alive. But at some point, most of the conducting cells will have cavitated; then the parenchyma cells produce antimicrobial compounds, fill the adjacent cells and themselves with aromatic, dark-colored chemicals, and then die. The outer living wood is sapwood; the inner dead wood is heartwood. Heartwood has many desirable properties: it has a rich color, often a pleasant aroma; it is especially strong and it resists decay. Heartwood of cedar is so rich in these chemicals it is used to make chests and closets that keep moths away from woolen cloth. In contrast, sapwood is pale and decays quickly. As a tree ages, on average one ring of sapwood converts to one of heartwood per year. A young tree trunk, branch, or root has only sapwood, but the trunk of an old tree is mostly heartwood. When loggers cut down large old trees, they can harvest a large amount of the desirable heartwood, but modern plantation-grown trees usually do not get old enough to form much heartwood before they are harvested; most of them provide lumber cut from sapwood, which is weak and decays quickly.

Sawed Lumber

Much of the wood from trees is used as sawed lumber for constructing houses, buildings, the inner framework of furniture, and so on. Much of this merely needs to be strong, not beautiful, because it will be covered by something else. Much of our lumber is softwood from conifers simply because most of our forests are dominated by conifers. Also, conifers usually have a single, straight trunk that is more suitable for sawing into long pieces of lumber than are the irregular trunks of dicot trees. Many conifers grow to be gigantic trees, for example, redwoods, Douglas fir, red cedar, and ponderosa pines. Most large dicot trees, at least in temperate areas, are much smaller; examples of large dicot trees would be maples, hickories, and oaks. Cottonwoods and willows can be big trees but their wood is too weak for most uses as sawed lumber. Other dicot trees such as cherry, pecan, walnut, and mesquite are too small to provide large beams, boards, or planks, but they do provide beautiful woods that are valued by woodworkers for making furniture and decorative pieces (FIGURE 16.27).

FIGURE 16.27. The heartwood of certain trees contains chemicals that give it especially rich color. These woods are prized for their beauty and are used to make objects of special esthetic value. (© Sergio Azenha/Alamy.)

Logs that are to be sawed are allowed to dry outdoors for up to a year, or they may be placed in a kiln and dried with heated, moving air. If lumber is cut from logs that have not been dried, the boards will warp, twist, and crack as they dry. The dry logs are carried into a sawmill and cut lengthwise in several passes through band saws or rotary saws with elaborate cutting schemes designed to maximize the number of high-value large boards and beams while minimizing the number of small, low-value sticks that are left over.

Some logs are not sawed but merely stripped of their bark and used as **round wood** for fence posts, poles, pilings for docks, and mine shaft supports. Round wood that is intended for use in soil or water must be treated with preservatives.

Engineered Woods

In the past, logs were sawed into lumber as described above, but with the invention of strong glues and resins, it is now possible to bond small pieces of wood together into large pieces that are as strong as a natural piece of wood. By altering the types and sizes of the wood pieces used, or the glues and other additives, the nature of the bonded wood can be more flexible, or stronger, or have completely novel grain, color, and texture. These are **engineered woods**. One of the first engineered woods was **plywood**: a log is rotated about its longitudinal axis as a long knife cuts off a thin peel (a **veneer**) of wood that is as wide as the log is long (FIGURE 16.28). Veneers can be cut as thin as 0.25 mm (0.01 in) in some species or as thick as 10 mm (0.4 in) in others, depending on the nature of the wood, its uniformity, and its tendency to splinter when being cut. The veneer is allowed to dry, and then it is cut into sheets that are stacked as the individual layers in plywood. In most plywoods, the grain of each layer is oriented at right angles to the layers above and below, which gives plywood equal strength in all directions and no tendency to curl or roll. Specialty plywoods are needed for curved surfaces, and those are made by aligning the grain of each layer parallel to each other. Because it is manufactured in sheets, plywood can be used in building walls, floors, and roofs more rapidly than when many small individual boards must be handled. Numerous types of plywood are manufactured, with softwood plywood (made of Douglas fir, spruce, pine, and fir) being the most common, and birch plywood being used where additional strength is needed. Most plywood is still used as basic building material that will later be covered by a finishing material, but some plywood is used in furniture, to display its own unique grain and figure. This decorative use of plywood, especially molded plywood, was pioneered by Charles and Ray Eames; an example that you have probably seen and used is the Eames Lounge Chair, designed in 1945 (FIGURE 16.29).

Many additional types of engineered wood have been developed recently. **Particleboard** is made by chipping logs into particles of various sizes, usually a centimeter or more (a half an inch or so), then gluing them into a solid block that can be cut into boards. By itself, this material can be broken rather easily, but if the exposed surfaces are covered with veneer, the particleboard becomes much stronger and resistant to breaking. The framework of many factory-made cabinets consists of particleboard made of sawdust.

Oriented strand board is a type of particleboard made from strips of wood oriented parallel to each other before being glued together. Oriented strand board has about the same strength as sawed lumber of the same size, but by varying the size, texture, and colors of the strips that are used, boards of unique beauty can be

FIGURE 16.28. This machine cuts a thin layer of wood, a veneer, from the entire length of a log as the log is turned. The machine has a knife as long as the log and which must be strong enough to withstand hours of cutting. The knives must be sharpened frequently. (© Neil McAllister/Alamy Images.)

FIGURE 16.29. Charles and Ray Eames pioneered many new styles of furniture; they are especially known for the use of molded plywood. The development of techniques to steam the wood to soften it and make it flexible was important, as was the invention of glues that are strong enough to maintain the wood's strength in its new shape. (Courtesy of David West and Will Klemm.)

FIGURE 16.30. The majority of oriented strand board (OSB) is known as plywood and is used in constructing houses and other buildings in which it is covered with a more attractive material such as brick, stone, or siding. Some OSB, however, is made to be beautiful by itself and is used for furniture such as tables and benches. (Courtesy of Tommy R. Navarre.)

constructed. These are often used in furniture and benches or as surface coverings because of their beauty (**FIGURE 16.30**).

By gluing together long, thick (several centimeters [an inch or two]) boards of wood, **laminated beams** can be produced that are much longer, thicker, and stronger than any single beam that could be cut from a living tree. By carefully choosing the types of boards used, the color and strength of the laminated beam can be designed to meet the aesthetic needs of the building. Laminated beams are often used to construct large rooms, such as auditoriums and gymnasiums, without internal pillars that would interrupt the space.

A benefit of engineered wood is that much more of each harvested tree is used rather than being wasted. In the past, when all lumber was sawed lumber, the small branches were discarded, as were branches that were too twisted to give a straight board at least 2 m (6 ft) long. These "scraps" can now be used for lumber, reducing the number of trees that need to be harvested. These engineered woods thus allow a degree of conservation of resources and reduce the need to harvest the largest possible old trees. Although there is still great demand for sawed lumber, there is also public pressure to conserve our forests and especially to conserve any stands of forest that have never been logged or cut; these are called **old-growth forests**. Some logging companies and lumber mills have developed an environmental awareness and do make efforts to minimize the ecological damage that lumber and paper production cause. Various programs and guidelines have been developed; examples are those by the Forest Stewardship Council, the Leadership in Energy and Environmental Design (LEED), and the Sustainable Forestry Initiative; all have certification programs that ensure that production and construction practices are sustainable. (These have interesting, informative websites that can be found with an internet search.)

The techniques for creating engineered wood are now being applied to a number of other plant materials that are fibrous but not woody. "Lumber" is now being made by gluing together straw of rice, wheat, and rye, or the stalks of hemp, kenaf, or sugar cane. Bamboo strips too are being used. These often do not have the strength of wood for building large structures, but they provide a unique beauty for small household items, wall coverings, and decorative pieces.

Artisan Woods

Many woods are valued for their beauty; these are **artisan woods**. They often have such exceptional color and figure that they are used for musical instruments, fine furniture, frames for art, bowls, lamps, or are carved into pieces of art themselves (FIGURE 16.31). The heartwood of artisan woods may be black (ebony), red (mahogany), white (maple and holly), and even purple (junipers and purpleheart). Also important is the alignment of the long cells in the wood, called its **grain**. In many woods, the grain is straight and parallel to the long axis of the wood, but in some it spirals around the trunk or branch, or it appears to undulate as if the wood had a wavy surface (bigleaf maple, koa). The grain is intricate and contorted and has a unique appearance in wood near knots and in burls. The total, overall appearance of wood, including its rays, the patterning of its earlywood and latewood, and the presence or absence of knots (especially tiny knots as in bird's eye maple, or larger ones as in knotty pine; FIGURE 16.32), is referred to as the wood's **figure**.

FIGURE 16.32. Bird's eye maple is an especially striking type of figure, consisting of many tiny dots that disrupt the otherwise smooth grain of the wood. It is not known what causes bird's eye figure, and not all maple trees have it; instead, it is discovered after a tree has been felled and is being processed. This is such a valuable type of wood that technicians always watch for it in certain species of maple. Bird's eye figure also occurs in Cuban mahogany and black walnut.

FIGURE 16.31. The unusual color and figure that cause some wood to be especially beautiful are often found in only small parts of a trunk, limb, or root. Thus the wood is used to make small objects such as bowls, platters, boxes, and so on. Because such combinations of color and figure are so rare, especially as compared to the large amounts of ordinary color and figure in wood used in lumber, artisan woods are extremely expensive.

Bark

Bark is the outer part of a shrub or tree and is a mixture of secondary phloem and cork. Bark texture varies depending on the nature of the components in both the phloem and the cork (see Chapter 3, Figures 3.38 and 3.39). The secondary phloem is produced by the vascular cambium to its exterior, and it always contains at least sieve elements. Almost always the secondary phloem also contains parenchyma cells, sclereids, fibers, and in some species, secretory ducts or chambers filled with resins, milky latex, or various other compounds.

Some of the phloem parenchyma cells divide in an orderly manner and organize themselves into a cork cambium. Most of the cells they produce differentiate into parenchyma cells with thin walls that become impregnated with suberin, which makes them impermeable to both liquids and gases. The cork cells then die. In some species, the cork cambium will produce a few layers of sclereids (stone cells) that alternate with layers of cork cells.

As branches and roots continue to grow, the bark is pushed outward and stretched tangentially, usually finally cracking and breaking. New cork cambia must arise in the newly produced secondary phloem; subsequent cork cambia arise every few years and produce cork for just a few weeks. Consequently, bark consists of alternating layers of secondary phloem and cork (or cork and sclereids; see Figure 3.38).

The nature of bark differs greatly from species to species. An extremely hard bark is produced if the secondary phloem contains fibers and the cork cambium makes sclereids. If the secondary phloem is mostly parenchyma and the cork cambium produces mostly cork, the bark will be soft and spongy. Numerous variations occur.

Almost all bark contains lenticels, radial patches of cells that have microscopic intercellular spaces. These spaces allow oxygen to diffuse past the otherwise gas-proof cork and penetrate deep into the branch or root, allowing the living cells of sapwood, vascular cambium, and inner bark to respire and not be suffocated by the bark.

FIGURE 16.33. This is a tree trunk of cork oak (*Quercus suber*) right after a layer of corky bark has been peeled from it. The surface you see here is still bark; it is not the vascular cambium or sapwood, so the tree is still protected from fungi, bacteria, and other problems.

Cork

Cork that is familiar to us as corkboards and stoppers for wine bottles comes from the bark of a species of oak tree, *Quercus suber* (**FIGURE 16.33**). Its bark is unusual in having thin layers of tough secondary phloem alternating with thick layers of cork. It has enough strength to hold its shape but is soft and elastic enough that it can be pushed into a bottle and then expand to form a tight seal. Cork oaks can be large trees that live for 200 years or more, and they must be about 25 years old before the first bark is harvested. Subsequent harvests occur at long intervals, about once every 9 or 10 years. The first two harvests of cork are usually too coarse to be used for wine stoppers, so trees must be almost 45 years old before they provide their first useful crop of cork. All harvesting is done by hand and with great care so as not to cut too deeply and damage the vascular cambium, which would harm the tree. Cork is pulled off in thick, curved sheets. Stoppers must be cut out of the bark vertically so that the lenticels run from side to side; if stoppers were cut out radially, the lenticels would allow oxygen to diffuse into the wine bottle, ruining the wine.

The bits and pieces of cork left over after stoppers are cut out are chipped into small pieces and glued together to make corkboards, fishing floats, buoys, and other things that must be light and elastic. One rather new product is cork flooring: the chips are bonded together in flat tiles that are glued to a floor to give a comfortable surface.

Obtaining cork from cork oaks is an example of low-impact agriculture. The trees are a long-term investment and must be protected by not overharvesting them: a sustainable harvest makes economic sense. The cork itself is easily recyclable and many stores accept wine corks so that they can be ground into cork particles to be made into new products (they cannot be reused as stoppers). Most cork oaks occur as managed natural stands rather than as orchards planted in rows, and grasses and shrubs are allowed to grow between the trees. Half of all commercial cork oaks are cultivated in Portugal and Spain, and cork oak forests have become an important habitat for the endangered Iberian lynx.

Mulch

Mulch is a protective material placed on soil around plants (FIGURE 16.34). It helps retain moisture in the soil, reduces erosion, keeps the soil cool in the summer, and suppresses germination of weed seeds. Sometimes plastic or fiberglass sheeting is used, but those are not biodegradable and they eventually end up in a landfill. Here we are concerned with mulches made of plant material. Just about any plant matter that does not contain seeds is acceptable. Pine bark is one of the most common for home gardeners to use in flower beds. When pine trees are harvested for lumber or fibers, the bark can either be sold as mulch or it is a waste produce that must be disposed of. Occasionally, the wood of small branches is also shredded and used for mulch along with the bark. Most cities now shred Christmas trees into mulch. Almost any kind of shell or husk (for example pecan shells and cotton seed "burs") can be ground to an appropriate particle size for mulch, as long as the material does not decompose too quickly or contain compounds that are toxic to plants. In farming areas, straw from cereal crops is often used to mulch between row crops; this retains moisture and minimizes the amount of tilling needed for weed suppression.

FIGURE 16.34. These leaves of spider lily (*Licoris*) are being produced by subterranean bulbs. The bulbs have enough stored starch to support the growth of the leaves upward through a thick layer of soil, which here is covered by a layer of mulch, the pieces of pine bark. The mulch has been applied for several reasons, such as shading the soil and keeping it cool in the summer, preventing this area from being muddy when it is watered, and its darkness prevents many weed seeds from germinating. The weeds that do germinate are easy to pull out because the mulch is so soft.

Important Terms

abacá
acid-free paper
artisan wood
bast fiber
bayberry wax
burlap
candelilla wax
carnauba wax
coir
copra
cotton
drying oil
engineered wood
figure
flax
geotextiles
grain
hard fiber
hemp fiber
jojoba wax
jute
kapok
kenaf
kraft method for wood fibers
laminated beam
linen
mulch
old-growth forest
oriented strand board
particleboard
phloem fiber
plywood
ramie
rayon
resin
retting
round wood
rubber
saponification
sisal
sizing
soft fiber
sulfate method for wood fibers
sulfite method for wood fibers
varnish
veneer
wax
wood fiber
xylary fiber

Concepts

- "Fiber" refers either to a single sclerenchyma cell or to a bundle of many cells.
- Only a fraction of cloth is used for clothing or household products; much industrial cloth can be coarse, as long as it is strong and resists being torn.
- Cotton fibers are actually plant hairs from the seed coats of cotton.
- Several species of fiber plants can be grown as annual crops and may be developed into sources of fiber that can reduce the need to cut down forests.
- Most wood fibers for paper are produced by the kraft method. Removing lignin and bleaching fibers are difficult and have been the main sources of pollution, but newer techniques have greatly reduced the pollution caused by pulp mills.
- Although much wood used for construction is still sawed lumber, an increasing amount is now engineered woods.
- Many reforestation projects are really just monocultures of firewood trees, not true forests with high biodiversity.

Ornamental Plants: Plants that Refresh Us

17

We humans differ from all other animals in that we appreciate certain plants simply for their beauty. We cultivate plants that provide neither food nor shelter, but that lift our spirits when we see them or smell their fragrance. Some plants help us express joy or friendship or sorrow; others we study because they are so different from us. These are the plants we keep close to us, in our gardens, in our homes; we keep them where we live. (Courtesy of Chris McCoy.)

Introduction

This chapter is about plants, people, beauty, fun, curiosity, and thoughtfulness. There is no hard and fast definition of what ornamental plants are; they are the plants we have and care for not to obtain food, fuel, fibers, or medicine but because we like the plants for themselves (FIGURE 17.1). We enjoy their foliage, their flowers, and fragrances, or with plants such as succulents and carnivorous plants, for their "otherness," for being different from other plants. We construct and tend gardens in which the plants are each beautiful as individuals and also contribute to the pleasing fabric of the entire garden (FIGURE 17.2). We watch plants grow, put on new leaves, make flower buds, and then anticipate the opening of the flowers; it is easy to take this for granted, but we should give ourselves credit: we watch plants grow and develop because we have an intellectual curiosity about these things that are so different from us. Few people consider weeding a garden or mowing a lawn to be enjoyable, but it is satisfying to make a cutting of a plant and watch it produce

FIGURE 17.1. Azaleas are cultivated just for their beauty. They do not provide food, fibers, medicines, or anything else of commercial value, but their springtime display of color and form is a tremendous boost to the spirit.

adventitious roots and grow, or to sow seeds and see them push up through soil. And a particular appeal of many ornamental plants is that they can be shared: cuttings can be made, bulbs can be divided, seeds collected and given to friends.

These plants are also used to express our emotions. Flowers are used on happy occasions like weddings, anniversaries, birthdays, and promotions, as well as sad ones such as funerals and memorials. We take flowers and potted plants to friends when they are sick or just to show friendship. Religious celebrations too would be incomplete without flowers or foliage plants (FIGURE 17.3).

A particularly human endeavor is to wonder what sets us apart from other animals: our use of tools, our capacity for abstract reasoning, and so on. To that list, we should add our appreciation of plants as individual living creatures.

FIGURE 17.2. This is just one of many types of gardens. Here, every bit of space is used for plants that provide vivid colors; other gardens might be planted to provide a more calming, tranquil effect, and still others would be dedicated to whatever plants the gardener wants. Some gardens are certainly planted more spontaneously, reflecting the desires of the gardener at the moment, while others are planned carefully. This particular garden is planted in the center of the mental hospital building where van Gogh had himself committed so that he could paint—with vivid colors—without interruption. (Courtesy of Chris McCoy.)

FIGURE 17.3. We often use plants to show feelings and sentiments that we have trouble expressing verbally. (Courtesy of Chris McCoy.)

Hobby Plants

Hobby plants are the plants we like to collect, to amass a set of plants that are either related to each other or have some feature in common that unites them and makes us pay special attention to them (BOX 17.1). Some of the plants that hobbyists like to collect are orchids, cacti, succulents, African violets, roses, bromeliads, bulbs, epiphytes, ferns, mosses, cycads, irises, and many more. For most of us, hobby plants need to be small enough that a collection of 10 or 20—or a few hundred—can be fit onto a window sill or home greenhouse, unless we are lucky enough to be fascinated by plants that can be cultivated outdoors where we live (for example, cacti in California and Arizona, orchids in Seattle, cycads in Florida). Other than that, hobby plants are just any particular set of plants that are especially appealing to someone.

For those of us who do like to collect and cultivate a particular set of plants, hobby plants offer many benefits, especially for our intellectual side. The first few plants of any specialist collection will almost certainly be similar to each other and representative of the group as a whole: the first cacti in someone's collection are always very typical, very obvious cacti, as is true of the first plants in a collection of orchids, ferns, and so on. But almost every plant group has several or many excellent books written about it, and now the internet is a superb resource for information on any group of plants. With just a little reading about the group, we quickly discover that every plant group has a few oddballs, a few species that differ from all the others in their body shape, flower type, or ecology. Anyone who has a collection of plants soon learns the true meaning of diversity within a genus, family, or any other clade. For example, many orchids do have the large, exquisite flowers and thick, leathery "pseudobulb" stems that live as epiphytes, but others have tiny flowers and grassy leaves and live in soil (FIGURES 17.4 and 17.5). Some cacti are small spherical spiny "pincushions," but others are tall and slender, some have real leaves, others can be gigantic, and many (especially those from the Andes in South America) can be grown outdoors even where there is an occasional light freeze in winter. And people who have collections

BOX 17.1. Using Your Plants and This Book to Study Plant Biology in Your Life

If you have a few plants in your home, or if you have a garden or yard, you can sit near those plants with this book and observe the plants as you read. If you don't have any plants at all in your life, then go buy one, or take this book to your favorite park or natural area, some place you visit—to sit and be quiet, to jog or bike, or play Frisbee. By paying attention to a set of plants that you visit regularly, most of the principles in this book will change from being abstract to real, and you will discover that you already know a lot of biology. It is just that most of us rarely take the time to stop and organize our thoughts or to verbalize our insights.

Diversity. This book emphasizes diversity and its importance in keeping ecosystems stable. A species is made up of individuals that resemble each other but that are not identical. Even if you have dozens of cuttings you have made of the same plant over several years, those cuttings will resemble each other in many ways, but their environment will cause some to have more leaves, more branches, a more upright habit, and so on. If you have a collection of several species in one genus, or several genera in one family, the diversity there will also be striking. Most petunias are easily recognizable as petunias, but the range of color and size is great. The same is true of lilies, geraniums, spice plants, everything. And if you are tempted to buy a few more specimens, you can consider the range of plants available throughout Kingdom Plantae: mosses, ferns, cycads, conifers, flowering plants, and many smaller groups not listed here. Diversity tends to be the hallmark of life; species that are becoming extinct often do so because they have too little diversity to cope with changing habitats.

Monoculture versus intercropping. If you have just a single plant on a window sill or balcony, that is definitely a monoculture. But if you also have a lawn of a single species, that too is a monoculture, as is a flower bed containing all the same species of a particular plant. These are beautiful and definitely have their merits, but monocultures can teach us that if something goes wrong, it goes wrong for every single plant and we may end up with most of our specimens either diseased or dead. A particular mite or fungus may one day arrive that can feed on our monoculture plants, and that invader will then be able to easily find more plants on which it and its offspring can feed. Disease spreads rapidly through a monoculture. But if we have the same number of plants widely dispersed among other plants in our garden (the same effect as intercropping), many may go unnoticed by the pest. It is surprising how far certain insects can fly and how acutely they can detect host plants, while at the same time completely missing similar host plants. Avoiding monocultures protects our plants at no cost or effort to us, and with no need to use pesticides.

Finite resources. Everybody who has plants quickly learns about finite resources. The window sill is full of plants and there is no more room; every nook and cranny of the garden has something planted in it. Before long, every gardener faces the choice of either expanding the garden or taking out some of the plants we know and love. Even if we decide to expand, we will at some point run out of room again. This is true of any ecosystem, and it is definitely true of Earth as an ecosystem: Earth only has so much room, and only a certain number of plants, animals, and other organisms can live here. Some people are fooled into thinking that we will be able to improve crop plants so that they have higher yield or that more can be grown per acre or that they can be modified to grow in marginal lands and waste lands, but . . . there is an absolute limit. The Earth's resources are finite, as is the number of organisms, including people, that can live here.

Erosion control and soil loss. This is easy to see in potted plants indoors: if we are a little careless when watering, dirt is splashed everywhere. And if a garden planting bed is not level when installed, it soon will be as dirt washes from high to low areas every time we water. Dirt moves downhill. We can see it with our plants every day, and we should think of the erosion that occurs in nature. A thick layer of vegetation holds much land in place, but landslides and avalanches occur even if a slope is vegetated. And on fallow cropland or areas left open by fire or deforestation, erosion is a serious problem. Next time you are biking or jogging or driving through the countryside, imagine the erosion that happens in your yard, and compare the size of your yard to the size of all the continents.

Essential elements and limiting factors. When we grow plants in pots, at some point we must fertilize them. Their roots will have taken up all the available nutrients in the soil, and without fertilizer, they will suffer slow growth and maybe a deficiency disease. For plants in a garden, fertilizers are often used to get plants to be a bit leafier (high nitrogen fertilizer) or to bloom better (high phosphorus), but it is not absolutely necessary. In a garden, roots are able to spread farther and deeper than when they are in pots. If we allow fallen leaves and twigs to stay in our garden, and if we mulch, that may allow nutrients to be recycled and avoid the need for fertilizer. On the other hand, if we are neat and keep the soil under our plants fastidiously raked and clean, then the debris that is raked up and hauled away is taking essential elements with it and leaving the beds less and less rich each year. You will notice that roses need lots of sun, and if planted in a shady area, they will not bloom even if watered and fertilized: light is the limiting factor. For many gardens in the sunny, hot southwestern United States, water is the limiting factor: plants grow only if given more water; they do not need more sun or more fertilizer.

Natural cycles. Taking care of plants is a superb way of learning about plant development and their seasonal cycles. Novice gardeners often panic in autumn as their plants appear to die, only to realize that it is just the onset of natural dormancy needed to survive winter. Most plants bloom only at a certain time or year, and either on new twigs or on last year's twigs. We soon realize that pruning, if done correctly, encourages flowering, but if done incorrectly it cuts off all the newly formed buds. Look at your garden throughout the year: where does each plant produce new leaves and flowers, and in what part of the season? And watch for weeds; it may seem as if they pop up all the time, but they don't, they germinate with a seasonality.

Carbon capture and sequestration. Any woody plant, whether in a flowerpot indoors or in a garden, captures carbon dioxide and locks it into wood. Wood takes carbon dioxide out of the air for a long time, and trees in a garden are just as effective as those in any faraway rainforest. In a xeriscape garden, the tough, leathery leaves of agaves and yuccas lock up carbon dioxide and store it away for many years because each leaf lives a long time and is slow to decay. In contrast, the carbon dioxide used to build roots, flowers, fruits, and nonsucculent leaves of most plants is only briefly locked up and then is released as soon as they decay; those carbon dioxide molecules are out of the atmosphere only a few months and the atmosphere's content of greenhouse gasses is not really affected.

Weather/climate change. We scientists had thought that climate change would be such a slow process that we ourselves would not have to face the consequences of our actions; it would be our children or grandchildren who would live in a hotter world. Even though Earth's climate is changing fast enough that we will have to suffer at least some, it will probably not change fast enough to affect our personal collections of plants. But variability of weather is a good substitute for showing what can happen when growing conditions change. All geographical areas have average years as well as extremely hot or cold years, wet or dry years, years with many insect pests and years with few. With bedding plants and annuals, we only have to worry about weather and other conditions for a few months, but with perennials, we need to consider not merely the lowest temperature in an average winter, but what is the severest winter we can except within 10 years, 20 years, or 100. The same with drought, heat, and pests. It is not unusual for a garden to thrive for years; then the area gets its coldest freeze in 20 years, and half the garden is dead. Cycads, palms, succulents, and other tender plants that had been thriving are suddenly dead or badly damaged. In a garden, these can be replaced by buying new specimens from a garden center, but stop and think about the effects that climate change will have on both plants in natural habitats and croplands. If altered conditions cause a forest or alpine meadow to be damaged, those plants will be replaced, but not by the same ones; instead, new species that can survive the new conditions will move in and the area's vegetation will be altered, affecting birds and other animals that depend on the plants.

Interactions with other organisms. Once we start to have a collection of plants, we automatically have insects. We sometimes have to use our biological training to remain convinced that spontaneous generation does not occur. Bugs appear as if by magic, as do weeds. In outdoor plants, butterflies, birds, and moths come to visit and feed, and so do squirrels, raccoons, mice, rabbits, possums, armadillos, and whatever other city-dwelling wildlife there is. At first our inclination is often to protect our plants at all costs and to spray pesticides, but unless the animals are killing our plants outright, why not think of the animals as an integral part of our garden? For the cost and effort of putting in plants, we also get to help and observe animals. Some animals do hurt plants by eating them or laying eggs in them, but others merely nest in them; and of course, they often bring organic fertilizer, which may be a bit messy, but ultimately benefits the plants. Before we bring out any pesticides, we should think of the many, many animals in our gardens that will be harmed just so that we can control one particular pest. Also remember that you and your family, your pets, and your friends will be in that garden in the next few hours and days.

The important thing to do with plants is to observe them, think about them, respect them, tend them, and enjoy them.

FIGURE 17.4. When most people think of orchids, species with large, showy flowers like this usually come to mind, and when most people start a collection of orchids, they choose something like this. But the orchid family is extremely diverse, and an enthusiastic collector may soon decide to include more diversity by acquiring an orchid like the one in Figure 17.5. (Courtesy of Chris McCoy.)

FIGURE 17.5. This *Spiranthes*, like many species in the orchid family, have just small flowers and grow in the soil. Most people do not recognize them as orchids right away, but the value of having a collection of plants is that it encourages us to learn more about our plants and their diversity, ecology, evolutionary history, pollinators, and so on. (© Martin Fowler/ShutterStock, Inc.)

of plants with bulbs soon learn that bulbs occur in many families, so they will differ greatly in their flowers, leaves, and the amount of cold or heat they need during their dormant period.

With almost every type of hobby plant, it is possible to propagate specimens in our collection, either because they form offsets (easily detachable side branches), or they produce seeds that geminate well. It is possible to learn a great deal of plant biology by propagating our plants and nursing them along while they are in vulnerable stages.

One of the few problems with hobby plants is that some specimen plants are still being collected from the wild. This is now illegal for virtually all species of orchids, bromeliads, and cacti, because overcollecting in the past has actually removed so many plants from natural populations that it has put certain species at risk. At present, there are nurseries that specialize in almost every type of plant, so it is possible to buy plants that have been grown from seed or made from cuttings of plants that have been in cultivation for many years. There is no longer any reason to collect wild plants except a greediness to have an unusually large or old plant.

Cultivation Methods

Many hobby plants can be grown just on a window sill or a patio, but others might require special housing. Greenhouses (also called glasshouses) have several functions. First, they are designed to enclose a cultivation area and protect the plants inside from wind, animals, rain, and low temperatures (FIGURE 17.6). Of course, almost any building could do these things, but greenhouses are also designed to admit the

FIGURE 17.6. Many companies sell prefabricated kits so that it is possible to build a greenhouse in a personal garden. A small one such as this will heat up well during the day, but will not hold its heat throughout a long cold winter night, so if it is used in areas with severe winters, supplemental heaters are necessary. A larger greenhouse would stay warm longer at night. All greenhouses need ventilation so that they do not overheat on warm days. (Gary Whitton/ShutterStock, Inc.)

FIGURE 17.7. Too much sunlight is a problem for many plants, so a shadehouse such as this may be necessary. This one uses narrow strips of wood (lathes) to block part of the sunlight, others use screens. Shadehouses often need maximum ventilation to prevent overheating, so they often do not have solid walls the way greenhouses do. (Mark Bolton Photography/Alamy.)

maximum amount of light so that plants can photosynthesize without having to rely completely on artificial lighting. Because of the greenhouse effect (see Chapter 11), a greenhouse also warms the enclosed space, which is an added benefit during winters in cold areas. But during summer, greenhouses usually must be opened so that wind can ventilate and cool them, or large fans must force air through the greenhouse to prevent overheating. In very sunny areas, greenhouses may need shades to reduce the amount of light and prevent sunburning if delicate plants are being cultivated. Greenhouses often have some electric light so that they can give plants artificially long days during winter if needed to make them flower or otherwise control their development. Greenhouses used for orchids or other tropical plants often have mist systems to spray fine droplets of water into the air to keep humidity high.

Shadehouses are made of wire or plastic screen that reduces the amount of sunlight the plants receive (FIGURE 17.7). Shadehouses are used to cultivate more delicate or shade-tolerant plants such as ferns and tropical understory plants. Shadehouses may have no walls at all, thus allowing maximum air flow and natural cooling, or they have screen walls to keep animals out. Unlike greenhouses, shadehouses do not automatically provide heat in winter, so they cannot be used for frost-sensitive plants in cold areas.

Cold frames are small areas of a garden that have low (about 1 m [3 ft]) walls that can be covered with glass in winter or on cold nights. Cold frames are used for plants that would grow well in an area except for an occasional, brief cold period.

Raised beds are narrow planting areas that have been built up several inches or more higher than the surrounding land (FIGURE 17.8). This allows rainwater to drain out of the beds quickly rather than staying wet or sodden for several days. Raised beds are used to cultivate plants in areas that are otherwise too wet. An obvious use is to cultivate desert plants in rainy areas, but raised beds are often used just for garden plants, to give the gardener more control over soil moisture.

FIGURE 17.8. This set of three raised beds with stony soil and coarse edging allows excess rainwater to drain out of the bed quickly so that the roots of these plants do not remain too wet and in too much danger of being attacked by moisture-loving fungi. By using raised-bed techniques, The Royal Botanic Gardens at Kew are able to successfully cultivate these desert plants on the outskirts of London. (Courtesy of Chris McCoy.)

FIGURE 17.9. This growth chamber is large enough to hold many racks of plants, with those on upper shelves receiving stronger light than those on the bottom shelves. The length of the light period is controlled by computer, as is temperature during light and dark. The air can be circulated to simulate windy conditions.

Growth chambers are large cabinets or small rooms in which multiple aspects of the environment can be controlled precisely (FIGURE 17.9). They have artificial lighting and heating so they can be programmed to match any type of day length/night length conditions that are needed. Most can also control relative humidity. Growth chambers are expensive and typically used only to carry out experiments, not to cultivate plants commercially or just for pleasure.

Propagation Methods

A **cutting** is, as its name suggests, a piece of a stem, root, or even a leaf that has been cut off from its parent plant and then is grown as a new individual genetically identical to its parent (see Figure B6.1a). Cuttings are very often used to propagate hybrid plants; even if the exact same parent plants are crossed several times, the hybrid progeny will vary in the combination of characters they have (unless each parent is homozygous for every gene, which is rare). Once a hybrid has been found that is especially appealing, the only way to guarantee that its descendants will have the same features is to propagate the plant clonally. One way to do this is by making cuttings: pieces of branches are cut off, allowed to form adventitious roots, then they are planted and tended. Either the shoot apical meristem or an axillary bud will become active and permit the cutting to grow as a new plant. In a few species, a piece of root or leaf can be used as a cutting, but only if they are able to make an adventitious shoot bud. Cuttings form new roots most easily if they are made from young, growing branches; older, woody branches, and cuttings from dormant plants are less likely to make adventitious roots and grow.

We often make cuttings without even realizing it. Most bulbs form clusters of new bulbs around their base, and it is easy to separate them by simply loosening the soil and pulling them apart. But the new bulbs are really branches formed from axillary buds, so by dividing a cluster, we have made cuttings of branches. Irises and gingers spread by rhizomes, and when their beds become overgrown, we dig them up, cut or break them into smaller pieces, and replant just the ones we need: by dividing the rhizome, we have made cuttings of the plant.

Some plants have surprisingly weak or sensitive root systems. The roots may grow poorly, even under natural conditions, or they do not recover after being transplanted. In these cases, plants may be more easily cultivated if their shoots are **grafted** onto the root system of another plant. The plant that supplies the root is called the **rootstock** (or just stock) and the grafted shoot is the **scion**. Rootstocks are often taken from seedling or sapling plants, while their shoots are vigorous and not very woody (FIGURE 17.10). The shoot is cut off, and the shoot of the scion is set on top of the rootstock's cut shoot, taking care to align the vascular cambia of the two as carefully as possible. Cell division in the vascular cambia produces new cells that differentiate into xylem and phloem and unite the two pieces. In some woody species, axillary buds are used as scions; a small incision is cut in the stump of the rootstock to open two flaps of bark, and then a bud is slipped into the incision. It becomes active and its new cells form a connection with new cells from the rootstock's vascular cambium. By using bud grafting, a single scion can produce as many grafted plants as it has buds.

FIGURE 17.10. To graft two plants together, the top is cut off of the one with the stronger root system (the stock), and the base is removed from the one with the desirable shoot characters (the scion); the two are then held together such that their vascular cambia contact each other. The two are next taped together, or sometimes held together with a waxy adhesive. The newly grafted plants must be kept in high humidity otherwise the scion will die before a good vascular connection develops. It is very rare that plants of different genera can be grafted together, and impossible to graft plants of different families. The rootstock of this kumquat is growing thicker than the scion is.

Horticultural techniques such as transplanting, grafting, and making cuttings put plants at risk of death by dehydration. The shoots cannot obtain adequate water from the roots, which have been damaged or have not formed yet. This risk is reduced partially by removing several of the leaves, and partially by misting the plants, spraying them frequently with extremely fine mist of water that cools the leaves and raises the air's relative humidity, both of which reduce water loss from the leaves. Once the root system is strong enough, misting is reduced over a period of days until the plants are able to survive natural conditions. It is easy to hesitate to snip healthy leaves off a cutting, but this actually helps the twig avoid water stress and almost any cutting will have more than enough reserve materials to make many new leaves.

When plants are propagated by cuttings, rooting them (inducing them to form adventitious roots) is critically important. Some plants such as those with bulbs, rhizomes, or runners form adventitious roots automatically even while the plants are intact, and are easy to propagate. Others, such as most cacti and euphorbias, form adventitious roots readily, within just a few days and most cuttings survive well. In most species, the formation of adventitious roots is slow or erratic but chances of success can be improved by treating the base of the cutting with a rooting compound, which usually contains indole acetic acid or an artificial auxin. But cuttings of some species, especially of old, woody stems of conifers, rarely form adventitious roots under any circumstances, and propagating these species by cuttings is difficult or impossible.

Sprouting seeds can be either fun or a challenge. Seeds of ornamental plants sold in nurseries and garden centers have been selected because they do germinate easily. For most, it is a matter of sowing the seeds on clean potting soil, covering them lightly with a layer of soil that is only one or two times thicker than the diameter of the seeds, then watering them and waiting for a few days. Before long, either the cotyledons or the new shoots show themselves. Seedlings are susceptible to **damping off** (being killed by fungi) and may need to be treated with fungicide, or for a more natural control, the seedlings may need to be allowed to dry slightly every 2 or 3 days. Newly sprouted fungal spores are more quickly killed by dryness than are most newly sprouted seedlings.

The seeds of many hobby plants have natural mechanisms that control when they germinate in nature. They may have a chemical inhibitor that must be washed out by rain, thus ensuring that they germinate only when the environment is very wet. Or seeds may have a tough seed coat that must be broken down by spending years with soil microbes or by passing through the digestive tract of a bird, deer, or some other animal. Gardeners might plant such seeds and never see any of them germinate. For those cases, it is best to consult an expert on the group, as the necessary techniques may have been discovered already. Otherwise, the gardener can experiment with various treatments such as giving the seeds a cold period or by scraping off part of their seed coat, and so on. The seeds of orchids are a special case. They

FIGURE 17.11. Seeds of orchids are extremely underdeveloped when released from the fruit. They are just a small mass of cells with no shoot, root, cotyledon, xylem, or phloem. Most will not develop further unless the proper fungus associates with them, but they can be encouraged to grow in sterile tissue culture by providing sugars, minerals, and numerous vitamins. This is not especially easy, but this can be done at home with kitchen utensils and bottles used for canning fruit or jelly. (Courtesy of Chad Husby.)

stop developing inside the fruit even before an embryo has formed, and in nature they must establish an association with a fungus before they will develop further. Hobbyists can cultivate orchids from seed by preparing sterile nutrient media in their kitchens. It is not easy, but it is possible and very satisfying when successful (FIGURE 17.11).

Outdoor Plants and Landscaping

Lawns and Groundcovers

Many gardens and landscapes are anchored by a large region of lawn that has many uses. In home gardens and public parks, the lawn is an area for living, an area where people walk, play, or sit (FIGURE 17.12). A lawn is an area to be used, and lawn plants must be strong enough to withstand being trampled, yet soft so that we can lie on them. A lawn also functions as either a point of beauty by being an expanse of uniform green, or it is a foreground that accentuates something just beyond it, such as a bed of flowers, a set of shrubbery, or a building. Lawns also keep an area from being dusty when dry or muddy when wet, and they prevent erosion.

The types of plants used for a lawn vary depending on the amount of traffic they must bear. Lawns that cover a playing field for football, soccer, baseball, and so on must be extremely tough, whereas those for parks and home gardens should be soft. Grasses are by far the most commonly used plants for lawns that will be walked on and that must be mowed to maintain uniformity. In cool areas, "cool season grasses" such as bluegrass (*Poa* species) and fescues (*Festuca* species) are used, and these grow from spring until autumn. In hot areas, "warm season grasses" such as *Zoysia*, Bermuda grass (*Cynodon dactylon*), and St. Augustine (*Stenotaphrum*) are used because they withstand the maximum heat of summer; they may turn brown in July and August (unless given large amounts of water) but they are not killed by the heat and

FIGURE 17.12. A grass used for a sports field lawn must be much tougher than the grasses used for homes and personal gardens. This lawn withstands the soccer of these youngsters now, but adults play on this field too, and concerts and kite festivals are held here; the grass must survive having thousands of people standing and running on it.

will recover in autumn. Lawn grasses spread by horizontal runners that lie close to the soil surface with just the leaves projecting upward; a lawn mower cuts off only the leaf tips without harming either the runners or the leaf basal meristems.

If an area will not be walked on, then a wide variety of low-growing plants can be used, including vines such as ivy (*Hedera*), mints (*Mentha*), and periwinkle (*Vinca*), or low-growing, spreading shrubs (FIGURE 17.13). In very moist areas, mosses are particularly beautiful (FIGURE 17.14).

FIGURE 17.13. This area of the campus of the University of Texas is home to a grove of ancient, historically important oak trees. Having people walk across their root zone would compact the soil and damage the trees, so a small number of walkways has been established, and the open spaces are planted with a vine-like ground cover that is difficult to walk through and not comfortable to sit on. It thus protects the soil around the roots and it also suppresses weeds.

FIGURE 17.14. The shoots of mosses are much too delicate to be walked on, but they provide a beautiful ground cover for areas of a garden away from where people walk or stand. Landscaping mosses that are capable of covering a large area of soil typically need frequent rainfall and high humidity; desert mosses can be used for landscaping too, but typically they form only small, discrete patches. (Courtesy of Annie Martin, www.mountainmoss.com.)

FIGURE 17.15. Turf grasses used for lawns and golf courses require large amounts of water, which causes significant problems in dry areas such as much of the southwestern United States. Lawns typically also require heavy use of pesticides, herbicides, fungicides, and fertilizer, all of which pollute nearby streams. (© Gary Whitton/ShutterStock, Inc.)

Technically, all the plants described above are groundcovers. As an alternative to living plants, groundcovers such as decorative stones and coarse bark mulch are becoming increasingly popular because they require no water and no weekly mowing.

Several concerns have been raised about lawns, especially high-maintenance grass lawns. Most require large amounts of water, and some need to be irrigated even in regions that are not considered dry (FIGURE 17.15). Many cities in the desert southwestern states have now banned lawns because of the amount of water needed to keep them green and healthy. Although some people have lawns consisting of two or three species, most lawns are a dense monoculture of just a single species, so they are ideal habitats for fungal diseases and insect pests, both of which can spread rapidly. Many homeowners and professional lawn care workers have used fungicides and pesticides freely at high concentrations, along with fertilizers. This has caused considerable environmental harm, not only due to the contamination by run off into surface water and seepage downward into groundwater, but also because birds and small mammals suffer from eating poisoned insects. A special concern is that small children and pets play on grassy areas and their bare skin is exposed to these chemicals. Many cities have established a policy of having their public parks be organic (no artificial chemicals).

Bedding Plants

Bedding plants are smaller plants that are planted in masses to provide color, texture, or line to a garden (FIGURE 17.16). These are the plants used in flower beds, and often a gardener will group dozens of the same species together. Petunias, snapdragons, coneflowers, lupines, larkspurs, and daylilies are familiar bedding plants that provide a mass of color when blooming and a mass of texture between a lawn and a background of shrubs or a building. Also, bedding plants can be arranged in lines to guide the viewer's eye to an important garden feature such as a fountain or gate.

Bedding plants are often herbaceous annuals or perennials whose shoots die back to a persistent base such as a bulb (lilies), a corm (*Gladiolus*), or a rootstock (columbine,

FIGURE 17.16. Bedding plants are small annual plants that are cultivated in large numbers close together to provide a line or block of color in a garden. Bedding plants are often considered temporary, and as soon as they are no longer at their peak, they are replaced by new plants, often of a different species that blooms a month or two later than the first set of bedding plants. This can be repeated throughout the growing season. This photo was taken during spring planting season; during autumn the nursery will stock winter-hardy plants.

phlox, asters). Most die back in the winter, but in warm areas many ferns and *Acanthus* die back in summer and revive in autumn then grow best through winter. Such plants remain small and easy to work with, so often a gardener will dig them up in late autumn or early spring, divide them, and then replant them. This gives the opportunity to change at least part of the garden each year. In areas with warm winters, a particular bed might be planted with one species of bedding plants in spring so that they flower through spring and summer, then those will be replaced by a different species that blooms in autumn and winter.

Because an important function of bedding plants is to provide color, it is worthwhile to consider some of the plant breeding and artificial selection they have experienced. For example, plant breeders have mutated plants and then selected for individuals that have an increased number of petals. If at least one extra set is present this changes the number from five to ten in many varieties, and the flower has twice the color and mass as the wildtype flower. Often very high numbers of extra sets of petals are obtained; roses are a familiar example: wild roses have five petals whereas modern hybrids have many times that number (FIGURES 17.17 and 17.18). A mutation that causes the formation of extra copies of a normal organ in a new position is called a **homeotic mutation**; these became famous and the object of intense study when they were discovered in insects, but plant breeders had been working with homeotic mutations for years before that. Other mutations alter petal shape, making them larger or rolled or undulate, as in cabbage roses and varieties of *Chrysanthemum*. Flowers on modern horticultural varieties are often much larger than those of their wild relatives (compare violets and pansies, both are *Viola*; see Figures 9.2 and 9.3), or more numerous, or they last longer. By selecting for mutations that reduce apical dominance, plants that are more highly branched and therefore bushier have been obtained. And many exotic leaf shapes have been produced by artificial selection. This is particularly important in foliage plants, plants cultivated for their dramatic leaves rather than their rather inconspicuous flowers. Examples are *Coleus*, *Begonia*, and many ferns (FIGURE 17.19).

FIGURE 17.17. The flowers of wild roses are similar to those of ancient roses, roses as they were before gardeners began artificially selecting for particular qualities. Wild roses have just a few petals, usually five, and a ring of numerous stamens.

FIGURE 17.18. Numerous varieties of domesticated roses now exist, each variety having been selected for particular traits. These cabbage roses have many more petals than just the five in wild roses, and flowers of cabbage roses typically do not open fully so the stamens remain hidden. (Courtesy of Tommy R. Navarre.)

FIGURE 17.19. This cultivar of *Begonia* has been artificially selected to have dramatically spiral-shaped leaves. Begonias are often cultivated as much for their foliage as for their flowers, if not more. Plant breeders have produced numerous variations in leaf size and shape in begonias.

Trees and Shrubs

Woody perennial plants are classified either as trees if they have a single trunk emerging from the soil, or as shrubs if they branch at soil level or below such that they have several more or less equal stems but no dominant trunk. In general, trees grow to be taller and larger plants than do shrubs, but some, such as ornamental maples, grow slowly and remain smaller than many shrubs of the same age. Being perennial, both trees and shrubs are more permanent elements of a garden or landscape than are bedding plants, and they remain the same year after year, providing continuity to the garden. Being so large, trees occasionally require the care of a professional arborist (FIGURE 17.20).

FIGURE 17.20. An arborist, a person especially trained to care for trees, must occasionally be relied upon to remove dead branches from trees that have been damaged by storms or age.

An important aspect of trees and shrubs is that they persist through the winter and thus provide beauty when other plants have withered or died back to the ground. Evergreens, as their name indicates, do not abscise their leaves in autumn and so provide a large green mass throughout winter. All conifers except three species (bald cypress [*Taxodium distichum*], larch [*Larix*], and dawn redwood [*Metasequoia*]) are evergreen, as are many broadleaf trees and shrubs, with hollies, live oaks, and magnolias being common examples (FIGURE 17.21). Large trees with a spreading crown are important shade trees: in warm regions, they keep a house cool in summer, and if they drop their leaves in autumn they allow sunlight through to warm the house in winter. In cool rainy areas, shade trees are not an advantage; in these areas the slender profiles of conifers is preferred because they do not block the warming light of the sun, which may be needed in both summer and winter.

Both trees and shrubs also offer protection from wind and noise. If planted upwind of a building or garden, they block wind and

FIGURE 17.21. Holly trees, in the genus *Ilex*, are especially prized in the garden in winter. Because they are evergreen, they keep their leaves throughout winter, providing a dramatic mass of dark green color, along with their distinctive red berries.

FIGURE 17.22. Many species of trees provide a mass of color when they bloom, and bring spectacular beauty to a garden. Flowering trees must be chosen with care: some flowering trees such as oaks have small, inconspicuous flowers that shed so much pollen it makes cars and outdoor furniture dirty; other flowering trees produce fruit that may be unwanted and either messy or unsightly. Some of the most popular flowering trees have been artificially selected to be sterile, producing only flowers but no pollen or fruit. (© Steven Russell Smith Photos/ShutterStock, Inc.)

direct it to the sides, which is especially important for protection against cold winter winds and the snow they often carry. In contrast, trees and shrubs can be aligned so as to direct summer winds toward a house to provide fresh air and cooling breezes. Dense plantings of shrubs block street noise and provide privacy without the need for a fence. Spiny shrubs also provide security: no one is going to trespass through a mass of roses or *Pyracantha* ("firethorn").

Last but not least, trees and shrubs are cultivated for their beauty. Cherry trees, magnolias, and dogwoods are spectacular in cooler areas, as are redbuds, mimosas, huisaches (*Acacia*), and mountain laurels (*Sophora*) in warm areas (**FIGURE 17.22**). In autumn, fall foliage and ripe berries add color just as other plants are fading, and throughout winter, deciduous trees reveal their dramatic architecture. Many shrubs, especially willows, display beautiful bark.

From an ecological point of view, as well as the concept of low-impact living, trees and shrubs are excellent. Many people either plant native trees or save the ones that are already present on their property rather than planting exotics. Consequently, most trees need no extra watering, and they are too large to be treated with pesticides or fungicides. Cultivating most trees contributes no pollution to the environment. Their wood is an excellent means of carbon capture and sequestration: the carbon dioxide captured by photosynthesis and made into a tree's wood will remain out of the atmosphere for at least as long as the tree is alive. Shrubs also sequester carbon, but most live only a few dozen years so they may die, decay, and release the carbon dioxide within the lifetime of the homeowner.

Specialized Gardens

Native Plant Gardens

Although the world has over 200,000 species of plants, just a small number dominate gardens, not only in the United States, but worldwide. Petunias alone account for about 40% of all plant sales at garden centers and other reliable favorites are roses, snapdragons, daffodils, and geraniums. Even among trees, gardens are dominated by a few species of maples, oaks, ashes, cherries, and magnolias. One of the reasons these are so widespread is that they are indeed attractive, beautiful plants. But another reason is that these are familiar plants: they were in our parent's gardens, and those of our grandparents also. Many of our favorite ornamental plants were already popular while most of the population of the United States was living on the east coast, or even earlier when many of our ancestors lived in northern Europe. As we spread westward across North America, not only did we take seeds for our standard food crops, but also seeds for our favorite garden plants. At present, many people spend considerable time, money, effort, and water to have gardens suited to our rainy, temperate past even if we now live in the hot, dry western states. Alternatives to this type of heritage gardening is to cultivate **native plants**, species that occur naturally in the areas where we live.

Gardening with native plants has many benefits. First, the plants are adapted to the region where they are being cultivated, so natural rainfall should be sufficient and the garden will need little or no irrigation. Second, a garden of native plants restores some of the habitat that was lost by the development of the city: a **native plant garden** will attract and help sustain birds and butterflies of the area, providing nectar, berries, seeds, nesting sites, and so on (FIGURE 17.23). One aspect of native plants is that many tend to be less compact and orderly than are standard garden plants; they are more open and sprawling, whereas many of our heritage garden plants have been bred to be compact with many leaves or flowers in a small, defined space. A native plant garden tends to have a more natural, unkempt look. This appeals to many people, but not to all. Part of the gardening mentality of many people is to control nature, to mow the lawn, trim the hedges, prune the bushes, and so on. Many of these activities require machines and energy, and they release greenhouse gases: gasoline-powered lawnmowers are a major source of pollution in cities. One of the concepts of low-impact living is to spend less time controlling our gardens and more time enjoying them, actually observing the plants as they grow and develop.

FIGURE 17.23. Pecan trees are native to most of the area drained by the Mississippi River, so they are authentic members of native gardens in that area. They are large trees with open lacy foliage that provides dappled shade. Their flowers are inconspicuous but they provide abundant fruit that is valued by squirrels and other wildlife as well as many homeowners. One problem with pecan trees in gardens is that shaded branches tend to die and fall off, so it is often necessary to pick up dead twigs and small branches.

Xeriscape Gardens

A **xeriscape garden** specializes in plants adapted to desert or semidesert areas and that therefore need little irrigation (FIGURE 17.24). Xeriscape gardens are becoming increasingly popular in the arid regions of the southwestern United State where water is both scarce and expensive. The plants that are used are cacti, of course, and succulents in other families, along with plants of *Agave*, *Yucca*, and desert trees and shrubs such as ocotillo (*Fouquieria*), acacias, mountain mahogany (*Cercocarpus*), and desert willow (*Chilopsis linearis*). Deserts are rich in spring annuals, so xeriscape gardens can have a spectacular variety of annual herbs, most of which will reseed themselves; examples are lupines, asters, globe mallows (*Sphaeralcea*), *Baileya*, larkspurs, *Coreopsis*, and many more

FIGURE 17.24. Plants in xeriscape gardens are adapted to dry conditions; they need little or no extra water, and often grow slowly enough that they need little pruning or thinning. By choosing carefully, the garden can have not only plants with dramatic shapes and textures, but also some that bloom abundantly and supply nectar and fruit to animals.

(FIGURE 17.25). Many bulbs are native to desert areas, so those can be used as well. Groundcovers in xeriscape gardens are usually pebbles, gravel, or decomposed granite, not grass. Xeriscape gardens tend to have the lowest impact on the environment as the plants typically need no water or fungicides, and rarely any pesticides. They are often planted in a natural, nonformal, nonlinear pattern, so weeding is done by hand rather than with chemicals.

Many of the plants in a xeriscape garden have interesting shapes even when not in flower, so the garden is attractive all year. Xeriscape plants generally tolerate only a light frost, and must be rather dry in the winter or whenever it is cold. If one attempts to plant a xeriscape garden in a rainy area, it is best to use raised beds with extremely good drainage. Xeriscape gardens can even be planted in areas that have cold winters, but then it is necessary to avoid the more succulent (frost-sensitive) plants and focus on species native to cold deserts (FIGURE 17.26).

Ponds, Streams, and Marsh Gardens

Wet areas provide wonderful opportunities for creating a garden that is out of the ordinary (FIGURE 17.27). Sites range from a small pond only a few feet in diameter to a lake, and from quiet water to a flowing stream. Moving water, whether as a pump-driven fountain or as a stream falling over rocks, provides a natural background sound to a garden. Plants used in water gardens are usually herbs rather than trees, but there is a great variety of attractive plants grouped into three basic categories: submerged plants, marginal plants, and floating plants.

Submerged plants are those that live completely underwater or that have their leaves floating on the surface. Waterlilies (both *Nymphaea* and *Nuphar*) are favorites because of their round, floating leaves and beautiful flowers with many petals. Leaves of *Nelumbo* project just above the water's surface, and their fruits are interesting pods with individual chambers for each seed (see Figure 7.26). Plants of water milfoil (*Myriophyllum*) are less well known but spectacular for being entirely submerged and having highly dissected, feathery leaves (FIGURE 17.28).

FIGURE 17.25. *Coreopsis* is one of many annual wildflowers that can be grown in xeriscape gardens. They can be sown in dense masses to provide dramatic color, or they can be sown sparsely among other, larger plants.

FIGURE 17.26. Xeriscape gardens are usually planted with desert-adapted plants such as cacti, agaves, yuccas, and grasses. Deserts are classified as being either hot (such as those in Arizona, California, and northern Mexico) or cold (such as those high in the Andes of South America). Many plants from cold deserts, such as this *Trichocereus pasacana* from northern Argentina and Bolivia, survive cold winters, and thus xeriscape gardens can be planted in northern states where snow and freezing nights are common. Most cacti from hot deserts would be killed by having snow and frozen rain on them.

FIGURE 17.27. Water gardens allow a gardener to explore many types of plants and gardening techniques that cannot be used in gardens based on soil. Water lilies and other plants may be rooted directly in the bottom of the pond, or very often they actually are in individual pots setting on the floor of the pond, which allows them to be rearranged easily so that the texture of the water garden can be changed without much trouble.

FIGURE 17.28. *Myriophyllum* is an unusual aquatic plant that will add a special touch to your water garden. It grows completely immersed in the water; its stems and feathery leaves are too weak to support themselves in air. If the garden has flowing water, shoots of *Myriophyllum* undulate gracefully in the current.

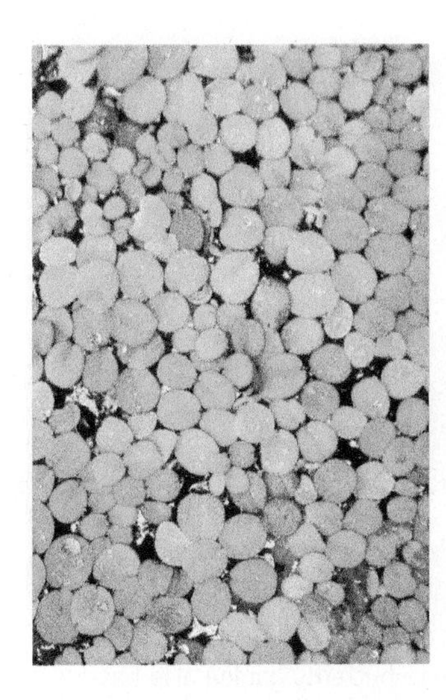

FIGURE 17.29. Plants of *Salvinia* resemble tiny leaves of water lilies with stout hairs just barely visible on their upper surface. Salvinias are aquatic ferns that float on the surface of quiet water, not rooted to the soil. They are such a novelty that many people are tempted to scoop some up into a plastic bag and take them home, but that is usually a mistake. Plants of *Salvinia*, as well as those of *Azolla*, *Pistia*, and *Eichhornia*, grow so prolifically that they can quickly cover the entire surface of your water garden, crowding out other plants, and leaving not even a small patch of open water. They quickly become weeds, and the only way to salvage your garden is to pick out every single plant: if even one remains, it will reproduce vegetatively and your problem will return.

Marginal plants are those that in nature grow along the edge of a pond or stream, with their roots always wet but with much or all of their shoots above water. In a garden pond, marginal plants are often planted in containers located just below the water's surface. Examples are *Iris*, rushes (*Scirpus*), cattails (*Typha*), elephant ears (*Colocasia*), arrowheads (*Sagittaria*), and pickerelweed (*Pontederia*).

Floating plants, as their name indicates, float on the surface of water with their roots dangling freely and not attached to anything. Most are tiny plants but form small masses that lend an interesting aspect to the pond or stream. Examples of floating plants are water hyacinth (*Eichhornia crassipes*), sea lettuce (*Pistia stratiotes*), water spangle (*Salvinia*; FIGURE 17.29), mosquito fern (*Azolla* species), and water clover (*Marsilea vestita*); the last three are ferns. Plants of water hyacinth have beautiful flowers, are extremely attractive, and grow easily.

Unfortunately, all of these floating plants grow too easily and reproduce too well: they are considered dangerous weeds that foul waterways. In many areas it is illegal to own these plants, and many states have programs to control or eradicate them. If they escape cultivation, or if they are carried from one body of water to another on a boat's hull, they proliferate rapidly and can cover the entire surface of a pond or slow-moving stream. Most will not survive cold winters that have many nights of freezing temperatures, but in the southern states with mild winters they survive year round. In Florida especially, water hyacinth and sea lettuce become so thick that boats cannot use waterways, and the layer of vegetation of the water's surface prevents light and oxygen from entering the water for fish, amphibians, and other aquatic animals.

A special benefit of water gardens is the rich variety of animal life that visits them. Frogs and other amphibians take up residence on their own, as will turtles. Birds come to drink and bathe, and if the water garden has fish, predators like herons and raccoons will be regular visitors. Unfortunately, mosquitoes will enjoy the moisture as well.

Butterfly Gardens

Some gardens are designed to attract and feed butterflies. Butterflies themselves feed only on flower nectar (or overripe fruit), and plants such as mist flower (*Conoclinium greggii*), purple cone flower (*Echinacea purpurea*), marigolds (*Tagetes*), salvias (*Salvia* species), and daisy fleabane (*Erigeron* species) are good choices (FIGURE 17.30).

FIGURE 17.30. The nectar of these flowers attracts butterflies, and by cultivating appropriate plants in your garden, you encourage butterflies to visit and remain in your garden. Check with local nurseries to see which species are best for your area. This specimen is a great purple hairstreak. (Courtesy of John C. Abbott.)

FIGURE 17.31. A complete butterfly garden also has plants that provide sites where butterflies can lay their eggs. The eggs hatch and the caterpillars feed on the leaves of the plant; many butterflies are coadapted to particular plants and will not lay their eggs on the wrong plant, so it is important to choose the correct plants for the butterflies of your area. This specimen is the caterpillar of the question mark butterfly. (Courtesy of John C. Abbott.)

During their caterpillar stage, however, butterflies feed only on leaves of completely different species, such as milkweeds (*Asclepias*), passionflowers (*Passiflora*), and many species related to cabbage and broccoli (*Brassica* and other members of the mustard family Brassicaceae), or dill and fennel (the family Apiaceae) (**FIGURE 17.31**). When planning a butterfly garden, it is best to think in terms of plants that are hosts for the eggs and caterpillars as well as plants that provide nectar for the adults. Butterflies also need open, moist ground where they can collect moisture.

House Plants

Houseplants are cultivated indoors, which is an extremely difficult environment. Even with large windows, rooms inside buildings tend to have very low light; they appear bright to us only because our eyes dilate to accommodate the low light, whereas when we are outdoors, our pupils are narrowed. Plants best suited to being houseplants are those adapted to shady forests as their natural habitat (**FIGURE 17.32**). The brightest area in most buildings is a south facing window, and if a plant is placed next to it, the plant usually quickly grows toward the window (positive phototropism) and must be turned every few days to prevent lopsided growth. Because we keep our rooms lighted until late into the night while we are awake, houseplants usually experience long days and short nights throughout the year, without ever experiencing the short day/long night cycles typical of winter. This interferes with many of their processes—especially flowering—that are controlled by day length.

Whenever a building is heated or cooled, these processes lower the relative humidity in the air, thus causing houseplants to transpire much more rapidly than they would if outdoors. Setting the plant's pot on a dish filled with pebbles and water can increase humidity near a plant, and some people have found it best to put their orchids in the bathroom where humidity from the shower helps them. The stress caused by dry air is compounded by the fact that houseplants are grown in containers that restrict their root systems. The soil around a potted plant's roots alternates between being completely saturated (when we water the plant) to dry (just before we water it the next time), and this cycle is much faster in pots than it is in natural soil outdoors.

Despite all these difficulties, many plants thrive as houseplants. African violets (*Saintpaulia*) grow and bloom profusely indoors, as do *Begonia*, *Coleus*, jade trees (*Crassula ovata*), and wax plants (*Hoya carnosa*). Other plants grow well but do not flower when cultivated indoors: corn plants (*Dracaena fragrans*), dumb cane (*Dieffenbachia*), *Philodendron*, rubber plant (*Ficus elastica*), and *Peperomia* are reliable. If kept in very bright windows, many cacti and succulents do well indoors, and species of *Mammillaria* and *Parodia* will even bloom.

Some houseplants are used for special occasions and typically are kept only temporarily. Poinsettias (*Euphorbia pulcherrima*), *Amaryllis*, and Christmas cacti (*Zygocactus*) are cultivated by nurseries under day/night conditions that induce them to bloom at Christmas. Red tulips (species of *Tulipa*) are likewise cultivated for Valentine's Day, and Easter lilies (*Lilium longiflorum*) for Easter (FIGURE 17.33).

FIGURE 17.32. Any number of plants qualify as house plants, and many are cultivated on window sills where they get the brightest light in the house. If kept on a kitchen or bathroom window, they will have higher humidity than they would receive in other rooms of the house.

FIGURE 17.33. Easter lilies are induced to bloom by photoperiod (the length of the day and night), and nurseries manipulate their lighting such that millions of Easter lilies bloom just in time for Easter, which occurs on slightly different dates each year. Many of us just throw the plants away when they stop blooming after Easter, but they are extremely hardy plants and can be planted outdoors in many part of the United States. They will thrive and bloom year after year with little care. (Courtesy of Chris McCoy.)

These plants are purchased, enjoyed, and then usually discarded. Even if we try to maintain them throughout the year, many of these just do not thrive once removed from the nursery.

Important Terms

bedding plant	growth chamber	raised bed
cold frame	homeotic mutation	rootstock
cutting	native plant	scion
damping off	native plant garden	shadehouse
grafting		xeriscape garden

Concepts

- Beyond being necessary for our fundamental metabolic needs (food, medicine) and survival (fuels, fibers, chemicals), plants are important for the intellectual and aesthetic aspects of our lives.
- Hobby plants are an excellent means of learning about plant diversity.
- Overcollecting specimen plants from natural habitats has endangered several species of ornamental plants; for many species, collecting wild plants or transporting them internationally is illegal.
- Specialized propagation methods allow plants to be cultivated with controlled amounts of light, temperature, and soil moisture.
- Groundcovers—such as lawns—control dust, mud, erosion, and weeds, and provide a beautiful area as a foreground or as a place to walk, sit, and play.
- Ornamental plants provide color, texture, line, and form to the gardens and landscapes that surround them.
- Cultivating long-lived ornamental trees is an excellent method by which individual gardeners can capture carbon dioxide and sequester it for many years.
- Many ornamental plants have been subjected to considerable artificial mutation, selection, and cross-breeding to obtain cultivated plants that now differ markedly from their wild ancestors.
- Fertilizers, pesticides, herbicides, and fungicides applied to lawns and gardens are often a significant source of pollution and of damage to nonpest birds and other animals.
- Native plant gardens, xeriscape gardens, water gardens, and butterfly gardens provide extra habitat, food, and shelter to local animals.

Algae and Fungi: Close (and Not-So-Close) Relatives of Plants

Introduction

Fungi and algae, the organisms in this chapter, are not considered to be plants by most biologists (FIGURES 18.1 and 18.2). Somewhat surprisingly, at least for biologists, is the conclusion based on DNA sequence data and cladistics that fungi are much closer to animals than either is to any other group (see Figure 9.14). That means that some population of early eukaryotes that had nuclei and unicellular bodies diversified into two lines of evolutions, one of which later became fungi, the other later became animals. That may seem strange if we compare ourselves to mushrooms, but we are not typical animals and mushrooms are not typical fungi. Most animals do not have as many derived, specialized features as we humans have; indeed, there are many animals with extremely simple bodies and metabolisms. Likewise, many fungi are much simpler than mushrooms, and if we compare the most primitive animals with the most primitive fungi, then the two groups do not look so different. Both clades have undergone tremendous evolutionary diversification and change since the time they each became distinct from their common ancestor and each other.

Fungi, such as this mushroom, are not plants but they are important to all organisms as agents of recycling. Fungi secrete enzymes that break down organic matter, such as the dead pine needles here, thus releasing the minerals and other nutrients and converting them to forms that plants can absorb and use. Within a few months, these pine needles will not exist but molecules from them will be enriching the soil, all due to fungi.

FIGURE 18.1. Mushrooms are familiar examples of fungi, but Kingdom Fungi is large with many species that differ greatly from mushrooms. The mushroom's cap produces tens of thousands of spores from gills on its underside, and most of the spores blow away and germinate into long, slender fungal cells called hyphae. No fungus ever carries out photosynthesis and none ever has chlorophyll, although some have a slight greenish color.

FIGURE 18.2. Two very different algae are growing together here. The green one is a green alga (Chlorophyta) and the brownish one with yellow "mid-veins" is a common and easily recognizable brown alga, *Fucus* (in the Phaeophyta). Although both are algae, many features of the green alga's cell biology (pigments, cell wall structure, the presence of starch) have more in common with true plants than they do with those of brown algae.

Algae are a bit more complex. Some, the green algae, are very similar to plants in their metabolism and cell structure. But other algae have less in common with plants and indeed, algae are an extremely diverse group in many fundamental features.

Lichens resemble plants but each is an association of fungal cells with algal cells (**FIGURE 18.3**). If you hold a lichen in your hand, you do not have a single organism

FIGURE 18.3. This is a lichen *Usnea cirrosa*, and it is not a plant. The body is produced only when a particular fungus grows together with a certain alga; when the two are growing together and influencing each other, the fungal cells form a structure like this and algal cells grow intermingled with the fungal cells inside the body. Lichens never produce flowers or cones or seeds, instead they reproduce when small clusters containing both fungal and algal cells break off and are dispersed. If a cluster contains only fungal cells or only algal cells, it will not grow into a lichen.

as if you were holding an animal, a plant, or a mushroom; instead you have hundreds of individual fungal cells and algal cells living together and behaving as if they were the cells and tissues of a single individual.

Although fungi, algae, and lichens are not plants, they are included in this plants and people book for several reasons. One is tradition: fungi and algae are almost always included in botany books and curricula. But the main reason is that they are interesting and we can learn something about both plants and people by studying relationships with algae and fungi. Hopefully, you have a deeper understanding of your own biology after studying the earlier chapters of this book. Now by examining the three new groups, we can explore more aspects of ourselves, of plants, and of biology in general.

Fungi, the Kingdom Fungi

The Major Types of Fungi

Fungi differ from plants in many ways. None is ever photosynthetic, none ever has plastids. Consequently all fungi, like all animals, are heterotrophic, they must obtain their energy and the carbon for their organic compounds by absorbing them from the environment, then respiring them. Unlike animals, fungi digest their food outside their body (they have extracellular digestion) by secreting enzymes onto wood, leaves, fruit, bread, and so on where the enzymes break nutrients down into their monomers, which are then absorbed into the fungus.

Fungal cells are long, slender tubes called **hyphae**, and each hypha has hundreds of thousands of nuclei. A mass of branched hyphae is a **mycelium**: the spots of mold we see on bread and cheese are mycelia (FIGURE 18.4, and see Figure 18.23). All fungi, like all plants and animals, reproduce and most are capable of both sexual and asexual reproduction. Fungi do not produce sperm and egg cells, but instead, when hyphae of two compatible mycelia grow together, they fuse and exchange nuclei (FIGURE 18.5). Hyphae that grow from this point of fusion have two types of nuclei; at some later time, each nucleus fuses with another nucleus of the other mating type, then the resulting diploid nuclei undergo meiosis with pairing of homologous chromosomes and crossing-over. Afterward, each new haploid nucleus forms a spore around itself, then blows away and germinates into a new hypha. Although fungi have no sex organs and never produce seeds, their life cycle includes the bringing together of two haploid genomes from two parents, then combing them into a diploid nucleus where the genomes are shuffled, creating greater genetic diversity.

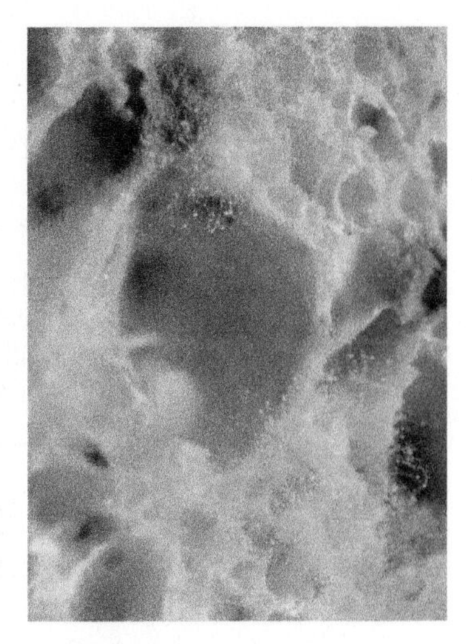

FIGURE 18.4. It is easy to see the structure of typical fungi by looking closely at bread mold. The cells are long and slender, and almost visible to the naked eye. Here, some are producing the dark blue-gray spores typical of bread mold, others are producing yellow spores. In this image, we can see only the fungal cells on the surface of the bread; all visible fungi have many more cells that penetrate into the substrate and that typically go unnoticed.

The three main groups of fungi differ from each other in many features, but the easiest way for us to recognize them is by their fruiting bodies, structures like mushrooms and morels (**TABLE 18.1**). Fruiting bodies are composed of hyphae that grow in an organized, interwoven network such that a body with predictable size, shape, and color is formed, as is true of mushrooms. The ordinary feeding mycelium grows more or less at random, gathering nutrients in the substrate; the size and shape of the feeding mycelium are determined by the richness of the substrate.

Basidiomycetes are the fungi that produce mushrooms, puffballs, brackets, and the pustules on cereal grains (FIGURE 18.6). The underside of a mushroom has many gills, each of which is composed of hyphae that adhere to each other. Along the surface of each gill, the tips of hyphae develop into club-like cells called **basidia**, each of which has four projections (FIGURE 18.7). The nuclei in each basidium fuse to become diploid then undergo meiosis; each of the four resulting nuclei migrates through one

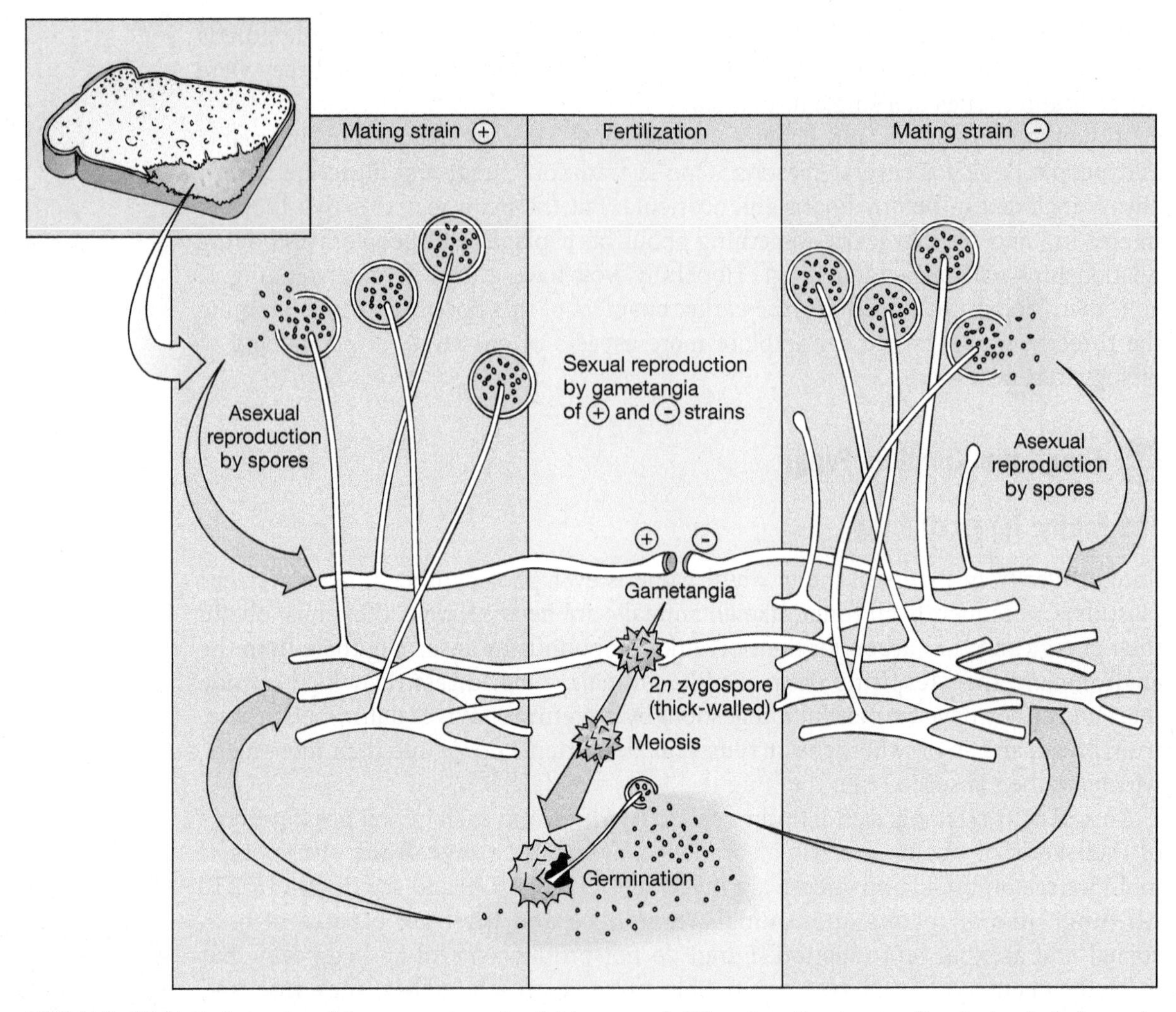

FIGURE 18.5. Each species of fungus produces haploid spores of different mating types, often just called plus and minus. Spores germinate and grow into long, slender, branched hyphae; the whole mass of hyphae is a mycelium (left and right panels). When hyphae of different mating types touch each other, they fuse and some nuclei are exchanged (middle panel). In Zygomycetes, a large spore is formed right at the point of fusion, and it uses its nutrient reserves to grow into a stalked structure that undergoes meiosis and then releases hundreds of haploid spores.

TABLE 18.1. The Main Groups of Fungi

Zygomycetes	Live in decaying plant and animal matter. Their body is a simple mycelium of branched, coenocytic hyphae; they never produce a complex fruiting body. Bread mold, *Rhizopus stolonifer*, is a familiar example.
Ascomycetes	Live in various substrates, including decaying plant matter. Many, such as morels (*Morchella*) and truffles (*Tuber*), have fruiting bodies. Yeasts and many others do not. Ascomycetes produce a saclike ascus in which meiosis occurs, resulting in four or eight haploid ascospores inside the ascus.
Basidiomycetes	Live in various substrates. Many have fruiting bodies such as mushrooms (many genera), puffballs (many genera), and bracket fungi (many species). Basidiomycetes produce a small club-shaped cell called a basidium with four small projections; each basidium is the site of meiosis and the production of haploid basidiospores.

of the projections, taking cytoplasm with it, and then each forms a haploid basidiospore. The spores abscise from the projections, drop downward between the gills, and fall out of the mushroom where they might be caught by wind and dispersed across a wide area. Bracket fungi are similar except their underside consists of hollow tubes lined with basidia.

Truffles and morels are the fruiting bodies of **ascomycetes** (FIGURE 18.8). This group of fungi is characterized by an **ascus**, the terminal cell of a hypha where meiosis occurs, producing four haploid spores all surrounded by the ascus cell wall (FIGURE 18.9). In many ascomycetes, a round of mitosis after meiosis results in eight ascospores rather than four in each ascus. Yeasts, such as those used in brewing (see Chapter 13), are unusual ascomycetes that almost always are unicellular and never form either a mycelium or a fruiting body.

Zygomycetes are the bread molds and their relatives (see Figure 18.5). They do not produce fruiting bodies; instead, they consist of a spreading mycelium that produces spores on almost any part of their body, at any time. Because so much of the mycelium is on the surface of the substrate, the spores can be carried away by wind or splashed away by raindrops. Because they produce no **fruiting body**, zygomycetes are more difficult to detect in the environment, but they are extremely common and widespread. Almost any sample of soil, mud, or decaying plant matter you might notice will contain hundreds of them.

A group called the oomycetes had been traditionally believed to be a group of fungi, but new evidence indicates that they are more closely related to brown algae, and will be treated with the rest of the algae later in this chapter.

FIGURE 18.6. These brackets (also called bear cookies) are the fruiting bodies of basidiomycete fungi, and are produced by an extensive feeding mycelium growing inside the tissues of the tree. Unlike mushrooms, which have gills, the undersides of brackets are made up of thousands of tubes: spores are produced by the sides of the tube then fall out and are blown away. It is necessary for the tubes to be vertical, and just like plants, bracket fungi sense gravity and respond to it.

Fungi, Breads, and Cheeses

Mix some water with wheat flour, spread it into a thin layer so that it dries quickly, and the result is—paste. Nutritious but not tasty. Start again but reduce the amount of water and knead the dough slightly, then spread it into a thick layer and cook it. Baked with a high heat and quick baking time, you will get a cracker; cooked on a griddle and depending on other ingredients, it will be a flatbread such as a tortilla, matzo, or lefse.

Things start to really change if we add a leavening agent, something to make the bread rise. Baking soda (sodium bicarbonate, $NaHCO_3$) reacts with acids in foods causing it to release bubbles of carbon dioxide. For baking soda to be effective the dough must contain an acidic ingredient such as buttermilk or vinegar. A similar leavening agent, baking powder, is a mix of both baking soda and an acid that is inactive as long as it is dry. Once moisture is added to flour and baking powder, the acid is activated and the baking soda starts to release carbon dioxide. Baking powder can be used to leaven foods that have little natural acidity, such as wheat flour.

Yeast too is a leavening agent that produces carbon dioxide. Surprisingly, the yeast used to produce carbon dioxide in bread is *Saccharomyces cerevisiae*, the same yeast used to produce ethanol when brewing beer and wine. As water is added to a mix of yeast and wheat flour, enzymes begin to depolymerize the starch into sugar, which the yeast then respire. Because bread dough is mixed and then kneaded, the yeast are well aerated and carry out aerobic respiration, no fermentation occurs as it does in brewing. Simultaneously, water hydrates a wheat protein called gluten (one of the

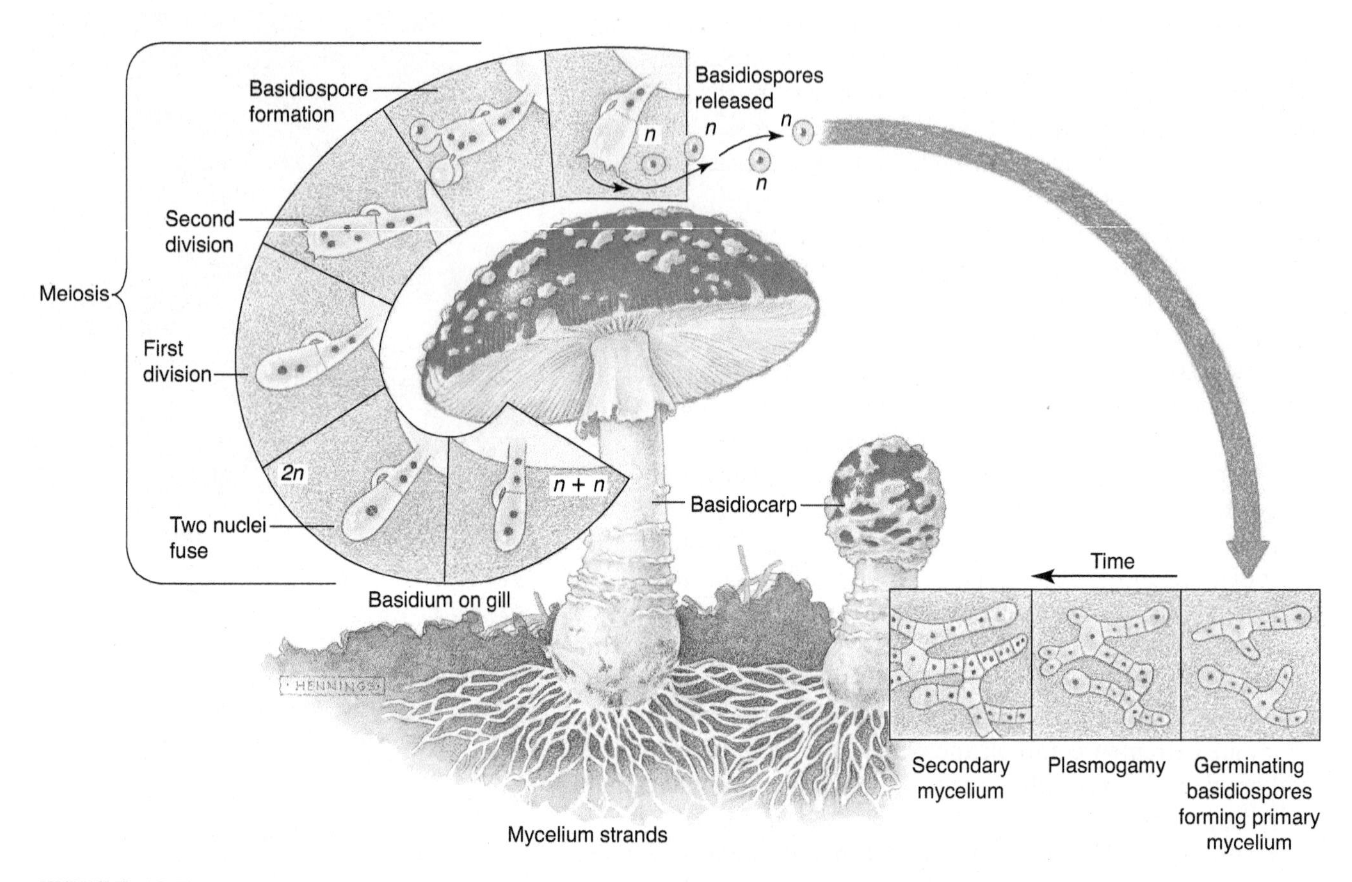

FIGURE 18.7. All mushrooms and other basidiomycete fruiting bodies are composed of binucleate hyphae, each cell in each hypha having one plus and one minus nucleus (curved panel to left of mushroom). In the terminal cells (called basidia) that line the surface of the cap's gills (or the tubes of a bracket), the nuclei fuse (so that one single cell becomes diploid), and it immediately undergoes meiosis. Each of the four resulting haploid nuclei, accompanied by some cytoplasm, migrates through a spike-like projection and forms a basidiospore. Two of the spores are plus, two are minus, and each blows away then germinates and grows into a new mycelium (set of three panels on lower right).

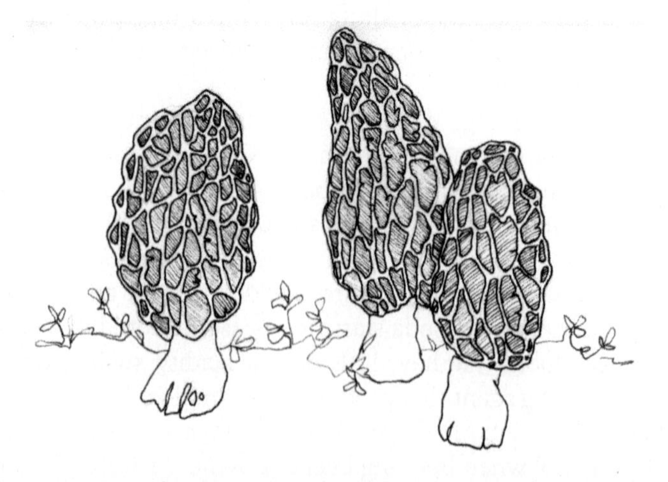

FIGURE 18.8. Morels are the fruiting bodies of certain species of ascomycete fungi. The upper part of each morel is composed of numerous deeply recessed indentations covered with hyphae tips that produce spores; thousands of spores are released from each indentation. The sporogenous upper part is elevated by a stalk, increasing the chances that the spores blow away rather than falling directly to the soil below.

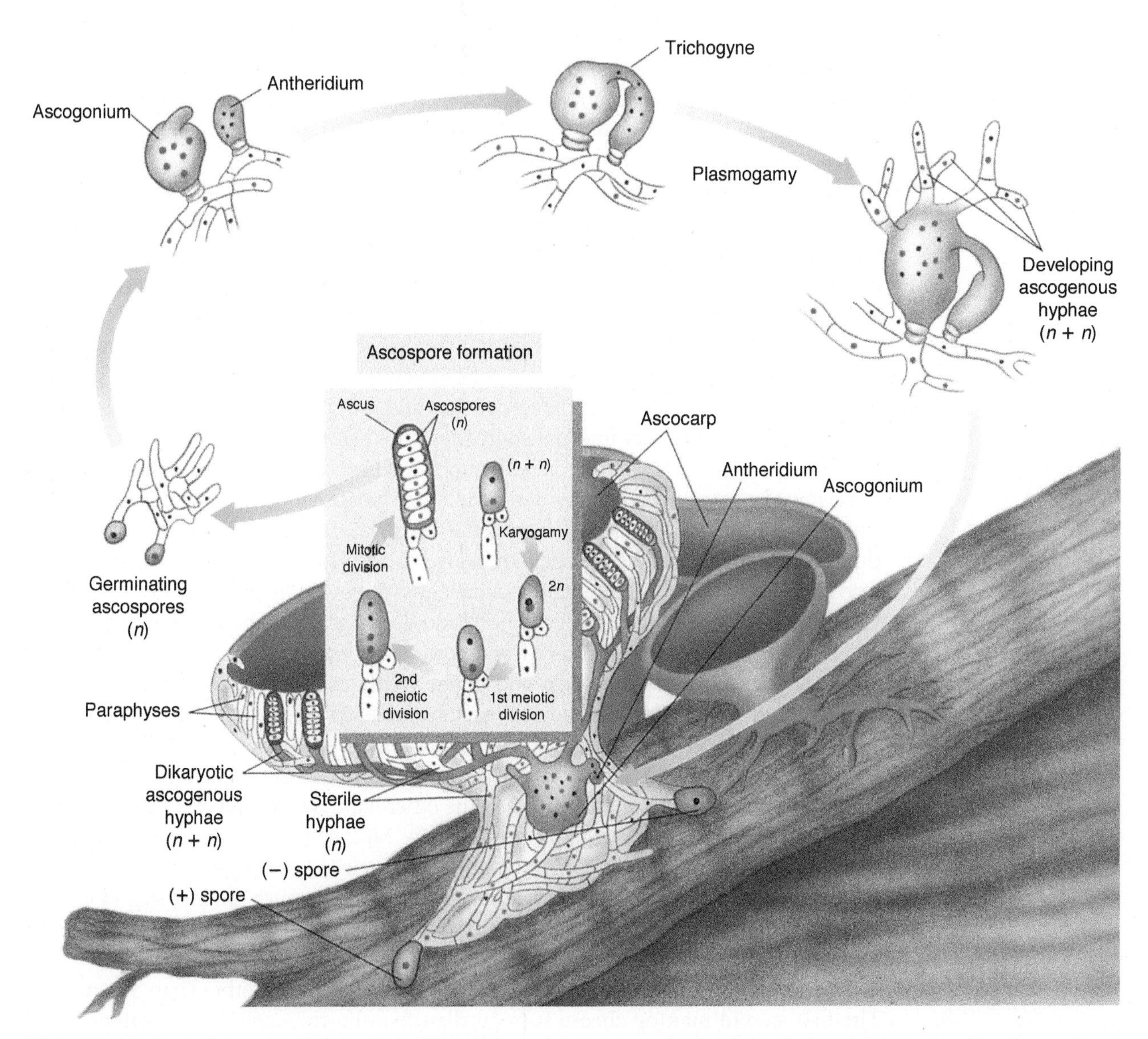

FIGURE 18.9. Similar to the situation in basidiomycetes, when two ascomycete plus and minus hyphae meet, they fuse and share nuclei (top of diagram). New hyphae, called ascogenous hyphae and which are binucleate, grow out and form the fruiting body. In the terminal cells (called asci) on the surface of the fruiting body, nuclei fuse to the diploid condition, then undergo meiosis and produce four or eight ascospores inside each ascus. The ascus cell then breaks down, releasing the ascospores, which blow away and germinate into new haploid hyphae.

longest proteins known), causing its molecules to extend as long, elastic filaments. As bread dough is kneaded, gluten molecules are pulled, extended, and become reoriented into a lattice that helps trap the bubbles of carbon dioxide produced by the yeast (see Figure 13.5). Without the gluten network, bread dough would not hold the carbon dioxide, would not rise, and would not be springy and elastic.

Both sodium bicarbonate and yeast are able to leaven bread by producing carbon dioxide. What is the difference between the two? Sodium bicarbonate only provides carbon dioxide, nothing more. Yeast, being a living organism, produces many more compounds, many of which contribute to the flavor and nutrition of bread. During the time bread rises, yeast cells grow, reproduce, and multiply so that when the dough goes into the oven, it has much more yeast than when the baker started. As yeast cells grow, they produce esters that give the bread a fruity flavor, and sulfur compounds with a flavor of egg. Yeasts also synthesize vitamins, enriching the bread's nutritional value. In general, bread dough that starts with only a small amount of yeast and thus must rise for a longer time will have a richer flavor.

The yeast so necessary for making our traditional breads is obtained from several sources. If a bit of flour is mixed with water and allowed to stand, the yeasts naturally present on its surface will thrive and cause the mix to bubble and be frothy in a few days. This can be used as the starter for sourdough bread. The risk and the benefit of this method are that many microbes are present on flour and many will grow, so the sourdough starter may be far from a pure culture of *Saccharomyces cerevisiae*. Sometimes this is beneficial because the starter will be especially flavorful and aromatic in a good way, other times it just stinks and has to be thrown away. Once a good starter is obtained, some of it is saved in the refrigerator and is maintained by periodically adding more flour and water. Commercial sources of pure *S. cerevisiae* cultivate it on molasses in large tanks. Some is harvested and compressed into cakes of fresh, active yeast cells; these must be kept refrigerated until use, and they have a short shelf life. Active dry yeast consists of granules of many dormant yeast cells. They must be reactivated by mixing them with warm water before they are added to the flour. Instant dry yeast has been dried very rapidly and aggregated into smaller particles that hydrate quickly. Instant dry yeast can be mixed directly with dry flour and does not need a period of reactivation.

The use of yeast in bread adds a few vitamins, but otherwise its real role is to make a food that is soft and easy to eat and digest. Without leavening of some sort, a mix of wheat and water bakes into a hard biscuit or cracker; these are long-lasting and are still a main food on long expeditions where food must be lightweight and durable. But for more pleasant eating on a daily basis, bread is what most people chose.

In the conversion of milk to cheese on the other hand, yeasts and other microbes play an absolutely essential role, that of preservation. Milk sours and becomes undrinkable within just a few days if not kept very cool, so in prerefrigeration days, converting milk to cheese allowed it to be stored for months or years. Also, because the first step of making cheese is to remove much of the water from milk, cheese is more lightweight and easily transported than raw milk.

The conversion of milk to cheese involves microorganisms other than *S. cerevisiae*. The first step in making cheese is to curdle the milk, thus collecting together milk's proteins and lipids as curd while leaving behind the water as whey. Curdling is done with rennet and various types of acid-producing bacteria. Rennet is a set of enzymes obtained from calf's stomach (calves must be killed to obtain the rennet); an alternative that does not involve sacrificing animals can be made from the fungus *Mucor miehei* or the thistle *Cynara*. The Indian cheese paneer is made by curdling milk with nothing other than acid; it forms a very soft curd and must be used immediately, but it qualifies as vegetarian cheese. Recently, bacteria have been genetically modified to produce rennet; by using rennet extracted from these bacterial cultures, cheese can be made without any need to kill calves. Many of us avoid the use of genetically modified organisms (GMOs), but in this case we can save the lives of thousands of calves by accepting the use of these bacteria.

For fermented milk products such as yogurt, sour cream, buttermilk, and crème fraiche, little further processing is needed other than curdling, and fungi are not added for flavor. The same is true for certain cheeses such as Swiss, Münster, and Limburger. Their texture and flavor develop due to the bacteria present as they age. But for other cheeses, fungi are added to give special flavor, aroma, and color. *Penicillium roqueforti* is, as you might guess, the fungus used for Roquefort cheese (and also is related to the fungus that gives us penicillin). This fungus, like bread yeast, carries out aerobic respiration, not fermentation; even though cheese is dense and compact, its interior still has at least 5% oxygen, enough for *Penicillium roqueforti* to grow aerobically (**FIGURE 18.10**). A related species, *Penicillium glaucum* is used to make Stilton and Gorgonzola cheeses. Both species break down part of the milk fats, converting them to short chain fatty acids, alcohols, and chemicals called methyl ketones, all of which contribute to the flavor.

Another *Penicillium*, *P. camemberti*, is used in making Camembert, Brie, and Neufchâtel; it does not produce blue pigment. Its enzymes break down milk proteins, giving the cheese its softer, creamier texture and producing aromas of ammonia, mushrooms, and garlic.

Tempeh is produced by fermenting soybeans. Soybeans are washed, cooked briefly, and then inoculated with a fermentation starter containing spores of the fungus *Rhizopus oligosporus*. They ferment for 1 to 2 days, allowing fungal enzymes to break down the soybean oligosaccharides that cause gas and indigestion in other soybean products. After fermenting, the tempeh is cut into cubes or chunks and can be fried or mixed in with other foods to replace meat. Because tempeh is made with whole soybeans, it is especially nutritious.

FIGURE 18.10. The conversion of milk to cheese occurs due to the action of bacteria, but this Roquefort cheese was also inoculated with the fungus *Penicillium roqueforti*. The fungal mycelium grows throughout the cheese, but in certain areas it produces the dark blue spores that give the cheese its distinctive color and aroma.

Mushrooms, Truffles, and Others

Fungal spores grow into extensive mycelia, but for the most part, we eat only fruiting bodies such as mushrooms (**TABLE 18.2**). Mushrooms are initiated by a basidiomycete mycelium that grows either underground or inside dead trees (see Figure 18.7). The mushrooms do most of their development while hidden. After nuclei have fused then undergone meiosis and formed spores, the hyphae absorb water, push up above ground, and then open their cap such that spores can fall from the gills. Mushrooms shed their spores quickly then the mushroom breaks down within a few days. Some actually digest themselves and turn into a liquid pulp (they deliquesce). Because they are ephemeral structures, mushrooms store few nutrients other than what is in the spores themselves; consequently, young mushrooms, still in the button stage or just after they have opened, are more nutritious than older mushrooms. However, their best flavor develops just after they open. Mushrooms are a low calorie food because we cannot digest many of their polymers, and they are rich in minerals and vitamin D (mushrooms that have been exposed to ultraviolet light are apparently the only natural vegan source of vitamin D).

Although there are about 20,000 fungi that produce fruiting bodies, only a few dozen can be cultivated (**FIGURE 18.11**). They are usually grown on composted straw, wood shavings, or a variety of other plant debris, often with composted horse manure added as

TABLE 18.2. Mushrooms and Other Fungi We Eat*

Button mushroom (also known as champignon)	*Agaricus bisporus*
Crimini mushroom	*Agaricus bisporus*
Oyster mushroom	*Pleurotus* species
Paddy straw mushroom	*Volvariella volvacea*
Portobello	*Agaricus bisporus*
Shiitake mushroom	*Lentinus edodes*
Tempeh starter	*Rhizopus oligosporus*
Tree ear fungus	*Auricularia polytricha*
Truffle	*Tuber*

*The term "toadstool" is often used for poisonous mushrooms or for those that have the characteristic cap-and-stem form. "Toadstool" is not a botanical term; it has no precise definition.

(a) (b)

FIGURE 18.11. Common commercial mushrooms, *Agaricus bisporus*, are grown on mushroom farms that consist of a set of unlighted trays that contain soil rich in organic matter and with high humidity. Harvesting the mushrooms does not harm the mycelium, which continues to grow and produce more mushrooms. These mushrooms are being cultivated at Kitchen Pride Mushroom Farms, Inc. All parts of this facility are kept scrupulously clean to prevent the introduction of other fungi that would act like weeds, outcompeting the desired species.

a source of minerals and nitrogen. Wood mushrooms are cultivated on short pieces of branches of the appropriate tree. The humidity in the cultivating rooms must be high, and usually the rooms are kept dark: fungi do not need light for photosynthesis, so darkness saves cost and it also prevents algae and ferns from growing as weeds in the cultivating beds. Mushrooms must be shipped, sold, and consumed quickly, otherwise they continue converting nutrients into indigestible cell wall materials and their nutritional value decreases.

The ascomycetes produce two of the most highly desired fruiting bodies, morels (*Morchella*; Figure 18.8) and truffles (*Tuber*). Morels resemble mushrooms in having a stalk and a cap, but the cap is spongelike or honeycombed, covered with cavities where the spores are produced. Like mushrooms, the fruiting body is supported by an extensive feeding mycelium that digests decaying plant matter. Morels do not have a symbiotic relationship with plants, but certain species tend to be found near certain trees; for example, black morels (*Morchella elata*) occur in oak and poplar forests. Like morels, truffles are another cultivated ascomycete that is highly valued. Their mycelium must form an association with living tree roots, so truffles are cultivated by tending the soil around trees. Acorns picked from forests known to have truffles can be planted in a new area with the proper soil, and as the oak trees grow, truffle mycelium develops and after several years fruiting bodies can be harvested. When searching for wild truffles, it is best to look near trees of beech, poplar, oak, hazel, and pine. Truffles never emerge above ground, so animals with acute sense of smell, especially pigs and dogs, are trained to find them. Truffles have a wonderful delicate, earthy flavor, and because they are rare and demand is high, truffles are always expensive. A product called "truffle oil" is sold as a less expensive means of obtaining truffle flavor, or of even enhancing the flavor of natural truffles, but truffle oil is just olive oil with a synthetic aroma added.

While gathering mushrooms, morels, and other fruiting bodies, the underlying feeding mycelium is left unharvested. It consists of extremely slender, delicate hyphae ramifying through decaying matter, so it cannot be gathered or cleaned. But in the 1990s, a method was developed to cultivate the pure mycelium of the ascomycete *Fusarium venenatum* in aerated liquid culture, in large tanks called fermenters (that is the general name of the device; because the culture is oxygenated, the mycelium is not growing by fermentation). Glucose, minerals, and vitamins are provided, and the fungus grows as a massive mycelium, never forming fruiting bodies. When large enough, the mycelium is gathered and the protein extracted. A process removes most of the RNA and DNA, because these are too concentrated in the mycelium for the human diet (if we consume excess amounts of nucleotides, our bodies convert them to uric acid, which causes gout). The processed protein is sold as "Quorn," and is used as a meat substitute; it is popular in vegetarian diets.

Fungi and Plants

Plants have several relationships with fungi, some beneficial to the plant, some harmful. In mycorrhizal relationships (see Chapter 4), fungi receive sugars and nutrients from roots while supplying the plant with phosphorus. Because both partners benefit, this is a mutualism. Other plants, formerly described as saprophytes (living on dead plant matter), are now known to be **mycotrophic**: they form an association with fungi and obtain nutrients from them. Most cases have not been studied in detail, but it seems that the plant parasitizes the fungus. The fungus is already living successfully by obtaining nutrients from some other plant at the time it contacts the mycotrophic plant; it appears that the mycotrophic plant simply absorbs material from the fungus, which obtains more from the other plant. There is no evidence that the mycotrophic plant supplies the fungus with anything beneficial. This plant-fungus relationship occurs in many orchids and a small number of other photosynthetic plants, as well as nonparasitic, nonchlorophyllous plants such as *Monotropa* and *Pterospora* (FIGURE 18.12).

FIGURE 18.12. These are just the clusters of flowers (inflorescences) of plants of Indian pipe, *Monotropa*. The vegetative body of the plant is below ground, drawing nutrients from fungi that have made contact with its base, thus it does not need to carry out photosynthesis and it has no chlorophyll. These mycotrophic plants were once thought to be parasites because they have no chlorophyll, but careful digging reveals that they do not have roots or haustoria that directly connect to another plant, so they cannot be parasites. Then it was thought that they lived by absorbing nutrients from dead, decaying plant material, so they were called saprophytes. Only recently have we realized their association with fungi; the term "mycotrophic" is new. (© Jupiterimages/Liquidlibrary/Thinkstock.)

One of the most common and most serious relationships between plants and fungi is that of pathogenicity by fungi. Fungi cause many plant diseases and may often kill plants completely. Fungi cause some diseases in humans (*Candida* and ringworm), but they cause thousands of diseases in plants; 70% of all crop diseases are due to fungi. Generally, plants are able to resist most fungi, but watch your local weather broadcast and notice the number of mold spores per cubic meter of air present every day. Plants are exposed to a constant barrage of fungal spores; thousands of new spores land on a plant's body every day. But danger occurs only when a plant receives a spore capable of attacking it. The spore must germinate and then be able to penetrate the plant's epidermis or bark, often by growing through a stoma, a lenticel, or a fresh wound. Most plants detect the growth of fungi once a hypha reaches a living plant cell, and if the plant is resistant, it can mount one of several types of responses that block the fungus. But if the plant is susceptible, it either does not detect the fungus or its response is inadequate and the fungus then grows deep into the plant, invading cell after cell. In some species, the mycelium reaches the xylem and then releases spores that are carried by vascular flow long distances throughout the plant. Other species might invade the phloem and then grow upward and downward through this soft, nutrient-rich tissue. Often a fungus kills a plant merely by attacking a small part; enzymes of some fungi convert hemicellulose or cellulose to gums, thus clogging xylem vessels, blocking water flow, and killing the plant by dehydration.

Plants and fungi have coevolved for hundreds of millions of years, waging a constant battle against each other. All members of a plant species might be resistant to a fungus until a gene in one of the fungus individuals mutates to make it pathogenic and able to overcome the plant's defenses. Now the whole plant species is in danger unless some members, just by luck, happen to be resistant to the new allele as well as the original one. After many years, the susceptible plants will have been killed; the plant species now consists of the descendants of the resistant few, and the fungus thus is no longer pathogenic. But this stalemate will not last forever; at some point, another fungal mutation will occur and the process will repeat itself. The survival of the plant species depends on it being genetically diverse, such that some members are resistant to most fungi. It is always possible that a fungus will mutate to a pathogenic form that can attack all individuals of a plant species and thus cause that species to go extinct. Even if the fungus doesn't kill all members of the plant species, it may reduce its population numbers so much that the surviving plants are so few and so widespread they can no longer interbreed or they are susceptible to local disturbances. Species of plants, animals, and all other organisms go extinct all the time, and there must always be a cause: disease is an important factor.

Many crop disasters have been caused by fungal diseases. In 1942, an outbreak of "brown spot" in rice, caused by *Drechslera oryzae* (an ascomycete), caused crop failure and the death of two million people in Bengal. And more recently, *Bipolaris maydis* (also known as *Helminthosporium maydis*, also an ascomycete) caused a widespread epidemic of southern leaf blight in the United States, destroying 15% of our corn crop.

FIGURE 18.13. *Puccinia graminis*, wheat rust, is one of the most destructive crop diseases. There are hundreds of millions of individuals of *P. graminis* in the world, and like any organism, its nuclear DNA periodically has mutations, so new types of *P. graminis* arise every day. Occasionally, one of the new types is able to attack varieties of wheat that had been resistant to previous types of the fungus. (Courtesy of Brian Hudelson, University of Wisconsin.)

Wheat is our most important crop, and all wheat-cultivating areas around the world have programs to watch fields for any sign that a plant or two might be infected with a new strain of pathogenic fungus. These centers maintain thousands of varieties of wheat and are constantly breeding new varieties that might be needed to resist the next new type of pathogenic fungus. Wheat is especially susceptible to *Puccinia graminis* subspecies *tritici*, which causes wheat rust (the subspecies name refers to wheat *Triticum*; other subspecies of this fungus attack other cereals). *Puccinia graminis* is a basidiomycete, but it never forms mushrooms or puffballs; instead it causes rust-colored patches on leaves and stems of wheat, weakening the plants and, when not killing them, lowering their yield (**FIGURE 18.13**). The life cycle of wheat rust is extremely complex, involving five separate types of spores, and also needing to attack barberry plants (*Berberis*) to complete its sexual life cycle. Part of the method of controlling wheat rust in Europe has been to eliminate barberry plants. Several states in the United States also prohibit the cultivation of barberry and its closest relatives.

Crop plants are not the only victims of fungal diseases. Few of us are old enough to remember what forests and cities in the eastern United States looked like before Dutch elm disease (*Ophiostoma ulmi*, an ascomycete) wiped out almost all the American elms (*Ulmus americana*). They are described as having been larger and more majestic than other trees, and they dominated whole forests. The virulent form of Dutch elm disease was noticed in Europe in 1919, and for 11 years North America was safe. Our trees were protected by the Atlantic Ocean, which is so wide that spores cannot be blown from Europe to North America without dying from desiccation or being killed by ultraviolet light in the upper atmosphere. But in 1930, our first outbreak occurred in Ohio, apparently infected by a shipment of contaminated logs. Whereas nature could not carry spores from Europe to the U.S., people could and did. Since that time, over 77 million elms have died and the nature and aspect of our eastern forests have been altered. Botanists occasionally find a few American elm seedlings, and there is hope that a resistant strain of elms can be bred, but so far, results have not been promising.

For a description of the most famous "fungal disease" of all, the one that caused the Irish potato famine, see "Algae as Disease Organisms in Plants" below; the disease organism *Phytophthora* was long believed to be a fungus, but now we realize it is an unusual alga.

Algae

The Major Groups of Algae

Algae have complex evolutionary relationships, and so it is easier to start by considering just the red and green algae. The earliest ancestor, a single-cell eukaryote, became photosynthetic when it engulfed a cyanobacterium but did not digest it. The cyanobacterium and the eukaryote formed a symbiotic relationship (an endosymbiosis, see Chapter 9) and gradually evolved so much that chloroplasts in land plant leaf cells today show little resemblance to their cyanobacterial ancestors. However, the "chloroplasts" of the first photosynthetic eukaryotes had many cyanobacterial characters, and the same is still true of the chloroplasts in red algae. For this and many other reasons, we believe red algae have a great number of ancestral features and, at least in cell organization and metabolism, very few derived ones.

After the red algae had continued evolving for millions of years and had diversified into many groups, one of its clades gradually evolved into green algae. This was not a sudden change like picking up a chloroplast and switching from being heterotrophic to autotrophic. Instead, many metabolic, structural, and organelle features changed bit by bit, one at a time, until now green algae have many more derived features and very few ancestral, relictual features. Looking at red and green algae, people rarely think of them as closely related, but DNA-based evidence indicates that the two are closer to each other than to the other algae. Later, one group of green algae underwent a further step of divergent evolution and gave rise to two new clades: true plants and a group of algae called the Charophyta (stoneworts; see Figure 9.16).

The other algae have turned out to have surprising relationships. All have chloroplasts, one of several reasons why they are considered algae instead of animals or fungi, but it turns out that their "chloroplasts" are actually cells of red algae! We had thought, for example, that one group of red algae had gradually evolved to become brown algae, and simultaneously their red algal chloroplasts had gradually evolved to become brown algal chloroplasts. This would have been an evolution similar to that which produced green algae. A similar process was believed to have produced the yellow-green algae and their chloroplasts, and so on. But now we realize instead that the other groups of algae began as plastid-free cells that existed at the same time as red algae. They engulfed red algae and then did not digest them, much the same way that red algae had originated by engulfing a cyanobacterium. Much to our surprise, endosymbiosis occurred several times in the early history of algae.

At present, the main groups of algae are:

Red algae (Rhodophyta; **FIGURE 18.14**) are often medium-sized individuals that are easy to see and pick up, although some grow as crusts on rocks along the seashore

FIGURE 18.14. This is a red alga whose body is composed of many cells interconnected by broad pits that run through thick, gelatinous walls. Whereas thick walls of plants are usually lignified and rigid, those of red algae contain chemicals called colloids that make the walls flexible and tough. (© Robert & Jean Pollock/Visuals Unlimited, Inc.)

FIGURE 18.15. Almost any stream, especially if it flows slowly and is in a warm area, will have green algae present, usually with bodies composed of long slender filaments, such as this *Spirogyra*. This image shows an especially abundant growth of algae that completely covers the stream's surface; at such times, the alga is then often called "pond scum" or "moss." A rich growth of algae like this usually indicates that the stream is so polluted that there are plenty of mineral nutrients to support the alga's growth. If the stream water were pure and clean, there would be too few nutrients to support this much algae.

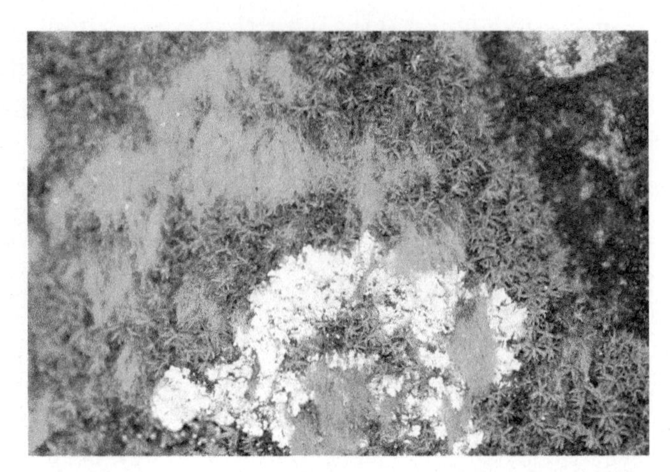

FIGURE 18.16. Brown and red algae are almost never found in freshwater or on land, but green algae are diverse in their habitats: some live in ocean water, others in freshwater, some in soil, others on bark or rock and so on. Some (called snow algae) even grow on snow. Shown here is *Trentopohlia* growing with lichens and mosses on the surface of rock in Austin, Texas. A rock face in a dry area is not a typical habitat for algae, but *Trentopohlia* is so well adapted that it not only thrives, it grows large enough to be easily visible. Despite having yellow pigments, it is a green alga.

FIGURE 18.17. These giant kelps (*Nereocystis*) are quintessential brown algae; their bodies consist of a "holdfast" attached to rocks on the bottom of the seacoast, a "stipe" (stalk, the main bulk of the body) that has a hollow, bulbous "float" at the top. Long leaflike blades extend from the float and carry out photosynthesis and reproduction. Kelps are complex internally as well, having a variety of highly specialized cells. (Courtesy of Claire Fackler/CINMS/NOAA.)

(almost all red algae are marine, only a few live in freshwater). Their bodies consist of aggregated filaments resembling parenchyma, and they store their extra carbohydrate as a polymer called floridean starch. An unusual feature is that red algae never have cells with flagella, not even their gametes.

Green algae (Chlorophyta) vary from unicellular organisms to individuals that consist of a single row of cells (filamentous algae) or a sheet of cells, the sheet being only one or two cells thick (laminar algae). Green algae occur in oceans but are more diverse and plentiful in freshwater where they are often called "pond scum" (FIGURE 18.15). You have probably noticed them on any warm pond or slow-moving stream. Being so abundant, they are an important part of the base of food chains in freshwater ecosystems. Microscopic green algae occur in many soils, often abundantly, and a few terrestrial green algae grow to be the size of an easily visible small herb (FIGURE 18.16). Green algae have the same photosynthetic pigments as true plants, and they also store reserve carbohydrates as amylose (starch) just as plants do; green algae do not produce floridean starch like red algae do. Green algae have sex cells that swim with flagella, and many have life cycles like those of plants, involving a diploid sporophyte that alternates with a haploid gametophyte.

Brown algae (Phaeophyta) might be familiar to you as kelp that grows along rocky coasts and has a brown or brown-yellow-green color (FIGURE 18.17; see also Figure 18.2). Many kelps are very large, up to 30 m (almost 100 ft) long and a few centimeters (an inch or two) in diameter, and they can be so abundant that they form "forests" along the rocky coast of California. They have no

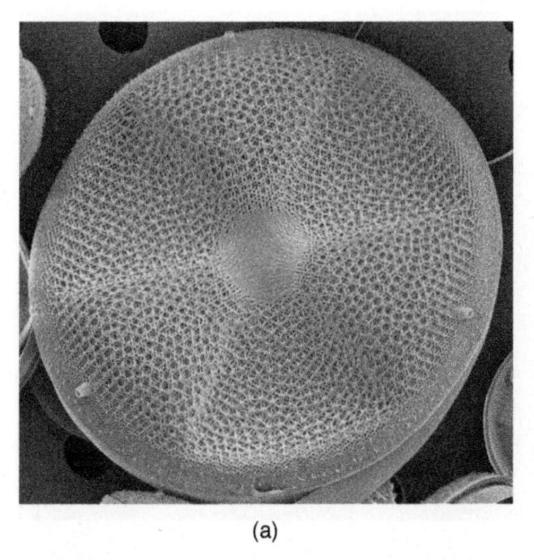

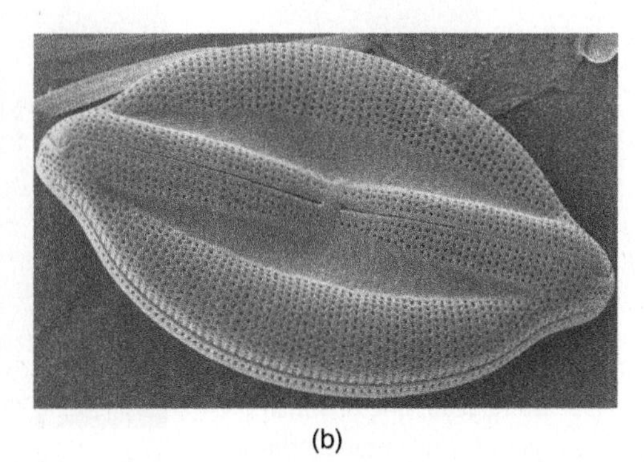

(a) (b)

FIGURE 18.18. The bodies of diatoms are almost always unicellular, only rarely do two or three cells remain attached to each other. Each cell has a two part wall (each part called a frustule) composed of silicon and pierced by hundreds of tiny holes. Diatoms are always microscopically small, but they are often so abundant in seawater that a single drop contains thousands of cells. Altogether, they carry out a considerable amount of photosynthesis. **(a)** *Actinoptychus*, a centric diatom from South Africa; **(b)** *Lyrella*, a pennate diatom from Guam. Both are marine and are shown at about 5000 times life size (x5000). (Courtesy of Elizabeth Ruck.)

collenchyma or sclerenchyma to give them elastic toughness, but they do have many unusual chemicals in their walls that provide strength and resiliency. Some brown algae are too small to see with the naked eye, but none is unicellular. They have unusual photosynthetic pigments, probably because their chloroplasts are descendants of red algal chloroplasts.

The division Chrysophyta contains several algae that probably are not familiar to most people: diatoms, golden-brown algae, and yellow-green algae. These are sometimes classified with brown algae, because many of their photosynthetic pigments are the same. Chrysophytes store extra carbohydrates as either oil or a polysaccharide called laminaran. Diatoms have a silicon-based wall consisting of two halves that fit together like the two parts of a Petri dish (FIGURE 18.18). Long slender diatoms are pennate; disk-shaped ones are centric. Diatoms are extremely abundant in both freshwater and marine ecosystems, and because they are such plentiful photosynthetic organisms, they are vitally important as the base of marine food chains.

Dinoflagellates (Pyrrhophyta) are unicellular algae whose bodies are covered in cellulose plates (FIGURE 18.19). Dinoflagellates have characters that seem to be similar to the very earliest stages of eukaryote evolution: their chromosomes have no histones, and when the nucleus divides, many aspects of mitosis are unusual, especially the fact that the nuclear envelope does not break down. Dinoflagellates store either starch or oil, and if they are disturbed they give off light (bioluminescence). At times, dinoflagellates undergo explosive population growth and suddenly an area of ocean will become red or rust colored because each liter of water contains up to 20 million dinoflagellates; this is a **red tide**, and it usually is toxic to fish and people (FIGURE 18.20). *Gymnodinium breve* causes massive fish kills by producing neurotoxins. Red tide organisms also damage gills of fish and suppress heart rate, causing fish to die of asphyxiation. Shellfish can accumulate the poisons without being harmed, but become so toxic that a single clam can kill a person.

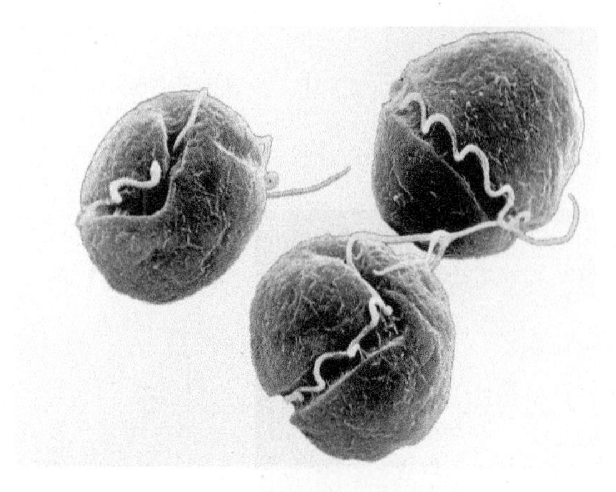

FIGURE 18.19. Dinoflagellates are microscopic single-cell algae. They have two flagella, one of which wraps around the body, the other of which propels the body forward. The body is covered by plates of cellulose. (© Dr. David Phillips/Visuals Unlimited, Inc.)

FIGURE 18.20. This is a red tide caused by dense populations of a dinoflagellate. In certain species of dinoflagellates, growth and cell division are so rapid that populations become so dense that seawater turns visibly red (such rapid growth is called an algal "bloom"): we cannot see individual dinoflagellates without a microscope, but we can see the populations when they are dense enough. (Courtesy of P. Alejandro Diaz.)

Euglenoids (Euglenophyta) appear to be another ancient line of algae. They are unicellular, some have chloroplasts and some do not; the ones without chloroplasts feed on bacteria and other forms of microscopic food, just as an animal would. Rather than a cellulose-based cell wall, their body is protected by a protein-based layer called a pellicle.

Oomycetes (Oomycota) are a special case of algae. Until just a few years ago, they were thought to be fungi because they have no plastids, and their long, slender cells resemble hyphae. But oomycetes have a variety of characters that are typical of algae, and it was suspected that they might be algae that had lost their chloroplasts and had adopted a heterotrophic life. DNA analysis has shown that theory to be correct. The oomycetes contain some of our worst plant pathogens, such as *Phytophthora infestans*, which caused the Irish potato famine (see Chapter 13 and **BOX 18.1**).

BOX 18.1. Using Integrated Pest Management to Control Microbial Diseases in Plants

Recall the potato blight discussed in Chapter 13. When a member of *Phytophthora infestans* mutated such that it could overcome the resistance of potatoes, the potato crop of Ireland was destroyed in the 1840s. More than one million people starved to death and one and a half million emigrated. This is not just a tragic story from the past, it is a continuing problem: *P. infestans* still occurs and still attacks potato crops. Some varieties of potato are somewhat resistant, but none is completely safe. Treating entire crops with algicide is more expensive than the crop is worth, so other methods of disease control are needed.

The most effective method is **integrated pest management**, a set of farming practices. One key of these techniques is *crop rotation:* potatoes are only cultivated in the same field for 2 years, then that land must be used for other crops for several years, long enough for any *Phytophthora* spores in the soil to die. Once the soil becomes clean enough, potatoes can be planted there again, but only for 2 years. Another important aspect is *breeding disease resistance*; whenever a pathogen attacks a crop, geneticists search for survivors, plants that may still be alive because they have an allele that confers resistance. If such plants are found, they may become the parents needed for breeding a new variety of resistant plants. *Disease forecasting* keeps track of where disease outbreaks are occurring, predicting wind distribution patterns of the spores, and then estimating which areas might be unsafe for planting next year, and so on. *Sanitation* is crucial; infected plants must not be shipped from one area to another, and even trucks, farm equipment, and so on should be cleaned after being in contact with a crop that might be infected. For example, the *P. infestans* that caused the Irish potato famine was just a single mating type called A1. Because only one mating type was present in Ireland (and the rest of Europe as well as North America), *P. infestans* could not undergo sexual reproduction, so its evolution was extremely limited. In 1950, a second mating type, A2 was discovered in Mexico, and in 1970, an unsanitized shipment of infected plants from Mexico to Europe brought the A2 mating type in contact with the A1, and now *P. infestans* undergoes sexual reproduction. New pathogenic allele combinations may arise much more rapidly now.

At present, a new oomycete disease is devastating forests on the west coast. "Sudden oak death" is caused by *Phytophthora ramorum* (related to *P. infestans*); since 1995 it has been infecting numerous species of oak as well as coast redwood and Douglas fir. At present it is confined to California and Oregon, but, unfortunately, infected camellias were discovered at a wholesale nursery in Los Angeles, and it is possible that infected plants have been shipped throughout the country.

Algae as Food

A variety of algae are eaten in Asia, but only a few enter the diet of North Americans. The most familiar is probably nori, the thin, black-red paper-like wrapper around sushi (**FIGURE 18.21**). Nori is made from several species of the red alga *Porphyra*, especially *Porphyra yezoensis*, and *P. tenera*. These algae grow as sheets or leaf-like blades attached to poles or nets suspended in the ocean, just off shore. As blades reach full size, they are harvested, washed and cleaned, then chopped into a fine slurry. This is poured over a screen to let water escape and retain the *Porphyra* as a large thin sheet, which is peeled off and dried. Converting *Porphyra* into nori is similar to making paper. Nori, as mentioned, is used as a wrap for sushi as well as a garnish for soups and noodles. It is extremely high in vitamins, iodine, iron, protein, and carotene. Most of the carbohydrates it contains are types we cannot digest, so it is low in calories.

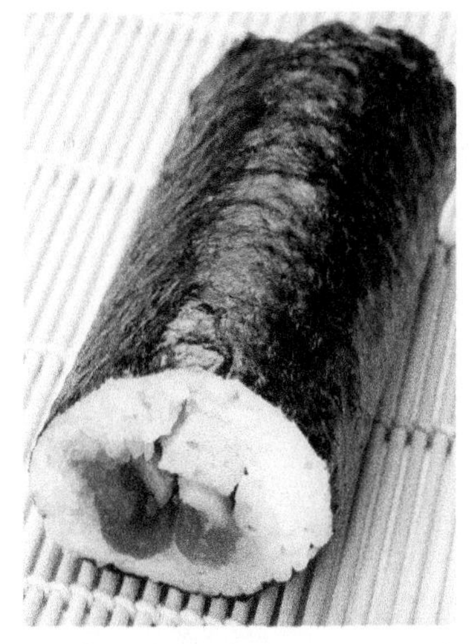

FIGURE 18.21. If sushi is served wrapped in a thin sheet of dark black-red material, that wrapping is called nori and it is made of the red alga *Porphyra*. Although the body of *Porphyra* is a thin sheet that resembles nori, the body is too delicate and irregular to be used in its natural form. Instead, nori is manufactured by shredding the bodies of Porphyra and then reforming them into the thin sheets of uniform size, shape and thickness. (Reika/ShutterStock, Inc.)

We can use nori to reexamine an important concept of plant biology. All true plants have a life cycle involving an alternation of heteromorphic generations. Recall that in flowering plants, the plants we see are diploid sporophytes. Some cells within the anthers undergo meiosis and become haploid microspores that develop into haploid microgametophytes called pollen grains, each with two sperm cells. Inside the ovules within the carpels, some cells undergo meiosis to become megaspores, and those develop into megagametophytes (the embryo sacs) that have an egg, synergids, and antipodals. After a sperm cell fertilizes an egg cell, the new diploid zygote develops into an embryo within a seed and later into a whole new sporophyte. This is a life cycle with an alternation of heteromorphic generations, and the same thing occurs in *Porphyra*. The blades that are harvested are the gametophyte phase, and each cell has a haploid nucleus. When they reproduce, the new plants do not grow into bladelike plants like their parents but instead settle onto seashells, attach to them, and grow into a set of branched filaments that are diploid. These branched algae had been known as a completely different genus, *Conchocelis*, until the whole life cycle was worked out in a laboratory. Spores of the branched form germinate and grow into new bladelike gametophytes used for nori. In seed plants, it is easy to know which gametophytes are parts of the same life cycle as sporophytes because the two are attached to each other at some point. But in *Porphyra*, the spores, gametes, and zygotes all float away from their parent and settle down, so if the generations are heteromorphic, it takes a great deal of research to establish which gametophytes are part of the same life cycle as which sporophytes. The same is true in ferns: fern sporophytes (the familiar leafy plants) release their spores that then grow into tiny green independent gametophytes. If you find a patch of ferns with several species of sporophytes and various gametophytes, it is impossible to tell which go with which unless you cultivate them through an entire life cycle. Once it was discovered that *Conchocelis* was actually the diploid generation of *Porphyra*, its name was changed to *Porphyra* also.

Two other edible seaweeds are wakame (*Undaria pinnatifida*) and kombu (*Saccharina japonica*, until recently known as *Laminaria japonica*), both brown algae. Both are sheetlike, and wakame is eaten directly as a leaflike vegetable in seaweed salad (**FIGURE 18.22**). Wakame is also an ingredient in miso soup and tofu salads. It is high in vitamins, minerals, and omega-3 fatty acids. Unfortunately, it has also been declared one of the 100 worst invasive species in the world. It is native to cold coastal waters along Japan and Korea, but has now spread to New Zealand, South America, the United States, and Europe where it displaces native algae. Kombu is an ingredient in a soup stock called dashi, and it is often pickled with sweet and sour flavorings and eaten as a snack. Rice that will be used in sushi may be flavored with kombu.

Spirulina is a cyanobacterium, not an alga, but it is mentioned here because it is often treated as an alga commercially. It forms corkscrew shaped long filaments and is dark blue-green in color. It is rich in protein, up to 55% to 77% of its weight is protein, and furthermore, it is complete protein, containing all essential amino acids. It is an especially good source of many essential lipids, as well as vitamins and minerals. *Spirulina* is sold in health food stores and is used as a food additive, but there is risk of contamination by toxin-producing cyanobacteria.

Many algae are used as animal feed. Species with large individuals, such as *Macrocystis*, *Ascophyllum*, and *Fucus*, are chopped into feed for cattle and other livestock.

FIGURE 18.22. This is a bowl of *Gracilaria parvispora*, a species of red alga that is eaten as a cold vegetable. It has a slightly crunchy texture, a delicious flavor, and is high in various nutrients. This is being cultivated in controlled conditions in tanks of seawater rather than being harvested from an ocean. (Courtesy of Lewis Weil, Austin Sea Veggies.)

Smaller, microscopic algae are gathered with filters, then mixed with regular animal feed as a supplement rich in protein and carotenoids. Animals, we humans included, cannot synthesize carotenoids even though they are essential for our retinas and our vision; we must obtain carotenoids in out diet. We obtain most from eating fruits and leafy vegetables, but we also obtain a significant amount from the yolk of chicken eggs, which will be especially rich if the chicken feed has been supplemented with algae.

Industrial Products Derived from Algae

Many algae contain substances that give their walls flexibility, elasticity, and strength. Cell walls of true plants are mostly cellulose and hemicellulose, and one cell is glued to another with an extremely thin middle lamella of calcium pectate. For extreme strength and elasticity, plants use lignified secondary walls. Algae do make cellulose, but no alga makes either a secondary wall or lignin; those occur only in true plants. Instead, algae use a variety of other compounds that give their bodies flexibility and a resistance to tearing that they need to withstand being buffeted by waves and water movement. These chemicals also hold water tightly and keep the algae moist if they are exposed to dry air. This is especially important for algae that grow attached to rocks in the intertidal zone; as sea level changes from high tide to low tide and back again, the algae can be exposed for hours and would die if they could not keep their bodies moist. Plants use an epidermis and cuticle to retain water; algae use cell wall compounds located throughout their bodies. The ability to bind water is extremely useful to us, and we use algal products every day, certainly when eating ice cream or candy bars, drinking beer or wine, using paints, inks, paper, and many more things.

The algal products are called **colloids**, substances that become suspended in water rather than dissolving in it. Think of any water solution such as coffee, tea, or fertilizer for feeding your plants. Compare the thinness of those solutions with the texture of chocolate milk, creamy soups, yogurt, pudding, and Jell-O. In the latter group, the liquid is thicker, creamier, and finally a stiff gel, all due to the presence of colloids. Just a little in chocolate milk, more in pudding. By changing the consistency of the liquid, the colloids change the sensation we get as we eat or drink the food. Now think of paints and stains. Staining solutions are thin and soak into wood quickly, leaving the surface smooth or rough, however it was before the stain was applied. But paints are creamier because they contain colloids, so they form a thick protective layer on the surface of the object; the elasticity of the colloids lets them smooth out brushstroke marks. Because colloids make liquids thicker, it is easier to suspend other materials in the liquid. We typically think of solutions but often we deal with suspensions. Certain components of paint, toothpaste, or cosmetics are fine particles that are suspended in the liquid base, not dissolved in it. Without colloids, they would gradually separate from the liquid and either sink to the bottom of the container or float to its surface. Any container that advises to "shake well" contains a suspension. But with the correct amount of the right colloid, the particles will remain suspended.

Algae provide three main colloids: alginates, carrageenans, and agars. **Alginates** are polysaccharides, long polymers of alginic acid, and they occur in brown algae. Alginates are harvested from giant kelps (*Macrocystis pyrifera*, *Ascophyllum nodosum*, and several species of *Laminaria*). Alginates absorb as much as 200 to 300 times their weight in water and form a colloid. As a gelling agent, alginates are used to thicken beverages, ice cream, and cosmetics. They can make such a flexible, tough gel that they are used to make impressions for dentistry and or prostheses. This same property is employed to make artificial foods such as nuggets of cat food, or some of the foods we eat that are made by grinding and processing meat, fish, or fruits and then reforming them into a shape that may or may not resemble the real food. Think about some of ready-to-bake fish and meat products, or inexpensive, mass produced fruit pies. Take a look at the texture next time you eat some of these; are they real or are they ground food molded into shape with alginates? Because of their ability to absorb water and swell greatly, dried alginates are added to some powdered diet supplements; they swell in your stomach and give you a full feeling. A thin colloid of alginate is usually added to beer and wine where it traps impurities; afterward, the colloid is filtered out, taking the impurities with it, leaving the liquid clear and unclouded.

Carrageenans are polysaccharides extracted from the red alga *Chondrus crispus*. They are good gelling agents and are vegetarian and vegan alternatives to gelatin (gelatin is derived from skin and bones of animals). Jellies and jelly candies are made with carrageenans, as are chocolate milk, eggnog, yogurt, and ice cream. Carrageenans are also used to make paper and fabric with a marbled pattern. Oil-based inks can be floated on top of dilute, liquid carrageenan, and then the inks are swirled and paper or other material is laid on top, picking up the ink.

Agar, extracted from any of several red algae (for example *Gelidium* and *Gracilaria*), is a mixture of chemicals called agarose and agaropectin. As a gelling agent, agar too is used in jellies, puddings, and custards, but its most famous applications are in biology laboratories. A 1% gel of agar in nutrient solution is used to culture bacteria, fungi, and cells of plants and animals (FIGURE 18.23). These can also be cultured in a liquid, but the fluidity of the nutrient solution causes all the cells to constantly mix together. By adding agar, the solution becomes solid, allowing the cells to sit quietly on its surface. As the cells divide and grow, all daughter cells stay together as a colony, and we know they all came from the same parental cell. Often a culture is started with a mix of cells, such as the many microbes that might be present in a diseased plant or in a sample of soil. By spreading a dilute solution of these cells across a plate of agar, each cell will probably be separated from all others, and as it grows, its colony of daughter cells will stay separate from those of other cells. By looking at the culture several days later, we can tell how many different microbes had been present in the sample, and by carefully taking just a few cells from a single colony, we can establish a pure culture of just one species (this is a monoculture).

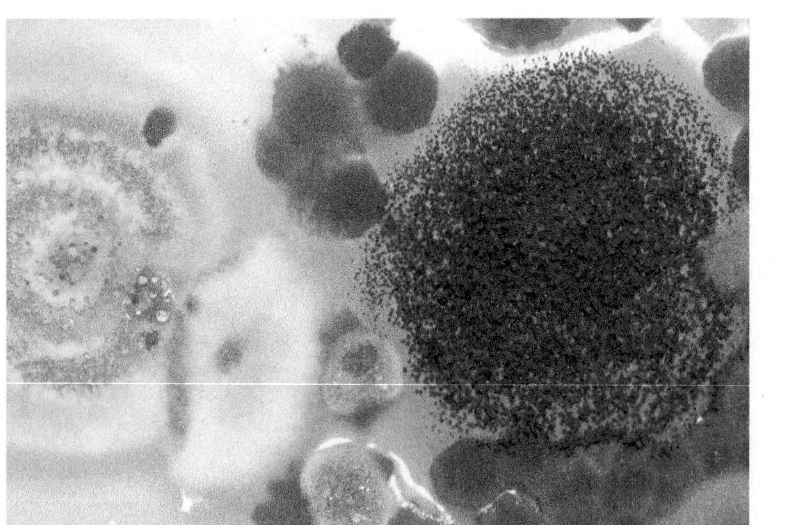

FIGURE 18.23. These colonies of fungi are growing on a nutrient medium solidified with agar, a colloid extracted from red algae. Each colony has grown from a single spore, and because the medium is solid, each colony and the spores each produces remain together in a single spot. If we want to experiment with just one colony or grow spores from only one, that is easy to do by picking up some of that colony with a fine needle or tweezers. If these same spores had been cultivated on a liquid medium without the agar, any sloshing or swirling of the medium would have mixed them all together and it would have been impossible to have kept them separate.

Agarose, one of the components of agar, is used in molecular biology in a technique called **gel electrophoresis**, which separates proteins based on their size. An agarose gel is prepared as a thin sheet or column, and a mix of proteins, usually an extract of a tissue, is added to one end. An electric current is then applied, which drives the proteins toward the other end of the agarose gel. The gel has pores so large that small proteins enter them and become temporarily trapped until they slip out and continue on, only to be trapped in another pore. The very largest proteins just slide by all the pores, and those of intermediate sizes slip into some, not into others. Consequently large proteins move quickly, small ones slowly. After the electrophoresis has been allowed to run for several hours, the gel can be stained, and the proteins show up as bands. At this point we know only how many proteins are present and what their relative sizes are, but if we have a stain that reacts with a particular protein of interest, that band can be identified and cut out of the gel, then the protein can be isolated by soaking it out of the gel. Agarose gel electrophoresis is also used for separating DNA and RNA.

An unusual product of diatoms is their silicified remains. Each diatom cell is surrounded by two cell walls called frustules that resemble a Petri dish and its cover. Frustules consist of silicon dioxide, and when a diatom dies, its frustules settle to the bottom of the ocean as the diatom's protoplasm decays. Certain areas have deposits of diatom frustules up to 300 m (almost 1000 ft) thick and extending for miles; these deposits are called **diatomaceous earth**. Each frustule, despite being microscopic, has hundreds of tiny holes, so diatomaceous earth is like finely ground glass with holes in it (see Figure 18.18). It is inert and can be used to filter out even extraordinarily small particles from a liquid. It is used in municipal water purification systems as well as almost every home swimming pool. Because its particles are so fine, it is an excellent polishing agent. It is also added to paint to reduce its glossiness.

Algae in the Food Chain

In lakes, streams, and oceans, algae are the main or only primary producers. When we go to the beach and see seaweed washed ashore or growing in the intertidal zone, that is only an inconsequential portion of the algae present. Most algae are microscopic, and any drop of water near a coastline contains thousands of algae, all photosynthetic, all producing oxygen, and all at some point ending up as the food for some heterotroph. The carbon that algae fix supports all the animals in the food chains of the lakes and oceans. Whales, dolphins, sharks, sea turtles, sea birds, and all other such creatures ultimately depend on the productivity of algae.

Ocean water near a coast supports the greatest concentration of algae. This water is enriched in nutrients washed in by rivers that have picked up minerals from all across their drainage basin, and virtually all precipitation that falls on land eventually flows into an ocean, bearing minerals with it. Rivers also pick up all the fertilizers, herbicides, pesticides, and industrial wastes that enter their watersheds and those too are carried to the oceans. The richness of the minerals washed into coastal waters supports large numbers of floating algae (called **phytoplankton**) as well as algae that grow attached to rocks or seashells (**benthic algae**). Far from shore, in the open ocean, there are fewer minerals and thus fewer algae. Minerals that had been present were used by algae, and as those were eaten or as they died and sank, the minerals were removed from the top layers of water. Waters at the bottom of the ocean are rich in nutrients but, because light does not penetrate far through water, there is no light for photosynthesis near the ocean floor and little or no primary productivity occurs there.

Surprisingly, the frigid waters near the North and South Poles have some of the highest concentrations of phytoplankton. Those areas are so cold that ice forms, taking water out of the seawater, and thus making the remaining minerals more concentrated and better for the algae. The low temperatures do not matter because the algae have evolved to have alleles that function best at such low temperatures.

Oceans cover most of our planet, and algae are extremely productive. Some of the most valuable items we obtain from algae are oxygen, reduced amounts of atmospheric carbon dioxide (a greenhouse gas), and the feeding and support of all the many sea creatures that share the planet with us.

Algae as Disease Organisms in Plants

Several algae cause significant disease in plants. See Box 18.1.

Lichens

Lichens occur all around us, growing on rocks, tree trunks, fences, gravestones, and soil. Many of them appear to be very plantlike, having what look like small shrubs with stems and oddly shaped leaves (**fruticose lichens**; FIGURE 18.24); other lichens are more of a crust on a surface (**crustose lichens**; FIGURE 18.25) and don't resemble plants, fungi, or animals; and some lichens look like bits of leaf attached to rock or wood at one side (**foliose lichens**; FIGURE 18.26). A very unplantlike feature of lichens is their color or pattern: it never looks quite right for a plant. Depending on the species, lichens may be dull gray-green, blue-green, yellow, red, black, smooth, spotted, dotted, splotchy, and so on. It turns out that lichens are not plants or animals; they are a mix of fungi and algae that grow together in a well-defined shape and morphology. If the fungus is cultured by itself, it usually just forms a nondescript mycelium that looks nothing at all like the lichen it was isolated from. And if the algae are cultured, they just form a smooth green film over the surface of agar in a Petri

FIGURE 18.24. Fruticose lichens, such as this *Teloschistes chrysothalmus*, have a body that somewhat resembles a leafy herb. But this is not a plant, it is a set of fungal cells growing together with algal cells. The large orange disks are one form of reproduction, but most reproduction occurs when small clusters of cells break off and grow into new lichens.

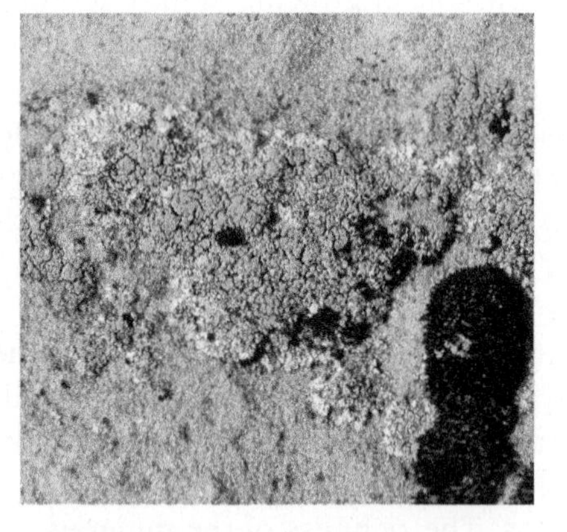

FIGURE 18.25. Crustose lichens are easy to recognize because they really do grow as a crust, usually on rocks or bark; the yellow lichen here is *Pleopsidium flavum* growing in Zion National Park. You would need a knife or a chisel to collect a piece of this intact, and lichenologists really do carry chisels on collecting trips. On this bare, sun-baked rock, neither the fungus alone nor the alga alone could survive, but together as a lichen they are well adapted to these severe conditions. The dark mass at the lower right is a desert-adapted moss, growing in full sun on bare rock; it would die if given shade and constant moisture.

FIGURE 18.26. Foliose lichens resemble pieces of foliage leaves attached to the substrate, which may be rock, bark or soil. This *Peltigera* is growing among mosses, and if it does not grow fast enough, it will be covered by them, shaded and killed. Even though some plants, animals and other organisms are so small we can barely see them, they still compete or cooperate and have the same types of interactions described for larger organisms. *Peltigera* is widespread and common, and you may be able to find it growing near where you live.

dish, again not resembling a lichen at all. But mix some of the two cultures together so that the fungi and algae interact again, and a new lichen will be produced. Fungi (called the **mycobiont**; "myco-" refers to fungi: the study of fungi is mycology) usually make up most of the biomass of the lichen and give it its shape, whereas the algal cells (called the **photobiont** because it is photosynthetic) often just sit between the fungal hyphae as a green layer (FIGURE 18.27). But it must be the algae that stimulate the hyphae into growing in an organized manner, resulting in a definite lichen body.

FIGURE 18.27. This lichen (*Ochrolechia africana*) has been sliced to show that the layers of algal cells (green) grow just below the upper surface of most lichens, in a position where they receive sunlight but are kept moist by a thin covering of fungal cells above them. Most of the fungal cells make up a spongy white mass between the algae and the substrate.

The relationship between the algae and fungi in a lichen is not known for certain, but it is suspected that the fungi are parasitizing the algae, because the algae grow easily and rapidly in culture whereas fungi need to have many nutrients in their culture medium. On the other hand, the algae by themselves could never grow on bare rock or tree trunks or many of the other severe microhabitats that lichens are adapted to. So even though the algae can and do grow by themselves in many places, they are able to inhabit more sites when incorporated into a lichen.

There are 13,500 species of lichens, although we have to be careful when using the word "species" because that is properly used to indicate groups of individual organisms, not associations of several species. Surprisingly, in 90% of all the lichens the algal partner is either a species of *Trebouxia* or *Trentopohlia* (both are green algae; see Figure 18.16) or *Nostoc* (a nitrogen-fixing cyanobacterium; see Figure B5.1). The fungus in most lichens is some sort of ascomycete, and we can only say "some sort" because they typically never form a fruiting body or anything else distinctive enough to identify which species each might be. We can only tell that most are ascomycetes by the nature of the occasional cross-walls in the hyphae.

Lichens are extremely hardy, being able to live on bare surfaces exposed to full sunlight and all its damaging ultraviolet radiation. They absorb water from rain, dew, mist, and even just high humidity in the air, and when the air is dry, the lichens dry out almost completely, but do not die. With a little moisture, they rehydrate quickly and begin growing again within minutes. Lichens reproduce by simply breaking into fragments that can grow into a new lichen. Many lichens also produce soredia, packets of algal cells surrounded by hyphae; the soredia are produced in masses, similar to the way a fungus produces masses of spores, and the soredia too can be distributed by wind, rain, or animals.

Lichens and people do not have many direct interactions the way that plants and people do. Lichens are never used as human food; we simply do not have the enzymes necessary to digest the many exotic carbohydrates and other compounds unique to lichens. There are reports of people eating lichens, but actually it is just that people wrap other foods, such as fish, in a mass of lichens and then steam the two together, then eat the fish and throw the lichen away. Ruminant animals (those with extra stomachs) such as caribou (in America) and reindeer (in Europe) do eat lichens; up to 90% of their winter diet is "reindeer moss" (various

FIGURE 18.28. It is easy to underestimate the importance of the lichens in an ecosystem. Here, they form only a thin crust on this rock whereas the surrounding grasses are much taller, and in this same area is a conifer forest. But even though the grasses and trees are tall, if you would measure all the leaf material directly above a patch of soil, it will never add up to more than the thickness of a few leaves, so despite the thinness of the lichen, it is harvesting light effectively. Many lichens are adapted to thrive on exposed rock faces in dry or cold areas such as deserts or high alpine areas, and the acids they produce help convert the rock to soil.

species of *Cladonia*) which is actually a lichen, not a moss. Ruminants do not digest the lichens either, but bacteria in their extra stomachs do.

Several indirect interactions between lichens and people are very important. Lichens hasten soil formation (**FIGURE 18.28**). Some species do grow on bare rock, and in the brief times when it is moist enough for them to be metabolically active, they secrete acids (all organisms do); this helps break the rock down faster than if it were just exposed to rain or dew. Also, the rough nature of most lichen surfaces traps tiny windborne soil particles, preventing the soil from blowing away. The branched, fluffy nature of many lichens makes them ideal seed beds for the tiny seeds of small flowering plants; in rocky or barren places, the only seed plants that become established are those whose seeds fall onto a mass of lichens instead of bare rock. Mosses too become established in lichen beds, and their acids also speed up soil formation. Some lichens have nitrogen-fixing cyanobacteria as their photobiont, so these add nitrogen compounds to the microhabitat, enriching it. Even some of the lichens based on green algae have special pockets called cephalodia that house nitrogen-fixing cyanobacteria.

In areas with even moderate amounts of rainfall, most of the photosynthesis of the region is done by vascular plants; that performed by lichens is just a small part of the productivity of the area. But in severe habitats, such as dry deserts or cold mountain tops and near the Arctic Ocean, vascular plants are sparse and lichens may dominate the landscape (see Figure 10.25). In those areas, lichens are the significant primary producers as well as the source of nitrogen. Also, in areas where lichens dominate, their color is important: many arctic lichens are gray or light green, so they reflect light away and do not contribute much to global warming. Other lichens are dark and absorb more heat, but they tend to be sparse.

Certain lichens have extremely bright colors, especially yellows, oranges, and reds. These have been used by indigenous people as dyes, and even today lichen pigments are used in the production of Scottish tweeds.

Important Terms

agar
alginate
ascomycete
ascus
basidiomycete
basidium
benthic alga
carrageenan
colloid
crop rotation
crustose lichen
diatomaceous earth
foliose lichen
fruticose lichen
fruiting body
gel electrophoresis
hypha
integrated pest management
mycelium
mycobiont
mycotrophic
photobiont
phytoplankton
red tide
zygomycete

Concepts

- Algae are extremely diverse evolutionarily, but some green algae are closely related to true plants.
- Fungi are more closely related to animals than they are to algae or plants.
- Lichens are associations of fungi and algae, sometimes fungi and cyanobacteria; they are not individuals in the same way that plants and animals are individuals. Lichens never grow from a fertilized egg.
- Most diseases of crop plants are caused by fungi or several algae (oomycetes) that resemble fungi.
- Many algae are eaten as food with little processing, but we mostly use algae by extracting their colloids for use as thickeners.
- Photosynthesis by algae supports almost all oceanic animals as well as many freshwater fish; algae are the primary producers of many food chains.
- Lichens are significant primary producers in environments that are too dry or too cold for plants.

Glossary

ABA Abscisic acid, a plant hormone.

abacá Fibers derived from *Musa textilis*, a relative of bananas, used to make paper for tea bags and manila envelopes.

abiological reproductive barrier Any nonbiological object or phenomenon that prevents organisms from interbreeding. Examples are mountain ranges and oceans.

abscisic acid (ABA) A plant hormone involved in water stress and dormancy.

abscission zone A region in a leaf petiole or flower stalk that allows the leaf or flower (or fruit) to separate from a plant in a controlled fashion, without damaging the plant.

absorption spectrum The various colors (or wavelengths) absorbed by a pigment.

accessory pigment A pigment that absorbs light and transfers it to chlorophyll during photosynthesis.

acid Any substance that gives off a proton (H^+) when dissolved.

acid rain Rain that is acidic because it has formed in or fallen through air contaminated with acidic chemicals.

acid-free paper Paper produced by the kraft method of separating and delignifying fibers; acid-free paper is durable and long-lasting.

action spectrum The various colors (or wavelengths of light) that cause a biological response to occur. By matching an action spectrum to the absorption spectra of various pigments, the pigment responsible for the biological response can be identified.

active site of enzyme The portion of an enzyme that binds to a substrate and catalyzes the reaction.

active transport The transport of a solute across a membrane by a membrane protein, using energy in the process and able to move the solute from an area where it is dilute to an area where it is more concentrated.

adaptive radiation The divergent evolution of a species into several new species after the founding species enters a new habitat or evolves to have a particularly beneficial feature.

adult phase The part of a plant's lifetime when it is able to reproduce; any plant that can produce flowers or cones is in its adult phase. Compare with juvenile phase.

adventitious root A root that forms in any organ other than another root.

aerobic respiration Respiration that consumes oxygen; most plants, animals, and fungi carry out aerobic respiration. Compare with anaerobic respiration or fermentation.

aflatoxins Poisonous compounds produced by certain fungi; if the fungi infect crops grown for food, the food becomes toxic.

agar One of several types of colloids extracted from algae.

albedo The general brightness of the sky.

algae Photosynthetic organisms that are not plants; most live in water, but some live in moist soil or bark.

algal bloom A sudden, rapid growth and reproduction of algae, much out of proportion to normal growth, usually caused by a body of water becoming polluted with nutrients that previously had been limiting.

alginate One of several types of colloids extracted from algae.

alkaline A condition of having more base present than acid, and having a pH above 7.0.

all-or-none response A developmental response that is not graded. If enough stimulus is present, the plant responds fully; if not enough stimulus is present, the plant does not respond at all.

allele A version of a gene. A gene can have different DNA sequences (different alleles) and they are considered to be versions of the same gene as long as the proteins they code for have similar functions.

alpine tundra Low, treeless vegetation that occurs in cold areas of mountains. Compare with arctic tundra.

alternation of generations Plant life cycles have a haploid gametophyte generation that alternates with a diploid sporophyte generation.

alternative energy source Refers to a variety of methods of generating energy other than by using traditional methods such as burning fossil fuels, damming rivers, or using nuclear reactors. Examples are solar energy, wind turbines, and biofuels.

amino acid A small organic acid that has an amino group; most are monomers of proteins.

amylopectin A polymer of glucose similar to amylose (starch) except that it has occasional side branches.

amyloplast A cell organelle that stores starch; a starch grain.

anaerobic respiration Respiration that does not consume oxygen; often called fermentation. Compare with aerobic respiration.

analgesic A compound that relieves pain.

analogous features Features that resemble each other but that have evolved from different ancestral features. They are the result of convergent evolution. Compare with homologous features.

anaphase The phase of mitosis in which the two chromatids of a replicated chromosome separate from each other and move to opposite ends of the mitotic spindle.

ancestral group The group of organisms that were the ancestors of another group. This can refer to the ancestral group that gave rise to just one species or to a set of species (such as a genus or family).

angiosperm A flowering plant.

annual plant A plant that is not perennial, that does not live for more than one year. Compare with biennial and perennial.

anoxia A condition in which no oxygen is present, so aerobic respiration is not possible.

anoxygenic photosynthesis Photosynthesis that does not use water as a raw material and does not produce oxygen.

anther The part of a stamen (an organ in a flower) in which some cells undergo meiosis and develop into pollen grains.

anthropogenic Something that is caused by people, such as carbon dioxide produced by factories, cars, and so on, as opposed to carbon dioxide produced by the decay of dead material.

antioxidant A chemical that neutralizes certain types of destructive oxygen compounds with an organism's body.

apical dominance The inhibition of axillary buds by an actively growing shoot apical meristem.

apomorphy A feature present in one (autapomorphy) or several (synapomorphy) derived members of a group, but which is not present in the ancestral members.

aquifer A subterranean layer of porous rock that allows water to flow through it.

Archaea One of the three domains of life. Compare with Bacteria and Eukarya.

arctic tundra Low, treeless vegetation that occurs in cold areas in the far northern parts of Earth. Compare with alpine tundra.

artificial selection In the breeding of organisms, selection done by humans to obtain organisms with particular traits. Compare with natural selection.

artisan wood A wood used for beauty rather than strength or low cost.

ascomycete A fungus that is part of the clade that produces asci, small sacs in which nuclear fusion and meiosis occur; yeasts, truffles, and morels are ascomycetes.

ascus (pl.: asci) In the ascomycete fungi, a small sac in which nuclear fusion and meiosis occur.

asexual reproduction During asexual reproduction, a single parent produces progeny that are genetically identical to itself.

atom A basic unit of matter; each atom has a nucleus and electrons, and each chemical element has its own particular atoms.

ATP Adenosine triphosphate, a small molecule that provides energy or phosphate groups to reactions.

autotroph An organism that obtains its carbon from inorganic sources; plants are autotrophs, obtaining their carbon from carbon dioxide and their energy from sunlight. Compare with heterotroph.

auxin One of the plant hormones, involved in growth and apical dominance.

axial portion of wood and secondary phloem The elongate cells aligned parallel to the long axis of the stem or root; for example, tracheids, vessel elements, fibers, and sieve tube members. They are produced by the fusiform initials. Compare with ray.

axillary bud A bud located in the axil of a leaf; axillary buds usually develop into flowers or branches or are inactive due to apical dominance.

Bacteria One of the three domains of life. Compare with Archaea and Eukarya.

bark The outermost covering of a woody plant, composed of secondary phloem and cork.

basal angiosperms The several groups of angiosperms that arose before monocots and eudicots. Examples are *Amborella* and water lilies.

base Any chemical that gives off an OH^- group when dissolved.

basidiomycete One of the main clades of fungi, basidiomycetes are characterized by producing sexual spores with a basidium; mushrooms and puffballs are basidiomycetes.

basidium (pl.: basidia) In basidiomycete fungi, the cell in which nuclear fusion and meiosis occur and which then produces basidiospores.

bast fiber A fiber extracted from the phloem of a plant, usually from bark.

bayberry wax Wax obtained from bayberry (*Myrica faya*) and used primarily to provide aroma to candles.

bedding plants Small herbaceous plants that are grouped into masses to provide a line or block of color in a garden.

beer An alcoholic beverage made by brewing sprouted barley; other cereal grains may also be added. The alcohol content is about 5%.

benthic algae Algae that live at or near the bottom of an ocean. Compare with phytoplankton.

biennial plant A plant that lives for two years, being vegetative its first year and reproductive its second.

binomial system of nomenclature The system for naming plants developed by Carl Linnaeus, in which each species has both a genus name and a species epithet.

biofuel A fuel made from plants or plant parts that were recently alive, as opposed to fossil fuels.

biogeography The study of the geographic distribution of organisms on Earth.

biological reproductive barrier Any aspect of an organism's physiology or behavior that prevents two members of a species from interbreeding. Compare with nonbiological reproductive barrier.

biome An extensive grouping of ecosystems, characterized by the distinctive aspects of dominant plants.

biparental inheritance A trait for which alleles are provided by both parents; all traits encoded by nuclear genes show biparental inheritance. Compare with maternal (uniparental) inheritance.

black tea Tea made from leaves of *Camellia sinensis* that have been fermented long enough to turn black. Compare with green tea and oolong tea.

boreal coniferous forest The extensive forests of conifers that occurs in the northern latitudes of Earth.

broad-spectrum (herbicide or **pesticide)** An herbicide or pesticide that affects a wide range of organisms, not just a specific one.

broadleaf plant Any flowering plant that is not a monocot; this term is used to distinguish eudicots and basal angiosperms from monocots and conifers.

bryophyte This term refers specifically to mosses, but it is very often used to refer to any of the three groups (mosses, liverworts and hornworts) that do not contain vascular tissue and in which the gametophyte is the dominant, independent generation.

BT plant A plant that has been genetically engineered to produce an anti-insect toxic protein naturally found only in certain bacteria (*Bacillus thuringiensis*).

bud scale A small leaf that protects buds while they are dormant.

bud trace A vascular bundle that transfers water and nutrients between a stem's set of vascular bundles and an axillary bud.

bulb A type of modified shoot that has short internodes, thick leaves, and is usually located underground.

bundle sheath A layer of cells that surrounds a vascular bundle. Monocots often have bundle sheaths of fibers; C_4 plants have bundle sheaths of special photosynthetic cells.

burlap Fabric woven from jute (several species of *Corchorus*) fibers.

C_3 cycle The basic type of photosynthesis in plants, algae, and cyanobacteria, in which the first reaction of photosynthesis produces a compound with three carbon atoms. Compare with C_4 photosynthesis.

C_4 photosynthesis A metabolism in which carbon dioxide is initially used to make a compound with four carbon atoms that is then transported into a bundle sheath cell where its carbon dioxide is used in C_3 photosynthesis.

cacao nibs Pieces of roasted cacao beans, from which chocolate is made.

caffeine An alkaloid present in coffee, tea, and chocolate, among other foods; it is a stimulant and diuretic.

callose A polymer of glucose involved in plugging damaged sieve elements and in incompatibility reactions between pollen and stigmas.

callose plug A layer of callose that forms instantly when a sieve element is damaged, and then seals the sieve areas, preventing loss of phloem sap.

callus A more or less irregular, unorganized proliferation of parenchyma cells; callus may form where a plant has been cut or damaged. When plant cells are grown in culture, they often form a callus.

Calvin/Benson cycle Another name for C_3 photosynthesis.

candelilla wax A hard wax derived from the desert shrub candelilla (*Euphorbia antisyphilitica*).

capsaicin The compound in *Capsicum* peppers that causes them to be pungent (hot).

carbon capture and sequestration The processes of collecting carbon dioxide, either from the atmosphere or as it is produced in factories that burn fossil fuels, and storing it permanently in some way that prevents it from reentering the atmosphere.

carbon fixation The addition of electrons to the carbon of carbon dioxide during photosynthesis, such that the carbon can then be used in organic compounds.

carbon neutral A process that removes as much carbon from the atmosphere as is put into it.

carbon offsets Processes in which the production of carbon dioxide or other greenhouse gases is reduced or eliminated in one area to offset the production of similar gases in another area.

carnauba wax A hard wax used for car polishes, derived from the palm *Copernicia prunifera*.

carnivory The eating of meat; many animals are carnivorous, but only a few plants are. Compare herbivory and omnivory.

carotenoid A type of pigment that occurs in chloroplasts and chromoplasts; carotenoids are red, orange, or yellow.

carpel The organ in a flower that contains ovules (with eggs), and that develops into a fruit as ovules develop into seeds.

carrageenan One of various colloids extracted from algae.

Casparian strip A layer of hydrophobic (water impermeable) material in the walls of endodermis cells; Casparian strips prevent uncontrolled diffusion of water-soluble material from the soil into the vascular tissues of roots.

cation exchange The respiration of roots produces carbon dioxide, which forms carbonic acid in soil water; the protons from the acid cause soil particles to release positively charged nutrients, which might then be absorbed by the roots.

cavitate The breaking of a column of water inside a vessel or tracheid that occurs when a plant is losing water to dry air and cannot replace it because the soil is too dry.

cell cycle The life of a cell from the time it is produced by cell division until it divides into two (mitosis) or four (meiosis) new cells.

cell plate The vesicle that forms during cell division and that is the site in which the new cell wall is synthesized.

cell wall The cell organelle that surrounds individual plant cells (except sperm cells) and most algal cells and that is composed of cellulose and other components.

cellulose An unbranched polymer of glucose molecules attached to each other by beta 1,4 bonds; cellulose is found in all plant cell walls.

central vacuole A plant organelle that consists of a semipermeable membrane (the vacuole membrane, or tonoplast) and a solution of water with material dissolved in it. In almost all mature plant cells, the central vacuole constitutes most of the cell volume and contributes to turgor pressure.

centromere A region of a chromosome where spindle microtubules attach during nuclear division. The two sister chromatids of a replicated chromosome remain attached to each other at the centromere until anaphase.

chaparral A shrubland biome found mainly in California, with wet mild winters and hot dry summers; it is dominated by shrubs and herbs that withstand wildfires.

chemiosmotic phosphorylation The production of ATP from ADP in mitochondria and chloroplasts by proton pumping across a membrane that is impermeable to protons but that has ATP synthase enzyme complexes. Compare with substrate level phosphorylation.

chloroplast A type of plastid (a plant organelle) that contains chlorophyll and carries out photosynthesis.

chocolate liquor The liquid produced when roasted cacao beans are ground; the liquor contains approximately equal parts cocoa solids and cocoa butter.

chromatid A single DNA double helix with associated proteins in a nucleus; a chromosome before replication has one chromatid and after replication has two.

chromoplast A type of plastid (a plant organelle) that that contains pigments other than chlorophyll; the red, yellow, and orange colors of many flowers and fruits are due to chromoplasts.

chromosome An organelle of inheritance; a chromosome consists of one or two DNA molecules, associated proteins, plus a centromere and two telomeres. A chromosome has one chromatid before replication, two afterward.

chromosome condensation The repeated coiling of a chromosome at the start of nuclear division (either mitosis or meiosis), during which the chromosome becomes short enough that the spindle can pull one chromatid away from the other.

circadian rhythm An endogenous rhythm that has a cycle about 24 hours long.

circumnutation A natural spiral growth pattern that is especially common in plants that twine around other objects.

cisgenic organism An organism that has been genetically modified using one or several of its own genes. Compare with transgenic organism. Also spelled *cis*-genic and *trans*-genic.

***cis*-unsaturated fatty acid** A fatty acid that has a carbon-carbon double bond in which the two hydrogens are on the same side of the molecule. Compare with *trans*-unsaturated fatty acid.

citric acid cycle A set of reactions in aerobic respiration that result in the oxidation of carbon in various acids to carbon dioxide, while simultaneously reducing several electron carriers. Also called the tricarboxylic acid (TCA) cycle.

clade A group consisting of all the descendants of a common ancestor.

cladistics A method of studying the evolutionary relationships of organisms by examining their shared derived characters.

cladogram A graphic representation of the evolutionary relationships of the members of a clade, usually consisting of a branched set of diverging lines. Also called a phylogenetic tree.

class A level of classification in systematics, a class is composed of orders.

clear cutting A method of harvesting trees, involving the cutting of all trees, even ones too small to be useful for lumber, to make it easier to harvest the big trees. Compare with selective cutting.

climacteric ripening A type of natural maturation of fruits in which changes occur slowly for many days or weeks, then very rapidly at the end. Usually controlled by the plant hormone ethylene.

coca Plants that are the source of cocaine, *Erythroxylum coca*.

cocaine One of the psychoactive alkaloids found in coca, *Erythroxylum coca*.

coccolithophorid A type of single-celled alga covered with plates of calcium carbonate. When the coccolithophorids synthesize their plates, they remove carbon dioxide from the atmosphere; when they die their plates sink, removing the carbon dioxide more or less permanently.

cocoa butter The oils released when roasted cacao beans are ground. Cocoa butter has no chocolate flavor.

cocoa solids The solid powder produced when roasted cacao beans are ground and the cocoa butter is removed.

coevolution A type of evolution in which two organisms each benefit from evolutionary changes in the other organism; for example, a plant and its pollinator might coevolve such that the plant's flowers are more suitable for the pollinator, and the animal becomes a more effective pollinator.

coir The fibrous mesocarp of coconut fruits; it is a coarse, strong fiber; coir is often used in door mats.

cold frame A small planting area surrounded by a low wall that can be temporarily covered with glass in cold or rainy weather.

coleoptile The first, tubular leaf of a grass seedling; oat coleoptiles have been used to study tropisms.

collenchyma A plant tissue that provides plastic strength; it is strong and can grow to a new size and shape. Compare with sclerenchyma.

colloid A chemical that binds a great deal of water, forming a thick, creamy liquid or a gel.

commensal relationship An interaction of two species in which one benefits and the other is unaffected.

common ancestor The group of organisms that have given rise to one or more descendant organisms; this term is often used in cladistics to describe the organisms represented by a point of divergence on a cladogram.

compact city The concept of building cities that minimize urban sprawl.

companion cell The phloem cell in angiosperms that controls the metabolism of a sieve tube member after it becomes enucleate.

compartmentation The isolation of a set of compounds or of metabolic reactions in a space where they do not interfere with, and are not influenced by, other compounds or reactions. Often used to describe the interiors of organelles, separated from the rest of the protoplasm by semipermeable membranes.

compatibility barrier Any physiological phenomenon that prevents two members of a species from interbreeding.

competition The interaction of two individuals or species that both need a particular resource; neither individual nor species survives as well as it would if the other species were absent.

complete dominance In genetics, the ability of a form of a gene (an allele) to completely mask the presence of another allele of the same gene.

complete flower A flower that has all four floral organs (sepals, petals, stamens, carpels). Compare with incomplete, perfect, and imperfect flower.

complete protein In a diet, a protein that contains all the amino acids in adequate amounts for the animal's metabolism.

complementary proteins In foods, two proteins can compliment each other if each contains amino acids that the other is missing.

compost Partially decomposed organic material used to enrich soil.

compound leaf A leaf that has several leaflets instead of a single blade. If all leaflets attach at the same point, it is a palmately compound leaf; if they attach in two rows, it is pinnately compound. Compare with simple leaf.

conifer A member of the Coniferophyta clade, trees that have strong wood and reproduce with cones rather than flowers. Compare with cycads.

contact face The portion of a cell wall that contacts another cell; the region where cells are glued together by their middle lamella.

contact herbicide A chemical that kills plants and which must be applied to their surface. Compare with systemic herbicides.

continental drift The movement of continents across Earth's surface as they float on the underlying liquid mantle.

contour plowing Plowing a field such that the furrows follow the contour of the land and are perpendicular to the slope; this reduces erosion caused by precipitation.

conventional farming The alternative to organic farming. In conventional farming, artificial pesticides, herbicides, and fertilizers can be used.

convergent evolution Evolution of two different species, metabolisms, or structures such that they eventually resemble each other; convergent evolution is usually the result of two separate organisms responding to similar environments. Compare with divergent and parallel evolution.

copra The white, sweet, solid edible part of coconut fruits.

cork Cork is usually part of the bark derived from the cork cambium and consisting of parenchyma cells with suberin in their walls.

cork cambium A set of meristematic cells whose daughter cells differentiate into cork cells (sometimes into sclereids).

cork cell A parenchyma cell in the bark that has suberin in its walls; cork cells prevent water loss and the entry of pathogens.

corm A short, vertical stem that is broad and has papery leaves; corms are typically subterranean.

cortex The region of a plant's stem or root located between the epidermis and the outermost vascular tissue.

cotton Cotton plants (*Gossypium hirsutum*) have seeds whose seed coats have extremely long trichomes that are spun into cotton thread and woven into cotton cloth.

cotyledon A lateral appendage on a seed, very leaf-like in some species, thicker in others. Monocots have one cotyledon, eudicots have two, and gymnosperms have many. The two halves of a peanut are each a cotyledon.

crassulacean acid metabolism (CAM) A set of metabolic reactions that temporarily attach carbon dioxide to an acceptor molecule, creating an acid. This occurs at night. The acid then breaks down during daylight, releasing the carbon dioxide such that it can be used in photosynthesis (the C_3 reactions).

critical night length For plants that respond to photoperiod, nights must be shorter than the critical night length to induce a response in short-night plants, and nights must be longer than this in long-night plants.

crop rotation Changing the type of crop grown in a field periodically, usually growing the same crop only 2 or 3 years successively. By putting in an alternate crop, pests that are dependent on the first crop should die while the host crop is not present.

cross In genetics, fertilization of one organism by another; this term usually describes reproduction controlled during an experiment.

cross-pollination The pollination of one plant by pollen of a different plant. Compare with self-pollination.

crossing-over During meiosis I, after homologous chromosomes have paired, chromatids break and parts of maternal chromatids are spliced to those of paternal chromatids.

crustose lichen A lichen (an association of fungi and algae) that grows as a thin, flat sheet or small particles, usually on a solid surface such as rock or bark.

cryptochrome A group of photoreceptor pigments that absorb blue light; they regulate processes such as germination, photoperiodism, and several other types of development.

cryptogam Any member of the groups of plants that never produce seeds; examples are mosses, lycopodiums, and ferns. Compare with spermatophyte.

cuticle A layer of cutin that makes the epidermis almost impermeable to water movement and that also hinders invasion by microbes and fungi.

cutin A hydrophobic polymer secreted by epidermis cells; cutin accumulates as a pure layer called the cuticle.

cutting Usually a piece of stem with several nodes and axillary buds, cut or pinched from a plant and then allowed to form adventitious roots and propagated.

cyanobacterium A clade of bacteria that contain chlorophyll *a* and that carries out oxygenic photosynthesis; a member of the cyanobacteria was the ancestor to chloroplasts.

cycad A member of the clade of seed plants that has soft wood and reproduces with cones; many are often called sago palms.

cytokinesis The division of the cytoplasm of a cell; nuclear division is karyokinesis.

cytokinin A plant hormone, involved in cell division and activation of dormant axillary buds.

cytoplasm All the parts of a plant cell excluding the cell wall and nucleus.

cytoskeleton The set of microtubules and microfilaments that provide a framework within a cell.

cytosol The liquid, nonparticulate portion of cytoplasm; all the cytoplasm excluding any organelles.

damping off The death of seedlings due to fungal attack while the seedlings are young and have a delicate cuticle.

day length The length of the lighted portion of a 24-hour period, important for plant responses that are affected by the length of the dark period; if day length is long, then night length is short in natural habitats.

day-neutral plant A plant in which flowering is not induced by the length of either day or night. Compare long-day plant and short-day plant.

deciduous forest A forest dominated by broadleaf (dicot) trees rather than by conifers.

deciduous leaf A leaf that is abscised and shed rather than being retained on the plant after the leaf dies.

deficiency disease A disease caused by a lack of, or an insufficient amount of an essential element.

dehiscent fruit A fruit that opens when it is mature. Compare with indehiscent fruit.

deoxyribonucleic acid (DNA) The molecule that stores genetic information in most organisms; genes are composed of deoxyribonucleic acid, which is a polymer of nucleotides.

desert A biome (a region of Earth) that receives very little precipitation; less than 10 inches of rain per year is often considered the critical amount.

desert island An island so low it does not cause air flowing across it to rise enough to cool and form rain.

determinate organogenesis The formation of a particular number of organs on an individual, the number being set by the organism's genes, not by its environment. Animals usually have determinate organogenesis whereas plants typically have indeterminate organogenesis.

diatomaceous earth Large accumulations of the silicified cell walls (frustules) of diatoms (a type of alga).

dicot This is an old term that described all the angiosperms (flowering plants) that were not monocots; however, a small number of the plants that were formerly called dicots are now called "basal angiosperms," the rest are called eudicots. The term "dicot" is often still used for convenience.

dictyosome An organelle involved in the packaging of material that will be secreted from a cell.

differential gene expression Cells differ in their characteristics because some express certain genes whereas other cells express other genes.

differential growth Unequal growth that causes an organ to bend or to have lobes.

diffuse growth Cell division and growth that occurs throughout an organ or organism. Compare with localized growth.

diffusion The random movement of atoms or molecules from areas where they are more concentrated to where they are less concentrated.

dihybrid cross A cross in which two characters are studied simultaneously.

diploid A cell or organism that has two complete sets of chromosomes. Compare with haploid and polyploid.

disaccharide A sugar composed of two simple sugars; sucrose (composed of one molecule of glucose and one of fructose) is the most common disaccharide.

distal This term is used to compare the position of two points or sites on a plant, the more distal point being farther away from the root/shoot junction. For example, a tulip flower is distal to the leaves of the plant, and a root cap is distal to the root hair zone of the same root. Compare with proximal.

diurnal "During the day;" diurnal flowers open during daytime. Compare with nocturnal.

divergent speciation Speciation in which one species gives rise to additional species.

division The first classification level below kingdom; the Kingdom Plantae is composed of divisions such as Angiospermophyta, Coniferophyta, and so on. In zoology, the term "phylum" is used instead.

DNA Deoxyribonucleic acid.

DNA probe A short single strand of DNA that is complementary to and will bind to a sequence of DNA that is being studied; the probe is synthesized to be radioactive or fluorescent such that it can be located easily.

DNA sequencing Techniques that reveal the sequence of nucleotides in a strand of DNA.

domestication The conversion of a species of natural, wild plants to a form that can be cultivated.

dominant allele A form of a gene that produces a phenotype that masks the phenotype of other alleles when they occur together in a diploid organism. Compare with incomplete dominance and recessive allele.

dosage-dependent response A developmental response whose strength or degree of expression depends on the amount of stimulus the organism received. Compare with all-or-none response.

double fertilization In angiosperms, one sperm nucleus of a pollen tube fuses with the egg nucleus and the second sperm nucleus fuses with the central cell nuclei.

dry fruit A fruit that is not fleshy at maturity, but instead is dry; dry fruits often are dehiscent. Compare with fleshy fruit.

drying oil A highly unsaturated (contains many double bonds) plant oil that polymerizes into a hard film when exposed to oxygen; many are used to protect wood and to make it shine.

duplication division An informal term for mitosis; the two daughter nuclei have the same ploidy level after duplication division. Compare with reduction division or meiosis.

earlywood The portion of a growth ring in wood produced in the first part of a growing season. Earlywood usually has more or wider conducting cells than does latewood.

ectomycorrhiza A root-fungus association in which the root obtains phosphate from the fungus; the fungus cells form a network around the exterior of the root. Compare with endomycorrhiza.

El Niño A weather phenomenon in which surface water in the eastern Pacific Ocean near the equator becomes warmer for several months. This causes disrupted rain patterns in various areas, even throughout the United States.

elastic strength Strength provided by sclerenchyma tissue, in which the cells can be deformed, but return to their normal size and shape after the deforming force stops. Compare with plastic strength.

electron carrier A small molecule that picks up electrons in one area and releases them in another; can be located either in a membrane of mitochondria and chloroplasts or dissolved in cell fluid (examples are NAD^+ and $NADP^+$).

electron transport chain A series of electron carriers located close to each other in a mitochondrion or chloroplast membrane; they transfer electrons in sequence and create a chemiosmotic gradient of protons.

embryo An immature sporophyte, developing from a zygote (fertilized egg); in seed plants, the embryo is located in the seed.

embryo sac Refers to the fact that in flowering plants, the megagametophyte (female gametophyte) develops as a multinucleate cell that only becomes cellular just before maturity.

endodermis In roots and some stems, a cylindrical layer of cells with Casparian strips that stops uncontrolled diffusion between the cortex and the vascular tissues enclosed by the endodermis.

endogenous rhythm A cyclic change in some metabolic state, the duration of the cycle being controlled by the organism's own internal metabolism, not by environmental factors.

endomycorrhiza A root-fungus association in which the root obtains phosphate from the fungus; the fungus cells invade the root cortex cells. Compare with ectomycorrhiza.

endoplasmic reticulum A net-shaped, tubular membranous organelle that transports material throughout cytoplasm.

endosperm In angiosperms, the triploid nutritive tissue formed by fusion of one sperm nucleus and two polar nuclei; the endosperm proliferates rapidly and nourishes the developing embryo. Monocots often have abundant endosperm in their mature seeds; dicot embryos often consume all the endosperm during development.

endosymbiosis A mutually beneficial relationship between two organisms in which one organism lives inside the other. Plastids and mitochondria arose by two cases of endosymbiosis between early eukaryotes and two types of bacteria.

engineered wood An artificial wood product made by gluing smaller wood pieces or fibers into a large shape.

enzyme A protein molecule that acts as a catalyst.

epicotyl The portion of a seed located above the cotyledons; the epicotyl is the embryo's shoot.

epidermis The outermost layer of the primary plant body (the herbaceous body); the epidermis is covered by cutin and wax, and has stomata in most leaves and some stems.

epiphyte A plant that lives upon another plant; it may merely be balanced on the host plant, or attached to it by twining around it or by adhesive roots, but most epiphytes are not parasites.

equally parsimonious In cladistics, this refers to alternative cladograms that are equally likely; the available data can be interpreted in multiple ways, each just as likely as the other.

essential amino acid An amino acid that an animal must obtain in its diet to survive; it either cannot make the amino acid at all, or not in sufficient quantities.

essential element An element such as nitrogen or phosphorus that an organism must obtain from its environment in order to survive, reproduce, and complete its entire life cycle. No organism can make any element.

essential fatty acid A fatty acid that an animal must obtain in its diet to survive; it either cannot make the fatty acid at all, or not in sufficient quantities.

ethanol Another name for ethyl alcohol, produced by yeasts during fermentation; it has only two carbon atoms. Ethanol is the alcohol in alcoholic beverages.

ethyl alcohol Another name for ethanol.

ethylene A gaseous plant hormone, involved in ripening of climacteric fruits and in responding to flooding.

etiolation The response of a plant to prolonged darkness or very low light; typically the internodes become long and slender, leaves remain small, and little or no chlorophyll is produced.

eudicots The clade of angiosperms that does not include the monocots or basal angiosperms; this is almost synonymous with the old concept of "dicots," except that "dicots" included the basal angiosperms.

eugenol An essential oil found in cloves (it is the aroma of cloves) and several other spices.

Eukarya One of the three domains of all organisms. Eukarya are the eukaryotes, the organisms that have a membrane-bounded nucleus; examples are plants, algae, fungi, and animals. The other two domains are Archaea and Bacteria.

eukaryote An organism with a membrane-bounded nucleus, a member of the domain Eukarya.

evergreen forest A forest dominated by trees that are never leafless; this often refers to the conifer forests of the northern hemisphere, but can also be applied to tropical forests that have no season in which most trees are leafless.

F_1 (first filial) generation In genetics, this is the progeny of an experimental cross between two individuals, or the progeny of individuals that have been self-pollinated. The progeny of a particular cross are often interbred for many generations so that the pattern of inheritance can be studied; each succeeding generation has a higher F number, for example F_1, F_2, F_3, and so on.

F_2 generation In genetics, this is the progeny of an experimental cross between two individuals of an F_1 generation, or the progeny of individuals that have been self-pollinated.

facultative aerobe An organism that uses oxygen when it is available but can survive without oxygen if it is absent; such organisms can also be called facultative anaerobes. Compare with obligate aerobe or anaerobe.

family In systematics, a level of classification composed of genera; a group of families constitutes an order.

fat An informal term for lipids, especially lipids that are solid at room temperature.

fatty acid A type of unbranched lipid molecule that consists almost exclusively of a carbon backbone with hydrogens attached, and having an acid group at one end.

fermentation Another term for anaerobic respiration, used especially for yeasts and the brewing of beer and wine.

fern A member of the monilophyte clade; ferns have vascular tissues and megaphyllous leaves (evolved from branches) but do not produce seeds. Members of *Equisetum* are now believed to be ferns.

fern ally A term used to refer to plants that, like ferns, have vascular tissue but do not produce seeds, and that have microphyllous leaves (evolved from enations). Fern allies are members of the Lycophyta.

Fertile Crescent The portion of the Near East that includes Iran and Iraq, and which was the site where the first cities and agriculture were established.

fiber In plant anatomy, a fiber is a long, slender cell with a secondary wall. In textiles, a fiber is anything that can be spun into thread or woven into fabric; it may be a trichome (cotton) or large groups of fiber cells (flax).

fibrous root system A root system that is not dominated by a single large taproot; the many adventitious roots of a monocot constitute a fibrous root system, as do the many more or less equal sized roots of many dicots.

field capacity The amount of water held by a soil after gravity has removed the unbound water.

figure The "look" of wood as the result of its color, the striping of the early- and latewood, and the size and texture of rays.

flax An herb (*Linum usitatissimum*) whose fiber bundles are extracted for an especially fine fabric called linen.

fleshy fruit A fruit that is soft and usually edible at maturity. Compare with dry fruit.

flower The structure in flowering plants (angiosperms) involved in sexual reproduction. Flowers usually have sepals, petals, stamens, and carpels, and after pollination, parts of a flower develop into seeds and a fruit.

foliose lichen A lichen (an association of fungi and algae) composed of thin, flat, somewhat leaflike bodies attached by one edge to the substrate.

forage Grasses and small herbs that herbivorous animals can eat.

fossil fuel Fuels such as coal, lignite, and petroleum derived from fossilized remains of plants that lived millions of years ago. Compare with biofuel.

founder The individual or group of individuals that establishes a population in a new area.

free-living organism An organism that is not dependent on another organism; most organisms are free-living. Compare with symbiotic relationship.

frond An informal, nontechnical word for the leaf of a fern, palm, or kelp.

fruit A carpel develops into a fruit after it has been pollinated and its ovules are developing into seeds.

fruticose lichen A lichen (an association of fungi and algae) whose body resembles a small shrub, with somewhat stemlike portions.

fruiting body In ascomycete and basidiomycete fungi, the structure in which meiosis occurs; examples are mushrooms, puffballs, truffles, and morels.

functional group In a molecule, a functional group is a small set of atoms that together have a specific chemistry, such as being an acid, a base, or an alcohol.

G1 (gap 1) phase In the cell cycle, G1 phase occurs after nuclear division and before DNA replication.

G2 (gap 2) phase In the cell cycle, G2 phase occurs after synthesis and before nuclear division.

gamete A haploid sex cell; a sperm or an egg cell.

gametophyte In a plant life cycle, the gametophyte is the haploid phase that produces the gametes (the sperms and eggs). Compare with sporophyte.

gasohol A mixture of gasoline and ethanol, usually in a 9:1 ratio.

gel electrophoresis A technique for separating proteins or nucleic acids based on their size by forcing them through agarose gel by an electric current.

gene activation Part of the concept of differential gene expression; the use of a gene to guide the formation of messenger RNA.

gene bank Computer databases that contain DNA sequences of genes.

gene family If a gene is duplicated during evolution, the copies can evolve to have different properties but are related evolutionarily.

gene flow The movement of genes from one area to another by means of movement of pollen, seeds, or other propagules.

gene pool All the alleles and all the genes of a population.

gene repression Part of the concept of differential gene expression; some genes are not transcribed into messenger RNA.

genetic drift In evolution, the genotype of a population may change not because of natural selection but just by random chance; genetic drift is more common in very small populations rather than in large ones.

genetic engineering The modification, by people, of genes and the genome of whole organisms, using recombinant DNA technology.

genetics The scientific study of genes and their alleles and of the characters they produce.

genome All the genes of something, for example, the genome of the plastids, the genome of the mitochondria, or of all the DNA in the nucleus.

genotype The set of alleles in one particular organism of a species.

genus In taxonomy, a level of classification above the species and below the family; a genus is made up of species, and one or several genera make up a family.

geotextiles Nets and coarse fabrics used to cover and stabilize bare ground after construction damage or a fire; geotextiles should decompose as plants become reestablished.

gibberellin (GA) A plant hormone involved in elongation of organs, in juvenile/adult transition, and in flowering, among many other things.

glacial age A period in Earth's history when the average temperature is so low that glaciers are present on many mountains.

glucose One of the most common simple sugars consisting of six carbons; glucose is used to make sucrose, cellulose, and amylose, among many other things.

gluten A protein found in the flour of wheat and several other cereals; it gives bread its elasticity and allows it to rise.

glycyrrhizin The compound with flavor or licorice, found in licorice root (*Glycyrrhiza glabra*). Glycyrrhizin is up to 50 times sweeter than sugar.

glycolysis The first step in respiration; if oxygen is present, glycolysis leads to the production of acetyl co-A and further respiration. If oxygen is not present, this leads to fermentation.

Gondwana The name given to the supercontinent made up of Africa, South America, India, Australia, and Antarctica. In older literature it is called Gondwanaland.

grafting The attaching of a bud or branch (scion) of one plant onto the roots or basal stem (rootstock) of another plant such that they grow together: the resulting plant is a combination of the two; usually done to give a shoot with desirable properties a stronger root system.

grain Typically used to refer to the fruits and seed of cereal grasses such as wheat and rice, although "grain" is often used to refer to any small, hard seed.

grassland An area dominated by grasses, with few or no trees and with few patches of open soil.

green building techniques Techniques of construction designed to reduce the environmental impact of the building both during its construction and use.

green tea Tea made from leaves of *Camellia sinensis* that have been fermented so briefly they retain their pale color. Compare with black tea and oolong tea.

greenhouse A structure made of a framework and glass that protects plants from weather and animals and heats itself by trapping infrared light.

greenhouse effect The trapping of energy by glass or atmospheric gasses that are transparent to visible light but which absorb at least some energy of infrared light.

greenhouse gasses Gasses that are transparent to visible light but that are partially opaque to infrared light, thus allowing energy from sunlight to reach Earth's surface but preventing part of that energy from being lost to space. The trapped heat causes Earth's temperature to rise.

growing season The season between the last killing frost of spring and the first one of autumn.

growth chamber A cabinet or small room for cultivating plants in conditions in which the light, temperature, and relative humidity are controlled.

growth ring In woody plants, the wood produced in one growing season by the vascular cambium.

guard cell In the epidermis, guard cells occur in pairs with a stomatal pore between them; guard cells swell and shrink, opening and closing the stomatal pore respectively.

habitat destruction Any processes that cause a habitat to be unable to support the species diversity it had when in a natural state.

hair A trichome; any outgrowth of the epidermis.

haploid Refers to having a single set of genes; a nucleus is haploid if it has a single set of genes, as occurs in gametes and gametophytes. Compare with diploid.

hard fiber A term used in the textile industry for fibers derived from leaves rather than bark or wood.

hardwood Refers to wood of broadleaf trees, the dicots, and basal angiosperms; it is called "hardwood" because in most cases it has wood fibers. Compare with softwood.

heartwood The central, older portions of woody stems and roots in which all the cells have died; usually, heartwood has antimicrobial compounds, is darker in color, and is more aromatic than the surrounding sapwood, which still has living cells. Compare with sapwood.

hemicellulose A set of branched polysaccharides that occur in plant cell walls and bind cellulose microfibrils together.

hemp fiber Fiber derived from hemp plants, *Cannabis sativa*.

herb (1) Any plant that never produces a vascular cambium or wood. All bryophytes, lycophytes, ferns, and most monocots are herbs. (2) A plant that is flavorful and used in cooking.

herbicide Any chemical that kills plants.

herbivory The eating of plants; animals that eat only plants are herbivores. Compare with carnivory and omnivory.

heroin An addictive drug synthesized from morphine, derived from the opium poppy, *Papaver somniferum*.

heteromorphic generations In plant life cycles, the generations are heteromorphic if the haploid gametophyte differs morphologically from the diploid sporophyte.

heterotroph An organism that obtains its energy and carbon from molecules in its food. Animals, fungi, and parasitic plants are heterotrophs. Compare with autotroph.

heterozygous Refers to having differing types of alleles present for a particular gene in an organism. Compare with homozygous.

high-energy bond In chemistry, a bond between two atoms in a molecule that, when broken, releases a large amount of energy.

histone One of a set of proteins that binds to DNA and stabilizes it, neutralizing the many positive charges on the nucleic acids. Histones allow DNA to be coiled and compacted into the small space of a nucleus.

homeotic mutation A mutation that causes organs to be produced in unusual places on a body, or in altered numbers. Flowers that have excess numbers of petals are the result of homeotic mutations.

Hominidae In systematics of animals, the family that contains humans, orangutans, gorillas, and chimpanzees.

Homo sapiens The scientific name for human beings.

homologous chromosomes In diploid organisms, each chromosome in the paternal set has an equivalent, homologous chromosome in the maternal set.

homologous features Features of organisms that have originated from the same ancestral feature; for example, leaves of seed plants and of ferns are homologous because both evolved from small branches in the common ancestor of the two groups.

homoplasy Convergent evolution; two features resemble each other but evolved from different ancestral features.

homozygous Refers to having identical alleles present for a particular gene in an organism. Compare with heterozygous.

hops Inflorescences of carpellate flowers of *Humulus lupulus*, used in brewing beer.

hormone A chemical produced by one part of a plant, often in response to a stimulus, and then transported to other parts where it induces a response.

hormone receptor A protein with an active site that binds a hormone, and which then is activated or inactivated when the hormone binds to the receptor.

hornwort A clade of plants in which the gametophyte is the dominant, independent phase, there is no vascular tissue produced, and the sporophyte (the "horn") grows by a basal meristem, the top part being mature and releasing spores while the bottom part is still growing and producing new spores.

hydrogenated oil An oil that has been artificially created from an unsaturated oil by forcing hydrogen onto some of the double bonds; this produces a mixture of fully hydrogenated oils, *trans*-unsaturated oils, and *cis*-unsaturated oils.

hydrophilic "Water-loving," this refers to compounds that are soluble in water. Compare with hydrophobic.

hydroponics A method of growing plants with their roots in a dilute water solution containing all essential elements but completely lacking soil.

hydroponic experiment An experiment in which plants are grown in a liquid nutrient solution; the objective is typically to examine the effect of nutrient concentration on plant growth and development.

hydrophobic "Water-fearing," this refers to compounds that do not dissolve in water; such compounds are usually lipid soluble, lipophilic. Compare with hydrophilic.

hydroxyl ion An ion consisting of OH^-.

hypha (pl.: hyphae) A long, slender cell or linear series of cells of fungi. All the hyphae of a fungus make up a mycelium.

hypodermis In plant anatomy, an outermost band of cortex cells, just interior to the epidermis, in which the cells are distinct from other cells of the cortex in any feature.

hypocotyl The portion of an embryo axis located between the cotyledons and radicle.

hypothesis (pl.: hypotheses) A model of a phenomenon constructed from observations of the phenomenon. It must make testable predictions about the outcome of future observations or experiments.

IAA (indole acetic acid) One of the plant hormones; also called auxin. Like all plant hormones, IAA is involved in many responses including general promotion of metabolic activity and unequal growth that causes curvature.

ice age A geologic age characterized by Earth having low temperatures; ice ages do not have to be cold enough to cause glaciers (a glacial age). We are in an ice age now.

imperfect flower A flower that lacks either stamens or carpels or both. Compare with perfect flower and with incomplete flower.

incomplete dominance In genetics, incomplete dominance occurs in heterozygotes when one allele does not completely mask the effect of the other allele. Compare with complete dominance.

incomplete flower A flower that lacks any of its lateral appendages, that is, it lacks sepals, petals, stamens, or carpels. Compare with complete flower and with imperfect flower.

indehiscent fruit A fruit that does not open spontaneously when it is mature, such as an orange or apple. Compare with dehiscent fruit.

independent assortment In genetics, this refers to the fact that alleles on one chromosome move to one daughter cell or the other independently of the alleles on the other chromosomes; there is no way to predict which alleles will move together, unless they are very close together on the same chromosome (they are linked). With crossing-over, even alleles that are far apart on the same chromosome assort independently of each other.

indeterminate organogenesis The formation of organs without a specific number of organs being determined by genes. Most plants have indeterminate organogenesis and continue to form new organs such as leaves, flowers, and roots throughout their lives. Animals typically have determinate organogenesis.

inferior ovary An ovary located below the sepals, petals, and stamens; those appendages are epigynous.

inflorescence A group of flowers on a plant, such as a raceme, an umbel, a head, etc.

infusion The result of soaking plants in hot water or oil to extract chemicals that are aromatic, flavorful, or medicinal; the plants are then discarded. Tea and coffee are infusions.

integrated pest management A set of techniques that controls pests with a minimum use of artificial chemicals. Examples are tilling soil to disrupt development of larvae, cultivating plants that encourage the natural predators of the pest, and intercropping to lower the density of susceptible plants.

insecticide Any chemical that kills insects.

intercellular space A space that occurs between cells in plants, usually caused as cells enlarge and become more rounded, the corners of a cell pull away from its neighboring cells.

intercropping A technique of farming in which two or more crops are grown simultaneously in one field.

internode In plant stems, the portion of a stem between two adjacent nodes (the sites where leaves are attached to stems).

interphase In the cell cycle, the phase between the end of one division (telophase) and the beginning of the next division (prophase); interphase consists of G1, S, and G2.

jasmonic acid One of the plant hormones; it is involved in defense responses when plants are attacked by pests.

jojoba wax A wax extracted from *Simmondsia chinensis* and used in cosmetics.

jute The tough, flexible bast fibers extracted from plants in the genus *Corchorus*; jute is often woven into burlap cloth.

juvenile phase In the life of a plant, the time between germination and the time when the plant is mature enough to be induced to flower. In many species, the juvenile phase lasts only a few weeks or months, but many tree species may grow as juveniles for 10 or 20 years before they become adults and are finally able to flower. Compare with adult phase.

karyokinesis Nuclear division, either mitosis or meiosis. Compare with cytokinesis.

kapok Fluffy fibers extracted from the seed pods of *Ceiba pentandra*, often used to stuff pillows and soft toy animals.

kenaf Fibers extracted from the plant *Hibiscus cannabinus*.

kinetochore The structure on a chromosome where spindle fibers attach.

kingdom In classification, the first subdivision of domains. All plants are in Kingdom Plantae; all animals in Kingdom Animalia. Kingdoms are composed of divisions in plant classification, phyla in animal classification.

kraft method for wood fibers A method of digesting fibrous materials to release the fibers so that they can be used for various purposes, most often for making acid-free paper.

lamina The blade of a leaf.

laminated beam A construction material made by gluing strips of wood together into a single beam. Laminated beams can be longer, thicker, and stronger than beams cut from intact trees.

land reclamation A process of restoring land to a condition as close as possible to its natural state before it was harmed by people for mining, farming, being flooded by a dam, and so on.

latewood In a growth ring, the wood formed in the later part of the growing season. Latewood typically has more cells that supply strength and fewer that supply conductivity than earlywood has. In conifers, latewood tracheids have very thick walls; in dicot wood, latewood has more fibers and fewer, narrower vessels.

Laurasia The former supercontinent that was composed of North America, Europe, and Asia. When Laurasia and Gondwana were united as Pangaea, Laurasia was the northern portion.

leaf One of the three basic organs of most plants; leaves tend to be flat and thin and carry out photosynthesis, but there are many types of modified leaves.

leaf axil The upper angle formed where a leaf attaches to a stem; leaf axils typically contain axillary buds.

leaf primordium A group of cells that will develop into a leaf; leaf primordia are produced by shoot apical meristems.

leaf scar The corky plate left where a leaf has abscised; leaf scars have precise organization and are produced by abscission zones. This does not refer to an irregular scar that might result if a leaf is torn off a plant.

leaf sheath In monocots, the basal part of a leaf that surrounds the stem for a distance of one or several internodes.

leaf trace The set of vascular bundles (either one bundle or several, depending on the species) that interconnect the vascular bundles of a leaf and a stem.

lethal allele An allele that causes the death of the organism.

lenticel An area in bark where cork cells have intercellular spaces, allowing oxygen to diffuse inward to the living cells of bark, cambium, and sapwood.

lichen An association of a fungus and an alga, the two growing together with a distinct, predictable morphology as if the association were an individual organism.

light-dependent reactions In photosynthesis, the reactions that occur only when chlorophyll is receiving light; these are the water-splitting and electron transport reactions in chloroplasts.

lignin A polymer of alcohols deposited by sclerenchyma cells in their secondary walls, making the walls especially strong and waterproof.

limiting factor The growth of any organism depends on the availability of many factors such as water, light, and various nutrients; the rate of the organism's growth is limited by whichever factor is in short supply.

linen The cloth woven from flax thread, made from fibers of *Linum usitatissimum*.

linked genes Genes that are located together on a chromosome; all the genes of one chromosome.

liverwort True plants in which the gametophyte is the photosynthetic, independent phase, and in which the sporophytes have a slender, ephemeral stalk. Similar plants are mosses and hornworts.

localized growth Growth by cell division and enlargement that occurs only in part of an organ or organism; shoot elongation growth is localized in the shoot apical region. Compare with diffuse growth.

long-day plant A plant that is induced to flower by exposure to days longer than some critical length of day. Compare with day-neutral plant and short-day plant.

macro essential element An essential element needed in relatively large quantities, such as nitrogen, potassium, and phosphorus. Compare micro essential element or trace element.

male sterile cultivar A cultivar that does not produce viable sperm cells, usually due to improper development of anthers, tapetum, or the sperm cells themselves; male sterile cultivars are unable to fertilize themselves, so are useful in genetic experiments because the scientist can control pollination.

malt Barley seeds that have germinated just long enough for the root to be easily visible, then dried and toasted; malt supplies the enzymes needed to convert starch to sugar in brewing beer.

map units A measure of the distance between two genes on a chromosome, based on the frequency of crossing-over events that occur between them.

mash In brewing beer, the mix of water and malted barley that is allowed to set so that natural enzymes break down starches into sugars. The sugar-rich liquid that results is the wort, which is then fermented.

maternal (uniparental) inheritance In genetics, traits coded by DNA in mitochondria and plastids are inherited just from one parent (usually the mother) because the sperm cell contributes only a nucleus but no organelles to the new zygote.

meiosis Reduction division, in which the total number of sets of chromosomes in each new daughter cell is only half the number present in the mother; it is usually a conversion from 2n to 1n, but can be from 4n to 2n. Compare with mitosis.

menthol An essential oil present in mints (*Mentha*); often used for flavoring foods, candies, and many other products.

meristem A small group of cells capable of undergoing cell division and producing new cells; typical plant meristems are the root and shoot apical meristems, leaf, and flower primordia.

mescaline The psychoactive alkaloid in peyote (*Lophophora williamsii*).

mesophyll In leaves, all the tissues other then the epidermis; all the inner tissues and cells of leaves.

messenger RNA (mRNA) RNA that carries information from a gene to a ribosome, where the mRNA's information is translated to protein structure.

metaphase One of the phases of mitotic nuclear division, during which chromosomes are aligned in the center of the spindle; the chromosomes, while located there are called the metaphase plate.

metaphase plate During mitotic nuclear division, the aggregation of all chromosomes in the center of the spindle.

methane clathrate An unstable combination of methane and water that results in an ice-like material. Methane clathrates usually occur in areas that are cold (the Arctic) or under pressure (the sea floor); if methane clathrates decompose (such as due to warming of the arctic tundra), the methane clathrates would release large amounts of methane, a strong greenhouse gas.

microfibril As adjacent molecules of cellulose are synthesized, they crystallize into a microfibril, which may be 10 to 25 nm wide.

microRNA This refers to any of many types of very short RNAs (less than a few dozen nucleotides) that affect the way DNA is read or the way other RNAs are processed.

microtubules Slender protein tubes inside cells that are involved in moving material around the cell; they are involved in preprophase bands, mitotic spindles, and flagella.

middle lamella A thin layer of calcium pectate that glues adjacent cells together in multicellular plants.

mid-ocean ridge Long mountain ranges near the middle of oceans where molten material rises and becomes part of Earth's crust, forming new seafloor and causing adjacent tectonic plates to push apart.

midrib Many leaves have a single largest vein located in the center of the blade, extending from the lamina base to the tip; if blades have several equally large veins instead of just one, they are considered to not have a midrib.

mineral nutrition The study of an organism's needs for inorganic chemicals such as iron, calcium, nitrates, and so on.

minimum tillage A farming technique in which ground is disturbed (by disking and plowing) as little as possible between crops, to conserve fuel and minimize erosion.

minor essential element An essential element needed in relatively small quantities, such as molybdenum, chromium, and chlorine; also called trace elements. Compare macro essential element.

mitochondrion (pl.: mitochondria) A cell organelle involved in aerobic respiration; the site of the tricarboxylic acid cycle and the mitochondrial electron transport chain.

mitosis (pl.: mitoses) Duplication division, in which the total number of sets of chromosomes in each new daughter cell is the same as the number present in the mother cell. Compare with meiosis.

molecule A group of atoms that have combined chemically into a single structure.

monocarpic A plant that undergoes sexual reproduction only once in its life; annuals and biennials are monocarpic, but only a few perennials are. Compare with polycarpic.

monocot A clade of angiosperms (flowering plants) characterized by having: flower parts in sets of three; narrow strap-shaped leaves with parallel venation; and vascular bundles located throughout the stem rather than in just one ring. Compare with basal angiosperms and eudicots.

monoculture A cultivation area that contains just one species; in microbiology or cell culture, a Petri dish or culture tube with only one species in it; in agriculture, a field with only one crop in it (although weeds, animals, and microbes may be present as well).

monohybrid cross In genetics, a cross between two parents in which only one single trait is analyzed. Even though many traits could be studied, only one is. Compare with dihybrid cross.

monomer A small molecule that combines with more monomers to make a polymer; amino acids are the monomers of proteins, nucleotides are the monomers of nucleic acids, monosaccharides are the monomers of polysaccharides, and so on.

monophyletic group A group in which all members share the same common ancestor, and none of the descendants of that ancestor are left out.

monosaccharide A simple sugar such as glucose or fructose; monosaccharides can be polymerized into disaccharides and polysaccharides.

monounsaturated fatty acid A fatty acid that contains only one carbon-carbon double bond.

montane Refers to habitats, conditions, and organisms that occur on mountains.

montane forest A forest that occurs on the lower levels of mountains, below the area of subalpine forests.

morphine One of the psychotropic alkaloids in opium poppies (*Papaver somniferum*); it can be chemically converted to heroin.

morphogenic response In plant development, a response that causes a plant to develop a new morphology, such as the transition from producing vegetative to reproductive organs, or the transition from growing with etiolated characters to those typical of exposure to sunlight.

moss True plants in which the gametophyte is the photosynthetic, independent phase and is leafy in all species. In contrast to liverworts, moss sporophytes have a slender, permanent stalk. Similar plants are liverworts and hornworts.

mulch A ground cover for gardens and fields, composed of plant material such as bark nuggets, straw, or leaves.

multiple alleles Many genes have undergone mutations at various times, so they exist in multiple forms known as multiple alleles.

mutagen Anything that causes a mutation in DNA, such as X rays, UV light, radioactivity, and certain chemicals.

mutagenic The ability to cause a change in DNA.

mutation Any change in the sequence of nucleotides in DNA; mutations may be the addition or loss of nucleotides, or the substitution of one nucleotide for another.

mutualism A symbiotic relationship in which two organisms live together and both benefit; each grows better when the other is present. The presence of nitrogen-fixing bacteria in root nodules and in many liverworts and hornworts are examples of mutualism.

mycelium (pl.: mycelia) The diffuse mass of hyphae that constitutes the vegetative body of a fungus.

mycobiont The fungus that is part of a lichen; the alga is the photobiont.

mycorrhiza (pl.: mycorrhizae) A symbiotic association of a fungus and plant roots. The fungus supplies the plant with phosphorus, the plant provide the fungus with sugar.

mycotrophic Some plants, formerly called saprophytes, have an association with fungi whereby the plants obtain some or many nutrients from the fungus. The plants typically lack chlorophyll and so cannot photosynthesize; it is suspected that the plants are parasitizing the fungi.

NAD^+ Nicotinamide adenine dinucleotide, an electron carrier in its oxidized state (without the electrons it can carry).

$NADH_2$ Nicotinamide adenine dinucleotide, an electron carrier in its reduced state (with the two electrons it can carry). Often written as NADH or as $NADH + H^+$.

$NADP^+$ Nicotinamide adenine dinucleotide phosphate, an electron carrier in its oxidized state (without the electrons it can carry).

$NADPH_2$ Nicotinamide adenine dinucleotide phosphate, an electron carrier in its reduced state (with the two electrons it can carry). Often written as NADPH or as $NADPH + H^+$.

nastic response A nongrowth response that is stereotyped and not oriented with regard to the stimulus.

native plant A plant that occurs naturally in the area being considered.

native plant garden A garden in which many or all the plants occur naturally in the area where the garden is located.

natural selection The preferential survival, in natural conditions, of those individuals whose alleles cause them to be more adapted than other individuals with different alleles.

natural system of classification A classification based on evolutionary, phylogenetic relationships.

negative feedback loop A system in which the product inhibits the action that produces it. This is a common control mechanism for enzymes. This also refers to nonbiological processes.

netted venation Synonym for reticulate venation, typically found in leaves of dicots.

nitrogen assimilation The incorporation of ammonium into organic compounds within an organism.

nitrogen fixation The conversion of atmospheric nitrogen into any compound that can be used by plants, typically either nitrate or ammonium.

nitrogen reduction The addition of electrons to nitrate or nitrite, converting these to ammonium, a form of nitrogen that can be used to construct amino acids and other organic compounds that contain nitrogen.

nocturnal Nighttime; a nocturnal plant opens its flowers at dusk. Compare with diurnal.

node Point on a stem where a leaf is attached.

nomenclature The branch of biology involved in giving names to objects, especially to species, genera, and higher groups.

nonpoint source pollution A source of pollution that is broadly distributed, such as an agricultural area, rather than arising from a single point, such as a factory.

nonpolar molecule A molecule that does not carry even a partial charge anywhere.

nonspecific (herbicide or **pesticide)** A chemical that kills plants (herbicide) or animals (pesticide) and that affects many species, not just members of a single species or genus. Also referred to as broad spectrum (herbicide or pesticide).

nucleic acid A polymer of nucleotides.

nuclear envelope A set of two membranes, the inner and outer nuclear envelopes, that surround the nucleus.

nucleus (pl.: nuclei) In eukaryotic cells, the organelle that contains DNA and is involved in inheritance, metabolism control, and ribosome synthesis.

nucleotide The monomer of nucleic acids; each nucleotide consists of a nitrogenous base, a sugar, and a phosphate group.

obligate (aerobe or **anaerobe)** An organism that absolutely must have (aerobic) or avoid (anaerobic) oxygen to live.

oil A type of lipid that is liquid at room temperature.

old-growth forest A forest that has never been cut down or logged; it has very old trees and a more or less natural mix of species.

omnivory The ability to eat either plants or animals. Compare with carnivory and herbivory.

oolong tea Tea made from leaves of *Camellia sinensis* that have been fermented longer than those of green tea but not as long as those of black tea. Compare with green tea and black tea.

opium The milky latex of opium poppy (*Papaver somniferum*); it is the source of morphine (which can be processed into heroin).

order In systematics, a level of classification composed of families; a group of orders constitutes a class.

organ A structure composed of several tissues; in plants, the three basic organs are roots, stems, and leaves; flower parts are considered to be modified leaves and stems.

organic farming An alternative to conventional farming, organic farming uses no synthetic pesticides, herbicides, or fertilizers.

organelle A functional subunit of a cell, such as a plastid, a mitochondrion, or a ribosome.

oriented strand board A type of fabricated building material made by gluing pieces of wood together into a larger mass; plywood is an example. Many fibrous materials other than wood are now used as well.

ortholog Genes in different species that evolved from the same ancestral gene. Compare with paralog.

osmosis The diffusion of a solute through a membrane.

osmotic effect The effect that solutes have on water's ability to do things; when solutes are added to water, some water molecules interact with them and hydrate them, and this prevents the water molecules from doing other things.

ovary The basal part of a carpel, the region that contains the ovules; an ovary develops into a fruit as the ovules develop into seeds.

ovule In seed plants, a small structure that contains a megaspore mother cell, which undergoes meiosis to produce megaspores and then megagametophytes (female gametophytes); an ovule develops into a seed after its egg is fertilized.

ovule parent In seed plants, the parent that provides the egg, the "female" parent. Compare with pollen parent.

oxidize In chemical reactions, to oxidize an atom or molecule is to increase the positive charge on it; the reaction does not need to involve oxygen. Oxidation reactions are always paired with reduction reactions. Compare with reduce.

oxygenic photosynthesis Photosynthesis that produces oxygen; plants, algae, and cyanobacteria have oxygenic photosynthesis; certain photosynthetic bacteria produce sulfur or compounds other than oxygen.

P_{fr} See *phytochrome.*

P_r See *phytochrome.*

p-protein A fibrous protein that occurs in sieve elements of phloem and seals them by forming a plug if the elements are damaged.

p-protein plug When sieve elements are broken open, the sudden drop in pressure sweeps p-protein to the site of damage, where it plugs the hole.

palisade mesophyll Elongate, sausage-shaped chlorophyllous cells in the interior of a leaf, usually located in one or two layers just below the upper epidermis. Palisade mesophyll is the main site of photosynthesis for most leafy plants.

palmately compound leaf See compound leaf.

Pangaea The ancient supercontinent composed of all the world's land, it existed in the late Paleozoic Era and consisted of Laurasia (north) and Gondwana (south).

parallel venation Almost exclusively in monocot leaves, a pattern in which all veins run approximately parallel to each other, either from the base of the leaf to its tip or from the midrib to the margin. Alternative: reticulate venation.

paralog Genes within single species that evolved from the same ancestral gene.

paraphyletic group A clade that contains an ancestral taxon and several but not all its descendants.

parasite An organism that obtains its nutrition from another living organism. Parasitic plants are either hemiparasitic (obtaining water and mineral nutrients from their host but carrying out their own photosynthesis) or holoparasitic (obtaining water, mineral nutrients, and sugars from their host, and being unable to carry out photosynthesis).

parenchyma Cells with only thin primary walls; all other features are highly variable from type to type. Alternatives: collenchyma and sclerenchyma.

parental generation In a genetic cross, the two individuals that provide the gametes for the cross.

parental type chromosome A chromosome that, after meiosis, has not undergone crossing over. Alternative: recombinant chromosome.

parsimony The concept of minimum complexity; the simplest hypothesis that explains several observations is the most parsimonious. In cladistics, a cladogram with the least number of steps is the most parsimonious.

partially hydrogenated oil An unsaturated oil (one with double bonds) that has had enough hydrogen forced on to it to convert some but not all the double bonds into single bonds.

particleboard A building material made by gluing small pieces (sometimes as fine as sawdust) of wood together to create large pieces; this allows the use of pieces that would otherwise be considered scrap or waste material.

pedicel The stalk of an individual flower.

pentose phosphate pathway A type of respiration in which glucose is converted either to ribose or erythrose. Synonyms: hexose monophosphate shunt and phosphogluconate pathway.

perennial plant A plant that lives for more than 2 years. Compare with annual and biennial plant.

perfect flower A flower that has both stamens and carpels. Alternative: imperfect flower.

perforation In a vessel element, the hole(s) where both primary and secondary walls are missing. Compare with pit.

pericarp The technical term for a fruit wall (not the seeds) that develops from the ovary after a flower is pollinated.

pericycle An irregular band of cells in the root, located between the endodermis and the vascular tissue.

permafrost A layer of permanently frozen soil that does not thaw, even during summer. Permafrost is extensive in far northern latitudes of Alaska, Canada, and Europe.

persistent (herbicide or **pesticide)** An herbicide or pesticide that does not break down quickly but instead lasts (persists) for many months or years in the environment.

pesticide A substance that kills animals, usually insects.

petal The appendages on a flower, usually pigmented and most often involved in attracting pollinators.

petiole The stalk of a leaf.

peyote A cactus (*Lophophora williamsii*) that contains a hallucinogenic alkaloid called mescaline.

pH A measure of the acidity of a solution. Acidic solutions have a pH of 7.0 or lower, alkaline solutions have a pH between 7.0 and 14.

phenotype The physical, observable characteristics of an organism. Alternative: genotype.

phloem The portion of vascular tissues involved in conducting sugars and other organic compounds, along with some water and minerals. Alternative: xylem.

phloem fiber A fiber located in phloem, often called a bast fiber or soft fiber.

phloem sap The liquid that is conducted through sieve elements, consisting mostly of sugars dissolved in water.

phospholipid A type of lipid containing two fatty acids and a phosphate group bound to glycerol.

photobiont In a lichen, the organism that carries out photosynthesis, either an alga or a cyanobacterium. The other partner is the mycobiont.

photoperiod In reference to cycles of light and darkness, the length of time that uninterrupted light is present.

phototropins Pigments involved in detecting the blue light that causes phototropism.

phototropism A plant response to a stimulus in which a part of a plant grows toward (positive phototropism) or away from (negative phototropism) light.

phragmoplast During cell division, the phragmoplast is a set of short microtubules oriented parallel to the spindle microtubules; it catches dictyosome vesicles and guides them to the site where the new cell wall (cell plate) is forming.

phyletic speciation The evolution of one species into a new species, such that the original species no longer exists. Alternative: divergent speciation.

phylogenetic tree A graphic representation of the members of a clade, usually consisting of a branched set of diverging lines. Also called a cladogram.

phyllotaxy The arrangement of leaves and axillary buds on a stem.

phylogeny The evolutionary relationships of a group of organisms.

phytochrome A family of pigments involved in many aspects of morphogenesis in which the stimulus is red light or the length of a dark period. P_{fr} absorbs far-red light, P_r absorbs red light.

phytoplankton Microscopic algae that float in the ocean, near enough to the surface to carry out photosynthesis.

pigment Any compound that absorbs light. Some pigments in plants protect other chemicals by absorbing light that is too strong or has damaging wavelengths; other pigments carry out a chemical reaction or a physical rearrangement that then causes the plant to respond to the light. Pigments in flowers and fruits signal to animals that the flower is ready for pollination or that the fruit is ready (or not) to be eaten.

pinnately compound See compound leaf.

pit In a sclerenchyma cell, an area where there is no secondary wall underlying the primary wall; material can pass into or out of the cell through pits.

pit membrane The set of two primary walls and the middle lamella that occurs between the two pits of a pit-pair.

pith The region of parenchyma located in the center of most shoots, surrounded by vascular bundles.

plant growth substance Term used for any hormone-like compound, whether natural or artificial.

plasma membrane The semipermeable membrane that surrounds the protoplasm of a cell. Synonym: plasmalemma.

plasmodesma (pl.: plasmodesmata) A narrow hole in the primary wall, containing some cytoplasm, plasma membrane, and a desmotubule; a means of communication between cells.

plastic strength A type of strength provided by collenchyma; it allows a cell to grow to a new size or shape and then maintain that new size or shape. Compare with elastic strength.

plastids A family of organelles within plant cells; proplastids are young plastids in meristematic cells. Chloroplasts have chlorophyll and perform photosynthesis; amyloplasts store starch; chromoplasts have pigments other than chlorophyll.

pleiotropic effect The multiple phenotypic expressions of a single allele whose activity affects various aspects of metabolism.

plywood A type of manufactured wood made by gluing thin sheets of veneer together, with the grain of each sheet being perpendicular to that of adjacent sheets so that the wood will not warp.

point source pollution Pollution that arises at a very localized area, such as a factory or a feedlot. Compare with nonpoint source pollution.

polar molecule A molecule that has a partial positive charge at one site and a partial negative charge at another site.

pollen In seed plants, the microspores and microgametophytes.

pollen parent In the sexual reproduction of seed plants, the pollen parent provides the pollen, which contains sperm cells. Often called the paternal parent.

pollen tube After landing on a compatible stigma or gymnosperm megasporophyll (cone scale), a pollen grain germinates with a tubelike structure that carries the sperm cells to the vicinity of the egg cell.

polycarpic A plant that undergoes sexual reproduction more than once in its lifetime. Most perennial plants bloom yearly and are polycarpic. Compare with monocarpic.

polyphyletic group A group of organisms, some of whose members have arisen from an ancestor different from that of other members; consequently, not all members of the group are closely related to each other. Compare with monophyletic group.

polymer A large compound composed of a number of subunits, monomers.

polyploidy Refers to a nucleus that contains three or more sets of chromosomes.

polysaccharide A compound made up of many simple sugars (monosaccharides).

polyunsaturated fatty acid A fatty acid that has more than one carbon-carbon double bond. Compare with monounsaturated and saturated fatty acids.

population All the individuals of a species that live in a particular area at the same time and can interact with each other.

positive feedback loop An enzyme system in which the product stimulates the action of one or several of the enzymes that produce it. The ripening of climacteric fruits is an example.

predation A relationship in which one species benefits and the other is harmed; not often used in botany.

preprophase band Prior to cell division, the preprophase band is a set of microtubules that encircles the cell just interior to the cell membrane and is located at the site where the new wall will attach to the preexisting wall.

primary cell wall A cell wall present on all plant cells except some sperm cells; it is formed during cell division and is usually thin, but some may be thick. See secondary wall.

primary growth The production of new cells by shoot and root apical meristems and leaf primordia. Alternative: secondary growth.

primary phloem Phloem that occurs in the primary plant body; it is the only phloem in herbs. Compare with secondary phloem, which occurs in woody plants.

primary pit field An area of primary cell wall that is especially thin and contains numerous plasmodesmata.

primary plant body The herbaceous body produced by apical meristems (roots, stems, leaves, flowers, and fruits). Alternative: secondary plant body.

primary structure of protein The sequence of amino acids in a protein.

primary tissues The tissues derived more or less directly from an apical meristem or leaf primordium; the tissues of the primary plant body. Alternative: secondary tissues.

primary xylem Xylem that occurs in the primary plant body; it is the only xylem in herbs. Compare with secondary xylem (wood).

primordium (pl.: primordia) A small mass of cells that grows into an organ such as a leaf or petal.

prokaryotes Organisms that have no true nucleus or membrane-bounded organelles. Prokaryotes are eubacteria, cyanobacteria, and archaea. Alternative: eukaryotes.

promoter That portion of a gene in which control molecules and RNA polymerases bind during gene activation and transcription.

prophase The initial phase of mitosis during which the nucleolus and nuclear membrane break down, chromosomes begin to condense, and the spindle begins to form.

proplastid See plastids.

protein A polymer of amino acids; many proteins are enzymes, others provide structure (such as microtubules), and others are part of membranes.

protoplasm All the substance of a cell, usually considered not to include the cell wall. The protoplasm of a single cell is the protoplast. See cytoplasm and cytosol.

protoplast All the protoplasm of a cell; the protoplast is surrounded by the plasma membrane.

proximal This term is used to compare the position of two points or sites on a plant, the more proximal point being closer to the root/shoot junction. For example, the base of a leaf is proximal to the tip of a leaf, and the base of a root is proximal to the root tip. Compare with distal.

Punnett square In genetics, a table listing all possible male gametes along one side and all possible female gametes along the top; used to determine all possible genotypes of progeny that might be produced in a cross.

pure-bred line A group of organisms that are homozygous for a particular trait of interest.

quantitative trait loci In genetics, most phenotype characters are the result of interactions of many genes (loci); QTL analysis determines just how much various genes contribute to the phenotype.

quantum (pl.: quanta) A particle of electromagnetic energy. Synonym: photon.

quaternary structure of protein The association of several proteins into a large structure such as a microtubule or synthetic complex.

radicle The embryo of a seed has an axis; one end is the embryonic root, the radicle, the other end is the embryonic shoot tip, the epicotyl.

rainforest A forest characterized by very high rainfall, usually defined as approximately 2000 mm (78 inches) of rain per year.

rain shadow The area on the downwind (leeward) side of a mountain that receives significantly less rain than the area on the upwind side.

raised bed A planting area that is raised several inches or more than the surrounding land and that has a permeable border. A raised bed has good drainage and prevents the planting area from being too wet in a rainy habitat.

ramie A bast fiber (phloem fiber) harvested from *Boehmeria nivea*.

ray Rays are sets of cells, usually just parenchyma, produced by the ray initials of the vascular cambium. Each ray is aligned radially. Compare to axial portion of wood and secondary phloem.

rayon A synthetic fiber made from chemically processed cellulose.

receptacle In flowers, the short portion of stem axis to which the lateral organs are attached.

recessive allele An allele whose presence is masked by a dominant allele.

recombinant chromosomes Chromosomes that are made up of pieces of various other chromosomes; they can be produced either artificially or by crossing-over in meiosis.

recombinant DNA technology The techniques based on several enzymes that cut DNA at specific locations and then attach various pieces together in new combinations.

red tide A bloom of algae in which the cells become so numerous as to give the water a reddish tint.

reduce In chemical reactions, to reduce an atom or molecule is to reduce the positive charge on it; a reduction reaction is always coupled to an oxidation reaction. Compare with oxidize.

reducing power In chemical reactions, the ability of an atom or molecule to place electrons onto another atom or molecule.

reduction division Meiosis; a nuclear division in which the daughter cells have half as many sets of chromosomes as the parent cell had. Compare with duplication division.

replicate In DNA synthesis, each newly synthesized strand of DNA is complementary to the original strand, so the new strand is a replica but not a duplicate.

resin A viscous, sticky hydrocarbon secretion, especially common in certain conifers. It is often called "pitch." Fossilized resin is amber.

respiration The set of metabolic reactions that converts ADP and phosphate into ATP; if the metabolism involves oxygen, it is aerobic respiration, without oxygen it is anaerobic respiration or fermentation.

resting phase In the cell cycle, all the cell cycle except nuclear and cell division; it consists of G1, S, and G2.

restriction endonuclease An enzyme that cuts DNA at some place other than its end, and always at a predictable short sequence of nucleotides; there are many restriction endonucleases, each differing in the sequence of nucleotides they recognize and cut.

restriction map A set of DNA fragments of varying sizes produced by treating the DNA with a particular type of restriction endonuclease.

reticulate venation In broadleaf plants, veins (vascular bundles) that occur in a net-like pattern. Compare with parallel venation.

retting The process of extracting fibers from plant material by placing the harvested plants in a shallow pool of water and allowing the soft parts of the plants to be decomposed by fungi and bacteria.

rhizome An underground stem that is usually thick, with short internodes and bract-like leaves.

ribosomal RNA (rRNA) RNA that is part of the structure of a ribosome, the organelle that translates messenger RNA into protein.

ribosome The organelle that translates messenger RNA into protein; ribosomes consist of both RNA and proteins.

ribulose-1,5-bisphosphate (RuBP) In photosynthesis, the acceptor molecule to which carbon dioxide is attached, resulting in two molecules of phosphoenolpyruvate.

root One of the three basic organs of most plants; roots absorb water and minerals, are usually subterranean, and anchor the plant to the substrate.

root cap The set of cells that cover and protect the apical meristem of a root.

root hair A trichome on a root; root hairs are unicellular and usually long and slender, being able to penetrate spaces in the soil that are too small for the root itself to enter.

rootstock In grafted plants, the rootstock is the lower portion that provides the root system. Compare with scion.

round wood Wood harvested and used for poles, posts, and round beams; it does not have to be sawed into planks.

RuBP carboxylase (RUBISCO) In photosynthesis, the enzyme that carries out the initial step, the addition of carbon dioxide to ribulose-1,5-bisphosphate. Now usually called by its abbreviation RUBISCO.

S phase The synthesis phase of the cell cycle, during which nuclear DNA is replicated (synthesized).

sake A fermented beverage made from rice.

salicylic acid The chemical basis of aspirin, extracted from willow trees (species of *Salix*).

salinization An increase in the saltiness of a soil due to water carrying minerals in; then the water evaporates leaving the minerals in the soil. Often caused by fertilizing low-lying cropland in valleys that have no natural drainage stream.

saponification The conversion of fatty acids to soap, typically by treating them with a strong base such as lye.

sapwood The light-colored, light-scented outermost wood of a trunk or branch; conduction is still occurring and many wood parenchyma cells are alive. Alternative: heartwood.

saturated fatty acid A fatty acid that has no carbon-carbon double bonds. Saturated fatty acids are stable and often solid at room temperature.

savanna An ecosystem dominated by grasses, but with occasional trees. Also called a grassland.

scientific method A means of analyzing the physical universe. Observations are used as the basis for constructing a hypothesis that predicts the outcome of future observations or experiments. Anything that can never be verified cannot be accepted as part of a scientific hypothesis.

scientific name The binomial name of a species, consisting of the genus name and the species epithet.

scion In grafting, the upper of the two members, the scion provides the shoot system. Compare with rootstock.

sclereid A sclerenchyma cell that is rather cubical, not long like a fiber. Masses of sclereids provide a tissue with strength and rigidity.

sclerenchyma Sclerenchyma cells have both a primary wall and an elastic secondary wall; if stretched to a new size or shape, the wall returns to its original size and shape after the deforming force is removed. Alternatives: parenchyma and collenchyma.

secondary growth Growth that occurs by means of either the vascular cambium or the cork cambium. It results in wood and bark, the secondary tissues. Compare with primary growth.

secondary phloem Phloem derived from the vascular cambium. Compare: primary phloem.

secondary plant body The wood and bark produced by the vascular cambium and cork cambium. Compare with primary plant body.

secondary structure of protein Short sequences of regular helix or regular pleating in a protein.

secondary tissues The tissues of the secondary plant body—those produced by the vascular cambium and the cork cambium. Compare with primary tissues.

secondary wall A cell wall present only in certain cells (sclerenchyma) and formed only after cell growth has been completed. When present, the secondary wall is located interior to the primary wall and is typically impregnated with lignin. See primary cell wall, sclerenchyma, and xylem.

secondary xylem Xylem derived from the vascular cambium. Compare: primary xylem.

seed In seed plants, seeds develop from ovules after the egg has been pollinated. The seed consists of the embryo, a seed coat, and endosperm in some (mostly monocots) but not in others (mostly dicots).

seed coat The protective layer on a seed; the seed coat develops from one or both integuments.

selective cutting A method of harvesting trees from a forest in which only the needed trees are cut and removed; unneeded trees are left as undisturbed as possible. Compare with clear cutting.

self-assembly The automatic assembly of a larger structure solely due to interaction of charges and hydrophobic/hydrophilic regions on the molecules.

self-pollination The pollination of one plant by pollen from the same plant. Compare with cross-pollination.

sepal In flowers, the outermost of the fundamental appendages, most often providing protection of the flower during its development.

sessile Refers to an organ that has no stalk but rather is attached directly to the stem or other underlying organ.

shadehouse A structure for cultivating plants, similar to a greenhouse but with material to provide the plants with shade when sunlight is too intense. It is often used to cultivate understory plants and ferns.

sheathing leaf base In many species of monocots, the leaf base extends down around the stem for one or several nodes as a tight-fitting sheath.

shelterbelt Rows of trees planted to protect buildings or crops from wind.

short-day plant A plant that is induced to flower by nights longer than the critical night length.

shrub A woody plant that has multiple stems arising from ground level, not with just one single trunk as is true of trees.

shrubland An area dominated by shrubs rather than trees, sometimes with such a high density of shrubs that grasses are rare.

sieve area In phloem, an area on a sieve element wall in which numerous sieve pores occur.

sieve pore In sieve elements, the holes (enlarged plasmodesmata) in the primary walls; sieve pores permit movement of phloem sap from one sieve element to another.

sieve tube In the phloem of angiosperms, a column of sieve tube members interconnected by large sieve areas and sieve pores.

simple leaf A leaf in which the blade consists of just one part. Alternative: compound leaf.

simple sugar A sugar that is not composed of smaller sugar molecules; the monomer for polysaccharides. Synonym: monosaccharide.

sink In phloem transport, the organ or site that receives sugars.

sisal A tough, coarse fiber extracted from the leaves of *Agave sisalana*.

sizing The addition of clay or other materials to threads or paper to fill in holes, so that the material has a smooth surface or so that it will accept dyes and inks more readily.

soft fiber A fiber extracted from the phloem, cortex, or bark of a plant; also called bast fiber.

softwood A term applied to both gymnosperms and their wood, because few gymnosperms have any fibers in their wood. Alternative: hardwood.

somatic mutation A mutation in a cell that is not a gamete and does not give rise to gametes.

source In phloem transport, the organ or site that donates sugars and loads sugars into the phloem.

species (pl.: species) A set of individuals that are closely related by decent from a common ancestor and can reproduce with each other but not with members of any other species.

species epithet The part of a binomial name for an organism that indicates the particular species within a genus. For example, the scientific name of people is *Homo sapiens*, but "sapiens" is the species epithet that distinguishes us from other species of *Homo*.

spermatophyte A plant that produces seeds. Alternative: vascular cryptogam.

spindle The framework of microtubules that pulls the chromosomes from the center of the cell to the poles during nuclear division.

spirit An alcoholic beverage that has an alcohol concentration higher than that of wine, higher than 22%. The high concentration can be achieved by distilling a beverage with lower alcohol content, or by adding concentrated ethanol to it.

spongy mesophyll Any part of leaf mesophyll in which the cells are not aligned parallel to each other and are separated by large intercellular spaces.

spore A single cell that is a means of asexual reproduction; it can grow into a new organism but cannot fuse like a gamete.

sporophyte A diploid plant that produces spores. Alternative: gametophyte.

stamen The organ of a flower involved in producing microspores (pollen).

statocyte A cell within the root cap that detects the direction of gravity.

statolith A type of starch grain that is so dense it sinks to the bottom of a cell's cytoplasm, indicating the direction of gravity.

stem One of the three basic organs of most plants; stems are produced by shoot apical meristems and always have nodes, internodes, and axillary buds. Stems always bear leaves. Stems have diversified in evolution such that various types now exist, for example, rhizomes, tubers, bulbs, corms, tendrils, spines, and others.

stigma (pl.: stigmas) In the carpel of a flower, the receptive tissue to which pollen adheres.

stolon An aerial, horizontal stem with elongate internodes; it establishes plantlets periodically when it contacts soil. Example: strawberry.

stoma (pl.: stomata) A word sometimes used to mean "stomatal pore," the intercellular space between guard cells through which carbon dioxide and water are exchanged. "Stoma" is sometimes used to mean "stomatal complex," the stomatal pore plus guard cells plus associated cells.

stomatal pore The intercellular space between two guard cells; carbon dioxide is absorbed through the pore and water is lost.

strict aerobe An organism that absolutely must have oxygen to live; without oxygen, strict aerobes soon die. Compare with facultative aerobe and strict anaerobe.

strict anaerobe An organism that is killed by oxygen; strict anaerobes must have a habitat free of oxygen to live. Compare with facultative anaerobe and strict aerobe.

style In the carpel, the tissue that elevates the stigma above the ovary.

subalpine forest The forest that occurs just below treeline in mountains; it is less dense than the forests located lower on the mountains.

suberin Lipid material that causes the hydrophobic properties of cork cell walls and the Casparian strip of the endodermis.

substrate-level phosphorylation The formation of ATP from ADP by having a phosphate group transferred to it from a substrate molecule.

substrate specificity The ability of an enzyme to distinguish one substrate from similar substrates.

sulfate method for wood fibers A method of digesting wood to pulp using compounds of sulfate; this is also known as the kraft method and it produces durable acid-free paper.

sulfite method for wood fibers A method of digesting wood to pulp using calcium sulfite. This produces a paper with acid in it, and which deteriorates with age.

superior ovary In flowers, an arrangement in which the carpels are located above sepals, petals, and stamens. Compare with inferior ovary.

sustainable agriculture Techniques designed with the goal of not damaging the soil or water around the agricultural area, so that the same level of cultivation can be carried out for years.

systemic herbicide (or pesticide) An herbicide (or pesticide) that is absorbed by the plant, such that it affects areas other than those to which it was applied. A systemic herbicide can be sprayed onto a plant's leaves, but it will be absorbed and transported throughout the plant, killing roots and other parts.

symbiotic relationship A relationship in which two or more organisms live closely together.

symplesiomorphy Shared relictual features; these are features that were present early in the evolution of a group so they do not show which members of the group are more closely related to other members.

synapomorphy Shared derived characters; these are present in closely related species because the character arose recently in evolution, in a recent common ancestor.

synapsis (pl.: synapses) The pairing of homologous chromosomes during zygotene of prophase I of meiosis. Synapsis precedes crossing-over.

syngamy The fusion of a sperm and an egg.

systematics The branch of biology that analyzes evolutionary relationships among organisms.

taiga Another word for boreal coniferous forest.

taproot system A root system that develops from the embryonic root, the radicle. Often, the radical is larger than the lateral roots (as in a carrot), but that is not necessary. Compare with fibrous root system.

taxon A term that refers to any taxonomic group such as species, genus, family, and so on.

tectonic plate The Earth's crust is composed of many pieces, each called a tectonic plate. The plates float on the hot, fluid mantle, and currents in the mantle cause tectonic plates to move relative to each other, resulting in continental drift.

telophase The fourth and last phase of mitosis, during which the chromosomes decondense, the nucleolus and nuclear envelope reform, the spindle depolymerizes, and the phragmoplast appears.

temperate Indicates that a habitat or ecosystem has freezing conditions at some time during the year. Compare with tropical.

temperate rainforest A forest that receives high amounts of rain (at least 2000 mm per year [almost 79 in.]) and that experiences at least occasional freezing conditions in winter.

tendril An organ that attaches a vine to a support by wrapping around it. It may be a modified leaf, leaflet, or shoot. Example: grape.

terminal bud A bud located at the extreme apex of a shoot; usually present only in winter as a dormant bud. Alternative: axillary bud.

tertiary structure of protein The overall three-dimensional shape of an entire protein molecule.

test cross A cross involving one parent known to be homozygous recessive for the trait being considered.

Tethys Sea An ocean that existed in the ancient past when Africa was not joined to Europe.

tetrahydrocannabinol (THC) The psychoactive chemical in marijuana (*Cannabis sativa*).

texturized vegetable protein (TVP) Plant protein that has been processed to resemble meat.

theory After a hypothesis has been confirmed by numerous observations or experiments, it is considered to be a theory.

trace essential element See minor essential element.

tracheid A xylem-conducting cell; tracheids tend to be long and tapered, and they never have a perforation (a complete hole in the primary wall) as vessel elements do.

***trans*-fat** See *trans*-unsaturated fatty acid.

***trans*-unsaturated fatty acid** An unsaturated fatty acid (it has one or more carbon-carbon double bonds), and the two hydrogens at the double bond are on opposite sides of the molecule. *Trans*-unsaturated fatty acids are synthetic, not natural, and they are not healthful. Compare with *cis*-unsaturated fatty acid.

transfer RNA (tRNA) RNA that carries amino acids to ribosomes.

transgenic organism An organism that has been genetically altered by having a gene from a different species inserted into its DNA. Compare with cisgenic organism.

trichome A plant hair; often restricted to structures that contain only cells derived from the epidermis.

triglyceride A type of lipid consisting of three fatty acids bound to one molecule of glycerol.

tropic response A growth response oriented with regard to the stimulus.

tropical Indicates that a habitat or ecosystem never has freezing conditions at any time during the year. Compare with temperate.

tropical rainforest A forest that receives high amounts of rain (at least 2000 mm per year [almost 79 in.]) and that never experiences freezing conditions.

true plant A member of the clade that is sister to the charophyte algae and that has multicellular sporangia with an outer layer of sterile cells. Bryophytes, lycophytes, monilophytes (ferns), and seed plants are true plants, but algae and fungi are not, nor are lichens.

tuber A short, fleshy, horizontal stem, involved in storing nutrients. Example: potato.

tundra The biome in which tree growth is impossible due to low temperatures; tundra is widespread near the poles and at high altitudes on mountains.

turgid Filled with water to such a degree that the surface of the cell or plant is firm.

tylosis (pl.: tyloses) After a vessel stops conducting because of cavitation, adjacent cells may push cytoplasm into the vessel through pits, plugging the vessel.

unsaturated fatty acid A fatty acid in which two of the carbons are joined by a double bond, so the fatty acid is not carrying as much hydrogen (is not saturated) as it theoretically could. The two types are *cis*-unsaturated and *trans*-unsaturated. Compare with saturated fatty acid.

vacuole membrane The membrane that surrounds a vacuole; if the vacuole is the central vacuole, the membrane can be called the tonoplast.

vanillin The flavor molecule of vanilla, most often extracted from the fermented and processed fruits of the orchid *Vanilla planifolia*, although it is present in small amounts in many plants.

varnish A solution of a drying oil, a resin, and often a thinner (usually turpentine); varnish is used to protect and give color to wood surfaces.

vascular bundle A set of vascular tissues, usually containing both xylem and phloem, but occasionally only one or the other.

vascular cambium The meristem in woody plants that produces secondary xylem (wood) and secondary phloem. It contains fusiform initials and ray initials.

vascular cryptogam A plant that has vascular tissue (xylem and phloem) but which does not make seeds. Ferns and lycophytes are vascular cryptogams.

vascular plant A plant that has vascular tissues; all plants other than bryophytes (mosses, liverworts, and hornworts) are vascular plants.

vascular tissue Conducting tissue in plants; xylem conducts water and whatever is dissolved in it; phloem conducts water, sugars, and sometimes hormones and other organic compounds.

venation The set of vascular bundles in an organ; in leaves, there is reticulate and parallel venation.

veneer A thin sheet of wood cut from a log on a lathe; veneers may be glued to each other to make plywood. Veneers cut from expensive wood may be glued to the surface of inexpensive wood to produce a construction material that is beautiful yet not expensive.

vernalization The treatment of a plant with low temperatures (just above the freezing point) necessary to induce certain plants, such as biennials, to be able to flower in the subsequent year.

vessel In xylem, a vessel is a set of vessel element connected end to end by means of perforations.

vessel member In xylem, dead, empty cells that interconnect by perforations (large areas with no cell wall) to form a vessel.

wax An extremely hydrophobic polymer of long-chain fatty acids; wax contributes to the water-retaining capacity of the epidermis.

wilted A plant or plant organ that has lost so much water that turgor pressure in the vacuoles of its cells is no longer enough to support the plant or organ.

wine An alcoholic beverage produced by fermenting sugary fruits, especially grapes.

wood Another term for secondary xylem, the water-conducting and supporting tissues produced by a vascular cambium. Wood is part of a plant's secondary body.

wood fiber A fiber extracted from wood; if wood of dicots are used, then the fibers are actually fibers in the botanical sense (single cells with narrow pits), but if conifer wood is used, then the "wood fibers" are actually tracheids.

woodland A geographic area that is covered by forest.

woody plant A plant that produces wood, secondary xylem. Compare with herb.

wort In brewing beer, after the mash (malted barley and water) has been allowed to sit for several hours, the resulting sugary liquid is wort. Wort is then fermented to make beer.

xeriscape garden A garden in which most of the plants are adapted to dry or arid conditions; xeriscape gardens typically do not need to be irrigated and so help to conserve water.

xylary fiber See wood fiber.

xylem The tissues that conduct water in vascular plants; the two types of conducting cells in xylem are tracheids and vessel elements.

zero population growth The concept that human population could remain the same rather than increasing as it has done. Related to the concept that Earth's resources are finite and cannot support an ever-increasing human population.

zone of elongation Refers to the portion of a root tip where the cells and the root are becoming longer; any part of this zone will be pushed forward through the soil by older cells that are elongating.

zygomycete A member of one of the major clades of fungi, characterized by producing zygospores but not fruiting bodies. Bread mold is a zygomycete.

zygote A fertilized egg; the zygote is the first cell of a new sporophyte generation in plant life cycles and the first cell of a new individual in animal life cycles.

Index

Note: Italicized page locators indicate photos/illustrations; tables are noted with *t*.

B

C

G